Advancing Artificial Intelligence Through Biological Process Applications

Ana B. Porto Pazos
University of A Coruña, Spain

Alejandro Pazos Sierra
University of A Coruña, Spain

Washington Buño Buceta
Cajal Institute (CSIC), Spain

MEDICAL INFORMATION SCIENCE REFERENCE

Hershey · New York

Director of Editorial Content: Kristin Klinger
Managing Development Editor: Kristin M. Roth
Assistant Development Editor: Deborah Yahnki
Editorial Assistant: Heather A. Probst
Senior Managing Editor: Jennifer Neidig
Managing Editor: Jamie Snavely
Assistant Managing Editor: Carole Coulson
Typesetter: Larissa Vinci
Cover Design: Lisa Tosheff
Printed at: Yurchak Printing Inc.

Published in the United States of America by
Information Science Reference (an imprint of IGI Global)
701 E. Chocolate Avenue, Suite 200
Hershey PA 17033
Tel: 717-533-8845
Fax: 717-533-8661
E-mail: cust@igi-global.com
Web site: http://www.igi-global.com

and in the United Kingdom by
Information Science Reference (an imprint of IGI Global)
3 Henrietta Street
Covent Garden
London WC2E 8LU
Tel: 44 20 7240 0856
Fax: 44 20 7379 0609
Web site: http://www.eurospanbookstore.com

Library of Congress Cataloging-in-Publication Data

Advancing artificial intelligence through biological process applications / Ana B. Porto Pazos, Alejandro Pazos Sierra and Washington Buno Buceta, editors.
p. cm.
Summary: "This book presents recent advancements in the study of certain biological processes related to information processing that are applied to artificial intelligence. Describing the benefits of recently discovered and existing techniques to adaptive artificial intelligence and biology, it will be a highly valued addition to libraries in the neuroscience, molecular biology, and behavioral science spheres"--Provided by publisher.
ISBN 978-1-59904-996-0 (hardcover) -- ISBN 978-1-59904-997-7 (e-book)
1. Artificial intelligence--Medical applications. 2. Artificial intelligence--Biological applications. 3. Neural networks (Neurobiology) I. Porto Pazos, Ana B. II. Pazos Sierra, Alejandro III. Buceta, Washington Buno.
R859.7.A78A382 2008
610.285'63--dc22

2007051645

British Cataloguing in Publication Data
A Cataloguing in Publication record for this book is available from the British Library.

Table of Contents

Section II
New Biologically Inspired Artificial Intelligence Models

Detailed Table of Contents

Foreword

In the last century, biologists have elucidated the fundamental basis of the simplest phenomena and mechanisms underlying life. We now know what the genetic material, the genetic code is and what are the mechanisms for its replication and expression in the living cells. We have also gained a high amount of knowledge on how molecules serve to cell differentiation and function, and on how cells interact to create organized tissues and organs. We also have started to have a glimpse on how these cellular phenomena lead to tissue and organ function and, finally, to animal behavior. Yet, a better understanding of the processes underlying the harmonized coordination of multiple cells to perform appropriate biological responses is one of the great challenges faced by current science. The nervous system function is probably the best paradigmatic example. It is an organized assembly of cells whose function in the organism is to receive, process and transmit information from the environment and from itself and to develop a biologically appropriate response. While a great amount of effort has been done to elucidate the function and behaviour of molecules and cells of the nervous system, we are far away from a clear understanding of how these events lead to the function of the nervous system as a whole.

There is no doubt that to solve such complex biological problems requires multidisciplinary approaches from different scientific disciplines. Artificial intelligence is one of these disciplines that, imitating some key and simplified features of biological process and implementing computational models, attempts to provide some key clues on the function of complex biological systems and to provide technical background to solve real world problems.

In this book, authors present and discuss recently reported experimental biological evidence regarding information processing. The book also presents novel artificial intelligence models based on biological process, and finally explores the benefits of these new models not only to artificial intelligence but also to other sciences, to propose examples of applications to solve real-life problems.

This book is organized in three main sections describing new advances in experimental biology about neuronal information processing, new biologically inspired artificial intelligence models and real-life applications of these models. Several aspects of the book will make it and important reference text. It provides a multidisciplinary vision, from biology to artificial neural network models, to finally present applications of models to solve real-life problems. Contributors are international experts in their respective field. This book will serve not only to students and postgraduates but also to senior researches in many different disciplines. Indeed, because the wide multidisciplinary fields covered, it will be of interest to experimental biologists, neuroscientists interested in information processing at cellular and systems levels as well as researchers in artificial intelligence and computational sciences.

Dr. Alfonso Araque Almendros
Instituto Cajal, CSIC
Madrid. Spain

Alfonso Araque Almendros was born in Madrid, Spain in 1965. He received an MS and PhD in biology from "Complutense" University of Madrid (1988 and 1993, respectively). Since 1990, he has worked with several research groups: "Laboratoire de Neurosciences Fonctionnelles" Unité de Sensori-Motricité Comparée et Comportements - Centre National de la Recherche Scientific, Marsella (France), Dept. of Zoology and Genetics, Iowa State University, Ames (Iowa), "Unidad de Neurofisiología del Instituto de Investigaciones Biológicas Clemente Estable" and Depto. of Phisiology of the Faculty of Medicine in Montevideo (Uruguay), etc. He is currently a "Titular" scientific of the "Consejo Superior de Investigaciones Científicas (CSIC)" in the Neural Plasticity Department, Cajal Institute in Madrid. His research interests covers neurophysiology, neural plasticity, glial system, etc. Dr. Araque is the author of more than 40 published papers from 1991 and is a member of honour societies including the Society for Neuroscience (EE.UU.), European Neuroscience Association, "Sociedad Española de Neurociencias", etc.

Preface

When reading in the book title the words "artificial intelligence", the image of a computer which possesses that amazing capacity characterizing human beings immediately springs to mind. This capacity is nothing but the result of several biological processes which faithfully follow the laws of physics and chemistry but which at present remain mostly unknown. A computer is, perhaps, the best device to represent technology and, besides, its use may contribute to the development of science.

Science and technology may be considered as two sides of the same coin. Human beings have always used devices which have helped them to expand their knowledge and they have used that newly-acquired knowledge to build new mechanisms. It is impossible to tell which of both aspects prevails in this necessarily symbiotic process. One of the greatest scientific challenges nowadays lies in understanding closely the functioning of biological systems, knowing the complex relations among their elements and reasons for the behavior of live beings, as well as their mutual relations. In this sense, technology assists human beings in their consecution of smaller or greater achievements. It also contributes to a gradual reduction of the time required to overcome this challenge.

In parallel, if a greater understanding of biological systems is achieved, numerous scientific fields may benefit from it, since their work models or systems are based on the behavior of the said biological systems. This is true because a great number of scientific fields, and not just Artificial Intelligence, are searching for solutions to complex problems in biological processes, due to the lack of success in reaching good solutions with other strategies. This happens in areas such as engineering, computing, bioinformatics, etc. Anyhow, this is no easy task, given that any biological process is characterized by having a notorious complexity and being conditioned by multiple parameters which may result in different behaviors. With this borne in mind, it is certainly a hard task to get computers to simulate or even imitate biological processes.

The types of problems that people try to solve by considering biological processes are usually characterized by the absence of a complete mathematical model of the relevant phenomena, the existence of a large number of variables to be adjusted and conditions to be simultaneously satisfied, the presence of high degrees of non-linearity, and the formulation of multi-purpose tasks with a combinatorial explosion of candidate solutions, among other challenging scenarios which become even more complicated due to the need to act in real time in most occasions. Another motivation to be highlighted is the realization that most complex problems have similar versions in the natural world. Adaptation, self-organization, communication, and optimization have to be performed by biological organisms, in parallel and in different scales and structural levels, in order to survive and maintain life. Nature is constantly developing efficient problem-solving techniques and these are now beginning to be explored by engineers, computer scientists and biologists for the design of artificial systems and algorithms to be applied to the greatest variety of problems.

There are many systems based upon biological processes. The initial section of the present book focuses on the description of biological processes whose behavior has very recently been discovered. Numerous models and techniques which imitate nature can be developed within the scope of Artificial Intelligence. These are portrayed in the second section of the book. Among all of them, we could highlight, e.g., connectionist systems as models of the human brain (McCulloch & Pitts, 1943), evolutionary computing which imitates the processes of species evolution (Holland, 1975, 1992; Koza, 1992; Whitley, 1999), biomolecular computation, which is a scientific discipline concerned with processing information encoded in biological macromolecules such as DNA, RNA or proteins, although the most important advances have been made using DNA (Adleman, 1994; Lipton, 1995; Rodríguez-Patón, 1999), swarm-intelligence systems based heavily on the emergence of global behavior through local interactions of components (Bonabeau et al., 1999), robotics, etc. Moreover, these systems once built can be applied to multiple scientific fields. The third section of this book presents some examples of these applications in a series of very interesting works developed by researchers from different centres across the five continents.

ARTIFICIAL INTELLIGENCE AND ITS RELATION TO OTHER SCIENCES

It may be generally stated that artificial intelligence is a branch of science whose aim is to build systems and machines showing a type of behavior which, if developed by a person, would be termed as "intelligent". Learning, the capacity to adapt to changing environments, creativity, etc., are skills usually related to intelligent behavior.

It is assumed that the origins of Artificial Intelligence lie in the attempts by human beings to increase their physical and intellectual capabilities by creating artifacts with automatisms, simulating their shape and skills.

The first written reference to an intelligent artificial system appears in Greek mythology, specifically in The Iliad, and it is attributed to Hephaestus, god of fire and forges, who made "two servants of solid gold with intelligent minds and the ability to speak".

In the Middle Ages, Saint Albert the Great built a "butler" who would open the door and greet visitors.

In the Modern Age (s. XVII) the Drozs, famous Central European watch-makers, built three androids: a boy who wrote another one who drew and a girl who played the organ and pretended to breathe. This achievement was based on clock-work devices and it was the reason for their arrest and imprisonment by the Inquisition.

As regards innovations in the XIX and XX centuries, we should note the works by Pascal, Leibnitz, Babbage and Boole. Also Ada Lovelace, Babbage's collaborator and Lord Byron's wife, who wrote in the well-known Lovelace's regime, that: "machines can only do everything that we can tell them how to do. Their mission is to help facilitate the already known". This is still in force for "conventional computing" but it was left behind by the progress made in artificial intelligence, thanks to automatic learning techniques, among other things.

The immediate origin of the concept and development criteria of artificial intelligence goes back to the insight by the English mathematician Alan Turing (1943), inventor of the "COLOSSUS" machine created to decipher the messages encrypted by Nazi troops through the "ENIGMA" machine, while the "Artificial Intelligence" term was coined by John McCarthy, one of the members of the "Dartmouth Group", which gathered in 1956 with funding from the Rockefeller Foundation in order to discuss the possibility to build machines which would not be limited to making pre-fixed calculations, but "intelligent" operations. Among the group members were: Samuel (1959), who had written a chequers program capable of learning from its own experience, therefore overcoming the already mentioned Lovelace's regime; McCarthy (1956), who studied systems capable of making common-sense reasoning; Minsky (1997), who worked with analogical geometry reasoning; Selfridge (1986), who studied computer-aided visual recognition; as well as Newell, Shaw and Simon, who had written a program for the automatic demonstration of theorems.

Two main artificial intelligence "schools" were born from this initial group: Newell and Simon (1963) were the leaders of the Carnegie-Mellon University Team, with the intention to develop human behavior models with devices whose structure resembled the brain as close as possible, which later on led to the "connectionist" branch and to the works on "artificial neural networks" also known as "connectionist systems".

McCarthy and Minsky integrated another team in the mythical Massachusetts Institute of Technology (MIT), focusing on the processing results having an intelligent nature, without being concerned whether the components functioning or structure are similar to those of human beings or not.

Nevertheless, both approaches correspond to the same priority goals of Artificial Intelligence: understanding natural human intelligence and using intelligent machines in order to acquire knowledge and to solve problems which are considered to be intellectually hard. But in traditional artificial intelligence (MIT) which was more successful in the first twenty years, researchers found out that their systems collapsed before the increasing length and complexity of their programming. Stewart Wilson, a researcher from the Roland Institute (Massachusetts), realised that something was wrong in the field of "traditional" artificial intelligence. He wondered about the roots of intelligence and was convinced that the best way to understand how something works in a human being is to understand it first in a simpler being. Given that, essentially, artificial intelligence tried to replicate human intelligence, Wilson decided that first he should replicate animal intelligence (Meyer & Wilson, 1991). This idea had never been popular among Artificial Intelligence researchers but he, together with others, soon turned it into a first informal principle of a new approach to Artificial Intelligence based on nature. From that point on, various Artificial Intelligence techniques based on biological principles were created and continue to surface.

Therefore, numerous researchers coming from different scientific fields have taken part in the creation and setting of the artificial intelligence foundations. Following what has been mentioned at the beginning of the present book, about what artificial intelligence suggests at first sight, we may find a clear interrelation between the various branches of science: computer science (computer) – neuroscience (intelligence). Other relations appear between branches of science which are, in principle, very different, such as: computer science – psychology, computer science – sociology, computer science – genetics, etc. All of these interrelations which gradually surfaced through time have led to innovations which benefit at least one of the two related scientific areas.

Consequently, although it cannot be determined which is the best way to proceed, it is clear that both in artificial intelligence and in its foundations or application areas, interdisciplinarity and dissemination of results do facilitate reaching relevant conclusions. If we tried to reach those conclusions only from computer science, they would have never been achieved. Therefore, it is obvious that science should not be based only on proofs obtained from a single perspective, since this would limit its scope of action.

Other viewpoints should also be considered, viewpoints based upon different theories which bring about ideas facilitating the generation of hypothesis about the phenomenon to be studied on each occasion.

ORGANIZATION OF THIS BOOK

This book contains 20 contributed chapters written by internationally renowned researchers and it is organized into three main sections. A brief description of each of the sections follows:

Section I presents recent advances in the biological processes related to information processing. Such advances refer to the area of neuroscience. The authors not only show the results of the study of biological phenomena, but they also propose several options for the later ones to be the basis of the elaboration of new Artificial Intelligence models and techniques.

Chapter I. Recent electrophysiological studies indicate the existence of an important somatosensory processing in the trigeminal nucleus which is modulated by the corticofugal projection from the somatosensory cortex. This chapter studies a new mathematical analysis of the temporal structure of neuronal responses during tactile stimulation of the spinal trigeminal nucleus.

Chapter II. Knowledge in invertebrate neuroethology has demonstrated unique advantages for engineering biologically-based autonomous systems. This chapter aims at presenting some basic neuronal mechanisms involved in crayfish walking and postural control involving a single key joint of the leg. Due to its relative simplicity, the neuronal network responsible for these motor functions is a suitable model for understanding how sensory and motor components interact in the elaboration of appropriate movement and, therefore, for providing basic principles essential to the design of autonomous embodied systems.

Chapter III. Reviews the underlying mechanisms and theoretical implications of the role of voltage-dependent dendritic currents on the forward transmission of synaptic inputs. The notion analysed brakes with the classic view of neurons as the elementary units of the brain and attributes them computational/storage capabilities earlier billed to complex brain circuits.

Section II illustrates some of the more recent biologically inspired Artificial Intelligence models involved in information processing. The biological processes that serve as inspiration for the models are related to the fields of neuroscience, molecular biology, social behaviour, etc. Not only the different approaches for the design and construction of these models are reviewed, but also their different areas of application are introduced.

Chapter IV. This is a quick survey of spiking neural P systems, a branch of membrane computing which was recently introduced with motivation from neural computing based on spiking.

Chapter V. This chapter presents an evolution of the recurrent ANN (RANN) to enforce the persistence of activations within the neurons to create activation contexts that generate correct outputs through time. The aim of this work is to develop a process element model with activation output much more similar to the biological neurons one.

Chapter VI. It presents the functioning methodology of the artificial neuroglial networks; these artificial networks are not only made of neurons, like the artificial neural networks, but also they are made of elements which try to imitate glial cells. The application of these artificial nets to classification problems is also presented here.

Chapter VII. This describes the experience gained when developing the path generation modules of autonomous robots, starting with traditional artificial intelligence approaches and ending with the most recent techniques of evolutionary robotics. Discussions around the features and suitability of each technique, with special interest on immune-based behavior coordination are proposed to meet the corresponding theoretical arguments supported by empirical experiences.

Chapter VIII. In this chapter, two important issues concerning associative memory by neural networks are studied: a new model of Hebbian learning, as well as the effect of the network capacity when retrieving patterns and performing clustering tasks.

Chapter IX. The contents a computational model which is inspired in the biologically morphogenesis ideas. This chapter contains the theoretical development of the model and some simple tests executed over an implementation of the theoretical model.

Chapter X. This shows the interrelations between computing and genetics, which both are based on information and, particularly, self-reproducing artificial systems.

Chapter XI. It discusses guidelines and models of mind from cognitive sciences in order to generate an integrated architecture for an artificial mind that allows various behavior aspects to be simulated in a coherent and harmonious way, showing believability and computational processing viability.

Chapter XII. The chapter presents a general *central pattern generator* (CPG) architecture for legged locomotion. Based on a simple discrete distributed synchronizer, the use of *oscillatory building blocks* (OBB) is proposed for the production of complicated rhythmic patterns. An OBB network can be easily built to generate a full range of locomotion patterns of a legged animal. The modular CPG structure is amenable to very large scale circuit integration.

Section III presents real-life applications of recent biologically inspired artificial intelligence models and techniques. Some works use hybrid systems which combine different bioinspired techniques. The application of these models to such different areas as mathematics, civil engineering, computer science, biology, etc., is described here.

Chapter XIII. This chapter tries to establish, the characterization of the multimodal problems and offers a global view of some of the several approaches proposed for adapting the classic functioning of the genetic algorithms to the search of multiple solutions. The contributions of the authors and a brief description of several practical cases of their performance at the real world will be also showed.

Chapter XIV. It focuses on the description of several biomolecular information-processing devices from both the synthetic biology and biomolecular computation fields. Synthetic biology and biomolecular computation are disciplines that fuse when it comes to designing and building information processing devices.

Chapter XV. Swarm-intelligence systems involve highly parallel computations across space, based heavily on the emergence of global behavior through local interactions of components. This chapter describes how to provide greater control over swarm intelligence systems, and potentially more useful goal-oriented behavior, by introducing hierarchical controllers in the components.

Chapter XVI. It proposes a bio-inspired approach for the construction of a self-organizing Grid information system and also describes the SO-Grid portal, a simulation portal through which registered users can simulate and analyze ant-based protocols. This chapter can foster the understanding and use of swarm intelligence, multi-agent and bio-inspired paradigms in the field of distributed computing.

Chapter XVII. It presents graph based evolutionary algorithms (GBEA). GBEA are a generic enhancement and diversity management technique for evolutionary algorithms. GBEA impose restrictions on mating and placement within the evolving population (new structures are placed in their parents' graph neighborhood). This simulates natural obstacles like geography or social obstacles like the mating dances of some bird species to mating.

Chapter XVIII. It aims to explain and analyze the connection between *artificial intelligence* domain requirements and the *theory of systems with several equilibria*. This approach allows a better understanding of those dynamical behaviors of the artificial *recurrent neural networks* which are desirable for their proper work, i.e., achievement of the tasks they have been designed for.

Chapter XIX. This involves an application of artificial intelligence in the field of civil engineering, specifically the prediction of longitudinal dispersion coefficient in rivers. Based on the concept of genetic algorithms and artificial neural networks, a novel data-driven method called GNMM (genetic neural mathematical method) is presented and applied to a very well-studied classic and representative set of data.

Chapter XX. The authors employ the fuzzy decision tree classification technique in a series of biological based application problems. A small hypothetical example allows the reader to clearly follow the included analytical rudiments, and the larger applications demonstrate the interpretability allowed through the use of this approach.

CONCLUSION

As time elapses, researchers are discovering the way living beings behave, the way their internal organs function, how intercellular substance exchange takes place and which is the composition of the smallest molecule. Most of these biological processes have been imitated by many scientific disciplines with the purpose of trying to solve different complex problems. One of these disciplines is artificial intelligence, where diverse research teams have analysed and studied from the late 19th century some biological systems implementing computational models to solve real life problems. Ramon y Cajal (1904), the Spanish Nobel prize winner, used the "daring" and currently used term Neural Engineering even before the 20th century started.

This book intends to be an important reference book. It shows recent research works of great quality about information processing. It presents biological processes, some of them recently discovered, which show the treatment of different life processes at distinct levels (molecular, neural, social, etc.) that can be applied to artificial intelligence. It also presents the benefits of the new techniques created from these processes, not only to Artificial Intelligence, but also to multiple scientific areas: computer-science, biology, civil engineering, etc.

REFERENCES

Adleman, L. M. (1994). Molecular computation of solutions to combinatorial problems. *Science,* 226, 1021-1024.

Bonabeau, E., Dorigo, M. & Theraulaz, G. (Eds.). (1999). *Swarm Intelligence: From Natural to Artificial Systems*. Oxford, UK: Oxford University Press US.

Holland, J. H. (1975). *Adaptation innatural and artificial systems.* Ann Arbor: The University of Michigan Press.

Holland, J. H. (1992). Genetic Algorithms. *Scientific American*, 267(1), 66-72.

Koza, J. (1992). *Genetic Programming. On the programming of computers by means of natural selection.* Cambridge, MA: MIT Press

Lipton, R. J. (1995). DNA solutions of hard computational problems. *Science*, 268, 542-545.

McCarthy, J. (1956). Inversion of Functions Defined by Turing Machines. *Automatica Studies*. Pricenton Universities Press.

McCulloch, W.S., & Pitts, W. (1943). A Logical Calculus of Ideas Immanent in Nervous Activity. *Bulletin of Mathematical Biophysics*, 5, 115-133.

Meyer, J. & Wilson, S. (Eds.). (1991). From Animals to Animats. *Proceedings of the First International Conference on Simulation of Adaptive Behavior*. Cambridge, MA: MIT Press

Minsky, M. (1967). *Computation: Finite and Infinite Machines, Englewood Cliffs*, N.J.: Prentice-Hall.

Newell, A. & Simon, H. A. (1963). GPS: A Program that Simulates Human Thought, in Feigenbaum, E.A. *& Feldman, J., Computers and Thought, McGraw-Hill.*

Ramón y Cajal, S. (1904). *Textura del Sistema Nervioso del Hombre y los Vertebrados. Tomo II.* Madrid.

Rodríguez-Patón, A. (1999). *Variantes de la concatenación en computación con ADN.* PhD thesis, Universidad Politécnica de Madrid.

Samuel, A.L. (1959). Some studies in machine learning using the game of checkers. *IBM Journal of Research and Development* 3 (3): 210-219.

Selfridge, M. (1986). A computer model of child language learning. *Artificial Intelligence*, 29:171-216.

Turing, A. (1943). *Computing machinery and intelligence*. Cambridge, MA: MIT Press.

Whitley, D. (1995). Genetic algorithms and neural networks. In *Genetic Algorithms in Engineering and Computer Science* (vol. 11, pp. 1-15). John Wiley & Sons Ltd.

Acknowledgment

The editors would like to acknowledge the help of all the people involved in the book, without whose support the project could not have been satisfactorily completed.

We wish to thank all of the authors for their insights and excellent contributions to this book. Deep appreciation and gratitude is due to Alfonso Araque, Ángel Núñez, Didier LeRay, George Paun, Juan Pazos, José Mª Barreiro, and Alfonso Rodríguez-Patón for your great interest from the beginning of this project.

Furthermore, most of the authors of chapters included in this book also served as referees for chapters written by other authors. Thanks go to all those who provided constructive and comprehensive reviews. Our acknowledgement is also due to whom asked to be part of the present book and whose works were not finally selected; we would like to highlight the good quality of those works despite not being fitted as far as the field approached in this book is concerned.

Special thanks also go to the publishing team at IGI Global, whose contributions throughout the whole process from inception of the initial idea to final publication have been invaluable. In particular to Ross Miller, Jessica Thomson, and Heather Probst, who continuously prodded via e-mail for keeping the project on schedule and to Jan Travers, Kristin Roth, and Mehdi Khosrow-Pour, whose enthusiasm motivated us to initially accept his invitation for taking on this project.

In closing, we also want to thank the resources and the support of the staff from RNASA/IMEDIR Laboratory (Artificial Neural Network and Adaptive Systems / Medical Informatics and Radiological Diagnosis Center) at the University of A Coruña, Spain and from the Electrophysiology Cellular Laboratory at the Cajal Institute in Madrid, Spain.

Ana Porto wants to thank her parents and grandparents Elvira and Carlos for their guidance, unfailing support, love and patience. She wants also to thank her boyfriend Gustavo for his love and constant encouragement and Mª Luisa Gómez de Las Heras for her inestimable help.

Alejandro Pazos wants to thank to Alejandro, Carla & Loly for their love and comprehension to his professional circumstances.

Washington Buño wants to thank to his wife, children and grandchildren for their love and support.

Ana Porto, Alejandro Pazos & Washington Buño
Editors
A Coruña & Madrid. Spain
May 2008

Section I
Recent Advances in Biological Processes Related to Information Processing

Chapter I
Corticofugal Modulation of Tactile Responses of Neurons in the Spinal Trigeminal Nucleus: A Wavelet Coherence Study

Eduardo Malmierca
Universidad Autónoma de Madrid, Spain

Nazareth P. Castellanos
Universidad Complutense de Madrid, Spain

Valeri A. Makarov
Universidad Complutense de Madrid, Spain

Angel Nuñez
Universidad Autónoma de Madrid, Spain

ABSTRACT

It is well know the temporal structure of spike discharges is crucial to elicit different types of neuronal plasticity. Also, precise and reproducible spike timing is one of the alternatives of the sensory stimulus encoding. This chapter studies a new mathematical analysis of the temporal structure of neuronal responses during tactile stimulation of the spinal trigeminal nucleus. We have applied the coherence analysis and the wavelet based approach for quantification of the functional stimulus - neural response coupling. We apply this mathematical tool to analyze the decrease of tactile responses of trigeminal neurons during the simultaneous application of a novel tactile stimulation outside of the neuronal receptive field (sensory-interference). These data suggest the existence of an attentional filter at this early stage of sensory processing.

INTRODUCTION

Repetitive synaptic activity can induce persistent modification of synaptic efficacy in many brain regions in the form of long-term potentiation (LTP) and long-term depression (LTD). Such synaptic plasticity provides a cellular mechanism for experience dependent refinement of developing neural circuits and for learning and memory functions of the mature brain. The precise timing of presynaptic and postsynaptic spiking is often critical for determining whether an excitatory synapse undergoes LTP or LTD.

This chapter proposes a new mathematical analysis to study the temporal structure of neuronal responses to sensory stimulation. We have applied the coherence on the wavelet based approach for quantification of the functional stimulus - neural response coupling and its modulation when two tactile stimuli appear simultaneously. This analysis reveals that modulation of sensory responses may imply an increase/decrease in the number of spikes elicits by a sensory stimulus and an increase/decrease in the temporal coherence of evoked spikes with the stimulus onset, as well.

Recent electrophysiological studies indicate the existence of an important somatosensory processing in the trigeminal nucleus which is modulated by the corticofugal projection from the somatosensory cortex. The somatosensory cortex may enhance relevant stimulus. Also, it may decrease sensory responses in the trigeminal nuclei when a novel (distracter) stimulus is applied. We interpret this decrease of the response as sensory-interference. The objective of the present chapter is to demonstrate that sensory interaction may occur in the first relay station of the trigeminal somatosensory pathway changing the number of spikes evoked by a tactile stimulus and temporal coherence with the stimulus onset. Data suggest the existence of an attentional filter at this early stage of sensory processing.

BACKGROUND

Somatosensory information coming from the face (including the mouth and the cornea) is collected, processed and finally sent to the thalamus by the trigeminal complex. For experimental study of the mechanism of information representation and processing the vibrissae sensory system of rodents is one of the most used models since it is particularly well organized and structured. Indeed, the large mystacial vibrissae of the rat are arranged in a characteristic, highly conserved array of five rows and up to seven arcs (Brecht et al., 1997; Welker, 1971). Rats use these facial whiskers to perform a variety of tactile discriminative tasks and behaviors (Carvell & Simons, 1990; Gustafson & Felbain-Keramidas, 1977). Sensory information from the vibrissae arrives to the trigeminal complex, which is organized in three sensory and one motor nuclei. The sensory trigeminal nuclei include: the principal nucleus (Pr5), the spinal nucleus (Sp5) and the mesencephalic nucleus (Me5). In turn Sp5 is divided into three subnuclei called oralis (Sp5O), interpolaris (Sp5I) and caudalis (Sp5C). In the trigeminal complex primary afferents and neurons form the "barrelettes", which replicate the patterned arrangement of the whisker follicles on the snout (Ma, 1991).

Three classes of morphologically and physiologically distinguishable neurons reside in the rat trigeminal nucleus: barrelette cells, interbarrelette cells, and GABAergic or glycinergic inhibitory interneurons (Ressot et al., 2001; Viggiano et al., 2004).

The Pr5 and Sp5 trigeminal nuclei are obligatory synaptic relays for somatic sensory information originated in the large mystacial vibrissae or "whiskers'' on one side of the face to the contralateral ventral posterior medial (VPm) nucleus of the thalamus (Peschanski, 1984; Smith, 1973). Pr5 projection neurons are characteristically described as having single-whisker receptive fields (RFs), whereas the rest of the population has RFs composed of multiple whiskers (Friedberg et al., 2004; Veinante & Deschênes 1999). In contrast, Sp5 thalamic neurons typically respond to more than four whiskers, and thus have larger RFs (Friedberg et al., 2004; Woolston et al., 1982). Thus, at the single-neuron level, there is a high degree of integration of tactile inputs from multiple whiskers in most trigeminal neurons.

In the contralateral thalamus (VPm) the information from the barrelettes is received by groups of neurons that are called berreloids (Sugitani et al., 1990). The sensory information arrives through the lemniscal pathway. Moreover, the paralemniscal pathway transports less precise information from Sp5 neurons to the contralateral posterior thalamic nucleus (Po), and also to VPm (Patrick & Robinson, 1987).

There is also a similar organization in the primary somatosensory (SI) cortex. The information of the vibrissae is organized in columns called barrels and each barrel receives information from a single or few vibrissae (Erzurumlu & Jhaveri, 1990; Hutson & Masterton, 1986).

There is also a feed-back projection from SI (and also from the primary motor cortex) to the trigeminal nuclei. This is a monosynaptic crossed projection going by the pyramidal tract with an extremely precise somatotopy (Dunn & Tolbert, 1982). The electrophysiological properties and the action of this projection are currently less known. Making a parallel, in the dorsal column nuclei this projection has shown to be relevant to modulation of somatosensory responses, being able to modify the sizes of the RF and increasing the acuity of the system (Malmierca & Nuñez, 1998, 2004, 2007; Mariño et al., 2000).

Synaptic Properties of Trigeminal Neurons

The peripheral and corticofugal inputs to trigeminal nuclei are glutamatergic (Feliciano & Potashner, 1995). There are also electrophysiological evidences of GABAergic activity within the trigeminal complex that control the RF size. GABAA antagonists increase the size of the RFs at the trigeminal nuclei, while the agonists reduce it (Takeda et al., 2000). Also glycinergic interneurons have been shown to inhibit the somatosensory responses at the trigeminal complex (Ressot et al., 2001).

Barrelette cells exhibit a monosynaptic EPSP followed by a disynaptic IPSP after stimulation of the trigeminal nerve. The disynaptic IPSP should be mediated by a feed-forward inhibitory circuit. GABAergic cells in the Pr5 most likely serve as inhibitory interneurons in this feed-forward circuit. An IPSP without the preceding EPSP can be evoked in barrelette cells by stimulating the trigeminal nerve, suggesting that there is a separate inhibitory circuit other than the feed-forward inhibitory circuit (Lo et al., 1999). Synaptic plasticity has been also described in the trigeminal complex. At day 1 after born (P1) when barrelettes are in their formative stage, high frequency stimulation of trigeminal ganglion afferents produces a long-term depression (LTD) in synaptic responses. Between P3–7 when barrelettes are consolidating, heightened activity of trigeminal ganglion axons leads to long-term potentiation (LTP) of the responses (Guido et al., 2004). This suggests that LTD can serve as a mechanism selectively eliminating multiple whisker inputs and LTP to consolidate inputs that are connecting.

Mathematical Tools to Analyze Neuronal Activity

Extracellular electrophysiological study of the neural activity results in simultaneous observation of several processes, e.g. tactile stimulation–neural response. Then the recorded activity can be considered as a multivariate process, where the variables available for analysis are the times of occurrence of spikes or local field potential. To address spike trains several statistical techniques to infer on the neuronal interaction have been developed both in time and in frequency domains.

The most common time-domain method to characterize association between two spike trains was and is still the cross-correlation histogram (Perkel et al., 1967), called peristimulus time histogram (PSTH) when neural response to stimulus events is studied. A waste amount of electrophysiological data show that neuronal spike trains may exhibit correlation; the occurrence of a spike at one time is not independent on the occurrence of spikes at other times, both within spike trains from single neurons and across spike trains from multiple neurons. The well extended opinion is that the presence of these correlations is a key element of the neural code. However, the absence of an event may be also considered as information bit. Such "negative logic" is less accepted by the neural community, in part due to difficulties of experimental and theoretical investigation. How to measure the absence of something? A possible solution is to use frequency domain methods.

In frequency domain, the Fourier transform estimates the power spectrum of the process providing a measure of the frequency content of the spike train. However, such transformation is known to have difficulties when dealing with point processes (Brillinger, 1978). To overcome some of them in the literature the use of multi-taper Fourier transform has been advocated (Jarvis & Mitra, 2001). Although the multi-taper Fourier transform usually provides a good estimate of the power spectrum, in the case of highly periodic spike trains (e.g. in experimental conditions of periodic sensory stimulation) it may fail to consistently represent the spectral density. A frequency domain analysis of a neural ensemble from spiking activity can be conducted by taking Fourier transforms of spike trains, and using these to compute the spectrum of individual trains and the cross-spectrum or coherence between pairs of spike trains. The spectral coherence is then a frequency-dependent measure of functional association (coupling) between two processes. It has two important advantages over its time domain counterpart: the normalization is not bin-size dependent, and it can be pooled across neuron pairs.

Although being useful, this method suffers from the fundamental assumption that the spike trains under study are generated by stationary stochastic processes. On the other hand, our understanding of the complicated interplay, communication, information exchange between cells and cell groups, and of the role of the intriguing dynamic phenomena they produce may stack in the framework of this hypothesis. To overcome this lack of balance between theory and experiment, first of all one requires a stronger emphasis on a systems oriented approach where mechanism-based modeling is used to establish a more complete and coherent picture. It also requires the use of concepts and methods from the rapidly growing fields of nonlinear dynamics and complex systems theory, and it requires the continuous development and improvement of tools that can help us interpret the information embedded in complex biological time series.

Relatively recently a new method of time series analysis that can, and has been designed to, cope with complex non-stationary signals has been introduced. The approach is based on the wavelet transform, technique providing high temporal resolution with a reasonably good balance in frequency resolution. Wavelet analysis is presumably one of the most powerful tools to investigate the features of a complex signal. The wavelet approach has shown its strength in connection with a broad range of application such as noise reduction, information compression, image processing, synthesis of signals, etc. The majority of its applications in neuroscience are in electroencephalographic recordings (see e.g. Alegre et al., 2003; Castellanos & Makarov, 2006; Goelz et al., 2000; Murata, 2005; Quiroga & Garcia, 2003; Schiff et al., 1994). However, there are few studies on synchronization between pair of spike trains (e.g. stimulus–neural response). In this direction the wavelet cross-spectrum has been used to analyze the phase-locked oscillations in simultaneously recorded spike trains in the motor area of rhesus monkeys trained to produce a series of visually guided hand movements according to changes in the target locations (Lee, 2002, 2003).

The first studies about the wavelet coherence are very recent (Grinsted et al., 2004; Klein et al., 2006; Lachaux et al., 2002; Le Van Quyen et al., 2001). The wavelet coherence, similarly to the spectral coherence, infers on the functional coupling between e.g. stimulus and neural response, but additionally it also provides the temporal structure of the coupling. Li et al. (2007) investigated the temporal interaction in CA1 and CA3 regions in rats with induced epilepsy using the wavelet coherence. In previous works (Pavlov et al., 2006, 2007) we advocated and illustrated the use of the wavelet transform for analysis of neural spike trains recorded in the trigeminal nuclei under tactile whisker stimulation.

MAIN THRUST OF THE CHAPTER

Previous results suggest that the temporal architecture of the spike response of a neuron is crucial to determine facilitation or depression of a postsynaptic neuron. To test this hypothesis we studied the response of trigeminal SP5 neurons to tactile stimulus and its modification when a novel, distracter stimulus appears simultaneously.

Methods

Data were obtained from 20 urethane-anaesthetized (1.6 g/Kg i.p.) young adult Wistar rats of either sex, weighting 180-250 g (from Iffa-Credo, France). Animals were placed in a stereotaxic device and the body temperature was maintained at 37°C. Supplemental doses of the anesthetic were given when a decrease in the amplitude of the EEG delta waves was observed. Experiments were carried out in accordance with the European Communities Council Directive (86/609/EEC).

Unit Recordings and Sensory Stimulation

Tungsten microelectrodes (2-5 MΩ; World Precission Instruments) were used to obtain single unit recordings in the SP5C (A: -14.3 mm, L: 3 mm from the bregma; H: 0.5 to 2 mm from the surface of the nucleus; according to the atlas of Paxinos and Watson 1986). Unit firing was filtered (0.3-3 KHz), amplified via an AC preamplifier (DAM80; World Precision Instruments) and fed into a personal computer (sample rate: 12 KHz) for off-line analysis using Spike 2 software (Cambridge Electronic Design, UK). Tactile stimulation was performed by an electronically gated solenoid with a probe of 1 mm in diameter that produced <0.5 mm whisker deflections. To avoid complex responses due to multiple deflections of the vibrissae, these were cut to 2 mm long, so the stimulation was reproducible and responses could be compared.

Experimental Protocol

Spontaneous spiking activity of the neuron was recorded during 30 s. Then neural response to deflections of the principal whisker was recorded in three following conditions:

a. **Control stimulation.** Tactile stimulation consisted in a sequence of 30 tactile pulses lasting 20 ms and delivered at the RF at 1 Hz rate was applied to the principal whisker.
b. **Ipsilateral sensory-interference.** During tactile stimulation of the principal whisker with the same characteristics as in (a) another tactile distracter stimulus was continuously applied with a hand-held brush to whiskers located outside of the RF of the recorded neuron.
c. **Contralateral sensory-interference.** The same as in (b) but the distracter stimulus was applied to the skin area that the recorded neuronal RF but located in the contralateral side of the body.

Data Analysis

Summed peristimulus time histograms (PSTHs) were calculated off-line with Spike 2 software running on a PC computer, using 2 ms bins. We considered a cellular response when the PSTH area in first 50 ms after the stimulus onset was at least 3 times bigger than the area corresponding to 50 ms preceding the stimulus. The latency of the sensory responses was measured as the time elapsing between the sensory stimulus onset and the largest peak in the PSTH. All data are shown as mean ± standard error.

Wavelet Transform of a Spike Train

The continuous wavelet transform (WT) of a signal $x(t)$ (e.g. spike train) involves its projection onto a set of soliton-like basis functions obtained by rescaling and translating along the time axis the so called "mother wavelet" Ψ:

$$W(p,z) = \frac{1}{\sqrt{p}} \int_{-\infty}^{\infty} x(t) \Psi^* \left(\frac{t-z}{p} \right) dt, \quad (1)$$

where parameters P and z define the wavelet time scale and localization, respectively. The choice of the function ψ depends on the research aim. To study rhythmic components of a signal the Morlet-wavelet is well suited:

$$\Psi(y) = \exp(j2\pi y)\exp(-y^2/2k_0^2), \quad (2)$$

where k_0 is a parameter, which can be tuned according to physical phenomena under study. In the wavelet transform (1) the time scale p plays the role of the period of the rhythmic component. Given a characteristic time scale (i.e., period) p the resolution of the wavelet in the time and frequency domains is given by

$$\delta t = ck_0 p, \quad \delta\omega = \frac{c}{k_0 p}, \quad (3)$$

where c is a constant of the order of unity. There is a trade-off between the frequency and time resolutions, small values of k_0 provide better time resolution, while using big k_0 improves frequency resolution. The commonly adopted value is $k_0 = 1$, and the limit $k_0 \to \infty$ corresponds to the Fourier transform. As we shall show further for our purpose $k_0 = 2$ is more suitable.

Since we deal with finite-length time series (spike trains) the evaluation of the wavelet spectrum (1) will have edge artifacts at the beginning and the end of the time interval. The cone of influence (COI) is the region in (p, z) plane where edge effects cannot be ignored. We define the size of the COI when the wavelet power is dropped by e^2 (Torrence & Compo, 1998), which gives $z = \sqrt{2}k_0 p$.

The spiking output (point process) of a neuron can be represented as a series of δ-functions at the times when action potentials occur:

$$x(t) = \sum_i \delta(t - t_i). \quad (4)$$

Representation (4) allows us to estimate analytically the wavelet-coefficients:

$$W(p,z) = \frac{1}{\sqrt{p}} \sum_i \exp\left(-j2\pi \frac{t_i - z}{a}\right) \exp\left(-\frac{(t_i - z)^2}{2k_0^2 p^2}\right) \quad (5)$$

Using the wavelet-transform (5) we can perform the time-frequency analysis of rhythmic components hidden in the spike train. Wavelet-coefficients can be considered as a parameterized function $W_p(z)$, where z plays the role of time.

Wavelet Power Spectrum and Coherence

The wavelet power spectrum of a spike train can be defined by

$$E(p,z) = \frac{1}{\sqrt{\pi} r k_0} |W(p,z)|^2 \quad (6)$$

where r is the neuron mean firing rate. The normalization factor in (6) ensures unit energy of the white-noise or "random" spike train (with equiprobable randomly distributed inter-spike intervals). Thus for a random spike train the energy is homogeneously distributed over all frequencies (periods) $\langle E(p) \rangle_z = 1$. Consequently, we quantify the power distribution in the train under study in units of the power of the random spike train with the same mean firing rate.

The global wavelet spectrum can be obtained from (6) by time averaging of the local (time dependent) spectrum:

$$E_G(p) = \int_0^T E(p,z)\,dz \qquad (7)$$

It provides an unbiased and consistent estimation of the true power spectrum (Percival, 1995).

Dealing with two spike trains N and M by analogy with the Fourier cross-spectrum we can introduce the wavelet cross-spectrum $W_{NM}(p,z) = W_N W_M^* / k_0 \sqrt{\pi r_N r_M}$. Then a normalized measure of association between two spike trains is the wavelet coherence (Grinsted et al., 2004):

$$C_{NM}(p,z) = \frac{\left| S\left(W_{NM}(p,z)/p\right)\right|^2}{S\left(E_N(p,z)/p\right) S\left(E_M(p,z)/p\right)} \qquad (8)$$

where S(•) is a smoothing operator (see for details: Torrence and Webster, 1998; Grinsted et al., 2004). The coherence definition (8) can eventually give artificially high values of coherence in the case of infinitesimally small values of the power spectrum of either or both signals (i.e., when $E(p,z) \sim 0$). To avoid this problem in numerical calculations we employ thresholding procedure setting to zero the coherence when either of the power values is below a threshold.

Two linearly independent spike trains have vanishing coherence, whereas $C(p,z) = 1$ indicates a perfect linear relationship between the spike trains at the scale p and localization z. Since we are interested in studying the coherence level (or functional coupling) between the stimulus events and neural response we focus on frequency band corresponding to the stimulus frequency, i.e., on $f = 1$ Hz, which corresponds to the scale $p = 1$s. To resolve well this frequency with minimal loosing in time resolution we set $k_0 = 2$. Then from (3) $\delta\omega \sim 1/2$ and $\delta t \sim 2$.

Although large amplitude of the coherence usually indicates the presence of a consistent phase relationship (coupling) between two spike trains in a given time interval, it is also possible that this may be a random casual variation in spike trains. Thus one should cross-check statistical significance of the observed coherence. The statistical significance of the wavelet coherence can be assessed relative to the null hypotheses that the two spike trains generated by independent stationary processes with given distribution of inter-spike intervals (ISIs) are not coherent. To evaluate the level of significance one can use surrogate data test (Theiler et al., 1992; Schreiber and Schmitz, 2000) with Monte-Carlo simulation to establish 95% confidence interval. The surrogate spike trains can be obtained from the original one by randomizing phase relations keeping intact other first order characteristics by means of, for instance, shuffling ISIs. To conclude positively on the connectivity between the stimulus train and neuronal response their coherence should be higher than the obtained significance level.

Stimulus Period Band and Power Spectrum of Ultraslow Oscillation of the Wavelet Coherence

The wavelet coherence allows studying temporal structure and variation of the functional coupling among stimuli and neural response. To quantify this variation we average the neural stimulus coherence over scales in a narrow band around the stimulus period. An estimate of the band limits can be obtain from (3): $p \in \left[(1+c/2\pi k_0)^{-1}, (1-c/2\pi k_0)^{-1}\right]$, which for $c = 2$ gives [0.86, 1.2] s. We shall refer to this band as the stimulus period band. Obtained this way coherence is a function of time $C(t)$, which then can be used to evaluate the power spectrum by the conventional Fourier transform.

RESULTS

Neuronal Population

Tactile responses were studied in the SP5C nucleus, whose participation in the sensory processing is less known. Twenty three single neuronal recordings in SP5C nucleus were included in this study. Selected neurons were silent under spontaneous conditions or displayed a low mean firing rate (0.9±0.89 spikes/s), and had a RF that corresponded to one or two whiskers. Tactile stimuli (pulses of 20 ms duration) deflecting the principal whisker evoked 2.1±0.4 spikes/stimulus in the control stimulation conditions with a mean latency of 20.1±0.9ms.

Sensory-Interference

Simultaneous application of another distractive tactile stimulus (a gentle tickling with a paintbrush for 30 s) on the contralateral whiskers reduced tactile responses from 2.1±0.4 spikes/stimulus to 1.7±0.3 spikes/stimulus (81% from control value; p= 0.02) in 16 out of 23 neurons (70%; Figure 1). We termed this decrease in tactile responses caused by the application of the distracter stimulus in the contralateral whiskers as contralateral sensory-interference, as it has been also described in the SI cortex (Alenda and Nuñez, 2004, 2007). The response of the remaining neurons was either not modified (n=2; 9%) or increased (n=5; 21%).

A similar sensory-interference effect was observed when the distracter tactile stimulus was applied on the ipsilateral whiskers during simultaneous stimulation of the principal whisker (ipsilateral sensory-interference). Tactile responses from the principal whisker were reduced to 1.6±0.5 spikes/stimulus (76% from control value; p=0.01) in 19 out of 23 neurons (82% of neurons; Figure 1). The remained four neurons were not affected. Figure 2 illustrates PSTHs of a typical neuron response in the three experimental

Figure 1. Percentage of I (sensory response Increase effect), D (sensory response Decrease effect) and No change in the cases for ipsilateral (black box) and contralateral (grey box) distraction with respect to control according the number of spikes evoked by each stimuli.

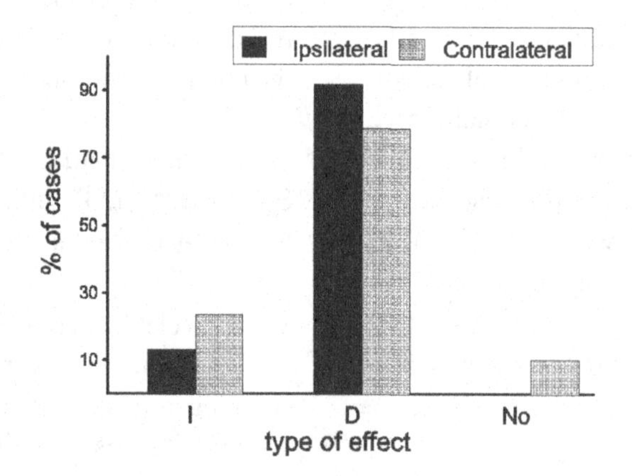

Figure 2. PSTHs of neural response to vibrissa deflection events in control conditions (A), and with parallel distraction in ipsilateral (B) and contralateral (C) sides

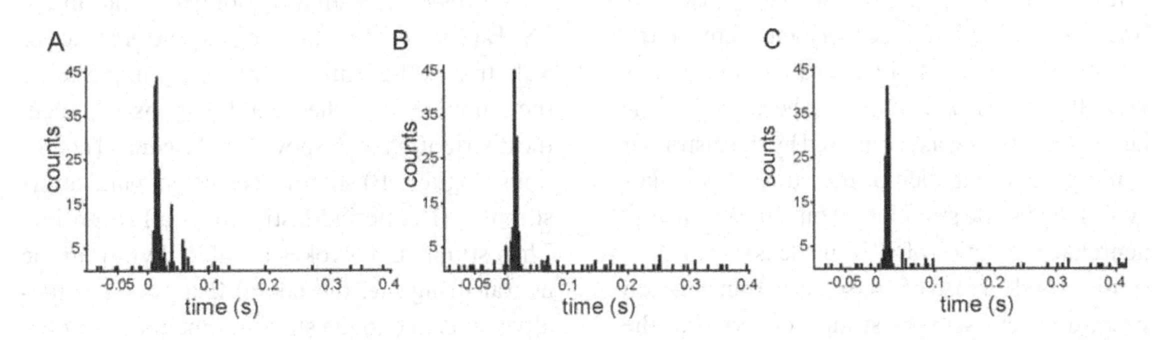

Figure 3. Neural spike trains (the same as used in Figure. 2) and corresponding wavelet power spectra. Color corresponds to the local spectral intensity. White area covers the cone of influence where the edge effects cannot be ignored. (A) Spontaneous activity. (B) Response to periodic vibrissa deflection events (1 Hz rate) at control conditions. (C, D) Response to tactile stimulation with ipsilateral and contralateral distraction, respectively. Tactile distraction reduces rhythmicity of the neural response to the vibrissa deflection.

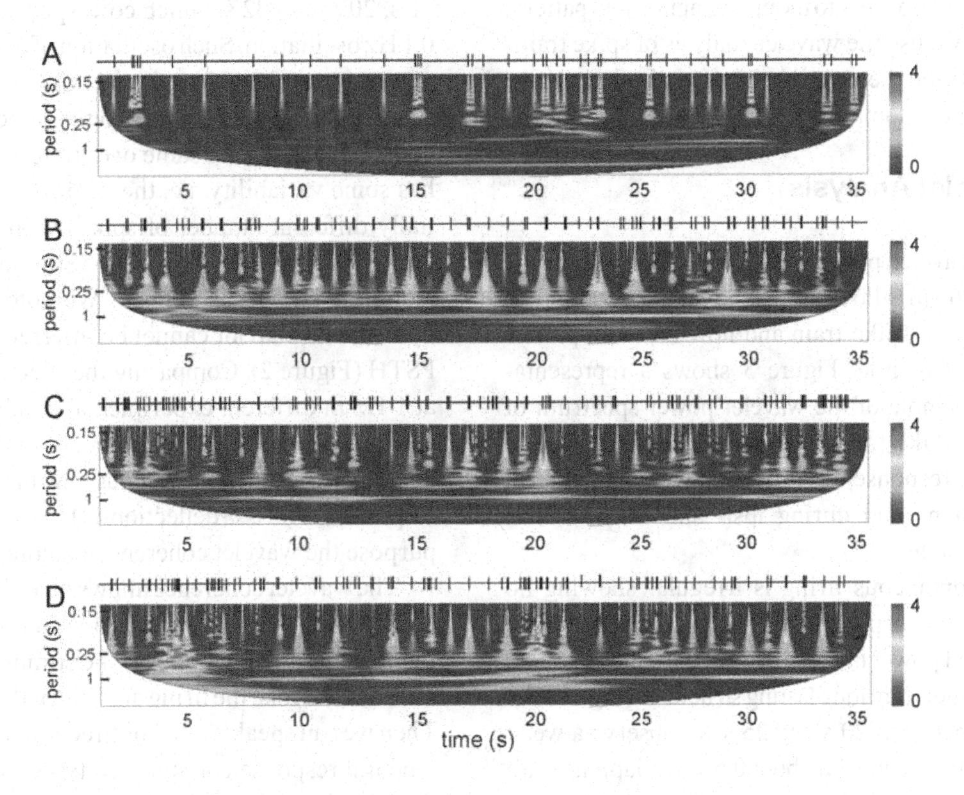

conditions showing a decrease of tactile responses during the application of a distracter stimulus. The spontaneous firing rate of trigeminal cells was not modified after sensory-interference trials compared to control values (0.9±0.9 spikes/s vs. 1±0.9 spikes/s). In all cases, the latency of the tactile responses was not altered by the distracter stimuli. This reduction of the number of spikes evoked by tactile stimulus (about 20%) in a large neuronal population of SP5C nucleus represents a significant decrease of the sensory transmission when different sensory stimuli occurred at the same time.

Our results demonstrate a clear tendency to decrease sensory responses in PSTHs during sensory-interference. However, changes in the spiking pattern evoked by the stimulus could not be studied with this method. The spiking pattern evoked by the stimulus is crucial in the sensory transmission because many studies have demonstrated that different types of sensory plasticity are elicited according to the presynaptic input pattern. Thus, we use the wavelet analysis of spike trains to describe the neural behavior in both time and frequency domains.

Wavelet Analysis

The wavelet power spectrum described in the equation (6) allows us to study the spectral properties of a spike train and how these properties change in time. Figure 3 shows a representative example of the wavelet power spectrum of neural spike-train in spontaneous condition and during response to vibrissa deflection in control conditions and during ipsi- and contra-lateral distractions.

Spontaneous firing is irregular showing no repetitive temporal patterns. Thus, maxima of the spectral power are localized at the spiking events and do not form long lasting structures (Figure 3A). Only between 20 s and 25 s we observe a weak rhythm with period about 0.5 s overlapping with 1.5 s period. Thus spontaneous spiking activity has no well defined dominant periodic activity. Tactile stimulation of the vibrissa leads to the neural response with a pronounced peak in the PSTH (Figure 2A). This suggests the presence of a rhythm at the stimulus frequency imposed by the stimulation in the neural response. Indeed, the distribution of the power in the control conditions (Figure 3B) shows a consistent band at the stimulus (1 s) period lasting over all recording. Thus stimulation evokes a stable rhythm in the neural firing, i.e. the neural firing is functionally associated to the stimulus and moreover the strength of this coupling is pretty constant over all 30 vibrissa deflections. Sensory distraction clearly influences the stability of the neural response to the vibrissa deflection (Figure 3C and 3D). The power of 1s period rhythm is decreased and becomes oscillatory, i.e. the peak amplitude is not persistent in time but instead exhibits a low frequency oscillation. For instance in Figure 3C (ipsilateral distraction) the power maxima are at 11 s, 20 s, and 32 s, which corresponds to about 0.1 Hz oscillation. Such oscillation of the spectral power are explained by the fact that during distraction the neural response to the same vibrissa deflections is not the same over time but instead has some variability, i.e. the neuron fires essentially different number of spikes with different inter-spike intervals to the same stimulus events along the stimulation epoch. We note that this dynamical behavior cannot be inferred from the PSTH (Figure 2). Comparing the spectral power at 1 Hz in different experimental conditions we can quantify the degree of influence of the tactile distraction on the stability (coupling) of the neural response to vibrissa deflections. However, for this purpose the wavelet coherence is more suitable.

The wavelet coherence allows quantifying the level of coupling between two systems analyzing their activity. In our case the stimulus onsets eventually evoke the firing activity of the neuron. Then we can speak about unidirectional stimulus – neural response coupling. To test the effect of the ipsi- or contralateral sensory-interference in

Figure 4. Wavelet coherences of the neural response to vibrissa deflection events evaluated in the stimulus period band [0.86-1.2] s. Color corresponds to the coherence level. Data correspond to Figures 2 and 3. (A) Coherence during control stimulation. (B, C) Coherence during stimulation with parallel ipsilateral or contralateral application of a sensory-interference (distractive) stimulus. Sensory distraction leads to an overall decrease of the coherence level and low frequency oscillations.

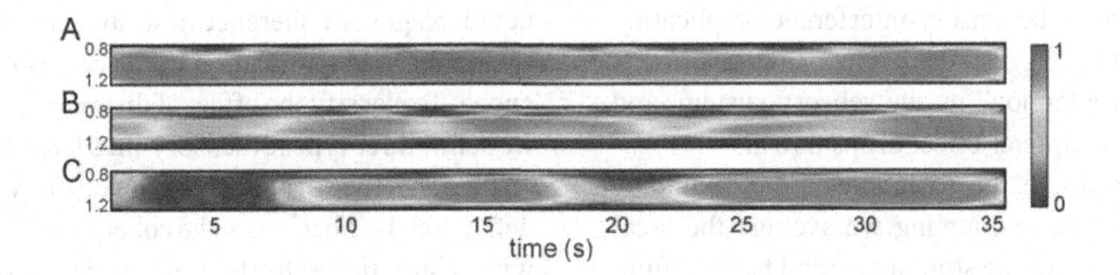

Figure 5. Averaged over stimulus period band wavelet coherence of the neural response to vibrissa deflection events for control, ipsilateral and contralateral epochs

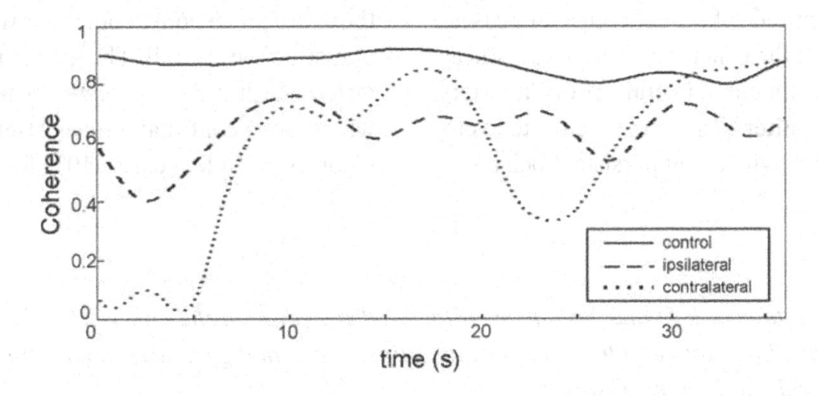

the stimulus processing we evaluate the wavelet coherence for each spike train in different epochs and compare coherences during sensory-interference with that found in the control conditions.

Figure 4 illustrates the wavelet coherence of the vibrissae stimuli and the evoked neural responses. Since the stimulation is periodic (1 s period) when speaking about the response coherence we shall refer to the stimulus period band only, i.e. [0.86, 1.2] s band. In control stimulation (Figure 4A), the neuron response was highly coherent to the sensory stimulation train. This evidences the presence of the stimulus-response association previously observed in the corresponding PSTH (Figure 2A) and in the power spectrum (Figure 3B). Moreover, as noticed above the strength of the sensory stimulus – neural response functional coupling is high and quite constant along the stimulation epoch.

Simultaneous application of the ipsilateral (Figure 4B) or contralateral (Figure 4C) distracter stimulus during tactile stimulation of the principal whisker decreases the coupling between the vibrissa deflection events and the sensory response of the trigeminal neuron. We also note that the sensory interference effect is not constant along the experiment. The maximum decrease of the sensory response coherence is observed at the beginning of the sensory-interference application period. This suggests that a novel stimulus attracts "highest attention" by the sensory system, and then the distracter effect drops down.

To examine the variation of the stimulus–neural response coupling we average the local coherence over the stimulus period band. Figure 5 shows obtained this way time series for the control conditions and during ipsi- and contralateral distractions. We observe that the mean level of the response coherence decreases when sensory-interference stimuli were applied and the oscillation amplitude of the coherence increases. The decrease of the coherence is more drastic at the beginning of the contralateral sensory-interference. However, after 10 s the coherence tends to recover although with a non persistent behavior. In this figure we can clearly observe slow oscillation appearing during sensory-interference and also note the high and constant coherent response to stimulus in the control condition.

Applying the above described analysis to each spike train we obtain characteristic values of the neural response coherence describing its reliability to vibrissae stimulation in the corresponding epoch. To classify the effect of distractive stimuli we define three types of sensory-interference types on the coherence of neural response to the vibrissa deflection: I-effect, when the coherence increases with distraction, D-effect, when the coherence decreases with distraction and No–effect when the coherence remains the same after distraction. Figure 6 shows percentage of cases for each type of the effect for ipsi- and contralateral distractions. There exists a clear tendency to decrease the neuronal response coherence with application of distractive stimuli. The ipsilateral sensory-interference diminishes coherence in 86% of cases, whereas the contralateral has D-effect in 70% of recordings. In few cases (10% for ipsilateral and

Figure 6. Percentage of I, D and No effects of tactile distraction on the coherence of neural response to vibrissa deflection relative to the control stimulation. Black and grey bars correspond to ipsilateral and contralateral distractions, respectively.

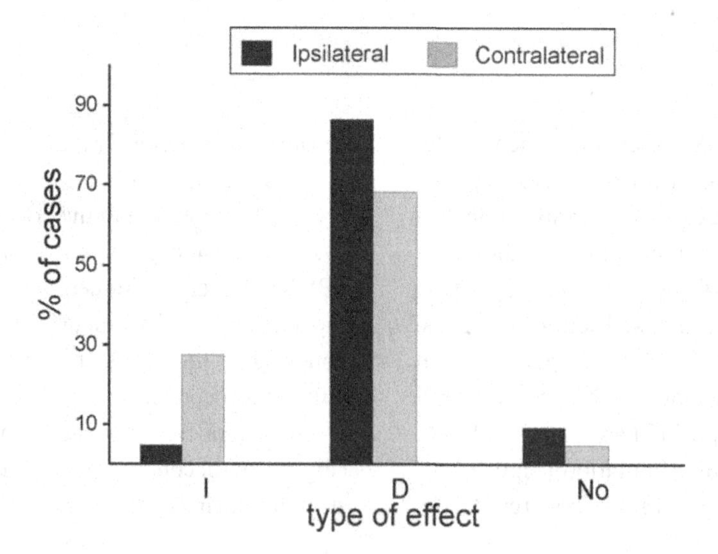

3% for contralateral) distraction had no effect on the neural response. Finally, in considerable number of cases (27%) distraction at the contralateral side led to the coherence enhancement. The same effect for distraction on the ipsilateral side has been observed in 4% cases only.

FUTURE TRENDS

Although some rough description of the somatosensory loop could be obtained on the basis of analysis of PSTHs, methods from the rapidly growing fields of nonlinear dynamics and complex systems theory can provide new insights on the underlying complex biological processes (Pavlov et al., 2006). In this line we have shown how the wavelet approach for analysis of non-stationary signals can be adapted to investigation of spike trains and can be used for quantification of the functional stimulus-neural response coupling. The coupling between two signals (i.e. stimulus and neural response) can be detected by means of the conventional spectral coherence. However, its Fourier transform origin limits its practical application to the neural spike trains. Here the wavelet coherence is better suited and can be directly evaluated from the spike trains. Similarly, we have studied the stimulus response coherence of the firing patterns in gracilis nucleus and its time evolution along stimulation epochs (Castellanos et al., 2007). This analysis has revealed that the coherence enhancement can be more relevant in sensory processing than an increase in the number of spikes elicited by the tactile stimulus.

For the optimal use of the wavelet coherence it is desirable: (1) to tune the Morlet parameter k according to the problem, and (2) to test coherence significance. For the constant frequency stimulations (1 Hz) we have found that the Morlet parameter $k = 2$ allows better resolving stimulus induced firing rhythm. The coherence significance affects our (statistical) believe at which extent the observed functional associations are not due to a chance. The latter can be tested by Monte-Carlo simulation with surrogate spike trains obtained by shuffling inter-spike intervals.

Neurons can communicate using time averaged firing rate (Shadlen & Newsome 1998). However, precise and reproducible spike timing is also frequently observed in the central nervous system, raising the likelihood that stimulus elicited firing patterns encode sensory information (Abeles et al. 1993; Bair & Koch 1996; Mainen & Sejnowski 1995). Thus, analysis methods studying the temporal coherence between evoked spikes and their stimulus may be an important tool to reveal synaptic plasticity in a neuronal network.

CONCLUSION

The somatosensory cortex may enhance relevant stimulus, as it occurs in the dorsal column nuclei (Malmierca & Nuñez, 1998, 2004) or may decrease sensory responses in the trigeminal nuclei when a novel (distracter) stimulus is applied (sensory-interference). This effect evoked by distracter stimuli applied outside of the RF of the recorded cell is observed even when the distracter stimulus is applied on the other side of the face. Consequently, these data suggest the existence of an attentional filter at this early stage of sensory processing. To evaluate this complex modulation of sensory responses we have created a new mathematical tool to reveal dynamic changes of synaptic interactions into a neuronal network.

By analysis of the wavelet power spectra we have shown that tactile stimulation of the neuron RF, as it was expected, strongly increases the spectral power in the stimulus frequency band. We have further demonstrated that the stimulus response coherence was decreased in about 86% for ipsilateral and in 70% for contralateral distractions with respect to the coherence measured in control stimulation. Moreover, the analysis showed that the coherence decay was larger immediately after the application of the distracter stimulus and then the stimulus coherence recovered but without reaching the value observed in the control conditions. According to previous results about synaptic plasticity, the decrease in the coherence has an important implication since it means a decrease of the temporal correlation between the tactile stimulus and the evoked spike train.

This decrease can be interpreted as sensory interference, as it has been previously described in the SI cortex (Alenda & Nuñez, 2004, 2007). In addition, it has been observed in the rodent cortex that deflection of ipsilateral whiskers suppresses cortical responses evoked by the contralateral whisker stimulation (Shuler et al., 2001). These data are consistent with the model where the evoked sensory activity from ipsilateral whiskers inhibits cortical responses to the stimulation of its RF that is located in the contralateral whiskers. According to this effect, the absence of ipsilateral whiskers to the somatosensory cortex increases the RF of the contralateral whiskers (Glazewski et al., 2007).

Sensory interference was observed as a decrease of the spike number elicited by a whisker deflection as well as a change in the temporal correlation between the stimulus onset and the evoked spike train (coherence). Both events are crucial in the sensory transmission since synchronized stimulus-driven discharges will induce EPSPs within a narrow temporal window in thalamic cells, increasing chances of generating spikes to be further transmitted to the somatosensory cortex.

Sensory interference evoked by ipsilateral distracter stimuli could be due to synaptic interaction into the trigeminal nucleus. However, the fact that a contralateral distracter stimuli also induce sensory interference in SP5C neurons, strongly suggests that this effect is mediated by corticofugal projections from the somatosensory cortex since there are no projections between trigeminal nuclei or thalamic nuclei of each side (Killackey, 1973).

Our results can be consistently interpreted under the assumption that somatosensory cortical neurons receive information from their contralateral RF and also from other RFs located in the contra- or in the ipsi- lateral side of the body. We propose that interhemispheric connections through the corpus callosum are a likely source of these sensory inputs. Callosal projections in somatosensory cortex are thought to be roughly homotopic (Olavarria et al., 1984) although other studies have indicated that there are few direct connections between barrel fields (Armstrong-James & George, 1988) and that callosal projections from the barrel field can be highly distributed over many different contralateral cortical areas, including SII and the motor cortex (Welker et al., 1988). Consequently, data suggest the existence of a complex sensory processing in the somatosensory cortex that modulates sensory responses in subcortical relay stations according to the appearance of novel sensory stimuli.

Resuming, cortical feedback enhances cortically relevant stimuli while decreases other sensory stimuli. This effect has been called "egocentric selection" (Jen et al., 2002) and play a pivotal role in the control of the sensory information that reaches the thalamus and cortex. Our results show that a distracter stimulus may reduce sensory responses in trigeminal neurons. For these reasons, a suggested role for the corticofugal feedback system is that it may contribute to selective attention (Malmierca & Nuñez, 2007). Selective attention is defined as the cognitive function that allows the focusing of processing resources onto the relevant sensory stimuli among the environmental information available at a given moment, while other irrelevant sensory information is largely ignored. Different studies have shown that, in addition to enhancing neuronal processing for attentionally relevant stimuli, selective attention also involves suppression of the sensory processing of distractive stimuli (Reynolds et al., 1999; Smith et al., 2000). Thus, corticofugal projections act as an attention filter that enhances relevant stimuli and reduces irrelevant ones according to the context and the behavioral task performed in the given time moment.

ACKNOWLEDGMENT

The authors are grateful to Dr. A. Pavlov for enlightening discussions. This research has been sponsored by the Santander-Complutense grant PR41/06-15058, by BFU2005-07486, FIS2007-65173, and by the Spanish Ministry of Education and Science under a Ramon y Cajal Grant.

REFERENCES

Abeles, M., Bergman, H., Margalit, E., & Vaadia E. (1993). Spatiotemporal firing patterns in the frontal cortex of behaving monkeys. *Journal Neurophysiology*, 70, 1629-1638.

Alegre, M., Labarga, A., Gurtubay, I., Iriarte, J., Malanda, A., & Artieda, J. (2003). Movementrelated changes in cortical oscillatory activity in ballistic, sustained and negative movements. *Experimental Brain Research*, 148,17–25.

Alenda, A., & Nuñez, A. (2004). Sensory-interference in rat primary somatosensory cortical neurons. *European Journal Neuroscience*, 19, 766-70.

Alenda, A., & Nuñez, A. (2007). Cholinergic modulation of sensory interferente in rat primary somatosensory cortical neurons. *Brain Research* 1133,158-167.

Armstrong-James, M., & George, M. J. (1988). Bilateral receptive fields of cells in rat Sm1 cortex. *Experimental Brain Research*, 70, 155-65.

Bair, W., Koch, C. (1996). Temporal precision of spike trains in extrastriate cortex of the behaving macaque monkey. *Neural Computation*, 8, 1185-202.

Brecht, M., Preilowski, B., & Merzenich, M. M. (1997) Functional architecture of the mystacial vibrissae. Behav *Brain Research*, 84, 81-97.

Brillinger, D. R. (1978). Developments in statistics. In *Comparative aspects of the study of ordinary time series and of point processes* (pp. 33–129). Orlando, FL: Academic Press.

Carvell, G. E., & Simons, D. J. (1990). Biometric analyses of vibrissal tactile discrimination in the rat. *Journal Neuroscience*, 10, 2638-2648.

Castellanos, N. P. & Makarov, V. A. (2006). Recovering EEG brain signals: artifact suppression with wavelet enhanced independent component analysis. *Journal Neuroscience Methods*, 158, 300-312.

Castellanos, N. P., Malmierca, E., Nuñez, A., & Makarov, V. A. (2007). Corticofugal modulation of the tactile response coherence of projecting neurons in the gracilis nucleus. *Journal Neurophysiology*, 98, 2537-2549.

Dunn R.C. Jr., & Tolbert, D. L. (1982). The corticotrigeminal projection in the cat. A study of the organization of cortical projections to the spinal trigeminal nucleus. Brain Research, 240, 13-25.

Feliciano, M., Potashner, S. J. (1995). Evidence for a glutamatergic pathway from the guinea pig auditory cortex to the inferior colliculus. *Journal Neurochemistry*, 65, 1348-1357.

Friedberg, M. H., Lee, S. M., & Ebner, F. F. (2004). The contribution of the principal and spinal trigeminal nuclei to the receptive field properties of thalamic VPM neurons in the rat. *Journal Neurocytology*, 33, 75-85.

Glazewski, S., Benedetti, B. L., & Barth, A. L. (2007). Ipsilateral whiskers suppress experience-dependent plasticity in the barrel cortex. *Journal Neuroscience*, 27, 3910-3920.

Goelz, H., Jones, R., & Bones, P. (2000). Wavelet analysis of transient biomedical signals and its application to detection of epileptiform activity in the EEG. *Clinical Electroencephalography*, 31, 181–191.

Grinsted, A., Moore, J. C., & Jevrejeva, S. (2004). Application of the cross wavelet transform and wavelet coherence to geophysical time series. *Nonlinear Processes in Geophysics*, 11, 561–566.

Gustafson, J. W., & Felbain-Keramidas, S. L. (1977). Behavioral and neural approaches to the function of the mystacial vibrissae. *Psychological Bulletin*, 84, 477-488.

Hutson, K. A., & Masterton, R. B. (1986). The sensory contribution of a single vibrissa's cortical barrel. *Journal Neurophysiology* 56, 1196-1223.

Jarvis, M. R. & Mitra, P. P. (2001). Sampling Properties of the Spectrum and Coherency of Sequences of Action Potentials. *Neural Computation*, 13, 717–749.

Jen, P. H. S., Zhou, X., Zhang, J., & Sun, X. (2002). Brief and short-term corticofugal modulation of acoustic signal processing in the bat midbrain. *Hear Research*, 168, 196-207.

Killackey, H. P. (1973). Anatomical evidence for cortical subdivisions based on vertically discrete thalamic projections from the ventral posterior nucleus to cortical barrels in the rat. *Brain Research*, 51, 326-331.

Klein, A., Sauer, T., Jedynak, A., & Skrandies, W. (2006). Conventional and wavelet coherence applied to sensory–evoked electrical brain activity. IEEE Trans. Biomed. Engineer. 53, 266-272.

Lachaux, J. P, Lutz, A., Rudrauf, D., Cosmelli, D., Le Van Quyen, M., Martinerie, J., & Varela, F. J. (2002). Estimating the time-course of coherence between single-trial brain signals: an introduction to wavelet coherence. *Neurophysiology Clinic*, 32, 157-174.

Le Van Quyen, M., Foucher, J., Lachaux, J. P., Rodriguez, E., Lutz, A., Martinerie, J., & Varela, F. J. (2001). Comparison of Hilbert transform and wavelet methods for the analysis of neuronal synchrony. *Journal Neuroscience Method,* 111: 83–98.

Lee, D. (2002). Analysis of phase-locked oscillations in multi-channel single-unit spike activity with wavelet cross-spectrum. *Journal Neuroscience Method*, 115, 67-75.

Lee, D., (2003). Coherent oscillations in neuronal activity of the supplementary motor area during a visuomotor task. *Journal Neuroscience,* 23, 6798-809.

Li X, Yao X, Fox J, Jefferys JG. Interaction dynamics of neuronal oscillations analysed using wavelet transforms. J Neurosci Meth 160: 178–185, 2007.

Lo, F. S,, Guido, W., Erzurumlu, R. S. (1999). Electrophysiological properties and synaptic responses of cells in the trigeminal principal sensory nucleus of postnatal rats. *Journal Neurophysiology,* 82, 2765-2775.

Ma, P. M. (1991). The barrelettes--architectonic vibrissal representations in the brainstem trigeminal complex of the mouse. I. Normal structural organization. *Journal Comparative Neurology,* 309, 161-199.

Mainen, Z. F., & Sejnowski, T. J. (1995). Reliability of spike timing in neocortical neurons. *Science.* 268, 1503-1506.

Malmierca, E. & Nuñez, A. (1998). Corticofugal action on somatosensory response properties of rat nucleus gracilis cells. *Brain Research*, 810, 172-180.

Malmierca, E. & Nuñez, A. (2004). Primary somatosensory cortex modulation of tactile responses in nucleus gracilis cells of rats. *European Journal Neuroscience*, 19, 1572-1580.

Malmierca, E. & Nuñez, A. (2007). Corticofugal modulation of sensory information. *Advances*

Anatomy Embryology Cell Biology, 187, 1-74.

Mariño, J., Canedo A. & Aguilar J. (2000) Sensorimotor cortical influences on cuneate nucleus rhythmic activity in the anesthetized cat. *Neuroscienc,e* 95, 657-673.

Martin, S. J., Grimwood, P. D., & Morris, R. G. (2000). Synaptic plasticity and memory: an evaluation of the hypothesis. *Annual Review Neuroscience,* 23, 649–711.

Murata, A. (2005). An attempt to evaluate mental workload using wavelet transform of EEG. *Human Factors*, 47, 498–508.

Olavarria, J., Van Sluyters, R. C., & Killackey, H. P. (1984). Evidence for the complementary organization of callosal and thalamic connections within rat somatosensory cortex. *Brain Research,* 291, 364-368.

Pavlov, A. N., Makarov, V. A., Mosekilde, E., & Sosnovtseva, O. V. (2006) Application of wavelet-based tools to study the dynamics of biological processes", *Briefings Bioinformatics,* 7, 375-389.

Pavlov, A. N., Tupitsyn, A. N., Makarov, V. A., Panetsos, F., Moreno, A., Garcia-Gonzalez, V., & Sanchez-Jimenez, A. (2007). Tactile information processing in the trigeminal complex of the rat. Proceedings of SPIE: Complex Dynamics and Fluctuations in Biomedical Photonics IV. Vol 6436, doi:10.1117/12.709155

Paxinos, G., & Watson, C. (1986). The Rat Brain in Stereotaxic Coordinates, Academic Publishing, New York.

Percival, D. P. (1995). On estimation of the wavelet variance. *Biometrika,* 82, 619–631.

Perkel, D. H., Gerstein, G. L., & Moore, G. P. (1967). Neuronal spike trains and stochastic point processes. II. Simultaneous spike trains. *Biophysical Journal,* 7, 419–440.

Peschanski, M. (1984). Trigeminal afferents to the diencephalon in the rat. *Neuroscience*, 12, 465-487.

Quiroga, R., Garcia, H. (2003). Single-trial event-related potentials with wavelet denoising. *Clinical Neurophysiology*, 114, 376–390.

Ressot, C., Collado, V., Molat, J. L., & Dallel, R. (2001). Strychnine alters response properties of trigeminal nociceptive neurons in the rat. Journal Neurophysiology 86, 3069-3072.

Reynolds, J.H., Chelazzi, L. & Desimone, R. (1999). Competitive mechanisms subserve attention in macaque areas V2 and V4. *Journal Neurosci*ence, 19, 1736-1753.

Schiff, S., Aldroubi, A., Unser, M., & Sato, S. (1994). Fast wavelet transformation of EEG. *Electroencephalography Clinical Neurophysiology*, 91, 442–455.

Schreiber, J. V, & Schmitz, A. (2000). Surrogate time series. *Physica D* 142, 646-652.

Shadlen, M. N., & Newsome, W. T. (1998). The variable discharge of cortical neurons: implications for connectivity, computation, and information coding. *Journal Neuroscience*, 18, 3870-3879.

Shuler, M. G., Krupa, D. J., Nicolelis, M. A. (2001). Bilateral integration of whisker information in the primary somatosensory cortex of rats. *Journal Neuroscience*, 21, 5251-5261.

Smith, A.T., Singh, K.D. & Greenlee, M.W. (2000). Attentional suppression of activity in the human visual cortex. *Neuroreport*, 7, 271-277.

Smith, R. L. (1973). The ascending fiber projections from the principal sensory trigeminal nucleus in the rat. *Journal Comparative Neurology*, 148, 423-436.

Song, S., Miller, K. D., & Abbott, L. F. (2000). Competitive Hebbian learning through spike-timing-dependent synaptic plasticity. Nature Neuroscience 3, 919–926.

Sugitani, M., Yano, J., Sugai, T., & Ooyama, H. (1990). Somatotopic organization and columnar structure of vibrissae representation in the rat ventrobasal complex. *Experimental Brain Research*, 81, 346-352.

Theiler, J., Eubank, S., longtin, A., Galdrikian, B., & Farmer, D. (1992). Testing for nonlinearity in time series: the method of surrogate data. *Physica D,* 58, 77–94.

Torrence, C., & Compo, G. P. (1998). A Practical Guide to Wavelet Analysis. *Bulletin of the American Meteorological Society,* 79, 61-78.

Tzounopoulos, T., Kim, Y., Oertel, D., & Trussell, L. O. (2004). Cell specific, spike timing-dependent plasticities in the dorsal cochlear nucleus. *Nature Neuroscience,* 7, 719–725.

Veinante, P. & Deschenes, M. (1999). Single- and multi-whisker channels in the ascending projections from the principal trigeminal nucleus in the rat. *Journal Neuroscience,* 19, 5085-095.

Viggiano, A., Monda, M., Viggiano, A., Chiefari, M., Aurilio, C., & De Luca, B. (2004). Evidence that GABAergic neurons in the spinal trigeminal nucleus are involved in the transmission of inflammatory pain in the rat: a microdialysis and pharmacological study. *European Journal Pharmacology*, 496, 87-92.

Welker, C. (1971). Microelectrode delineation of fine grain somatotopic organization of SM1 cerebral neocortex in albino rat. *Brain Research*, 26, 259–275.

Woolston, D. C., La Londe, J. R., & Gibson, J. M. (1982). Comparison of response properties of cerebellar- and thalamic-projecting interpolaris neurons. Journal Neurophysiology 48, 160-173.

ADDITIONAL READING

Armstrong-James, M., & Fox, K. (1987). Spatiotemporal convergence and divergence in the rat S1 "barrel" cortex. *Journal Comparative Neurology, 263*, 265-281.

Bell, C. C., Han, V. Z., Sugawara, Y., & Grant, K. (1997). Synaptic plasticity in a cerebellum-like structure depends on temporal order. *Nature 387,* 278–281.

Bliss, T. V. & Lomo, T. (1973). Long-lasting potentiation of synaptic transmission in the dentate area of the anaesthetized rabbit following stimulation of the perforant path. *Journal Physiology* (London) 232, 331–356.

Brillinger, D. R., Bryant, H. L. & Segundo, J. P. (1976). Identification of synaptic interactions. *Biological Cybernetic 22*, 213–229.

Dahlhaus, R., Eichler M., & Sandkühler J. (1997). Identification of synaptic connections in neural ensembles by graphical models. *Journal of Neuroscience Method, 77*, 93-107.

Donoghue, J. P. (1995). Plasticity of adult sensorimotor representations. *Current Opinion Neurobiology 5*, 749-754.

Dykes, R. W. (1983). Parallel processing of somatosensory information: A theory. *Brain Research Review 6*, 47-115.

Egger, V., Feldmeyer, D., & Sakmann B. (1999). Coincidence detection and changes of synaptic efficacy in spiny stellate neurons in rat barrel cortex. *Nature Neuroscience, 2*, 1098–1105.

Eichler, M., Dahlhaus, R., & Sandkühler J. (2003). Partial correlation analysis for the identification of synaptic connections. *Biological Cybernetics, 89*, 289-302.

Ito, M. (1985). Processing of vibrissa sensory information within the rat neocortex. *Journal of Neurophysiology 54*, 479-90.

Kaminiski, M. J., & Blinowska, K. J. (1991). A new method of the description of the information flow in the brain structures. *Biological Cybernetics 65*, 203-210.

Kaminski, M., Ding, M., Truccolo, W. & Bressler S. (2001). Evaluating causal relations in neural systems: Granger causality, directed transfer function and statistical assessment of significance. *Biological Cybernetics 85*, 145-157.

Korzeniewska, A., Manczak, M., Kaminski, M., Blinowska, K., & Kasicki, S. (2003). Determination of information flow direction among brain structures by a modified directed transfer function (dDTF) method. *Journal of Neuroscience Method 125*, 195-207.

Levy, W. B. & Steward, O. (1983). Temporal contiguity requirements for long-term associative potentiation/depression in the hippocampus. *Neuroscience 8*, 791–797.

Malenka, R.C. & Siegelbaum, S. A. (2001). *Synaptic plasicity.* In: Synapses, edited by Cowan WM, Sudhof TC, and Stevens CF. Baltimore, MD: Johns Hopkins Univ. Press.

Nicolelis, M. A. L., Baccala, L. A., Lin, R. C. S., & Chapin, J. K. (1995). Sensorimotor encoding by synchronous neural ensemble activity at multiple levels of the somatosensory system. *Science, 268*, 1353-1358.

Sameshima, K. & Baccalá, L. A. (1999). Using partial directed coherence to describe neuronal ensemble interactions. *Journal of Neuroscience Methods, 94*, 93-103.

Sjostrom, P. J., Turrigiano, G. G., & Nelson, S. B. (2001). Rate, timing, and cooperativity jointly determine cortical synaptic plasticity. *Neuron, 32*, 1149–1164.

Sumitomo I, Iwama K. Neuronal organization of rat thalamus for processing information of vibrissal movements. *Brain Research, 415*, 389-92, 1987.

White, E. L. (1979). Thalamocortical synaptic relations: a review with emphasis on the projections of specific thalamic nuclei to the primary sensory areas of the neocortex. *Brain Research Review, 1,* 275-311.

Chapter II
Neural Mechanisms of Leg Motor Control in Crayfish: Insights for Neurobiologically-Inspired Autonomous Systems

Didier Le Ray
Université de Bordeaux, Lab, MAC, France

Morgane Le Bon-Jego
Université de Bordeaux, Lab, MAC, France

Daniel Cattaert
Université de Bordeaux, CNIC, France

ABSTRACT

Computational neuroscience has a lot to gain from invertebrate research. In this chapter focusing on the sensory-motor network that controls leg movement and position in crayfish, we describe how simple neural circuitry can integrate variable information to produce an adapted output function. We describe how a specific sensor encodes the dynamic and static parameters of leg movements, and how the central motor network assimilates and reacts to this information. We then present an overview of the regulatory mechanisms thus far described that operate at the various levels of this sensory-motor network to organize and maintain the system into a dynamic range. On the basis of this simple animal model, some basic neurobiological concepts are presented which may provide new insights for engineering artificial autonomous systems.

INTRODUCTION

In the second half of the last century, diverse sets of computer or robotic models have appeared in an attempt either to explain or reproduce various features of behavior. Although the majority of models first arose from mathematical approaches, the use of ideas originating from biological studies are now pre-eminent in the construction of artificial organisms designed to achieve a given task. However, although such robots perform quite well (*e.g.*, the salamander robot by Ijspeert *et al.*, 2007), the actual animal's performance always appears much more efficient and harmonious. This probably results from the continuous interaction between motor activity and sensory information arising from the surrounding environment (Rossignol *et al.*, 2006), in association with the perfect use of the biomechanical apparatus. It is thus primordial to understand fully the neural mechanisms involved in such a dynamic interaction to be able to implement realistic integrative algorithms in the design of computational models.

In this context, knowledge in invertebrate neuroethology has demonstrated unique advantages for engineering biologically-based autonomous systems (*e.g.*, Schmitz *et al.*, 2001; Webb, 2002). Although invertebrates are able to generate complex adaptive behaviors, the underlying neuronal circuitry appears quite simple compared to vertebrates. This chapter aims at presenting some basic neuronal mechanisms involved in crayfish walking and postural control involving a single key joint of the leg. Due to its relative simplicity, the neuronal network responsible for these motor functions is a suitable model for understanding how sensory and motor components interact in the elaboration of appropriate movement and, therefore, for providing basic principles essential to the design of autonomous embodied systems. In walking legs for example, sensory information is provided by simple sensory organs associated with each joint, and integrated by relatively small populations of neurons within the central ganglia (see Cattaert & Le Ray, 2001). Here, we describe, the encoding signals generated by a specific sensor that monitors both upward and downward leg movements, as well as the integrative mechanisms used to process these signals within the central nervous system (CNS). The many ways of tuning sensory-motor processes are then presented, which allow the system to adjust perfectly the motor command and, consequently, to generate fully adapted behaviors (Clarac et al., 2000). Some possible applications and future research directions will conclude this chapter.

MULTI-SENSORY CODING OF LEG MOVEMENTS

Crayfish possess an external skeleton that allows movements only at the various joints, the movement of each joint being coded by simple sensory organs. Among these, the leg coxopodite-basipodite chordotonal organ (CBCO) plays a pivotal role in the control of locomotion and posture, since it monitors vertical leg movements (Figure 1). This proprioceptor consists of an elastic strand of connective tissue in which sensory cells are embedded and whose function is comparable to that of joint receptors in mammals (Clarac *et al.*, 2000). The CBCO strand is stretched during opening of the second, unidirectional joint and released during closure, which corresponds respectively to downward and upward movements of the leg. The sensing apparatus of the CBCO is composed of 40 neurons that are equally divided into 20 stretch-sensitive and 20 release-sensitive neurons that code depression and levation of the leg, respectively (see Cattaert & Le Ray, 2001).

Coding Movement Parameters

Joint movement can be monitored either as a displacement of one segment relative to the other (dynamic parameter) or in terms of the relative position of both segments (static parameter). In

Figure 1. Sensory-motor network controlling leg movements in crayfish. (A) Location of the coxo-basipodite chordotonal organ (CBCO) that encodes vertical movements of the crayfish walking leg. (B) In vitro preparation of the locomotor nervous system consisting of the last thoracic (T3-T5) and first abdominal (A1) ganglia, together with the levator and depressor nerves (Lev n, Dep n), and the sensory nerve (CBCO n) that can be recorded extracellularly. Simultaneous intracellular recordings can also be made from motoneurons (e.g., Dep MN) and CBCO terminals. A mechanical puller mimics natural vertical movements of the leg by stretching and releasing the CBCO strand.

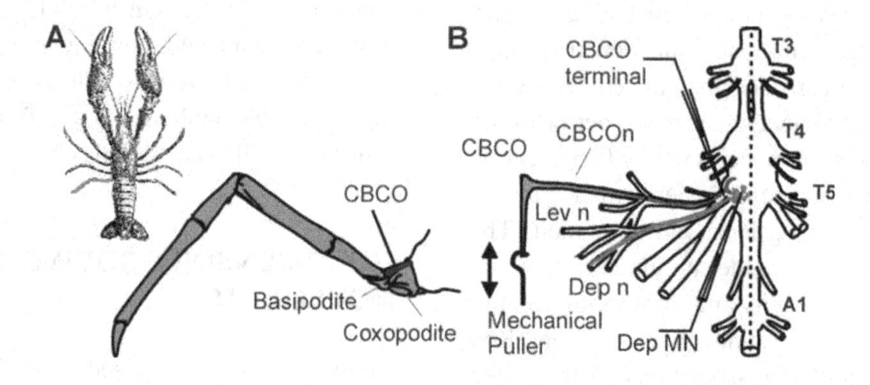

order to analyze in detail the coding capabilities of sensory afferents, a succession of 'ramp and plateau' stimulations were cyclically applied to the CBCO organ that had been isolated *in vitro* together with the crayfish CNS (see experimental arrangement in Figure 1B). In either release- or stretch-sensitive afferents recorded intracellularly the same three kinds of encoding were observed during monitoring of each joint movement parameters (Le Ray *et al.*, 1997a): only release-sensitive sensory neurons will be described here.

- Phasic afferents provide a purely dynamic coding characterized by velocity-sensitive, high-frequency bursts of action potentials during release ramps and silence during ramp plateaus (Figure 2A1). These afferents thus provide signals of limb movement and may play a key role in fast responses to external disturbances.
- Phaso-tonic afferents (Figure 2A2) are characterized by both phasic velocity-sensitive bursts during release ramps and tonic discharges during plateaus following a release ramp. In addition, both the phasic and tonic components are influenced by position, with their instantaneous frequency increasing linearly with the release position from which the release ramp is applied. Combining movement and position coding may confer on phaso-tonic sensory neurons an important role in the determination of joint movement relative to limb and overall body position.
- The third kind of release-sensitive CBCO afferents fires continuously whatever the angular position of the joint (Figure 2A3), but also combines the properties of phaso-tonic neurons. Indeed, the overall firing frequency of this latter group increases during release ramps and is correlated with the starting position of the imposed release movement. In the same way, the discharge frequency of the continuous tonic firing largely increases during the plateaus that follow a release ramp. Considering their

Figure 2. CBCO coding of leg movements. (A) Representative intracellular recordings of (1) a phasic release-sensitive CBCO neuron, (2) a phaso-tonic release-sensitive neuron, and (3) a continuously firing release-sensitive neuron. (B) Examples of discharge patterns of four extracellularly-identified sensory units in response to sinewave mechanical stimulation of the CBCO. (C) Position on the CBCO nerve of two chronic electrodes (1) used to discriminate orthodromic and antidromic spikes (2). (D) Analysis of orthodromic discharge in a levation-sensitive (Lev), depression-sensitive (Dep) and a tonic firing sensory unit, along with an integrated transform (Integ) of levator muscle activity (EMG) during real walking. Top traces are instantaneous frequency of corresponding units during locomotion. Bottom: Event histograms of sensory unit discharge in relation to the onset/offset of the Lev EMG burst, and cumulative sum curves, which highlights changes in occurrence probability, as determined from the event histograms. Black sections of the curves show significant increased or decreased activity, while stars indicate the level of significance.

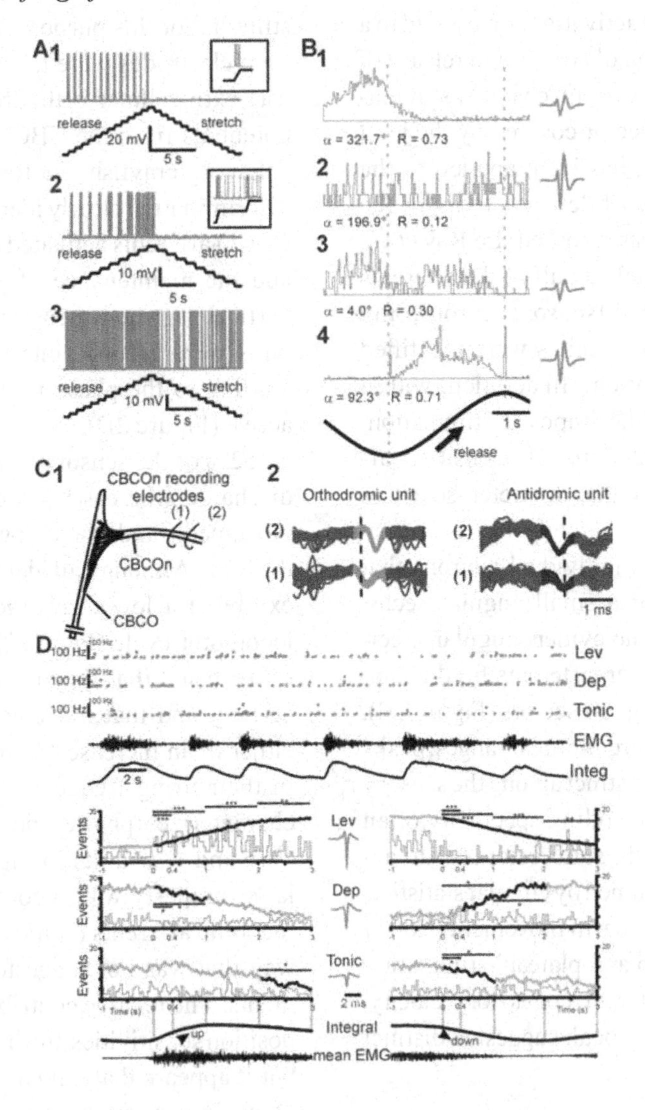

tonic background discharge, these afferents may supply continuous information on joint angular position. In addition, when a movement is imposed, they also perform powerful angular movement and position detection.

Encoding Imposed Leg Movement-Like Stimulation

Although stance control may be mimicked by a 'ramp and plateau' mechanical stimulation, the smoothness of real leg movements during locomotion makes the sensory activation correspond to a sinewave-like repetition of stretch and release of the CBCO. Thus, sinewave mechanical stimulation over the angular sector covered by the joint during actual walking has been applied to the CBCO strand *in vitro*, while the activity of the whole sensory nerve was recorded (Le Ray *et al.*, 1997a). Surprisingly, only a half of the afferents was activated (*i.e.*, of the 40 sensory neurons, only about 20 active neuron profiles were identified using spike discrimination). In addition, within the angular range of the imposed stimulation, the number of activated release-sensitive afferents was about twice that of stretch-sensitive afferents.

More astonishingly, individual neurons that specifically encode for a small angular sector were never found, and no evident angular specificity emerged, but rather, afferents fired action potentials over wide angular sectors (Figure 2B). However, within these large angular ranges, peaks of firing frequency are distinct among the sensory neurons, with the main coding specificity of an afferent resulting from the sum vector of all its action potentials as determined by circular statistics, as described in humans harm movement (Roll *et al.*, 2000). Since 'ramp and plateau' stimulation showed that every afferent codes for velocity, differences in frequency peak suggested distinct position sensitivities.

Coding Actual Leg Movements in Freely Behaving Crayfish

Isolating a sensory organ *in vitro* necessarily suppresses an ensemble of perturbing and modulatory influences that normally have substantial effects on its coding properties. For example, the neuromodulator serotonin modifies CBCO encoding in a dose-dependent manner (Rossi-Durand, 1993). Determining the actual coding properties of a sensory receptor thus requires an *in situ* analysis of its discharge evoked by real stimuli. For this purpose (Figures 2C1, C2), two wire electrodes were used *in vivo* to record and sort extracellular orthodromic (sensory) action potentials from the CBCO neurogram of freely behaving crayfish (Le Ray *et al.*, 2005). Spike discrimination clearly identified an average of 15 CBCO afferents activated during actual walking and the maintenance of posture. Compared to vertical leg movements, the identified sensory units were grouped into three main groups according to the phase in which they were most active (Figure 2D).

52% of the sensory profiles exhibited a tonic discharge, whereas 43% were sensitive to upward movements and 5% to downward movements of the limb. Although all identified sensory neurons exhibited a low discharge rate throughout the locomotor cycle (Figure 2D, top), a finer analysis revealed that they express distinct peaks of activity and that movement coding may result either in an increase or, surprisingly, a decrease in their firing frequency. The former probably characterizes phasic and phaso-tonic afferents (*Lev* and *Dep* units in Figure 2D), whereas the latter property was encountered in one third of the tonic afferents (*Tonic* unit in Figure 2D) in a way that was not suggested by *in vitro* experiments. The relative contributions of velocity and position sensitivities are difficult to assess *in vivo*, but it appears that only a minor fraction (<14%) of movement-sensitive neurons present purely phasic properties.

Sensory Coding in Autonomous Embodied Systems

One of the major challenges in studying the function of a sensory circuit is to understand what features of natural stimuli are encoded in the spiking of its neurons (Barlow, 1972). Movements comprise both dynamic (displacements) and static (successive positions) parameters that are encoded into phasic and tonic discharge, the latter playing also a role as an internal reference. Combining the two components allows a rapid (phasic) and referenced (tonic) spatial representation of joint position and movement. In the crayfish, phasic coding is a property of all CBCO sensory neurons, and their firing frequency during actual movement is always higher than during stationary plateaus, indicating that the major information conveyed by sensory afferents is dynamic rather than static. Thus, motor control depends largely on dynamic parameters, constituting a more economic system from a computational point of view. Indeed, when a perturbation occurs, a system based only on position encoding would require an incompressible integration time to measure positional changes for comparison with a precise pre-determined scheme, before counteracting the perturbation. In contrast, a motor system in which detection is based on joint movement will be able to adjust faster the ongoing movement or posture to compensate perturbations. However, since purely phasic coding only gives information about joint movement without accounting for joint position, a corrective system based exclusively on movement detection would result in position shifting. Nature solved this problem by adding two types of afferents that combine movement and position monitoring, thereby providing the CNS with both the dynamic and static parameters of joint movement relative to the whole body.

In vivo, CBCO coding appears much more complicated and less precise than *in vitro* studies have suggested. Although the population of sensory afferents obviously remains as diverse as *in vitro*, similarities of activation patterns among the population suggest that movement monitoring is based on a collegial sensory encoding of the various biomechanical parameters. Direct electrical synaptic connections between sensory afferents coding for the same direction (El Manira *et al.*, 1993) could play a substantial role in this collegiality. As recently proposed (Rowland *et al.*, 2007), it is likely that such multi-sensory information enhances network computation in such a way that the response latency is substantially shortened.

SENSORY-MOTOR INTEGRATION IN PARALLEL PATHWAYS

To simplify, the basic output of a neural network that generates a behavior results from the use of sensory inputs through a combination of circuit characteristics and intrinsic neuronal properties. For the control of the CB joint in crayfish, sensory-motor integration is mainly achieved through parallel pathways that convey CBCO sensory information to the motor centers. The organization of this sensory-motor network has been studied extensively (Le Bon-Jego & Cattaert, 2002; Le Bon-Jego *et al.*, 2004; and see Cattaert & Le Ray, 2001) and involves an extremely well organized and specific wiring.

CBCO release-sensitive afferents (sensing upward leg movements) excite monosynaptically the depressor motoneurons (Dep MNs), and stretch-sensitive afferents (sensing downward leg movements) directly stimulate the levator motoneurons (Lev MNs). These monosynaptic connections are therefore responsible for negative feedback reflexes. Two parallel polysynaptic pathways, comprising a dozen interneurons, also participate in integrating the sensory inflow to the motor command in either a resistance or an assistance reflex mode. Added to the forty CBCO afferents, this sensory-motor network thus comprises a small number of neurons, which are all accessible to intracellular recordings *in vitro*.

Network Properties

Resistance and Assistance Reflexes

During walking, sensory-motor pathways incorporate movement-generated information into the centrally programmed motor program in a cycle-dependent manner (see Clarac *et al.*, 2000). For example, during the stance phase, proprioceptive inputs mainly oppose gravity by strongly activating the extensor motoneurons, whereas during the swing phase the same inputs activate principally the flexor motoneurons in a way that the sensory information reinforces limb flexion. This means that the phase of the locomotor cycle determines the sign of the reflex response.

When isolated *in vitro*, the crayfish CNS either remains quiescent or produces an automatic rhythmic locomotor activity (termed 'fictive locomotion'). In the former, a sinewave mechanical stimulation of the CBCO elicits a typical negative feedback reflex (Figure 3A), devoted to postural control, resulting in the cyclic contraction of muscles that counteract the imposed movements (similar to the vertebrate stretch reflex; Le Ray & Cattaert, 1997). So, stretching the CBCO strand (mimicking leg downward movements) stimulates the Lev MNs, whereas a CBCO release activates the Dep MNs. By contrast, during pharmacologically-evoked fictive locomotion, the reflex response reverses to an assistance reflex that helps the ongoing movement.

Figure 3. Sensory-motor integration. (A,B) Network properties. (A) Levator and depressor nerve activity in resistance and assistance reflexes. (B) Schematic wiring diagram summarizing the different network components (sensory neurons, interneurons, motoneurons) and the principal connections within this sensory-motor circuit. C. Motoneuronal reflex responses to imposed ramp movements of the CBCO strand of a phasic (1) and a phaso-tonic Dep MN (2).

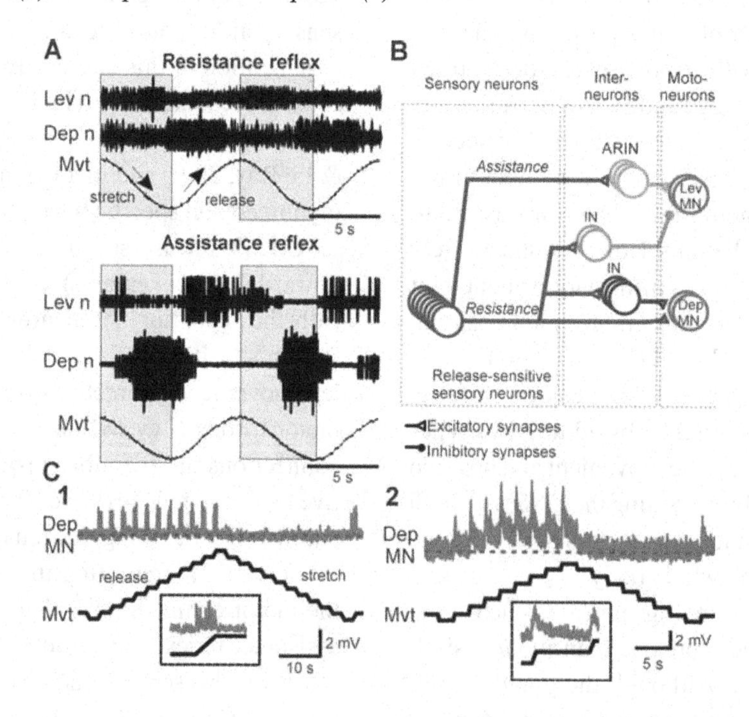

Sensory-Motor Synaptic Connections

Resistance reflexes evoked by CBCO stimulation are mediated by parallel, direct and indirect, excitatory and inhibitory connections between primary sensory afferents and motoneurons (El Manira *et al.*, 1991; Le Bon-Jego & Cattaert, 2002). For instance, releasing the CBCO activates release-sensitive sensory neurons that elicit both mono- and polysynaptic excitatory postsynaptic potentials (EPSPs) in the motoneurons that would oppose the imposed movement (Dep MNs in this case). In parallel, the same sensory neurons inhibit the antagonistic motoneurons (here, Lev MNs) through a polysynaptic GABAergic pathway (Figure 3B). Extensive studies of the monosynaptic sensory-motor connections (El Manira *et al.*, 1991; Le Ray *et al.*, 1997b) have shown that each Dep or Lev MN receives information from four to eight CBCO afferents.

This implicates that (i), the motoneurons themselves possess strong integrative capabilities, and (ii), the proprioceptive information is distributed to the two motoneuronal populations that control the joint, whatever the phase in locomotor cycle. This therefore requires that specific mechanisms regulate cyclically the transmission of sensory inputs toward their postsynaptic motoneurons (see below).

Mechanisms of Reflex Reversal

When the central pattern generator (CPG) is active (during actual or fictive locomotion), the sign of the reflex is cyclically reversed, which consists of a phase-dependent switch from a resistance (negative feedback) to an assistance (positive feedback) reflex. In crayfish, a key interneuron is specifically involved in this switch for the Dep MN population (Le Ray & Cattaert, 1997). This 'assistance reflex interneuron' (ARIN; Figure 3B) is accessed directly by up to eight velocity-coding, stretch-sensitive CBCO neurons and is likely to project directly onto Dep MNs. During active leg movement, the polysynaptic assistance pathway is concomitantly recruited to activate agonistic motoneurons while the monosynaptic resistance pathway is inhibited (see below), thereby silencing the antagonistic motoneurons. In contrast, during stance the monosynaptic connections prevail, and the resistance reflex is now expressed.

ARIN is a nonspiking interneuron that generates graded depolarizations in response to increasing CBCO stretches, which produce graded long-lasting EPSPs in Dep MNs. Considering that the ARIN only receives inputs from velocity-coding afferents, the activation of the assistance pathway thus depends on dynamic movement parameters. However, because ARIN gradedly and long-lastingly activates motoneurons, the result will not consist of a phasic motoneuronal activation, but rather of additional excitation to the motoneurons carrying out the movement. In addition, the ARIN receives phasic disynaptic inhibitory inputs from stretch-sensitive afferents via another interneuron, likely to prevent the over-excitation of the motoneurons involved in the ongoing movement. Thus, the CPG not only recruits different sets of motoneurons, but also controls the sensory-motor pathways in such a way as to finely adapt their inputs to the prevailing motor behavior.

Motoneuronal Reflex Responses

Because a major part of sensory-motor processing is achieved through direct afferent-to-motoneuron connections (El Manira *et al.*, 1991), motoneurons integrate the various parameters of CBCO information. Interestingly, among the pool of 12 Dep MNs (Le Ray & Cattaert, 1997), three motoneurons show no monosynaptic response, whereas eight others display a characteristic resistance reflex response (*i.e.*, they are activated by CBCO release). Surprisingly, one Dep MN always exhibits a direct assistance reflex response (*i.e.*, activation by proprioceptor stretch). The population of resistance Dep MNs can be further divided into two groups according to the shape of their response to 'ramp and plateau' CBCO stimulation (Figure 3C; Le Ray *et al.*, 1997b): one pool generates phasic bursts of EPSPs during release (upward) ramps (Figure 3C1) while the second develops an increasing depolarization throughout the CBCO release, on top of which phasic EPSP barrages are generated during ramps (Figure 3C2). Although compelling, it has not been possible to correlate the response shape with the type of sensory input (phasic, phaso-tonic or tonic) that the motoneuron receives, nor with its intrinsic electrical properties. In contrast, simulation experimentation suggested that the localization of sensory inputs along the non-homogeneous postsynaptic dendritic tree of a motoneuron determines the shape and time constant of EPSPs. Consequently, this may (or not) allow the development of a tonic depolarizing response: proximal synapses generate fast and large EPSPs involved in phasic reflex responses, while distal synapses produce slow and smaller EPSPs responsible for tonic depolarization. Nevertheless, although the motoneuron electrical properties appear to play a minor role in the monosynaptic response to passive movements, locomotor circuit activation confers active properties on these motoneurons, thereby modifying their integrative abilities (Cattaert *et al.*, 1994a,b).

Sensory-Motor Integration

In a mirror image of sensory coding, motoneuronal reflex responsiveness is mainly dynamic, allowing fast reactions to passive joint movements. However, a static component is also present in some motoneurons in order to prevent shifting or saccadic responses that would, finally, produce dis-equilibrium. Curiously, a purely phasic assistance Dep MN exists whose function within the network remains unclear. Because motoneurons within a pool are recruited progressively during walking, one can imagine that this particular assistance Dep MN is recruited close to the transition phase between returnstroke (where it preserves muscular tonicity and initiates downward leg movement at the end of the swing phase) and powerstroke (where resistance Dep MNs counteract gravity during stance and propel the body forward). Moreover, the interconnection of Dep MNs via electrical tight junctions (Chrachri & Clarac, 1989) also permits a role for the assistance motoneuron in the automatic recruitment of the other Dep MNs at the transition between swing and stance phases, thereby initiating the powerstroke. In addition to this role, electrotonic connections between motoneurons may substantially organize the output motor response by allowing all coupled motoneurons of a given pool to share their sensory input information.

How does the CNS coordinate the responses of the various motoneurons that receive distinct inputs? Bernstein (1967) proposed that the CNS elaborates motor synergies to prevent the overuse of central resources that would require the processing of all components independently. Evidently, the control of leg movement in crayfish relies on the 'simple' organization of two parallel sensory-motor pathways that are activated according to the phase of the locomotor cycle: one resists imposed displacements, while the other participates in the ongoing movement. Indeed, distinct mono- and polysynaptic connections convey sensory information toward motoneuron pools involved in either the resistance to, or assistance of movement. In the active phase, the CPG selects specific polysynaptic pathways that activate a large proportion of, if not all, motoneurons within a pool. Moreover, it unmasks active properties in motoneurons that, through gap junctions, may propagate to neighboring motoneurons and participate in the synergistic activation of the pool involved in a given motor task. In contrast, in the reflex mode, the response has to be directly proportional to the perturbation, and sensory inputs constitutes the only source of motoneuronal activation, individual motoneuronal integration shaping the unitary motor response according to this specific input. In addition, the wiring of the sensory-motor resistance pathway alone allows for the sequential recruitment of Dep MNs during postural correction (Le Ray *et al.*, 1997a,b). However, the switch between postural and locomotor modes involves the engagement of the CPG that not only cyclically selects parallel sensory-motor pathways, but also plays an important role in shaping the properties and efficacies of all the synapses involved.

PLASTICITY AND MODULATION OF SENSORY-MOTOR INTEGRATION

The sensory-motor circuits controlling the stretch reflex of vertebrates, or the resistance reflex of arthropods, operate like automatic controllers (Clarac *et al.*, 2000). Therefore, mechanisms should allow these circuits to adapt to the animal's ongoing behavior. In the crayfish, the CBCO terminal-to-motoneuron synapse involved in the resistance reflex is monosynaptic and should therefore, be constantly active. This feature has two consequences: (1) the strength of the synapse, which directly controls the gain of the reflex, should vary to match the activation level of postsynaptic motoneurons. If not, the gain would be too high in resting animals and insufficient in very active ones; (2) depending on the activity engaged, a monosynaptic negative feedback control may be inappropriate (see above). During the past fifteen years, several adaptive functional mechanisms of this sensory-motor circuit have been highlighted (Cattaert *et al.*, 1990; El Manira *et al.*, 1991; Cattaert *et al.*, 1992; El Manira *et al.*, 1993; Cattaert & Le Ray, 1998; Cattaert & El Manira, 1999; Le Ray & Cattaert, 1999; Cattaert *et al.*, 2001; Cattaert & Bévengut, 2002; Le Bon-Jego *et al.*, 2004; Le Ray *et al.*, 2005; Le Bon-Jego *et al.*, 2006; and for reviews, see Clarac *et al.*, 2000; Cattaert & Le Ray, 2001; Cattaert *et al.*, 2002), and we now present three classes of such mechanisms.

Automatic Gain Control Achieved by Postsynaptic Motoneurons

From Resting to Active State: 'Awakening' of the Negative Sensory-Motor Feedback

A crayfish may spend a large part of its life in a more or less totally immobile state. During such a resting state, the negative sensory-motor feedback controlling leg movements must be profoundly depressed. However, the synapse involved is capable of rapid functional restoration and capable of a long-lasting transformation to a high level of synaptic transmission (Le Ray & Cattaert, 1999). The underlying mechanisms have been deciphered in experiments in which paired intracellular recordings were made from a CBCO sensory terminal and a postsynaptic motoneuron (Figure 4A). After a long period of motoneuronal inactivity, sensory spikes evoked small unitary EPSPs (<0.2 mV) in the motoneuron (left inset in Figure 4B). At this point, the postsynaptic motoneuron was intracellularly stimulated with depolarizing current pulses that each evoked a few spikes (circle inset) over several minutes in the absence of sensory stimulation. Within ten minutes after such motoneuron activation, the unitary EPSPs were potentiated up to 200% of their control amplitude (right inset) for at least three hours.

Presynaptic activation of the sensory terminal alone (up to 100 Hz) never induced any monosynaptic EPSP enhancement in the absence of postsynaptic activity. Thus, the motoneuron activation was necessary and sufficient to obtain specific long-term potentiation (LTP) of its sensory-motor synapses. During activation, glutamate (the motoneuronal excitatory neurotransmitter in crayfish - Van Harreveld, 1980) is released and activates metabotropic receptors located on the presynaptic sensory terminals (Figure 4C). A quantal analysis of the EPSP amplitude distribution before and

Figure 4. Motoneuronal control of sensory synapses. (A) Paired intracellular recordings from a CBCO sensory terminal and a depressor motoneuron. (B) Postsynaptic induction of LTP at a sensory-motor synapse. (C) Cellular mechanisms of LTP induction. (D) Typical intracellular resistance reflex response to CBCO mechanical stimulation (mvt). During the course of the recording, the Dep MN spontaneously depolarized from -65 mV to -55 mV (dashed lines), and the reflex response increased in amplitude whereas the sensory firing frequency did not change. (E) Functional role of the muscarinic component of the sensory-motor EPSPs. 1. At rest, the central drive to motoneurons is weak (thin line), as is the efficacy of the reflex loop (thin arrow). The cholinergic proprioceptive feedback involves mainly nicotinic receptors (n). 2. When the CPG strongly excites motoneurons (thick line), muscarinic receptors (m) are recruited and considerably enhance sensory-motor transmission to maintain it proportional to the central drive. (F) Intracellular recording of slowly-developing primary afferent depolarizations (sdPADs) from a CBCO terminal during strong locomotor bursts. (G) Cellular mechanisms of sdPADs-mediated presynaptic inhibition.

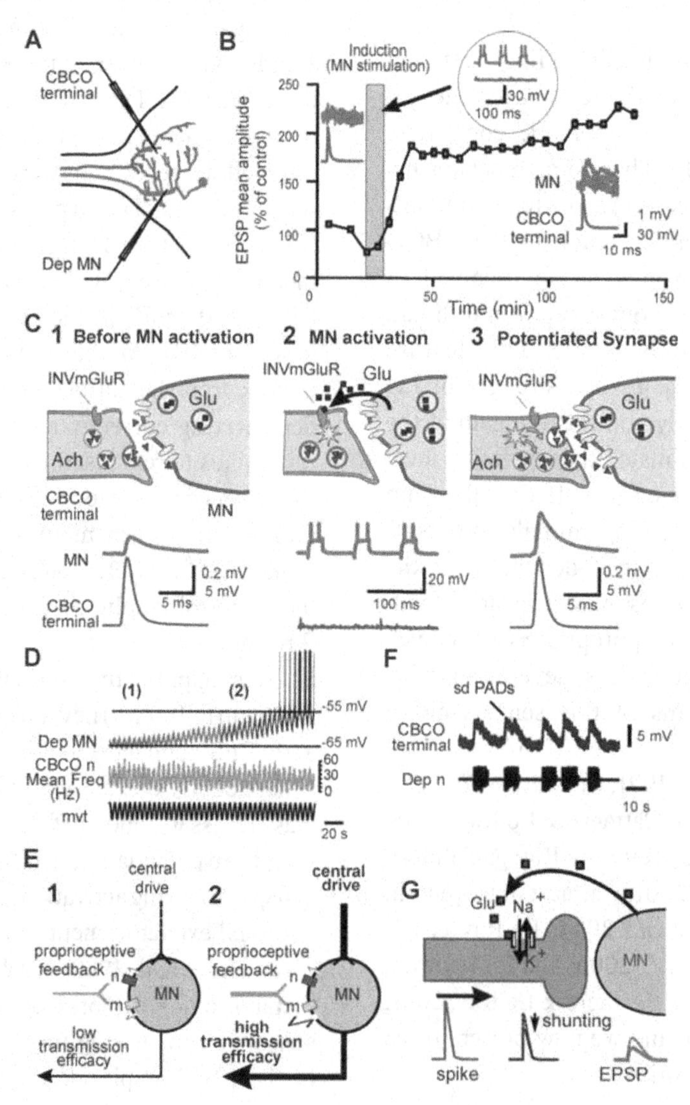

after LTP induction demonstrated that the EPSP amplitude enhancement resulted from a strong increase (>2 fold) in the probability of neurotransmitter release by the sensory terminal, without any changes in the quantal size or total number of quanta. This result and the lack of effect of LTP on motoneuron properties (input resistance, membrane time constant) strongly suggested that this new form of LTP affects exclusively the presynaptic site.

Automatic Gain Adjustment of the Sensory-Motor Synapse with Increasing Central Activity

Applying repeated sinewave stimulation to the CBCO evokes resistance reflex responses in an intracellularly recorded Dep MN (Figure 4D) which consists of a cyclic depolarization/repolarization for each release/stretch of the strand. During this time, if the motoneuron gradually depolarizes (either spontaneously or by intracellular injection of a ramp of depolarizing current), the amplitude of the reflex response dramatically increases (up to 300%; compare (1) and (2) on Figure 4D), without any change in the CBCO mean spike frequency. In the range of membrane potentials studied (-65 to -55 mV), the motoneuron input resistance decreased, and no active slow depolarization occurred (although spike bursts occurred when the membrane potential reached spike threshold). So, the voltage-dependent increase of the reflex response was not due to intrinsic motoneuronal electrical properties *per se*. On the other hand, the response enhancement persisted in the presence of a high divalent cation solution, indicating that no polysynaptic pathways are involved. Rather, the synapse between the sensory terminal and the motoneuron is directly responsible for this phenomenon.

The monosynaptic connections from CBCO sensory neurons to Dep MNs are cholinergic and include both nicotinic and muscarinic components (Figure 4E; Le Bon-Jego *et al.*, 2006). Whereas at resting potential (-65 mV) the decay time of sensory-triggered unitary EPSPs is rapid (5 ms), at a more depolarized potential (-55 mV), it increases markedly (>15 ms), and consequently, the summation of EPSPs is higher during a burst of sensory spikes. By contrast, in the presence of the muscarinic receptor antagonist scopolamine, the unitary EPSP shape is no longer voltage-dependent and is, thus, the same size and shape at -65 and -55 mV. Our results therefore indicate that the voltage-dependent enhancement of the reflex response involves the muscarinic synaptic component, and suggest that at higher depolarizing potentials, this component would be responsible for a strong reinforcing factor in the negative feedback control.

What could be the functional role of such reinforcement of the feedback loop? Motoneurons receive two types of inputs (Figure 4E): central commands that organize the motor pattern, and sensory cues that adapt the command to external constraints. However, we can imagine that during powerful central commands (or any source of motoneuron activity increase) the proprioceptive correction will be proportionally less effective. Our results indicate that the sensory feedback is indeed capable of maintaining a variable level of control adapted to the ongoing activity, by altering the efficacy of the sensory-motor synapse in a manner that is inversely proportional to the level of motoneuronal activity.

Overload Protection Mechanism Exerted by MNs on their Sensory Afferents

During intense motor activity (such as observed *in vitro* when the walking CPG is activated by oxotremorine, a muscarinic acetylcholine receptor agonist, or when picrotoxin blocks central inhibitory chloride-mediated synapses), intracellular recordings from a CBCO terminal display small amplitude, slowly developing primary afferent depolarizations (sdPADs – Figure 4F;

Cattaert & Le Ray, 1998). These sdPADs differ from classical picrotoxin-sensitive GABAergic PADs (see below) because sdPADs are related strictly to motoneuron activity, and their amplitude is much smaller. Since the addition of the two glutamate antagonists, CNQX and DNQX, to the picrotoxin bath significantly reduced the amplitude of sdPADs, these were likely to be mediated by glutamate. Their effect on sensory neurons was tested by direct micro-application of glutamate in the vicinity of the small end branches of CBCO terminals (or by direct stimulation of a motoneuron pool by antidromic stimulation of its motor nerve). Glutamate directly activates a presynaptic receptor-channel permeable to both Na^+ and K^+ ions, but not Cl^- ions (Figure 4G), which evokes a slowly depolarizing response in the sensory terminal, accompanied by a 50% decrease in input resistance that shunts by 20% the amplitude of sensory spikes throughout the depolarization.

Thus, it appears that crayfish motoneurons are able to exert directly a glutamate-mediated control of their input synapses. However, because this presynaptic inhibition requires high levels of motoneuronal activity, such a mechanism probably plays a protective role against overloaded motoneuron activity rather than a continuous control exerted over a whole range of activity levels.

Gain Control by the CPG

During motor tasks, proprioceptive input information may be inappropriate, and mechanisms exist that block the signal in the sensory terminal before it reaches its postsynaptic targets. In crayfish this occurs during tail flip or active leg movements. During a tail flip, the abdomen is rapidly flexed to propel the body rearward, and the legs are rapidly moved forward to minimize reaction against the water, thereby optimizing the escape movement. Such fast forward movements of the legs cause strong activation of chordotonal organs (and particularly the CBCO) that would result in a very powerful and inappropriate resistance reflex, in the absence of sensory input blockade. In this case and, more generally, during any active leg movements (such as during locomotion) resistance reflex circuitry must, therefore, be deactivated.

Primary Afferent Presynaptic Inhibition During Tail Flip

The crayfish tail flip is triggered by giant fibers that conduct spikes very rapidly to elicit a quick escape reaction. In the *in vitro* preparation of the walking system, medial giant fibers (MGF) can be electrically stimulated while a CBCO terminal is recorded intracellularly (Figure 5A1; El Manira & Clarac, 1994). When a spike is triggered in the MGF, a depolarizing response (PAD) is recorded in the CBCO terminal, accompanied by a marked decrease in its input resistance (Figure 5A2, A3). This shunting mechanism decreases the amplitude of sensory spikes in the terminal and thereby decreases the amplitude of the corresponding EPSPs in the postsynaptic motoneuron. The MGF-triggered PADs are mediated by both histamine and GABA since they are partially blocked both by the histaminergic antagonist, cimetidine, and picrotoxin.

Primary Afferent Presynaptic Inhibition During Walking Activity

Primary afferent presynaptic inhibition during locomotion also involves GABAergic synapses. Three complementary mechanisms are progressively recruited as the amplitude of GABA-mediated PADs increases.

- The first mechanism of GABAergic presynaptic inhibition occurs during moderate levels of central activity. At rest, intracellular recordings from CBCO terminals reveal the presence of a series of small (several millivolts) PADs. Such small PADs inhibit sensory-motor synaptic transmission by

Figure 5. CPG control of sensory inputs. (A) In response to a medial giant fiber (MGF) stimulation, an intracellularly-recorded CBCO terminal (1) exhibits a large depolarization (2) accompanied by a strong decrease in membrane input resistance (3). (B) During fictive locomotion, bursts of PADs time-locked to depressor bursts occur in an intracellularly-recorded sensory neuron. (C) Superimposed intracellular recordings showing the amplitude of the sensory spike and its motoneuronal EPSPs in the absence or presence of a PAD. (D) Inhibition of antidromic spikes and large PADs under picrotoxin (PTX) perfusion. (E) Motoneuronal responses to orthodromic and antidromic CBCO spikes. (F) Occurrence of three CBCO antidromic units (AU) during free behavior, with AU1 discharging preferentially in the absence of locomotor activity, and AU2 and AU3 being exclusively active when the animal moves. Arrowheads designate stepping levator activity, and horizontal bar indicates postural adjustments.

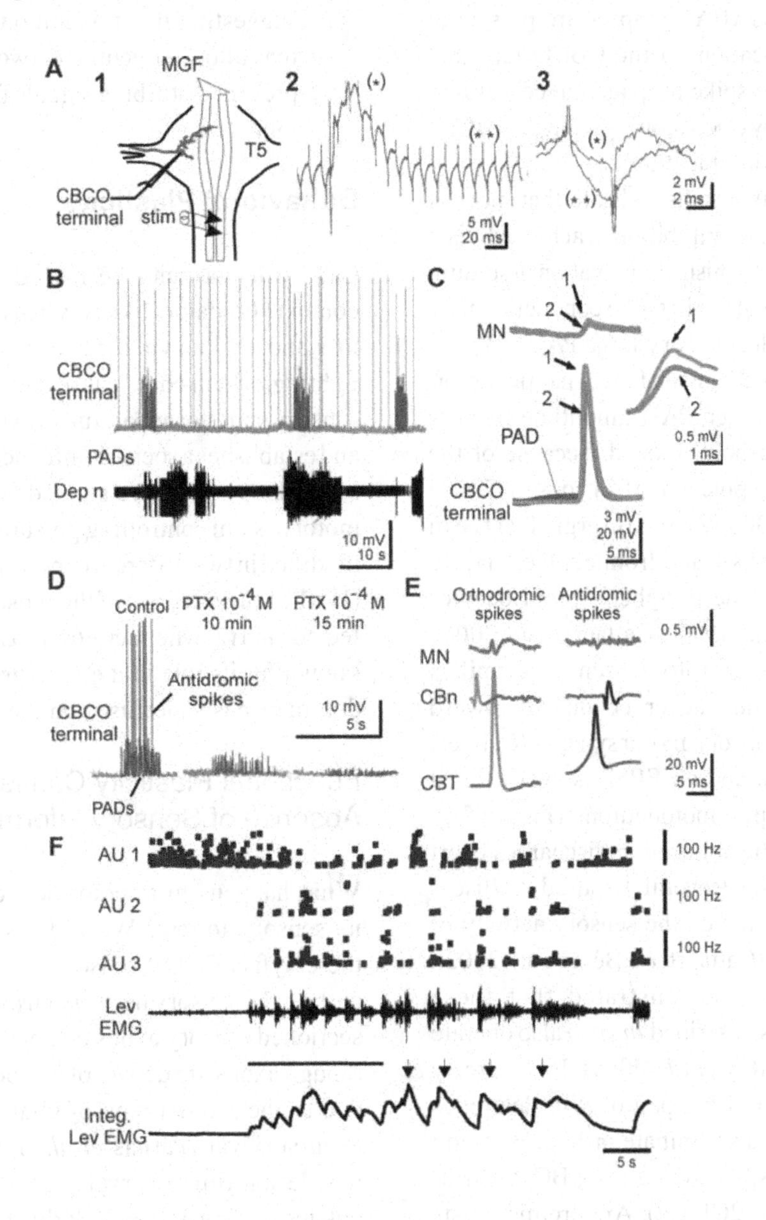

exerting a continuous and gradual shunting effect on passive spike propagation (Cattaert *et al.*, 1992): the higher the PAD frequency, the greater shunting of spike conduction. During fictive locomotion (Figure 5B) bursts of larger amplitude (20 mV) PADs occur, which also exert an inhibitory shunting effect on spike amplitude and, consequently, on EPSP amplitude (Figure 5C). To ensure that this shunting mechanism is functionally inhibitory, GABA synapses are present at a special location on the CBCO terminal where active spike propagation ceases and is replaced by electrotonic propagation (Cattaert & El Manira, 1999).

- Although our results indicate that most of the presynaptic inhibition is achieved via a shunting mechanism, inactivation of sodium channels involved in spike conduction may also occur during very large PADs.
- The third mechanism of presynaptic inhibition occurs when PAD amplitude reaches spiking threshold. Indeed, because of the high reversal potential of Cl^- ions in CBCO terminals (-30 mV), GABAergic PADs can generate bursts of antidromic spikes that are conveyed to the peripheral proprioceptor (Cattaert *et al.*, 1994c; Cattaert *et al.*, 2001). In contrast, picrotoxin-sensitive spikes (Figure 5D) are never conducted toward the central sensory-motor synapse (Cattaert *et al.*, 2001), and no EPSP is ever evoked in postsynaptic motoneurons (Figure 5E). Moreover, the antidromic discharges exert a long-lasting (several hundred milliseconds) inhibition on the sensory activity of the CBCO (Cattaert & Bévengut, 2002). Recently we demonstrated that these mechanisms described *in vitro* also operate *in vivo* (Le Ray *et al.*, 2005). In the freely behaving crayfish, a pair of wire electrodes was used to discriminate orthodromic and antidromic spikes from the CBCO neurogram (Figure 2C1, C2). Antidromic bursts in a single CBCO unit may reach 100 Hz during active locomotion (Figure 5F) and totally suppress any orthodromic sensory activity for 200-400 ms. This mechanism therefore exerts the strongest blockade of sensory activity and allows active movement to be achieved without reflex co-contraction of the antagonistic muscle. Moreover, we demonstrated that antidromic discharges are also present during joint immobilization, suggesting that this inhibitory mechanism may allow the central network effectively to prevent disturbing signals from the blocked joint.

Behavioral Plasticity

Can such plasticity be placed in a behavioral context? Predators like crayfish live in a community and are known to be very aggressive. Thus, fighting, which could cause damage to either the sensory or motor apparatus, is extremely frequent and establishes a social dominance hierarchy (Issa *et al.*, 1999). We have studied how the sensory-motor system controlling postural adjustments is modified in two different experimental situations: (1) when the integrity of the sensory system is affected, and (2) when serotonin, a neuromodulator known for its role in the establishment of social dominance is superfused on the CNS.

Functional Plasticity Caused by the Absence of Sensory Information

What happens to a motor network deprived of its sensory inflow? We addressed this issue in the crayfish CBCO sensory-motor system after cutting the sensory nerve *in vivo*. In crustaceans, sectioned sensory axons do not degenerate (even though axons are devoid of cell bodies), probably due to the proliferation of glial cells that allow axon survival (Parnas *et al.*, 1998). At various post-lesion time intervals, CNS were dissected out for testing intracellularly the Dep MN re-

Figure 6. Behavioral plasticity. (A) Post-lesional plasticity. 1. Time course of the decline in sensory-evoked PSPs at days (D) after the CBCO nerve had been cut in vivo. 2. Sequential restoration of the IPSPs (1) and then EPSPs (2) by a tonic electrical stimulation (>1h) of the sensory nerve. (B) Serotonergic neuromodulation. In the presence of serotonin (5HT), the resistance reflex response evoked by mechanical stimulations of the CBCO strand in the intracellularly recorded Dep MN is increased (1), as well as the motoneuron input resistance (2) and the excitability of the resistance reflex interneuron (IN, 3). This increase in interneuron excitability leads to an increase in the polysynaptic EPSPs evoked in the Dep MN by a single identified CBCO unit (4).

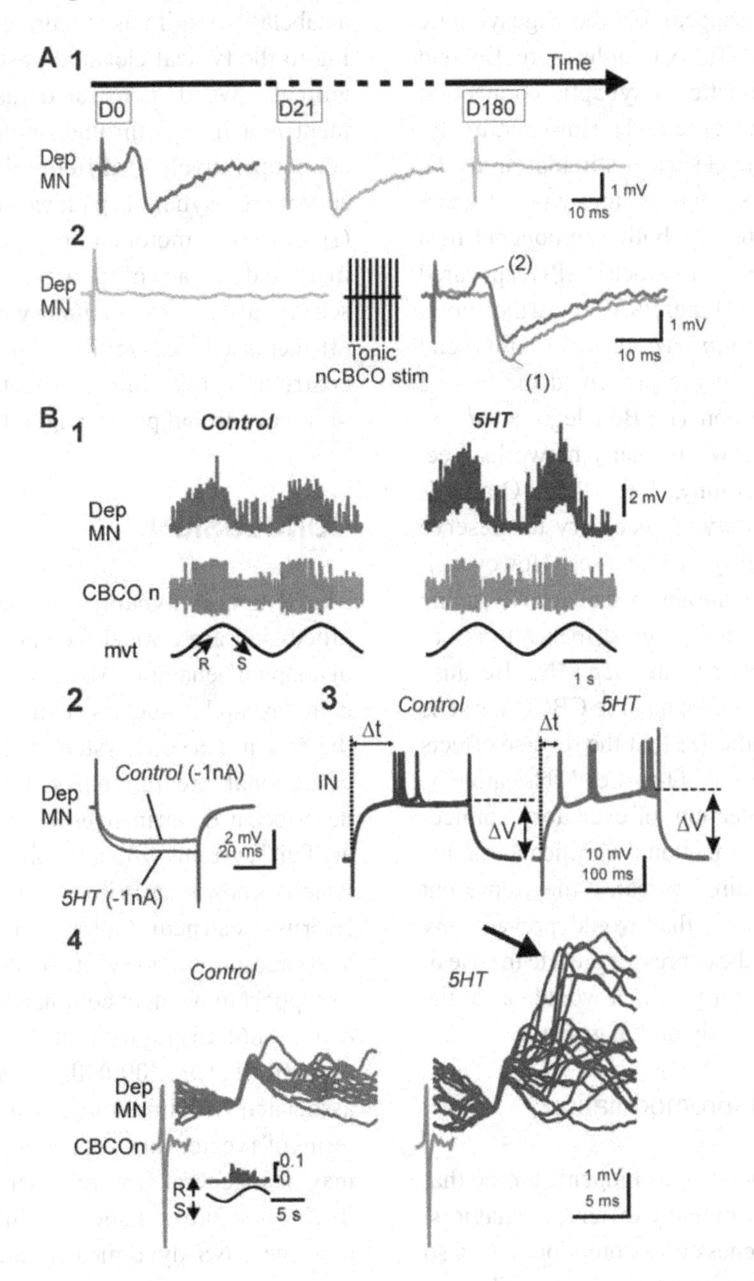

sponsiveness to electrical stimulation of the cut sensory nerve (mechanical stimulation being no longer possible). In control animals, this evokes a typical biphasic response through the recruitment of both monosynaptic EPSPs and oligosynaptic PSPs (see above). Within 21 days following proprioceptor ablation, no changes were observed in the motoneuronal response. Thereafter, the monosynaptic EPSPs disappear, but the oligosynaptic IPSPs persist for up to six months more. Beyond this period, no functional synaptic connection seems to remain (Figure 6A1). However, applying *in vitro* a tonic electrical stimulation to the sensory nerve for several minutes was sufficient to progressively restore both components in a characteristic sequence in which IPSPs reappeared first, then EPSPs (Figure 6A2). Furthermore, chronic stimulation *in vivo* of the proximal end of the cut sensory nerve prevented the loss of synaptic transmission (Le Bon-Jego *et al.*, in prep). Thus, as shown in many networks (see: Davis & Bezprozvanny, 2001; Perez-Otaño & Ehlers, 2005) activity is necessary to preserve the full functionality of synapses. However, it also appears that in the absence of activity, some compensatory mechanisms exist, mainly to assure the inhibitory pathway onto Dep MNs. Because the leg is kept elevated when the CBCO nerve is cut, we may hypothesize that the reverse effects are produced in the CBCO-to-Lev MNs pathway, including the 'protection' of excitatory connections. Although the functional significance of this latter process remains unclear, it may represent a control mechanism that would prevent any over-activation of the depressor muscle in case of sensory (CBCO) injury, which would cause the leg to trail continuously on the ground.

Serotonergic Neuromodulation

Serotonin is an endogenous biogenic amine that plays critical roles in many different behaviors, such as aggressiveness or locomotion, and also in the establishment and maintenance of social status (Issa *et al.*, 1999). In crayfish, social status is established by fighting, the victor becoming dominant and the vanquished a subordinate. Because both social states are characterized by distinct postures, we analyzed the central effects of serotonin on the CBCO sensory-motor system *in vitro* (Le Bon-Jego *et al.*, 2004). Serotonin was found to enhance resistance reflex responses in a subclass of animals (Figure 6B1), corresponding to the typical elevated posture of dominant animals. We demonstrated that this enhancement is achieved through multiple effects that act cooperatively at different levels in the sensory-motor synaptic pathway (Figures 6B2-4): (1) increased motoneuron input resistance, (2) increased efficacy of synaptic transmission from sensory afferents to excitatory interneurons, and (3) increased excitability of these interneurons contributing to an improvement of the amplitude of motoneuronal polysynaptic EPSPs.

CONCLUSION

The temporal dynamics of sensory feedback processing are critical factors in the generation of adapted behaviors. While difficult to tackle in complex models such as mammals and especially the human nervous system, the main principles of neuronal integration have been identified and described in detail in invertebrates. Consequently, artificial systems will probably continue to gain a lot from knowledge in invertebrates and, especially, crustacean neurobiology. For example, twenty years ago, population vector coding was proposed to support movement commands (Georgopoulos *et al.*, 1986). In crayfish, like human skilled hand tasks (Roll *et al.*, 2000), the sensory information associated with limb movement is also coded in terms of a vector sum, suggesting that this process may be a general feature of sensory coding. On this basis, a future challenge will be to understand how the CNS dynamically integrates multiple sensory vectors into particular motor command vectors to perform adapted behaviors.

Models of neuronal activity are based on the notion of threshold, a neuron discharging when excitation exceeds a particular threshold value. While this is true for most neurons, it is not the case for non-spiking neurons, often encountered in invertebrates (such as the ARIN) and in some vertebrate systems (*e.g.*, retina; Victor, 1999). In most models studied so far, non-spiking neurons play a substantial role in integrative processes by adding plasticity into the functional architecture of the network. But the real effects of such a cell type on the activity of a neuronal network, and how can the activity of such neurons be implemented into an artificial system remain to be clearly determined.

An auto-organizational capability is a feature of many neuronal networks which allows an adapted treatment of any incoming information. Various adaptive mechanisms exist at every single stage of the sensory-motor loop involved in the postural and locomotor control of crayfish legs. Such diversity permits specific control of integrative processes, depending either on specific elementary constraints (*i.e.*, motoneuronal control of afferents) or general demands (*i.e.*, CPG control or behavioral plasticity). However, although an isolated neuronal network has some ability to auto-organize, it remains unclear whether such mechanisms are expressed in the intact, free-behaving animal. Indeed, higher command structures generally delegate 'simple tasks' to downstream networks (*e.g.*, locomotor CPG), and it is conceivable that the latter possess a certain degree of freedom. Nevertheless, all components of an autonomous system continuously interact each other, which may largely influence the functional organization of a given network. Thus, one of the main questions to be addressed in future research will be to understand how the CNS conjugates individual neuronal command to enable expression of a fully adapted behavior. Resolving such issues would help considerably in the elaboration of 'behavioral algorithms' that would allow the full autonomy of artificial systems.

FUTURE RESEARCH DIRECTIONS

Deciphering a sensory-motor neuronal network only constitutes a first step towards understanding how information is processed in biological models. In invertebrates such as the crayfish, 'simple' neuronal assemblies encode inputs, process information and shape the adequate network output, and such biological models allow researchers to acquire strong knowledge on how a given neuronal network works. Although high levels of neuronal redundancy within single networks add complexity in vertebrate models, the study of neuronal circuits dedicated to 'simple' function has gained a lot from invertebrate researches, and many modes of information processing and output organization are now also well understood in these more complex systems. Nevertheless, in whatever the system studied, this knowledge is generally limited to the functioning of a single, more or less complex neuronal network dedicated to one or, sometimes, a little number of elementary actions.

The real challenge in the next future will, thus, be to understand how the so many functions that allow animal survival are coordinated and how the so many underlying neuronal assemblies interact dynamically. For this purpose, motor systems may again provide interesting research models since motor functions are tightly intermingled. For example, performing a simple pointing task requires controlling in a coordinated manner the arm movement, the body posture and the arm muscle stiffness in order to realize the perfect trajectory with the adequate arm speed. More simply, in every species the respiratory rhythm is always automatically adapted to the locomotor gait (in this case two different functions are coupled to allow the best performance). Recent evidences clearly demonstrate that this coupling partly relies on central interactions between locomotor and respiratory networks, and that sensory cues apparently dedicated to one function dramatically affect the other. Furthermore, many peripheral mechanisms (biomechanical, biochemical and hormonal processes…) also modulate network interaction. Indeed, a central nervous system is embodied and must take into account the dynamically changing body characteristics as well as all the other, non-neuronal processes (like those cited above, but also genetic, developmental…) that occur during the body life.

Neuroscience and artificial system researches will now need to incorporate all these integrated phenomena before we could have a real insight of what an embodied biological or artificial network must compute to ensure full autonomy.

REFERENCES

Barlow, H.B. (1972) Single units and sensation: a neuron doctrine for perceptual psychology? *Perception, 1*, 371-394.

Bernstein, N.A. (1967) *The co-ordination and regulation of movements*. Pergamon Press, London.

Cattaert, D., Araque, A., Buño, W., & Clarac, F. (1994a) Motor neurones of the crayfish walking system possess TEA^+-revealed regenerative electrical properties. *Journal of Experimental Biology, 188*, 339-345.

Cattaert, D., Araque, A., Buño, W., & Clarac, F. (1994b) Nicotinic and muscarinic activation of motoneurons in the crayfish locomotor network. *Journal of Neurophysiology, 72*, 1622-1633.

Cattaert, D., & Bévengut, M. (2002) Effects of antidromic discharges in crayfish primary afferents. *Journal of Neurophysiology, 88*, 1753-1765.

Cattaert, D., & El Manira, A. (1999) Shunting versus inactivation: analysis of presynaptic inhibitory mechanisms in primary afferents of the crayfish. *Journal of Neuroscience, 19*, 6079-6089.

Cattaert, D., El Manira, A., & Clarac, F. (1992) Direct evidence for presynaptic inhibitory mechanisms in crayfish sensory afferents. *Journal of Neurophysiology, 67*, 610-624.

Cattaert, D., El Manira, A., & Clarac, F. (1994c) Chloride conductance produces both presynaptic inhibition and antidromic spikes in primary afferents. *Brain Research, 666*, 109-112.

Cattaert, D., El Manira, A., Marchand, A., & Clarac, F. (1990) Central control of the sensory afferent terminals from a leg chordotonal organ in crayfish in vitro preparation. *Neuroscience Letters, 108*, 81-87.

Cattaert, D., Le Bon, M., & Le Ray, D. (2002) Efferent controls in crustacean mechanoreceptors. *Microscopy Research and Techniques, 58*, 312-324.

Cattaert, D., & Le Ray, D. (1998) Direct glutamate-mediated presynaptic inhibition of sensory afferents by the postsynaptic motor neurons. *European Journal of Neuroscience, 10*, 3737-3746.

Cattaert, D., & Le Ray, D. (2001) Adaptive motor control in crayfish. *Progress in Neurobiology, 63*, 199-240.

Cattaert, D., Libersat, F., & El Manira, A. (2001) Presynaptic inhibition and antidromic spikes in primary afferents of the crayfish: a computational and experimental analysis. *Journal of Neuroscience, 21*, 1007-1021.

Chrachri, A., & Clarac, F. (1989) Synaptic connections between motor neurons and interneurons in the fourth thoracic ganglion of the crayfish, *Procambarus clarkii*. *Journal of Neurophysiology, 62*, 1237-1250.

Clarac, F., Cattaert, D., & Le Ray, D. (2000) Central control components of a 'simple' stretch reflex. *Trends in Neuroscience, 23*, 199-208.

Davis, G.W., & Bezprozvanny, I. (2001) Maintaining the stability of neural function: a homeostatic hypothesis. *Annual Review of Physiology, 63*, 847-869.

El Manira, A., Cattaert, D., & Clarac, F. (1991) Monosynaptic connections mediate resistance reflex in crayfish (*Procambarus clarkii*) walking legs. *Journal of Comparative Physiology A, 168*, 337-349.

El Manira, A., Cattaert, D., Wallen, P., DiCaprio, R.A., & Clarac, F. (1993) Electrical coupling of mechanoreceptor afferents in the crayfish: a possible mechanism for enhancement of sensory signal transmission. *Journal of Neurophysiology, 69*, 2248-2251.

El Manira, A., & Clarac, F. (1994) Presynaptic inhibition is mediated by histamine and GABA in the crustacean escape reaction. *Journal of Neurophysiology, 71*, 1088-1095.

Georgopoulos, A., Schwartz, A.B., & Kettner, R.E. (1986) Neuronal population coding movement direction. *Science, 233*, 1416-1419.

Ijspeert, A.J., Crespi, A., Ryczko, D., & Cabelguen, J.M. (2007) From swimming to walking with a salamander robot driven by a spinal cord model. *Science, 315*, 1416-1420.

Issa, F.A., Adamson, D.J., & Edwards, D.H. (1999) Dominance hierarchy formation in juvenile crayfish *Procambarus clarkii. Journal of Experimental Biology, 202*, 3497-3506.

Le Bon-Jego, M., Cabirol-Pol, M.-J., & Cattaert, D. Activity-dependent plasticity induced in a sensory-motor network by the loss of a proprioceptive afferent in crayfish. (in prep)

Le Bon-Jego, M., & Cattaert, D. (2002) Inhibitory component of the resistance reflex in the locomotor network of the crayfish. *Journal of Neurophysiology, 88*, 2575-2588.

Le Bon-Jego, M., Cattaert, D., & Pearlstein, E. (2004) Serotonin enhances the resistance reflex of the locomotor network of the crayfish through multiple modulatory effects that act cooperatively. *Journal of Neuroscience, 24*, 398-411.

Le Bon-Jego, M., Masante-Roca, I., & Cattaert, D. (2006) State-dependent regulation of sensory-motor transmission: role of muscarinic receptors in sensory-motor integration in the crayfish walking system. *European Journal of Neuroscience, 23*, 1283-1300.

Le Ray, D., & Cattaert, D. (1997) Neural mechanisms of reflex reversal in coxo-basipodite depressor motor neurons of the crayfish. *Journal of Neurophysiology, 77*, 1963-1978.

Le Ray, D., & Cattaert, D. (1999) Active motor neurons potentiate their own sensory inputs via glutamate-induced long-term potentiation. *Journal of Neuroscience, 19*, 1473-1483.

Le Ray, D., Clarac, F., & Cattaert, D. (1997a) Functional analysis of the sensory motor pathway of resistance reflex in crayfish. I. Multisensory coding and motor neuron monosynaptic responses. *Journal of Neurophysiology, 78*, 3133-3143.

Le Ray, D., Clarac, F., & Cattaert, D. (1997b) Functional analysis of the sensory motor pathway of resistance reflex in crayfish. II. Integration of sensory inputs in motor neurons. *Journal of Neurophysiology, 78*, 3144-3153.

Le Ray, D., Combes, D., Dejean, C., & Cattaert, D. (2005) In vivo analysis of proprioceptive coding and its antidromic modulation in the freely behaving crayfish. *Journal of Neurophysiology, 94*, 1013-1027.

Parnas, I., Shahrabany-Baranes, O., Feinstein, N., Grant, P., Adelsberger, H., & Dudel, J. (1998) Changes in the ultrastructure of surviving distal segments of severed axons of the rock lobster. *Journal of Experimental Biology, 201*, 779-791.

Perez-Otaño, I., & Ehlers, M.D. (2005) Homeostatic plasticity and NMDA receptor trafficking. *Trends in Neuroscience, 28*, 229-238.

Roll, J.P., Bergenheim, M., & Ribot-Ciscar, E. (2000) Proprioceptive population coding of two-dimensional limb movements in humans: II. Muscle-spindle feedback during "drawing-like" movements. *Experimental Brain Research, 134*, 311-321.

Rossi-Durand, C. (1993) Peripheral proprioceptive modulation in crayfish walking leg by serotonin. *Brain Research, 632*, 1-15.

Rossignol, S., Dubuc, R., & Gossard, J.P. (2006) Dynamic sensorimotor interactions in locomotion. *Physiological Reviews, 86*, 89-154.

Rowland, B.A., Quessy, S., Stanford, T.R., & Stein, B.E. (2007) Multisensory integration shortens physiological response latencies. *Journal of Neuroscience, 27*, 5879-5884.

Schmitz, J., Dean, J., Kindermann, T., Schumm, M., & Cruse, H. (2001) A biologically inspired controller for hexapod walking: simple solutions by exploiting physical properties. *Biological Bulletin, 200*, 195-200.

Van Harreveld, A. (1980) L-proline as a glutamate antagonist at a crustacean neuromuscular junction. *Journal of Neurobiology, 11*, 519-529.

Victor, J.D. (1999) Temporal aspects of neural coding in the retina and lateral geniculate. *Network, 10*, 1-66.

Webb, B. (2002) Robots in invertebrate neuroscience. *Nature, 417*, 359-363.

ADDITIONAL READING

Ausborn, J., Stein, W., & Wolf, H. (2007) Frequency control of motor patterning by negative sensory feedback. *Journal of Neuroscience, 27*, 9319-9328.

Blanchard, M., Verschure, P.F., & Rind, F.C. (1999) Using a mobile robot to study locust collision avoidance responses. *International Journal of Neural Systems, 9*, 405-410.

Büschges, A. (2005) Sensory control and organization of neural networks mediating coordination of multisegmental organs for locomotion. *Journal of Neurophysiology, 93*, 1127-1135.

Chiel, H.J., & Beer, R.D. (1997) The brain has a body: adaptive behaviour emerges from interactions of nervous system, body and environment. *Trends in Neurosciences, 20*, 553-557.

Crochet, S., Chauvette, S., Boucetta, S., & Timofeev, I. (2005) Modulation of synaptic transmission in neocortex by network activities. *European Journal of Neuroscience, 21*, 1030-1044.

De Schutter, E., Ekeberg, O., Kotaleski, J.H., Achard, P. & Lansner, A. (2005) Biophysically detailed modelling of microcircuits and beyond. *Trends in Neurosciences, 28*, 562-569.

Devilbiss D.M., Page, M.E., & Waterhouse, B.D. (2006) Locus ceruleus regulates sensory encoding by neurons and networks in waking animals. *Journal of Neuroscience, 26*, 9860-9872.

Ijspeert, A.J. (2002) Locomotion, vertebrate. *The handbook of Brain theory and neural networks*, Second edition, Ed: M. Arbib, MIT Press, pp. 1-6.

Kullander, K. (2005) Genetics moving to neuronal networks. *Trends in Neurosciences, 28*, 239-247.

Liu D., & Todorov, E. (2007) Evidence for flexible sensorimotor strategies predicted by optimal feedback control. *Journal of Neuroscience, 27*, 9354-9368.

McCrea, D.A., & Rybak, I.A. (2007) Organization of mammalian locomotor rhythm and pattern generation. *Brain Research Reviews* (in press).

Oswald, A-M.M., Chacron, M.J., Doiron, B., Bastian, J., & Maler L. (2004) Parallel processing of sensory input by bursts and isolated spikes. *Journal of Neuroscience, 24*, 4351-4362.

Parkis, M.A., Feldman, J.L., Robinson, D.M., & Funk G.D. (2003) Oscillations in endogenous inputs to neurons affect excitability and signal processing. *Journal of Neuroscience, 23*, 8152-8158.

Pearson, K., Ekeberg, O. & Büschges, A. (2006) Assessing sensory function in locomotor systems using neuro-mechanical simulations. *Trends in Neurosciences, 29*, 625-631.

Rose, J.K., Sangha, S., Rai, S., Norman, K.R., & Rankin, C.H. (2005) Decreased sensory stimulation reduces behavioral responding, retards development, and alters neuronal connectivity in *Caenorhabditis elegans. Journal ofNeuroscience, 25*, 7159-7168.

Rybak, I.A., Stecina, K., Shevtsova, N.A., & McCrea, D.A. (2006a) Modelling spinal circuitry involved in locomotor pattern generation: insights from the effects of afferent stimulation. *Journal of Physiology, 577*, 641-58.

Ryckebusch, S., Wehr, M., & Laurent G. (1994) Distinct rhythmic locomotor patterns can be generated by single adpative neural circuit: biology, simulation, and VLSI implementation. *Journal of Computational Neuroscience, 1*, 339-358.

Song, W., Onishi, M., Jan, L.Y. & Jan, Y.N. (2007) Peripheral multidendritic sensory neurons are necessary for rhythmic locomotion behaviour in *Drosophila* larvae. *Proceedings of the National Academy of Science, 104*, 5199-5204.

Vinay, L., Brocard, F., Clarac, F., Norreel, J-C., Pearlstein, E., & Pflieger, J-F. (2002) Development of psture and locomotion: an interplay of endogenously generated activities and neurotrophic actions by descending pathways. *Brain Research Reviews, 40*, 118-129.

Vogelstein, R.J., Tenore, F., Etienne-Cummings, R., Lewis, M.A., & Cohen, A.H. (2006) Dynamic control of the central pattern generator for locomotion. *Biological Cybernetics, 95*, 555-66.

Chapter III
Forward Dendritic Spikes: A Mechanism for Parallel Processing in Dendritic Subunits and Shifting Output Codes

Oscar Herreras
Cajal Institute, Spain

Julia Makarova
Cajal Institute, Spain

José Manuel Ibarz
Hospital Ramón y Cajal, Spain

ABSTRACT

Neurons send trains of action potentials to communicate each other. Different messages are issued according to varying inputs, but they can also mix them up in a multiplexed language transmitted through a single cable, the axon. This remarkable property arises from the capability of dendritic domains to work semi autonomously and even decide output. We review the underlying mechanisms and theoretical implications of the role of voltage-dependent dendritic currents on the forward transmission of synaptic inputs, with special emphasis in the initiation, integration and forward conduction of dendritic spikes. When these spikes reach the axon, output decision was made in one of many parallel dendritic substations. When failed, they still serve as an internal language to transfer information between dendritic domains. This notion brakes with the classic view of neurons as the elementary units of the brain and attributes them computational/storage capabilities earlier billed to complex brain circuits.

ARE NEURONS THAT SIMPLE?

The quintessential feature allowing any processing of electrical signals within individual neurons is that their narrow elongated dendrites make them electrically non-uniform, enabling long electrical distances between different parts of the same neuron. Whatever electrical events take place in the dendrites, they are not the same as in the soma or the axon. These compartments are not totally isolated and signals travel from one another as in an electrical circuit. From early studies, neurophysiologists concluded that neurons were essentially input/output devices counting synaptic inputs in their dendrites, a numeric task that transform into a temporal series of action potentials or spikes, each one triggered when a certain voltage threshold is reached by temporal summation of the synaptic currents at a specialized region near the soma-axon junction. This simple working scheme enables the transformation of myriads of inputs scattered in a profusely branched dendritic arbor into a temporal sequence of binary axonal spikes that can be read by target cells in terms of frequency. Indeed, frequency modulation of spikes may perform with equivalent accuracy as the fine tuning capabilities of graded communication that prevail in non neuronal cells. For a while, researchers felt this was good enough for a single cell, and the uncomplicated view of dendritic trees as not-very-clever receiving black boxes settled in for decades. This classical picture is, however, incomplete. Such a basic scheme holds only in a few neuron types (if any).

Typically, neurons receive several thousand synaptic contacts, each conveying information from as many afferent cells. In order to understand how such a huge amount of inputs are integrated, it is useful to examine the output first. Lets have a look to it. The output range across neuron types in the Nervous System goes from regularly firing (pacemaker-like) to almost silent neurons. The later type fire spikes at extremely low rates lacking any apparent temporal structure. In fact, it constitutes a most interesting case as it applies to several neuron types in the cortex of mammals. One may argue that the few action potentials fired by these near-silent neurons have a strong informative load and they are signaling the recognition of particular combinations of inputs bearing a strong physiological meaning. While this is reasonable, some questions follow. If only some input combinations produce outgoing spikes, do neurons consider irrelevant the vast majority of their inputs, which cause no output? Would that mean that the fate of most presynaptic spikes is to become useless noise in brain circuits? If so, why bother that much to produce them anyway? And, is the postsynaptic cell structure or the specific assortment of electrogenic machinery genetically assembled to make the neuron recognize these critical combination of inputs that initiate a few outgoing *superspikes*? From the point of a working neuron, these and many other similar questions can be reduced to two major questions. First, how the thousands of inputs are selected within dendrites to end in a few outgoing spikes? And second, is there any benefit or function in the remaining inputs failing to produce output? The answers have to be found in the computational operations performed in the dendritic arborization of the neurons. Formerly viewed as a black box, we shall try to introduce ourselves within the dendritic apparatus to unveil the secrets of this miniature communication center.

CHANNELS IN DENDRITES: A CONCEPTUAL REVOLUTION

Based on modeling studies, some authors, suggested that the overall firing pattern is somehow engraved in the architecture of the dendritic tree (Mainen et al., 1995). Indeed, some architectonic dendritic prototypes have been preserved in specific brain nuclei spanning long periods of brain evolution. Although this may be indicative of a common integrative function being preserved to carry out a specific task, the large number of exceptions clearly argue against this hypothesis. Certainly, the physical structure of the dendritic arbors has a notable influence in the processing of inputs, and even some computational capabilities arise from it (see Segev and London, 1999). Nevertheless, most dendrites are *active* neuron elements, not mere passive cables. The number and variety of ion channels in their membranes modulate synaptic currents with a strength and complexity largely overcoming that of the cable structure. These "modulatory" dendritic currents have the same voltage-dependent nature as those in the soma and axon, and have been called *intrinsic currents* to differentiate them from the synaptic currents that co-activate and co-localize with them.

The presence of active electrogenic machinery in dendrites was first reported by founder of modern electrophysiology Lorente de Nó (1947) in the first half of the past century. He observed active backpropagation of action potentials into dendrites. But it was not until early nineties that researchers began to grasp the magnitude of the presence of V-dependent channels in sites other than the soma and axon. Today, we know that intrinsic dendritic currents participate in basic neuron functions as the mentioned conduction of action potentials from soma to dendrites and the plastic events born out of their local interaction with synaptic currents. This issue has been a major object of study in the late years and we refer the readers to specific literature on the issue (Johnston et al., 1996; Reyes, 2001).

Our focus here is on the role of dendritic channels on the forward transmission of inputs, an issue that has been poorly studied due to technical difficulties in recording from very thin dendrites. It is however highly relevant as it concerns to the very mechanisms of synaptic integration. An important move forward was the recognition of the participation of intrinsic dendritic currents in the reshaping of postsynaptic potentials, which has been proved in all cell types studied so far (Canals et al., 2005; Stuart and Sakmann, 1995). This finding has shaken vigorously the theoretical basis upon which neuronal integration is based. According to the classic doctrine of passive dendritic cables, the synaptic currents propagate along dendrites with a constant decay produced by the filtering properties of the membrane capacitance and resistance. Far from this, synaptic potentials are constantly modified in their propagation to the trigger zone/s. The most common role attributed to intrinsic currents is the amplification of excitatory postsynaptic potentials generated in remote sites to compensate for cable filtering. In some cells (but not others), this mechanism serves to equalize the impact of synapses located at different distances from the axon (Magee and Cook, 2000). This way, the traditional view that relegated distal inputs to a mere modulatory role of the proximal ones is no longer accepted. Therefore, the site of an input is not necessarily related to a genetically determined weight in deciding output. In fact, a delicate balance between opposing intrinsic currents determines the varying shape of synaptic potentials on their spread along the dendritic cable. It is also possible that the same synaptic input causes a totally different voltage envelope at the soma/axon trigger under different physiological situations due to the interference of multiple third parties. Though the actors (intrinsic currents) are relatively new, the concept is not. Let's bring to mind that inhibitory synapses trade with excitatory inputs in a similar way.

From the point of view of synaptic integration in dendrites one might think that little has changed with the discovery of dendritic channels since their currents should produce fixed alterations of the synaptic inputs. The intrinsic dendritic currents would only be translating the incoming binary messages into an internal graded code. However, that being the case, the non-linear properties of intrinsic currents would already multiply the computational possibilities of integration: among others, boosting, thresholding, coincidence detection, frequency-dependent filtering are functions out of the reach of synaptic currents integrating in a passive cable (Mel, 1999).

But things get here even much more exciting. In the late years we have learnt that intrinsic currents are not stable components of dendritic membranes. In fact, they can be up or down regulated by inhibitory synaptic currents, internal metabolites, extracellular effectors, and genetic mechanisms (Colbert and Johnston, 1998; Tsubokawa and Ross, 1996). The time windows used by such diverse modulators range from submillisecond to weeks or longer. As a consequence, the internal processing of synaptic inputs by intrinsic currents may, and in fact does change with numerous supracellular physiological variables, such as the metabolic state, previous activity, and age. Although we are only at the beginning of a major field of research, we can already say that the deterministic view of dendritic integration can be left to rest.

These breakthrough observations bear important theoretical implications. We are used to think of neurons as having one type of input (synaptic currents) and one type of output (axonal spikes), while inhibitory synaptic currents are sometimes classified as negative messages and others as modulators of excitability (global or local). At present, the overall view can be quite different. Excitatory inputs are not the only events carrying the informative load during synaptic integration. Lets think for a while on the computational significance of the fact that intrinsic currents can be modulated. One might well think of their modulating factors as additional input to the computing dendrite complementary to synaptic currents. Thus, dendrites would have several types of inputs acting on different sites and with different time scales. Even if we abide to the simplest of the schemes, we must now admit that synaptic integration is, at least, a game for three partners, excitatory, inhibitory and, the new boy in town, the intrinsic currents. To realize of their relevance in the behavior of neurons let's pay attention to this well known fact: when synaptic inhibition is blocked in a brain nucleus, even silent neurons begin to fire regularly at very high frequency. This continuous output is meaningless for target cells (epileptiform behavior), a disaster that results whenever excitatory and inhibitory inputs loss balance beyond the physiological ranges. Such meaningless and intense firing is brought about by the continuous uncontrolled activation of intrinsic dendritic currents unleashed by disinhibition. Since dendritic channels are all throughout the dendritic surface, we may envisage this intrinsic machinery as the actual target of synaptic inputs, which would operate as local switches turning on and off specific subsets of a complex distributed program in the dendritic tree, as if parallel subroutines in a large multipurpose software package.

DENDRITIC SPIKES

Are They Real Entities Or Lab Products?

In the previous section, slow subthreshold intrinsic currents activated or recruited by synaptic inputs are credited as important elements defining some computational properties of dendrites. Yet, a more intriguing function is their participation in the initiation of *local* action potentials or spikes in the dendrites of many neurons, so called because they generally fail to propagate to the axon. Dendritic spikes were initially considered an anomalous behavior of dendrites caused by improper observation or illness. It recently become clear that local spikes can be recorded under physiological conditions, but their significance still remains controversial. The conflict arises, at least in part, from the dramatic implications their presence would have on the very core of the synaptic integration doctrine. Indeed, local spikes break up with the classic idea of a single trigger zone at the axon initial segment.

The current thinking lingers between acceptance and skepticism. As recordings are more efficient and harmless to neurons the presence of local dendritic spikes is recognized in more cell types. Let's dig on it by reviewing the main findings in one of the most paradigmatic cell type, the pyramidal cell of rodents. For a rapid qualitative comprehension, always bear in mind that the physiological (and computational) meaning of the interaction between synaptic and intrinsic currents depends on whether the later contribute to the initiation or merely conduction of spikes, and whether these reach the axon or not (i.e., cell output).

The Data from Experiments

In pyramidal cells, active dendritic currents/potentials have been shown either as slow components or as local or fully propagated spikes (Andreasen and Nedergaard, 1996; Golding and Spruston, 1998; Herreras 1990; Masukawa and Prince, 1984; Wong and Stewart, 1992). Most researchers consider these active dendrites as *weakly excitable* based on the dominant observation that most action potentials initiate in the axonal trigger zone. This term, however, is somewhat gratuitous as it intends to open gap with *fully excitable* axonal membranes. The issue is far from clear. We and other groups have reported a sequential activation of synaptic, slow subthreshold and fast spike-like inward currents in proximal apical dendrites propagating toward the soma and axon (Herreras, 1990). We later reported that specific damage directed to the apical dendrites, or the local blockade of their depolarizing intrinsic currents, was enough to stop axonal firing upon large and synchronous synaptic excitation (Canals et al., 2005; Herreras and Somjen, 1993). An important concept can be derived from these observations. Thus, intrinsic currents constitute the amplifying mechanism by which the proximal main dendrite becomes a necessary relay station for remote synaptic inputs to reach the axon and cause cell output. While these are population studies that report the average behavior of a large number of cells, most single-cell studies, which are performed *in vitro*, point to the axon as the preferred site of action potential initiation. There seems to be a considerable experimental variability in pyramidal cells where it is claimed that dendritic spiking precedes or follows axonal firing (forward or backpropagation, respectively). While in the later case dendrites are only notified of cell output, in the former they become alternative zones for output decision (Stuart et al., 1997). The challenge is served!

For a long time, the main approach to study neuron activity was to evoke a response by stimulating their afferent pathways. In highly dynamic and non-linear systems as neuron dendrites, this method has some disadvantages, as it puts the emphasis on the strength and synchronization of the input that may not reflect actual ongoing behavior. The studies using synchronous activation of synaptic inputs showed that increases in the intensity of the stimulus (i.e., the number of activated synapses) produced a shift in the initial locus of the outgoing spike from the axon initial segment to a dendritic site, usually recorded in the main dendrite (Turner et al., 1991). This led some researchers to consider dendritic spike initiation as an abnormal response due to excessive input synchronization, far beyond what might be expected in spontaneous activity. However, the strength of the input is neither the only nor the main factor in determining the site of initiation. Other researchers found subthreshold responses, aborted spikes or fully propagated dendritic spikes using identical excitation (Golding and Spruston, 1998). Additional factors, such as the frequency of afferent input and local inhibition may also shift the initial locus of the spike (Canals et al., 2005).

Interpretations, Errors, and Other Considerations

Experimentalists are bound to draw conclusions from real data, but in the heat of the arguments we often neglect the limitations of our techniques.

Available data have been obtained from recordings in the soma and the thick main apical dendrite (i.e., the neuron sites where recordings are more accessible). The erroneous assumption is that these two recording sites mirror the electrical events in two other regions, the axon and smaller dendrites, respectively, which are both hard to access. In our opinion, this is a dangerous oversimplification feeding an artificial dispute. The four compartments (axon, soma, main dendrite, and thin dendrites) each have their own electrical identity. Let's analyze more carefully this important issue. While cell firing in the soma reflects faithfully the axonal firing, it does not provide information as to the decision zone (axon *vs.* dendrites). For example, there is experimental evidence that a spike coming from the main dendrite jumps to the axon and only then travels back to invade the soma, an effect caused by the large surface of the soma that requires much higher capacitive load than the slender axon connected to it (López-Aguado et al., 2002). Thus, the soma always fire after the axon, regardless of the axonal or dendritic origin of the spike. Because of this possibility, a simultaneous recording of the soma (as reflecting the axon activity) and main dendrite is required. Yet, there is also evidence that a dendritic spike may fail to propagate continuously through the main dendrite and still be able to fire the axon, so called pseudo-saltatory propagation. In this case, the output decision has also been made in dendrites. Time ago, some researchers suggested that specialized dendritic loci with a high concentration of channels (*hot spots*), possibly in dendritic bifurcations, may promote this type of spike propagation (Spencer and Kandel, 1961), although geometrical factors can also account for it (Ibarz et al., 2006).

Further on, the main point too often neglected is that recordings made in the thick primary dendrites do not mirror the electrical events in the secondary and tertiary dendritic branches, where the majority of inputs arrive and interact with local intrinsic currents. Instead, the behavior of thick dendrites is closely matching that in the soma, as explained long time ago by pioneer electrophysiologists (Eccles, Lorente de Nó). They already noted that the soma and its connected main dendrites with large diameter work as one, since the electrical distance is very small. This forgotten lesson take us to the core of the problem. When the main dendrite is very long, as in cortical pyramids, the function changes somewhat and it behaves as a collector of multiple independent signals originated in parallel dendritic domains (the lateral dendritic branches collecting the synaptic inputs). The questions: Are each of these dendritic branches capable of generating spikes? If so, do these spikes enter the main dendrite and propagate all the way to the soma/axon? Unfortunately, little is known on how action potentials are initiated in thin dendritic branches, how are they integrated locally, and which is the impact on cell output. There are some indications that local dendritic spikes outnumber the somatic spikes, and some are phase locked with them, suggesting that a few may reach the axon (Kamondi et al., 1998). Curiously, when *in vitro* preparations are treated as to reproduce more accurately well known properties of their *in vivo* mates, the initiation and forward propagation of distal dendritic spikes is regularly found (Williams, 2004). Based on these observations, the answers to the above questions gain special relevance as we may be on the wake of a totally new mechanism of synaptic integration based on the interplay of local spikes originated in dendritic substations. The main handicap for the experimental study is that active dendrites have a strong non-linear behavior that makes them highly sensitive to factors hard to control during experiments. We already mentioned that up- or down regulation of dendritic excitability depends on intrinsic factors as well as on the influence of local networks. Furthermore, many of these factors are likely to change during ongoing activity, or under experimental conditions, due to tissue manipulation and recording or simply through the choice of preparation.

LESSONS FROM THE COMPUTATIONAL MODELS

The rich repertoire of computational capabilities endowed by the multiple properties arising from membrane channels constitutes the greatest challenge for experimental studies. Some modern high-resolution techniques require excessive tissue manipulation or dissection, while this is strongly unadvised to study highly non-linear systems. Too often, the experimental approach cannot discriminate between real and fake mechanisms unveiled by an excessive perturbation of the system. Biophysical computation is at present the best approach. Next, we will review the computational studies on active properties of pyramidal cell dendrites in relation to the present issue and will use our own computational work to describe the most relevant mechanisms.

First Models and the Reliability of Experimental Data

The extensive knowledge of channel composition and subcellular distribution as well as the fine anatomical details obtained in the late years have facilitated the construction of mathematical single cell models with a high degree of realism. However, the single-cell models leading the way used *ad hoc* channel assortments and distributions in order to reproduce experimental data. In those, a single axonal trigger in the axon initial segment was assumed, and typically assembled with a huge density of channels while dendrites were given too few. The expected outcome was that dendrites never initiated spikes. These began in the axon and invaded dendrites in a decremental manner to extinction. Later on, as dendritic recordings were refined and the channels therein were identified and quantified, the biophysical models become more realistic and reproduced gross active properties of dendritic electrogenesis as observed in experiments, including the reshaping of synaptic currents and the initiation of dendritic spikes upon intense synchronous activation. In our opinion, these model studies had limited application as they were devoted to reproduce partial experimental data, i.e., they came in support of the dominant idea by the experimenters at the time (Hoffman et al., 1997; Mainen et al., 1995; Rapp et al., 1996). In general, a poor scrutiny of critical parameters yields rigid models, which are ineffective for reproducing any other electrical behavior than those they were designed for. We want to rise here a criticism to such a frequent practice, as non-specialized readers, for whom complex mathematical models are impenetrable jumble, may take them as a definitive demonstration of a certain mechanism, when it merely supports its possibility.

As an example of how the poor reliability of essential experimental data affects models, we can take the reproduction of the most emblematic event, the somatic action potential. By the time single-cell models became accessible to physiologists in mid-nineties, they were using *in vitro* preparations to study intensely the backpropagation of somatoaxonal action potential into dendrites, which by the way were thought to have too weak excitability to initiate and conduct forward spikes. The rule was to built the soma and the main apical dendrite with an homogeneous low density of Na channels that enables partial dendritic invasion of the spike but not its initiation. The prototype somatic spike chosen for modeling was that recorded *in vitro*. In fact, much higher densities are needed to generate realistic action potential waveforms at the soma as recorded *in vivo,* where it is much faster (Varona et al., 2000). Surely, the difference arises from the known depression of Na^+ currents in non intact or immature preparations (Spigelman et al., 1992), which may have only moderate consequences at the soma, but certainly it entails dramatic changes in local dendritic excitability (Shen et al., 1999; Varona et al., 2000).

Parametric Studies: Diving Deep into Dendrites

More beneficial are the models aiming to explore the impact of a given parameter on the electrogenesis of the different neuron domains, regardless of whether they reproduce or not specific experimental data. Some have been dedicated to study the effects of specific dendritic architectures, the fine geometry of dendrites, or the inhomogeneous distribution of channels on spike propagation. The exhaustive analysis of the parametric space is necessary since different combinations of somatodendritic channel types and distributions can reproduce the physiological events of interest at a single locus (e.g., Cook and Johnston, 1999), without this implying that the events in other neuron sites are even closer to reality. That is, one may reproduce faithfully the electrical behavior of a small neuron domain at the expenses of getting out of range everywhere else.

In addition, when some parameters are fixed, the number of possible solutions for the model is limited enormously. While it may be advantageous when the choice is correct, it may be misleading in other cases. For instance, using a fixed channel assortment in different prototype model neurons, Vetter et al. (2001) showed that dendritic architectures may define the mode of spike initiation and propagation. When this particular channel assortment was implemented in pyramidal-like cells, dendritically initiated spikes do not propagate forwardly. However, there are already numerous experimental observations showing this is not true (Larkum et al., 2001; Williams, 2004). Obviously, different cells have dissimilar channel assortment, surely because it is the function and not the structure what has been preserved by evolution.

One important issue is the computational role of the main apical dendrite. Since the initial experimental observations favored the idea of *back* over *forward* propagation of spikes along this cell element, the model studies also appeared to support such functioning. The particular concept of aborted forward propagation of distally initiated spikes did not represent any challenge since it was well known from early theoretical studies in axons that flaring cables produce the blockade of spike propagation (Goldstein and Rall, 1974). A similar conclusion was reached in some theoretical studies applied to active dendrites (Tóth and Crunelli, 1998). However, opposing to this widespread notion, we found that the moderate flare of the apical shaft in pyramidal cells actually works as a loudspeaker amplifying inward currents and fostering the forward propagation of distally initiated spikes (López-Aguado et al., 2002). The mystery is not other than the optimum balance between capacitive load and electroregenerative capability of a given segment of dendrite. Even with the moderate excitability attributed to this cell element, the small flare cannot counterbalance the increasing current density produced by forward spikes (Figure 1).

In other studies, we have used a realistic pyramidal cell model calibrated with *in vivo* and *in vitro* results to explore how some well known factors govern the synaptic initiation of dendritic action potentials and we have determined their ability to influence cell output. Among others, we have cross-analyzed the influence of the synaptic loci and strength, the role of different dendritic channels, the relative excitability of the axon and dendrites, and inhibition. When considering only the axon *vs.* the main dendrite, we found several types of spiking, namely, aborted dendritic spikes, full forward and pseudosaltatory propagation, and backpropagation. The number and variety of partial conclusions on dendritic functioning is very large (Ibarz et al., 2006). Some follow, just to name a few. Clustered dendritic inputs are more effective in producing output decision in dendrites than spatially dispersed inputs. The higher ratio of the axon-to-dendrite excitability was correlated with spike initiation at the axon for inputs in the middle part of the dendritic tree, but this effect was ill-defined for distal inputs.

Figure 1. Forward propagation of spikes: Loudspeaker effect of moderately flaring active cables. When the spike travels in a cable of constant diameter (left) the current remains stable. In flaring active cables (right), the spike initiated in the narrow edge travels forward with increasing transmembrane current that ensures propagation.

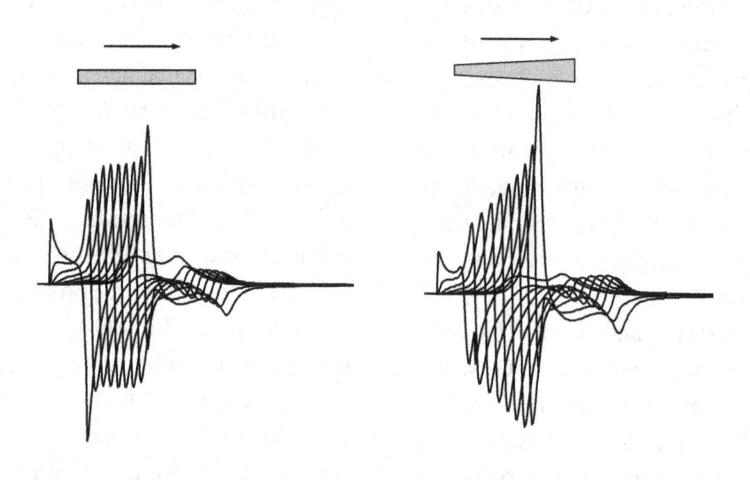

The precise distribution of some channels also had a profound effect on the site of initiation that depended on third variables.

Maybe, the most important conclusion of our study is that there are too many variables with a strong influence on the site of spike initiation within their recognized experimental ranges of variation. In fact, the activation mode varied extremely when two or more variables were cross-analyzed, becoming rather unpredictable when all the variables were considered. An example is illustrated in Figure 2 that shows multiple and variable activation modes in a cross-analysis of four variables, input strength, input locus, strength of inhibition, and dendritic gradient of potassium channels.

One notable observation is that spike initiation in the apical shaft does not necessarily lead to cell output when certain parameters are combined, a result verified by experimental observations using synchronous activation (Canals *et al.*, 2005; Golding & Spruston, 1998). In the model, the spike could either be aborted within the apical shaft or "jumped" into the axon, backpropagating from there into the apical dendrite, i.e. the so called pseudosaltatory mode. It should be noted that this mode of conduction may easily go undetected in experiments. Even with dual intracellular recordings from the soma and the apical shaft, data can be erroneously interpreted as a continuous forward propagation or backpropagation, depending on whether the dendrites were impaled at the firing or the refractory zones of the apical shaft. Examples of even more complex modes of forward propagation consisting of a spatial sequence of decremental and regenerative spike have been shown in neocortical pyramidal cells that have longer apical shafts (Larkum et al., 2001).

Figure 2. Interaction of several parameters makes unpredictable the spiking mode between dendrites (thin) and axon/soma (thick traces). A. Cross-analysis of inhibition, excitatory input and dendritic potassium conductance. B. Cross-analysis of inhibition, excitatory input location and dendritic potassium conductance. Note that for a fixed value in any one variable, the spiking mode can be totally different upon changes in third variables. N: no firing; Ax: axon fires first (backpropagation); D: dendrite first (full forward); ps: pseudosaltatory forward; d: aborted dendritic spike; =: simultaneous axo-dendritic firing. D-M: dendrite fires first and is followed by multiple firing. Ax-M: axon-first and multiple firing. Modified from Ibarz et al., 2006.

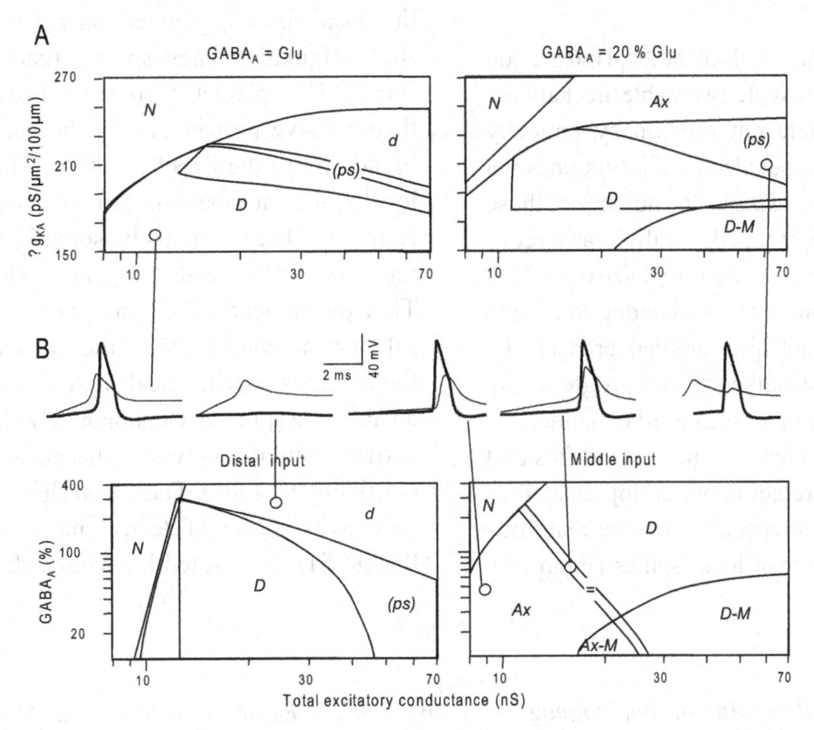

Small Dendrites Talking To Parent Dendrites

The different firing modes mentioned above to classify the dendritic apical shaft-to-axon coupling are derived from the variations in the spatiotemporal distribution of firing branches within the overall dendritic structure. We examined this pattern in order to gain insight of local integration, and the relationship between spiking on secondary branches and the main shaft. A small number of synaptically activated local spikes was sufficient to initiate a forward spike, although the minimum number varied enormously depending upon the different conditions. The difference between conditions promoting full-forward, pseudosaltatory, or backward propagation between apical shaft and the axon were neither simple nor obvious. The number of local spikes and the temporal scattering constituted the best predictive element for both cell output and the mode of dendrite-to-axon firing. As many as 6-10 lateral spikes (out of a total of 27 possible in our model) were sufficient to produce a cell output under most initial conditions and input locations. Temporal scattering between local spikes ≤ 1 ms was observed for input patterns that resulted in full-forward propagation, while local spikes that were scattered over 2 ms or more yielded pseudosaltatory conduction or aborted apical shaft spikes.

Dendritic Chattering by Cross-Spiking: An Internal Language Between Dendrites

A closer look into the details of the local integration of dendritic spikes revealed a notable mechanism. Local spikes initiate not only in synaptically activated branches, but also in adjacent ones via apical shaft (*cross spiking*). The number of these branches that fired "sympathetically" varied considerably depending on the initial settings. This mechanism facilitates forward conduction due to the successive current injections delivered into the apical shaft by local spikes at sites progressively nearer to the soma in a cascade-like manner.

The mode in which the lateral dendrites and the apical shaft interact is better appreciated in an example of subthreshold activation that initiated a small number of local spikes (Figure 3). The synaptically excited branch (as16a) fired a local spike that passively propagated into the apical shaft (vertical dashed line 1) originating a step-like increase (arrow a) in the EPSP envelope. Although the apical shaft failed to initiate an AP, it still carried enough current to initiate a local spike in a non-excited branch (as12) because of the lower threshold of thinner branches. In turn, this local spike fed current back into the apical shaft originating a new spikelet (vertical dashed line 2). The spikelets were smoothed out during their passive propagation to the soma (arrows b), adding to the EPSP envelope. Thus, though local spikes at positions distant from the soma cannot be discriminated by somatic recordings, they contribute notably to the EPSP envelope. This phenomenon facilitates branch-to-branch saltatory activation toward the soma, even when the common parent apical shaft was conducting an aborting spike. An analogous conduction was earlier postulated between adjacent active spines (Miller *et al.*, 1985). The apical shaft would thus serve as a reservoir of electric charge enabling the firing of non-activated branches, which in turn

Figure 3. Cross-spiking: An internal language between dendrites. Explanation in the text. Modified from Ibarz et al., 2006.

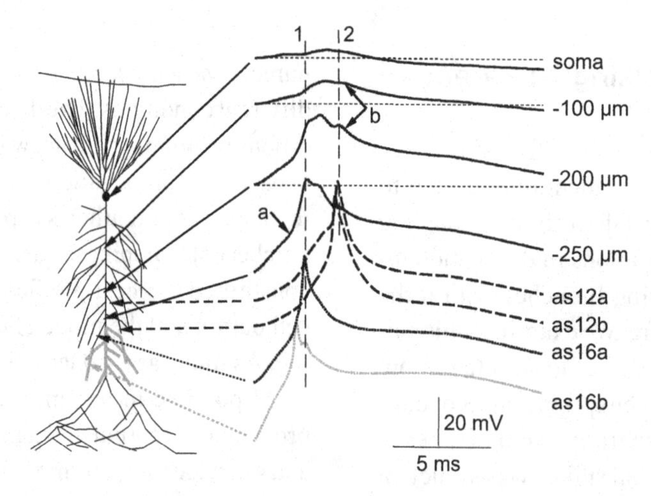

feed back more current into the parent dendrite. The successive current injections contribute to uphold the depolarizing envelope in the apical shaft, which may be critical to set the pseudosaltatory forward AP conduction. Even lateral firing alone without an apical shaft spike may still be considered pseudosaltatory forward conduction when it determines the immediate cell output.

OUTPUT DECISION IN DENDRITES: WHAT FOR?

So far, the dendritic electrogenesis described in the previous section remains in the category of the theoretical possibilities, as it has not been shown in actual experiments. Unfortunately, these have not attained enough technical resolution to record from intact dendritic branches without strongly modifying their electrical behavior. Some preliminary studies (Gasparini et al., 2004) apparently support the gross features.

We may ask what advantages can be obtained from multisite dendritic decisions contributing to cell output? One possibility is that local dendritic spikes can be used as an internal language between different cell subregions so that cell output becomes an elaborated response arising from the cross-talk between somato-axonal and dendritic spike generators. Such chattering between two separated action potential trigger zones (axon and apical dendrite) has already been shown in cortical pyramidal cells to delineate the final output pattern trough the axon (Larkum et al., 2001).

In the beginning of this chapter, we mentioned that many neuron types fire very poorly and irregularly. In general, these correspond to neurons dealing with inputs of varied nature, i.e., neurons ranking high in the canonical arbor of complex associations. Their irregular firing is compatible with output decision made in dendrites. Some theoreticians have indicated that irregular firing is not consistent with random input, as one might think. Instead, they postulated the synchronous arrival of inputs (Stevens and Zador, 1998; Softky and Koch, 1993). An irregular pattern of synchronous inputs can thus be translated in the occasional firing of local spikes in a dendritic branch/domain, collected by an excitable main dendrite that conveys all of them up to the soma/axon. The final temporal structure of the output that can be read in the axon is the summation of spikes generated in all dendritic substations (Figure 4).

A generic conclusion that can be drawn is that dendritic integration of local spikes is extraordinarily varied and subtle. The vast parametric space within which the dendritic trees appear to work is probably on the basis of the apparent unpredictability of cell output. Whether this is true in real cells awaits future studies. Recent work emphasizes the importance of homeostatic mechanisms that maintain firing regimes of a particular cell after severe changes of some influential parameters, as if neurons self-adapt the weight of intrinsic electrogenic machinery in response to adverse or unexpected conditions (Turrigiano, 2007). Being that so, the range of some parameters, as found in experiments, is much narrower than currently thought, and large variability only results from the variable perturbation of highly disturbing recording techniques.

Alternatively, the unpredictability of the output may reflect the subtle processing of inputs before an output decision is taken. Most input patterns would thus be insufficient to directly set an output. Still, we speculate that the apical tree as a whole is "tuned" to make immediate output decisions for specific sequences or combinations of inputs activated within a narrow temporal window. In essence, that would be the main computational difference as compared to the traditional view of synaptic integration based on spatiotemporal summation of slow synaptic potentials. A few spikes initiated in lateral branches will have an output efficiency equivalent to hundredths of synaptic potentials in absence of local spikes. Besides, integration based on local spikes has the chance

Figure 4. Modes of synaptic integration. A: traditional mechanism based on spatiotemporal summation of EPSPs and exclusive axonal firing. B: Proposed mechanism based on firing of spikes in dendritic substations that sum up in the axon. C: Multiplexed language. Both mechanisms may co-exist. Also burst firing and isolated spikes.

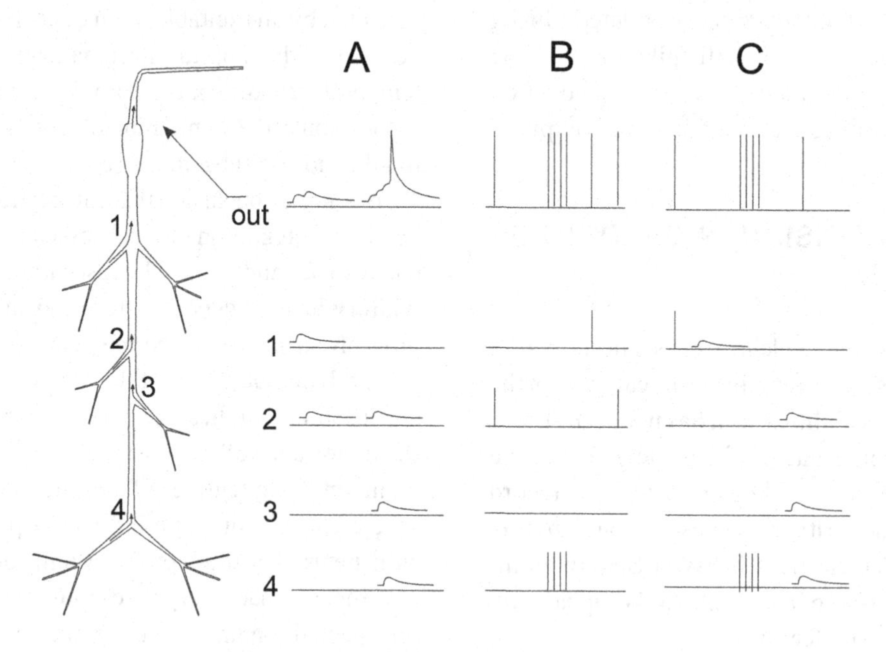

to maintain much better the temporal structure of selected inputs.

Although we advocate for this novel form of synaptic integration, it certainly does not exclude the functionality of the traditional mechanism based on spatiotemporal summation of inputs. As a matter of fact, the two may operate as a multiplexed language in the axon (Figure 4). It is very common that irregularly firing cells display isolated action potentials mixed up with occasional burst-like firing. The dendritic origin of the later has been proven in some cell types. The mixed patterns should not present any conceptual handicap. Although there is a single axon stemming out of the soma, most bifurcate one or several times and/or give off collaterals that contact different cell types in different brain nuclei. Even if a mixed message gets to different target cells, the specialized postsynaptic machinery of each one may be variably tuned to decode the incoming message and extract the interesting parts of it, so that only listen to either the bursts or the isolated action potentials. Thus, what might be considered input noise in some circuits, may actually be positive messages in others, all in just one delivery.

FUTURE RESEARCH DIRECTIONS

Through the chapter we tried to highlight the importance of getting reliable experimental data to feed the biophysical models. Small changes in some parameters that physiologists may not deem as critical, are revealed by modeling as crucial for determining the pattern of cell firing. A continuous effort should be made to develop experimental procedures that facilitate the observation of dendrites in intact preparations.

Obviously, disclosing the fine details of local dendritic integration is a must. So far, dendrites have been considered a black box. It is time to recognize their extraordinary computational capabilities, classify them, and describe their physiological ranges. Knowing the computational character that each channel and combination of channels confer to dendrites will also help to understand why have they been preserved in evolution, and will also provide new mathematical tools that can be applied in artificial networks and the real life.

An opening field of research addresses the homeostatic mechanisms of cell activity. There is not yet full awareness as to how the findings in this area may condition future research and the interpretation of data obtained so far. It may well be that some neuronal mechanisms established by past research are actually functioning only under the extreme conditions for adequate observation, i.e., activated by excessive perturbation of an object of study that self-adapts to the presence of our exploring devices/strategies.

An evident area of strong impact and fast development in neuroscience is the use of realistic mathematical models of single neurons. It not only helps to understand experimental data, but in cases like the examined here, it may advance the possible mechanisms that operate in systems that cannot be accessed experimentally. In the future, simplified models should only be used to explore the potential applications of basic neuronal principles in artificial systems. Those aiming to help physiologists must be detailed to the most and thoroughly scrutinized for safe parameter restriction.

ACKNOWLEDGMENT

This work was supported by grants BFU 2005/8917 from the Spanish Ministry of Science and Education, S-SEM-0255-2006 from the Comunidad Autónoma de Madrid, and I3/200620I186 from the CSIC.

REFERENCES

Andreasen, M., & Nedergaard, S. (1996). Dendritic electrogenesis in rat hippocampal CA1 pyramidal neurons: functional aspects of Na+ and Ca2+ currents in apical dendrites. *Hippocampus*, 6, 79-95.

Canals, S., López-Aguado, L., & Herreras O. (2005). Synaptically-recruited apical currents are required to initiate axonal and apical spikes in hippocampal pyramidal cells: modulation by inhibition. *Journal of Neurophysiology*, 93, 909-918.

Colbert, C. M., & Johnston, D. (1998). Protein kinase C activation decreases activity-dependent attenuation of dendritic Na+ current in hippocampal CA1 pyramidal neurons. *Journal of Neurophysiology* 79, 491-495.

Gasparini, S., Migliore, M., & Magee, J.C. (2004). On the initiation and propagation of dendritic spikes in CA1 pyramidal neurons. *Journal of Neuroscience,* 24, 11046-11056.

Golding, N. L. & Spruston, N. (1998). Dendritic sodium spikes are variable triggers of axonal action potentials in hippocampal CA1 pyramidal neurons. *Neuron*, 21, 1189-1200.

Goldstein, S. S., & Rall W. (1974). Changes of action potential shape and velocity for changing core conductor geometry. *Biophysical Journal,* 14, 731-757.

Herreras, O. (1990). Propagating dendritic action potential mediates synaptic transmission in CA1 pyramidal cells in situ. *Journal of Neurophysiology*, 64, 1429-1441.

Herreras, O., & Somjen, G, G., (1993). Effects of prolonged elevation of potassium on hippocampus of anesthetized rats. *Brain Research,* 617, 194-204.

Hoffman, D. A., Magee, J. C., Colbert, C. M., & Johnston, D. (1997). K+ channel regulation of signal propagation in dendrites of hippocampal pyramidal neurons. *Nature* 387, 869-875.

Ibarz, J. M., Makarova, I., & Herreras, O. (2006). Relation of apical dendritic spikes to output decision in CA1 pyramidal cells during synchronous activation: a computational study. *European Journal of Neuroscience,* 23, 1219-1233.

Johnston, D., Magee, J. C., Colbert, C. M, & Christie, B. R. (1996). Active properties of neuronal dendrites. *Annual Review of Neuroscience* 19, 165-186.

Kamondi, A., Acsady, L., & Buzsaki, G. (1998) Dentritic spikes are enhanced by cooperative network activity in the intact hippocampus. *Journal of Neuroscience,* 18, 3919-3928.

Larkum, M. E., Zhu, J. J., & Sakmann, B. (2001). Dendritic mechanisms underlying the coupling of the dendritic with the axonal action potential initiation zone of adult rat layer 5 pyramidal neurons. *Journal of Physiology (London),* 533, 447-466.

López-Aguado, L., Ibarz, J. M., Varona, P., & Herreras, O. (2002). Structural inhomogeneities differentially modulate action currents and population spikes initiated in the axon or dendrites. *Journal of Neurophysiology*, 88, 2809-2820.

López-Aguado, L., Ibarz, J. M., Varona, P. & Herreras, O. (2002). Structural inhomogeneities differentially modulate action currents and population spikes initiated in the axon or dendrites. *Journal of Neurophysiology*, 88, 2809-2820.

Lorente de Nó, R. (1947). Action potential of the motoneurons of the hypoglossus nucleus. *Journal Cell Comparative Physiology*, 29, 207-288.

Magee, J. C., & Cook, E. P. (2000). Somatic EPSP amplitude is independent of synapse location in hippocampal pyramidal neurons. *Nature Neuroscience*, 3, 895-903.

Mainen, Z.F., Joerges, J., Huguenard, J.R., & Sejnowski, T.J. (1995). A model of spike initiation in neocortical pyramidal neurons. *Neuron* 15:1427-1439.

Mainen, Z. F., & Sejnowski, T.J. (1996). Influence of dendritic structure on firing pattern in model neocortical neurons. *Nature* 382, 363-66.

Masukawa, L. M., & Prince, D. A. (1984). Synaptic control of excitability in isolated dendrites of hippocampal neurons. *Journal of Neuroscience*, 4, 217-227.

Mel, B. (1999) Why have dendrites? A computational perspective. In G. Stuart, N. Spruston and M. Häusser (Eds.), *Dendrites* (pp 271-289). OUP, New York.

Miller, J. P., Rall, W., & Rinzel, J. (1985). Synaptic amplification by active membrane in dendritic spines. *Brain Research*, 325, 325-330.

Rapp, M., Yarom, Y., & Segev, I. (1996). Modeling back propagation action potential in weakly excitable dendrites of neocortical pyramidal cells. *P.N.A.S.* 93, 11985-11990.

Reyes, A. (2001). Influence of dendritic conductances on the input-output properties of neurons. *Annual Review of Neuroscience,* 24, 653-675.

Segev, I., & London, M. A. (1999) Theoretical view of passive and active dendrites. In G. Stuart, N. Spruston and M. Häusser (Eds.), *Dendrites* (pp 205-230). OUP, New York.

Shen, G. Y., Chen, W. R., Midtgaard, J., Shepherd, G. M., & Hines M. L. (1999). Computational analysis of action potential initiation in mitral cell soma and dendrites based on dual patch recordings. *Journal of Neurophysiology*, 82, 3006-3020.

Softky, W. R., & Koch, C. (1993). The highly irregular firing of cortical cells is inconsistent with temporal integration of random EPSPs. *Journal of Neuroscience*, 13, 334-350.

Spencer, W.A., & Kandel, E.R. (1961). Electrophysiology of hippocampal neurons. IV. Fast prepotentials. *Journal of Neurophysiology* 24, 272-285.

Spigelman, I., Zhang, L., & Carlen, P. L. (1992). Patch-clamp study of postnatal development of CA1 neurons in rat hippocampal slices: membrane excitability and K^+ currents. *Journal of Neurophysiology* 68, 55-69.

Stevens, C. F., & Zador, A. M. (1998). Input synchrony and the irregular firing of cortical neurons. *Nature Neuroscience*, 1, 210-217.

Stuart, G., & Sakmann, B. (1995). Amplification of EPSPs by axosomatic sodium channels in neocortical pyramidal neurons. *Neuron*, 15,1065-76.

Stuart, G., Schiller, J. & Sakmann, B. (1997). Action potential initiation and propagation in rat neocortical pyramidal neurons. *Journal of Physiology (London)*, 505, 617-632.

Tsubokawa, H., & Ross, W. N. (1996). IPSPs modulate spike backpropagation and associated [Ca2+]i changes in the dendrites of hippocampal CA1 pyramidal neurons. *Journal of Neurophysiology*, 76, 2896-2906.

Turner, R. W., Meyers, D. E. R., Richardson, T. L., & Barker, J. L. (1991). The site for initiation of action potential discharge over the somatodendritic axis of rat hippocampal CA1 pyramidal neurons. *Journal of Neuroscience*, 11, 2270-2280.

Turrigiano, G. (2007). Homeostatic signaling: the positive side of negative feedback. *Current Opinion in Neurobiology*, 17, 1-7.

Varona, P., Ibarz, J. M., López-Aguado, L., & Herreras, O. (2000). Macroscopic and subcellular factors shaping CA1 population spikes. *Journal of Neurophysiology*, 83, 2192-2208.

Vetter, P. H., Roth, A., & Haüsser, M. (2001). Propagation of action potentials in dendrites depends on dendritic morphology. *Journal of Neurophysiology* 85, 926-937.

Williams, S. R. (2004). Spatial compartmentalization and functional impact of conductance in pyramidal neurons. *Nature Neuroscience*, 7, 961-967.

Wong, R. K. S., & Stewart, M. (1992). Different firing patterns generated in dendrites and somata of CA1 pyramidal neurones in guinea-pig hippocampus. *Journal of Physiology (London)*, 457, 675–687.

ADDITIONAL READING

Chen, W. R., Mitgaard, M., & Shepherd, G. M. (1997) Forward and backward propagation of dendritic impulses and their synaptic control in mitral cells. *Science*, 278, 463-468.

Joyner, R.W., Westerfield, M., & Moore, J. W. (1980). Effects of cellular geometry on current flow during a propagated action potential. *Biophysical Journal*, 31, 183-194.

Poirazi, P., Brannon, T., & Mel, B. W. (2003). Pyramidal neuron as two-layer neural network. *Neuron*, 37, 989-999.

Rall , W., Burke, R. E., Holmes, W. R., Jack, J. J. B., Redman, S. J., & Segev, I. (1992). Matching dendritic neuron models to experimental data. *Physiological Reviews*, 72, 159-186.

Softky, W. R. (1994) Sub-millisecond coincidence detection in active dendritic trees. Neuroscience, 58, 13-41.

Spruston, N., Schiller, Y., Stuart, G. & Sakmann, B. (1995) Activity-dependent action potential invasion and calcium influx into hippocampal CA1 dendrites. *Science*, 268, 297-300.

Wei, D. S., Mei, Y. A., Bagal, A., Kao, J. P. Y., Thompson, S. M., & Tang, C. M. (2001). Compartimentalized and binary behavior of terminal dendrites in hippocampal pyramidal neurons. *Science*, 293, 2272-2275.

Section II
New Biologically Inspired Artificial Intelligence Models

Chapter IV
Spiking Neural P Systems: An Overview

Gheorghe Păun
Institute of Mathematics of the Romanian Academy, Romania

Mario J. Perez-Jimenez
University of Sevilla, Spain

ABSTRACT

This chapter is a quick survey of spiking neural P systems, a branch of membrane computing which was recently introduced with motivation from neural computing based on spiking. Basic ideas, examples, some results (especially concerning the computing power and the computational complexity/efficiency), and several research topics are discussed. The presentation is succinct and informal, meant mainly to let the reader having a flavour of this research area. The additional references are an important source of information in this respect.

THE GENERAL FRAMEWORK

Learning computing ideas (data structures, operations with them, ways to control these operations, computing architectures, new types of algorithms—in general, of heuristic ways to search for fast solutions to complex problems, and so on) from biology was a permanent concern for computer science, but in the last decades this became a real fashion. Genetic algorithms, in general, evolutionary computing, neural computing, DNA computing, and several other research directions are already well established theoretically, sometimes also with numerous applications (this is especially the case of genetic algorithms). All these areas form what is now called *natural computing*.

Membrane computing is one of the youngest branches of natural computing. It has been initiated in (Păun, 2000) and soon became a "fast emerging research front of computer science", as Thomson Institute for Scientific Information, ISI, called it – see http://esi-topics.com. In a few words, the goal is to start from the living cell, as well as from the way the cells cooperate in tissues, organs, or other structures, and to devise computational devices. *Membrane systems,* currently called *P systems,* were introduced in this context, taking as basic architecture the structure of a cell, with chemical objects evolving in compartments delimited by membranes. In a cell, the membranes are hierarchically arranged (hence placed in the nodes of a tree), but also P systems with membranes placed in the nodes of a general graph were considered.

The (mathematical) theory of membrane computing is well developed, and a lot of classes of P systems were introduced and investigated – details can be found at the web site from http://psystems.disco.unimib.it, as well in the monograph (Păun, 2002). The main directions of research concern the computing power of P systems and their usefulness for solving computationally hard problems in a feasible time, by making use of the parallelism intrinsic to the model. Thus, most classes of P systems were proved to be Turing complete, and, in the cases where an exponential workspace can be produced in a linear time (e.g., by dividing membranes – *mitosis*), polynomial solutions to **NP**-complete problems were devised.

Recently, a series of applications of P systems were reported, expecially using membrane computing as a framework for building models of biological phenomena (approached at a micro level, thus contrasting with the models based on differential equations, which approach the reality at a macro level). The above mentioned web page contains information in this respect; details can be also found in the volume (Ciobanu et al., eds., 2006),

Also the neural cell was considered as an inspiration source for membrane computing. The present chapter is a quick introduction to so-called *spiking neural P systems* (in short, *SN P systems*), a class of P systems defined in (Ionescu et al., 2006), starting from the way the neurons cooperate in large neural nets communicating by electrical impulses (*spikes*). Neurons are sending to each others electrical impulses of identical shape (duration, voltage, etc.), with the information "encoded" in the frequency of these impulses, hence in the time passed between consecutive spikes. For neurologists, this is nothing new, related drawings already appear in papers by Ramon y Cajal, a pioneer of neuroscience at the beginning of the last century, but in the recent years "computing by spiking" became a vivid research area, with the hope to lead to a neural computing "of the third generation" – see, e.g., (Gerstner and Kistler, 2002), (Maass and Bishop, eds., 1999).

For membrane computing it is natural to incorporate the idea of spiking neurons (already neural-like P systems exist, based on different ingredients, efforts to compute with a small number of objects were recently made in several papers, using the time as a support of information, for instance, taking the time between two events as the result of a computation, was also considered), but still important differences exist between the general way of working with multisets of objects in the compartments of a cell-like membrane structure – as in usual P systems – and the way the neurons communicate by spikes. A way to answer this challenge was proposed in (Ionescu et al., 2006): neurons taken as single membranes, placed in the nodes of a graph corresponding to synapses, only one type of objects present in neurons, the spikes, with specific rules for handling them, and with the distance in time between consecutive spikes playing an important role (e.g., the result of a computation being defined either as the whole spike train of a distinguished output neuron, or as the distance between consecutive spikes). Details will be given immediately.

What is obtained is a computing device whose behaviour resembles the process from the neuron nets, meant to generate strings or infinite sequences (like in formal language theory), to recognize or translate strings or infinite sequences (like in automata theory), to generate or accept natural numbers, or to compute number functions (like in membrane computing). Results of all these types will be mentioned below. Nothing is said here, because nothing was done so far, about using such devices in "standard" neural computing applications, such as pattern recognition. Several open problems and research topics will be mentioned below (a long list of such topics, prepared for the Fifth Brainstorming Week on Membrane Computing, Seville, January 29-February 2, 2007, can be found in (Păun, 2007)), but probably this is the most important one: connecting SN P systems with neural computing, more generally, looking for applications of SN P systems.

AN INFORMAL PRESENTATION OF SN P SYSTEMS

Very shortly, an SN P system consists of a set of *neurons* (cells, consisting of only one membrane) placed in the nodes of a directed graph and sending signals (*spikes*, denoted in what follows by the symbol a) along *synapses* (arcs of the graph). Thus, the architecture is that of a tissue-like P system, with only one kind of objects present in the cells. The objects evolve by means of *spiking rules*, which are of the form $E/a^c \rightarrow a;d$, where E is a regular expression over $\{a\}$ and c, d are natural numbers, $c \geq 1$, $d \geq 0$. The meaning is that a neuron containing k spikes such that a^k belongs to the language $L(E)$, identified by the expression E, $k \geq c$, can consume c spikes and produce one spike, after a delay of d steps. This spike is sent to all neurons to which a synapse exists outgoing from the neuron where the rule was applied. There also are *forgetting rules*, of the form $a^s \rightarrow \lambda$, with the meaning that $s \geq 1$ spikes are removed, provided that the neuron contains exactly s spikes. We say that the rules "cover" the neuron, all spikes are taken into consideration when using a rule.

The system works in a synchronized manner, i.e., in each time unit, each neuron which can use a rule should do it, but the work of the system is sequential in each neuron: only (at most) one rule is used in each neuron.

One of the neurons is considered to be the *output neuron*, and its spikes are also sent to the environment. The moments of time when a spike is emitted by the output neuron are marked with 1, the other moments are marked with 0. The binary sequence obtained in this way is called the *spike train* of the system – it might be infinite if the computation does not stop.

In the spirit of spiking neurons, in the basic variant of SN P systems introduced in (Ionescu et al., 2006), the result of a computation is defined as the distance between consecutive spikes sent into the environment by the (output neuron of the) system. In the initial paper, only the distance between the first two spikes of a spike train was considered, then in (Păun et al., 2006a) several extensions were examined: the distance between the first k spikes of a spike train, or the distances between all consecutive spikes, taking into account all intervals or only intervals that alternate, all computations or only halting computations, etc.

Systems working in the accepting mode were also considered: a neuron is designated as the *input neuron* and two spikes are introduced in it, at an interval of n steps; the number n is accepted if the computation halts.

Two main types of results were obtained: computational completeness in the case when no bound was imposed on the number of spikes present in the system, and a characterization of semilinear sets of numbers in the case when a bound was imposed (hence for finite SN P systems).

Another attractive possibility is to consider the spike trains themselves as the result of a computation, and then we obtain a (binary) language generating device. We can also consider input neurons and then an SN P system can work as a transducer. Such possibilities were investigated in (Păun et al., 2006b). Languages – even on arbitrary alphabets – can be obtained also in other ways: following the path of a designated spike across neurons, or using *extended* rules, i.e., rules of the form $E/a^c \rightarrow a^p;d$, where all components are as above and

$p \geq 1$; the meaning is that p spikes are produced when applying this rule. In this case, with a step when the system sends out i spikes, we associate a symbol b_i, and thus we get a language over an alphabet with as many symbols as the number of spikes simultaneously produced. This case was investigated in (Chen et al., 2006a).

The proofs of all computational completeness results known up to now in this area are based on simulating register machines. Starting the proofs from small universal register machines, as those produced in (Korec, 1996), one can find small universal SN P systems. This idea was explored in (Păun and Păun, 2007).

In the initial definition of SN P systems several ingredients are used (delay, forgetting rules), some of them of an unrestricted form (general synapse graph, general regular expressions). As shown in (Ibarra et al., 2007), rather restrictive normal forms can be found, in the sense that some ingredients can be removed or simplified without losing the computational completeness. For instance, the forgetting rules or the delay can be removed, both the indegree and the outdegree of the synapse graph can be bounded by 2, while the regular expressions from firing rules can be of very simple forms.

There were investigated several other types of SN P systems: with several output neurons, with a non-synchronous use of rules, with an exhaustive use of rules (whenever enabled, a rule is used as much as possible for the number of spikes present in the neuron), with packages of spikes sent along specified synapse links, etc. We refer the reader to the additional bibliography of this chapter, with many papers being available at the membrane computing web site mentioned above.

A FORMAL DEFINITION

We introduce now the SN P systems in a general form, namely, in the extended (i.e., with the rules able to produce more than one spike) computing (i.e., able to take an input and provide an output) version.

A computing extended *spiking neural P system*, of degree $m \geq 1$, is a construct of the form $\Pi = (O, \sigma_1, \ldots, \sigma_m, syn, in, out)$, where:

1. $O = \{a\}$ is the singleton alphabet (a is called *spike*);
2. $\sigma_1, \ldots, \sigma_m$ are *neurons*, of the form $\sigma_i = (n_i, R_i)$, $1 \leq i \leq m$, where:
   a. $n_i \geq 0$ is the *initial number of spikes* contained in σ_i;
   b. R_i is a finite set of *rules* of the form $E/a^c \rightarrow a^p;d$, where E is a regular expression over $\{a\}$ and c, p, d are natural numbers, $c \geq 1, p \geq 0, d \geq 0$;
3. *syn* is a subset of $\{1, 2, \ldots, m\} \times \{1, 2, \ldots, m\}$ with $i \neq j$ for all (i, j) in *syn*, $1 \leq i, j \leq m$ (*synapses* between neurons);
4. *in, out* are elements of $\{1, 2, \ldots, m\}$ indicating the *input* and the *output* neurons, respectively.

An SN P system having only rules with $p = 1$ (they produce only one spike) or of the form $a^s \rightarrow \lambda$ is said to be of the *standard* type (non-extended).

The rules are applied as follows. If the neuron σ_i contains k spikes, a^k belongs to the language $L(E)$, and $k \geq c$, then the rule $E/a^c \rightarrow a^p;d$ from R_i can be applied. This means consuming (removing) c spikes (thus only $k - c$ spikes remain in σ_i (this corresponds to the right derivative operation $L(E)/a^c$ suggested by the writing of the rule), the neuron is fired, and it produces p spikes after d time units (a global clock is assumed, marking the time for the whole system, hence the functioning of the system is synchronized). If $d = 0$, then the spikes are emitted immediately, if $d = 1$, then the spikes are emitted in the next step, etc. If the rule is used in step t and $d \geq 1$, then in steps $t, t+1, t+2, \ldots, t+d-1$ the neuron is *closed* (this corresponds to the refractory period from neurobiology), so that it cannot fire and cannot receive new spikes (if a neuron has a synapse to a closed neuron and tries to send a spike along it, then that particular spike is lost). In the step $t + d$, the neuron spikes and becomes again open, so that it can receive spikes (which can be used starting with the step $t + d + 1$, when the neuron can again apply rules). Once emitted from neuron σ_i the spikes reach immediately all neurons σ_j such that (i, j) is in *syn* and which are open, that is, the p spikes are replicated and each target neuron receives p spikes; spikes sent to a closed neurons are "lost".

Note that rules as above, of the extended form, also cover the case of forgetting rules: when $p = 0$ we consume c spikes and produce no spike (hence in the extended SN P systems the forgetting rules are used under the control of regular expressions). In the standard case, the rule forgets all spikes from the neuron. If $E = a^c$, the rule is written in the simplified form $a^c \rightarrow a^p;d$; if all rules are of this form, then the system is called *bounded* (or *finite*), because it can handle only finite numbers of spikes in the neurons.

In each time unit, if a neuron σ_i can use one of its rules, then a rule from R_i *must* be used. Since two rules, $E/a^c \rightarrow a^p;d$, $E'/a^e \rightarrow a^q;f$ can have $L(E) \cap L(E')$ non-empty, it is possible that two or more rules can be applied in a neuron, and in that case, only one of them is chosen non-deterministically.

Thus, the rules are used in the sequential manner in each neuron, at most one in each step, but neurons function in parallel with each other. It is important to notice that the applicability of a rule is established based on the *total* number of spikes contained in the neuron.

The initial configuration of the system is described by the numbers $n_1, n_2, ..., n_m$, of spikes present in each neuron, with all neurons being open. During the computation, a configuration is described by both the number of spikes present in each neuron and by the state of the neuron, more precisely, by the number of steps to count down until it becomes open (this number is zero if the neuron is already open). Thus, $(r_1/t_1,..., r_m/t_m)$ is the configuration where neuron σ_i contains $r_i \geq 0$ spikes and it will be open after $t_i \geq 0$ steps, $i = 1, 2,..., m$; with this notation, the initial configuration is $C_0 = (n_1/0,..., n_m/0)$.

A computation in a system as above starts in the initial configuration. In order to compute a function $f: \mathbf{N}^k \rightarrow \mathbf{N}$, we introduce k natural numbers $n_1,...,n_k$ in the system by "reading" from the environment a binary sequence $z = 10^{n1-1}10^{n2-1}1... 10^{nk-1}1$. This means that the input neuron of Π receives a spike in each step corresponding to a digit 1 from the string z and no spike otherwise. Note that we input exactly $k + 1$ spikes, i.e., after the last spike we assume that no further spike is coming to the input neuron. The result of the computation is also encoded in the distance between two spikes: we impose the restriction that the system outputs exactly two spikes and halts (sometimes after the second spike), hence it produces a train spike of the form $0^{b1}10^{r-1}10^{b2}$, for some $b_1, b_2 \geq 0$ and with $r = f(n_1,..., n_k)$ (the system outputs no spike a non-specified number b_1 of steps from the beginning of the computation until the first spike).

The previous definition covers many types of systems/behaviours. If the neuron σ_{in} is not specified, then we have a generative system: we start from the initial configuration and we collect all results of computations, which can be the distance between the first two spikes (as in (Ionescu et al., 2006)), the distance between all consecutive spikes, between alternate spikes, etc. (as in (Păun et al., 2006a)), or it can be spike train itself, either taking only finite computations, hence generating finite strings (as in (Chen et al., 2007), etc.), or also non-halting computations (as in (Păun et al., 2006b)). Similarly, we can ignore the output neuron and use an SN P system in the accepting mode: a number introduced in the system as the distance between two spikes entering the input neuron is accepted if and only if the computation halts. In the same way we can accept input binary strings or strings over arbitrary alphabets. In the second case, a symbol b_i is taken from the environment by introducing i spikes in the input neuron.

SOME EXAMPLES

Not all types of SN P systems will be discussed below, and only two of them are illustrated in this section.

Figure 1 recalls an example from (Ionescu et al., 2006), and this also introduces the standard way to represent an SN P system (note that the output neuron, σ_7 in this case, is indicated by an arrow pointing to the environment).

In the beginning, only neurons σ_1, σ_2, σ_3, and σ_7 contain spikes, hence they fire in the first step – and spike immediately. In particular, the output neuron spikes, hence a spike is also sent to the environment. Note that in the first step we cannot use the forgetting rule $a \rightarrow \lambda$ in σ_1, σ_2, σ_3, because we have more than one spike present in each neuron.

The spikes of neurons σ_1, σ_2, σ_3 will pass to neurons σ_4, σ_5, σ_6. In step 2, σ_1, σ_2, σ_3 contain no spike inside, hence will not fire, but σ_4, σ_5, σ_6 fire. Neurons σ_5, σ_6 have only one rule, but neuron σ_4 behaves non-deterministically, choosing between its two rules. Assume that for $m \geq 0$ steps we use here the first rule. This means that three spikes are sent to neuron σ_7, while each of neurons σ_1, σ_2, σ_3 receives two spikes. In step 3, neurons σ_4, σ_5, σ_6 cannot fire, but all σ_1, σ_2, σ_3 fire again. After receiving the three spikes, neuron σ_7 uses its forgetting rule and gets empty again. These steps can be repeated arbitrarily many times.

In order to have neuron σ_7 firing again, we have to use the rule $a \rightarrow a;1$ of neuron σ_4. Assume that this happens in step t (it is easy to see that $t = 2m + 2$). This means that at step t only neurons σ_5, σ_6 emit their spikes. Each of neurons σ_1, σ_2, σ_3 receives only one spike – and forgets it in the next step, $t + 1$. Neuron σ_7 receives two spikes, and fires again, thus sending the second spike to the environment. This happens in moment $t + 1 = 2m + 2 + 1$, hence between the first and the second spike sent outside have elapsed $2m +$

Figure 1. A standard SN P system generating all even natural numbers

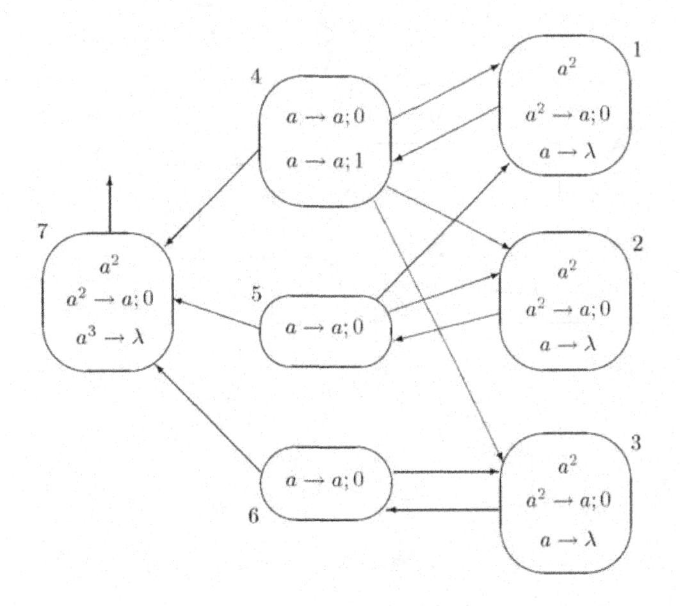

2 steps, for some $m \geq 0$. The spike of neuron σ_4 (the one "prepared-but-not-yet-emitted" by using the rule $a \rightarrow a;1$ in step t) will reach neurons σ_1, σ_2, σ_3, and σ_7 in step $t + 1$, hence it can be used only in step $t + 2$; in step $t + 2$ neurons σ_1, σ_2, σ_3 forget their spikes and the computation halts. The spike from neuron σ_7 remains unused, there is no rule for it. Note the effect of the forgetting rules $a \rightarrow \lambda$ from neurons σ_1, σ_2, σ_3: without such rules, the spikes of neurons σ_5, σ_6 from step t will wait unused in neurons σ_1, σ_2, σ_3 and, when the spike of neuron σ_4 will arrive, we will have two spikes, hence the rules $a^2 \rightarrow a;0$ from neurons σ_1, σ_2, σ_3 would be enabled again and the system will continue to work.

The next example, given in Figure 2, is actually of a more general interest, as it is a part of a larger SN P system which simulates a register machine. The figure presents the module which simulates a SUB instruction; moreover, it does it without using forgetting rules (the construction is part of the proof that forgetting rules can be avoided – see (Ibarra et al., 2007)).

The idea of simulating a register machine $M = (n, H, l_0, l_h, R)$ (n registers, set of labels, initial label, halt label, set of instructions) by an SN P system Π is to associate a neuron σ_r with each register r and a neuron σ_l with each label l from H (there also are other neurons – see the figure), and to represent the fact that register r contains the number k by having $2k$ spikes in neuron σ_r. Initially, all neurons are empty, except the neuron associated with the initial label l_0 of M, which contains one spike. During the computation, the simulation of an instruction l_i: (OPP(r), l_j, l_k) starts by introducing one spike in the corresponding neuron σ_{li} and this triggers the module associated with this instruction.

For instance, in the case of a subtraction instruction l_i: (SUB(r), l_j, l_k), the module is initi-

Figure 2. Module SUB (simulating li: (SUB(r), lj, lk))

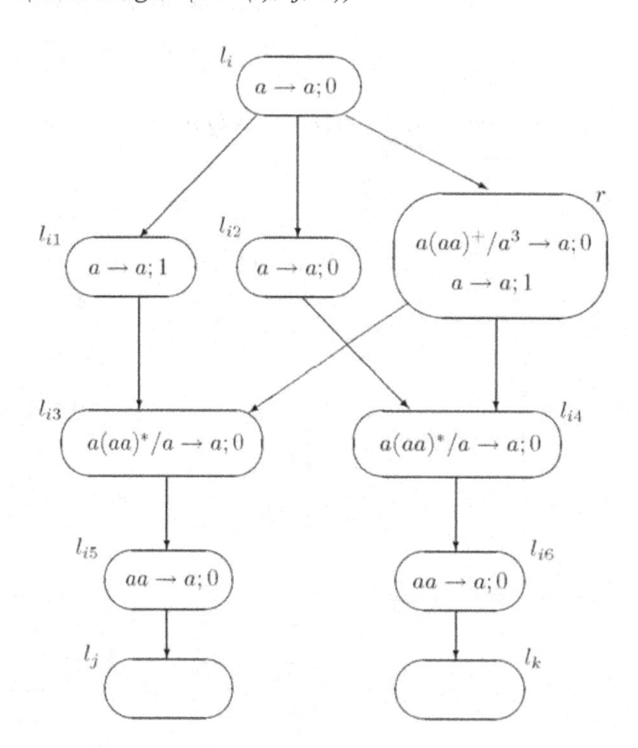

ated when a spike enters the neuron σ_{li}. This spike causes neuron σ_{li} to immediately send a spike to the neurons σ_{li1}, σ_{li2}, and σ_r. If register r is not empty, then the rule $a(aaa)^+/a^3 \rightarrow a;0$ will be applied and the spike emitted will cause neurons σ_{li3}, σ_{li5}, and finally neuron σ_{lj} to spike. (In this process, neuron σ_{li4} has two spikes added during one step and it cannot spike.) If register r is empty, hence neuron σ_r contains only the spike received from σ_{li1}, then the rule $a \rightarrow a;1$ is applied and the subsequent spikes will cause neurons σ_{li4}, σ_{li6}, and finally neuron σ_{lk}, to spike. (In this process, neuron σ_{li3} has two spikes added during one step and does not spike.) After the computation of the entire module is complete, each neuron is left with either zero spikes or an even number of spikes, allowing the module to be run again in a correct way.

The third example deals with an SN P system used as a transducer, and it illustrates the following result from (Păun et al., 2006b): *Any function* $f: \{0,1\}^k \rightarrow \{0,1\}$ *can be computed by an SN P system with k input neurons (also using further* $2^k + 4$ *neurons, one being the output one).*

The idea of the proof of this result is suggested in Figure 3, where a system is presented which computes the function $f: \{0, 1\}^3 \rightarrow \{0, 1\}$ defined by

$$f(b_1, b_2, b_3) = 1 \text{ if and only if } b_1 + b_2 + b_3 \neq 2.$$

The three input neurons, with labels in_1, in_2, in_3, are continuously fed with bits b_1, b_2, b_3, and the output neuron will provide, with a delay of 3 steps, the value of $f(b_1, b_2, b_3)$. The details are left to the reader.

SOME RESULTS

Mainly the power of SN P systems was investigated so far, but also their usefulness for solving computationally hard problems. We start by briefly presenting some results of the first type.

Computational Power

There are several parameters describing the complexity of an SN P system: number of neurons, number of rules, number of spikes consumed, produced, or forgotten by a rule, etc. Here we consider only some of them and we denote by $N_2SNP_m(rule_k, cons_p, forg_q)$ the family of all sets $N_2(\Pi)$ computed as specified in a previous section by standard SN P systems with at most $m \geq 1$ neurons, using at most $k \geq 1$ rules in each neuron, with all spiking rules $E/a^r \rightarrow a;t$ having $r \leq p$, and all forgetting rules $a^s \rightarrow \lambda$ having $s \leq q$. When any of the parameters m, k, p, q is not bounded, it is replaced with *. When we work only with SN P systems whose neurons contain at most s spikes at any step of a computation (finite systems), then we add the parameter $bound_s$ after $forg_q$. (Corresponding families are defined for other definitions of the result of a computation, as well as for the accepting case, but the results are quite similar, hence we do not give details here.)

By *NFIN, NREG, NRE* we denote the families of finite, semilinear, and Turing computable sets of (positive) natural numbers (number 0 is ignored); they correspond to the length sets of finite, regular, and recursively enumerable languages, whose families are denoted by *FIN, REG, RE*. We also invoke below the family of recursive languages, *REC* (those languages with a decidable membership).

The following results were proved in (Ionescu et al., 2006) and extended in (Păun et al., 2006a) to other ways of defining the result of a computation.

Theorem 1
(i) $NFIN = N_2SNP_1(rule_*, cons_1, forg_0) = N_2SNP_2(rule_*, cons_*, forg_*)$.

(ii) $N_2SNP_*(rule_k, cons_p, forgq_0) = NRE$, *for all* $k \geq 2, p \geq 3, q \geq 3$.

(iii) $NSLIN = N_2SNP_*(rule_k, cons_p, forg_q, bound_s)$, *for all* $k \geq 3, p \geq 3, q \geq 3, s \geq 3$.

Figure 3. Computing a Boolean function of three variables

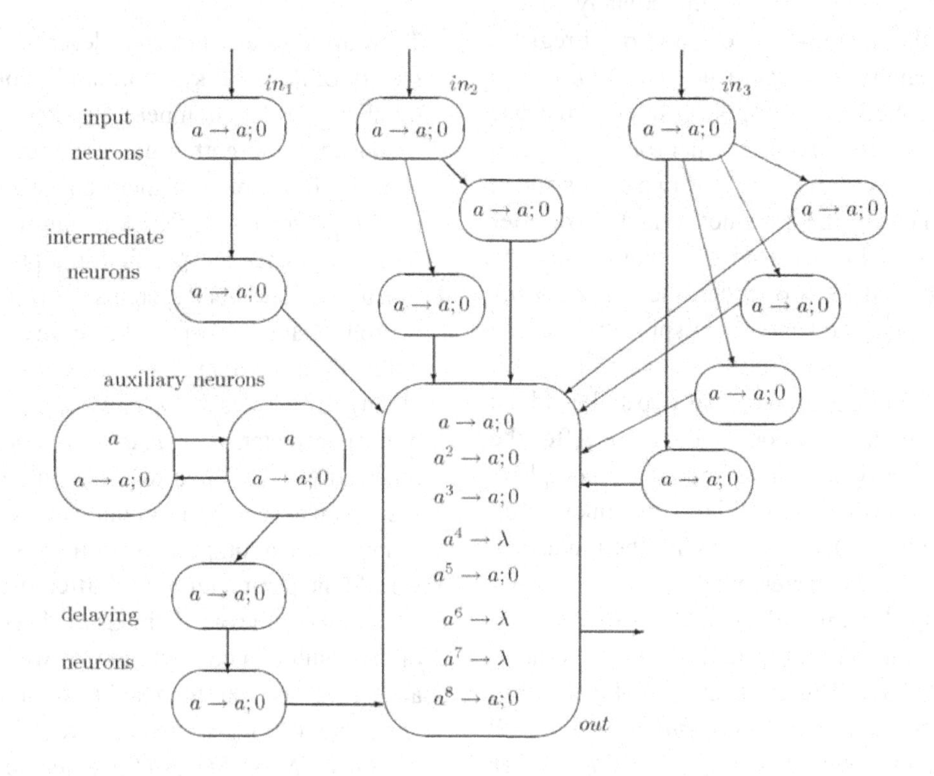

Point (ii) was proved in (Ionescu et al., 2006) also for the accepting case, and then the systems used can be required to be deterministic (at most one rule can be applied in each neuron in each step of the computation). In turn, universality results were proved in (Ionescu et al., 2007) and (Cavaliere et al., 2007) also for the exhaustive and for the non-synchronized modes of using the rules, respectively, but only for extended rules. The universality of standard systems remains *open* for these cases.

Let us now pass to mentioning some results about languages generated by SN P systems, starting with the restricted case of binary strings, (Chen et al., 2007). We denote by $L(\Pi)$ the set of strings over the alphabet $B = \{0, 1\}$ describing the spike trains associated with halting computations in Π; then, we denote by $LSNP_m(rule_k, cons_p, forg_q)$ the family of languages $L(\Pi)$, generated by SN P systems Π with the complexity bounded by the parameters m, k, p, q as specified above. When using only systems with at most s spikes in their neurons (finite), we write $LSNP_m(rule_k, cons_p, forg_q, bound_s)$ for the corresponding family. As usual, a parameter m, k, p, q, s is replaced with * if it is not bounded.

Theorem 2

(i) *There are finite languages (for instance, $\{0^k, 10^j\}$, for any $k \geq 1$, $j \geq 0$) which cannot be generated by any SN P system, but for any finite language L in B^+, the language $L\{1\}$ is in $LSNP_1(rule_*, cons_*, forg_0, bound_*)$, and if $L = \{x_1, x_2, \ldots, x_n\}$, then the language $\{0^{i+3}x_i \mid 1 \leq i \leq n\}$ is in $LSNP_*(rule_*, cons_1, forg_0, bound_*)$.*

(ii) *The family of languages generated by finite SN P systems is strictly included in the family of regular languages over the binary alphabet, but for any regular language L over an alphabet V there is a finite SN P system Π and a morphism* $h : V^* \rightarrow B^*$ *such that* $L = h^{-1}(L(\Pi))$.

(iii) $LSNP_*(rule_*, cons_*, forg_*)$ *is included in REC, but for every alphabet* $V = \{a_1, a_2, ..., a_k\}$ *there are two symbols b, c not in V, a morphism* $h_1 : (V + \{b,c\})^* \rightarrow B^*$, *and a projection* $h_2 : (V + \{b,c\})^* \rightarrow V^*$ *such that for each language L over V, L in RE, there is an SN P system Π such that* $L = h_2(h_1^{-1}(L(\Pi)))$.

These results show that the language generating power of SN P systems is rather eccentric; on the one hand, finite languages (like {0, 1}) cannot be generated, on the other hand, we can represent any RE language as the direct morphic image of an inverse morphic image of a language generated in this way. This eccentricity is due mainly to the restricted way of generating strings, with one symbol added in each computation step. This restriction does not appear in the case of extended spiking rules. In this case, a language can be generated by associating the symbol b_i with a step when the output neuron sends out i spikes, with an important decision to take in the case $i = 0$: we can either consider b_0 as a separate symbol, or we can assume that emitting 0 spikes means inserting λ in the generated string. Thus, we both obtain strings over arbitrary alphabets, not only over the binary one, and, in the case where we ignore the steps when no spike is emitted, a considerable freedom is obtained in the way the computation proceeds. This latter variant (with λ associated with steps when no spike exits the system) is considered below.

We denote by $LSN^eP_m(rule_k, cons_p, prod_q)$ the family of languages $L(\Pi)$, generated by SN P systems Π using extended rules, with the parameters m, k, p, q as above.

The next counterparts of the results from Theorem 2 were proved in (Chen et al., 2006a).

Theorem 3

(i) $FIN = LSN^eP_1(rule_*, cons_*, prod_*)$ *and this result is sharp, as* $LSN^eP_2(rule_2, cons_2, prod_2)$ *contains infinite languages.*

(ii) $LSN^eP_2(rule_*, cons_*, prod_*)$ *is included in REG and REG is included in* $LSN^eP_3(rule_*, cons_*, prod_*)$; *the second inclusion is proper, because* $LSN^eP_3(rule_3, cons_4, prod_2)$ *contains non-regular languages; moreover,* $LSN^eP_3(rule_3, cons_6, prod_4)$ *contains non-semilinear languages.*

(iii) $RE = LSN^eP_*(rule_*, cons_*, prod_*)$.

It is an *open problem* to find characterizations or representations in this setup for families of languages in the Chomsky hierarchy different from *FIN, REG, RE.*

We close this section by mentioning the results from (Păun and Păun, 2007):

Theorem 4

There is a universal computing SN P system with (i) *standard rules and 84 neurons and with* (ii) *extended rules and 49 neurons, and there is a universal SN P system used as a generator of sets of numbers with* (iii) *standard rules and 76 neurons and with* (iv) *extended rules and 50 neurons.*

These values can probably be improved (but the feeling is that this improvement cannot be too large).

Tool-kits for handling strings or infinite sequences, on the binary or on the arbitrary alphabet, are provided in (Păun et al., 2006b) and (Chen et al., 2006c). For instance, in this latter paper one gives constructions of SN P systems for computing the union and concatenation of two languages generated by SN P systems, the intersection with a regular language, while the former paper shows how length preserving morphisms (codings) can be computed; the problem remains *open* for arbitrary morphisms, Kleene *, inverse morphisms.

Computation Efficiency

Membrane computing is much investigated as a framework for devising polynomial solutions to computationally hard problems (typically, **NP**-complete problems). So far, only a few results of this type are known for SN P systems – and they are briefly mentioned here.

A first interesting result is reported in (Chen et al., 2006c): SAT can be decided in constant time by using an arbitrarily large pre-computed SN P system, of a very regular shape (as synapse graph) and with empty neurons, after plugging the instance of size (n, m) (n variables and m clauses) of the problem into the system, by introducing a polynomial number of spikes in (polynomially many) specified neurons. This way of solving a problem, by making use of a pre-computed resource given for free, on the one hand, resembles the supposed fact that only part of the brain neurons are active (involved in "computations") at each time, on the other hand, is not very common in computability and request further research efforts. What kind of pre-computed resource is allowed, so that no "cheating" is possible? How the given resources should be activated? Define and study complexity classes for this framework.

A more traditional investigation is carried in (Leporati et al., 2007a) and (Leporati et al., 2007b). In the first paper, it is proved that the Subset Sum problem can be decided in constant time by an extended SN P system working in a non-deterministic manner. (The Subset Sum problem asks whether, for given integers $V = \{v_1, v_2, \ldots, v_n\}$ and S there is a subset B of V such that the sum of the integers in B equal S. The problem is known to be **NP**-complete.) The idea is simple: rules of the forms $a^{vi} \rightarrow \lambda$, $a^{vi} \rightarrow a^{vi}$; 0 are placed in n neurons; non-deterministically, each neuron uses one of these rules and sends the produced spikes to a collecting neuron which spikes if and only if it receives in total S spikes. The system is constructed in a semi-uniform way, i.e., starting from the instance to solve. Moreover, with respect to the size of the problem representation, the system needs an exponential number of spikes to initialize.

These drawbacks are removed in (Leporati et al., 2007b). First, a module is provided which encodes the input in binary (and this is done in polynomial time). However, a new way to use the rules is needed in this phase, namely the maximally parallel use of rules in each neuron (it is also proved in (Leporati et al., 2007b) that without the maximal parallelism this encoding cannot be done in polynomial time). Then, an SN P system is constructed, *in a uniform way*, for solving the Subset Sum problem, also using the idea from (Leporati et al., 2007a). Therefore, the obtained system contains both neurons which work in a deterministic way and neurons which work in a non-deterministic way, as well as neurons using the rules in the sequential way and neurons using the rules in the maximally parallel way. Improving in unifying the types of neurons from these points of view is a research topic, and progresses are expected in this active area of research.

FUTURE TRENDS

The theory of spiking neural P systems is rather young, so many research topics wait to be addressed in this area. Among them, we mention here only some important ones: enlarge the model with further biological facts (for instance, including the astrocytes, which were already recently considered); explore in more details the complexity issues, which are related both to the (supposed) brain functioning and the possible applications of SN P systems; link this area to the traditional neural computing, bringing to the SN P systems framework such questions as pattern recognition, training/learning, etc.; develop dedicated hardware for simulating SN P systems, both with didactical goals (e.g., checking examples or proofs) and perhaps useful in applications; after having an enhanced model, faithful to neuro-biological reality, and efficient software tools, look for biological or bio-medical applications; try to learn computer science relevant ideas from the computing by spiking. All these research directions were already preliminarily addresses, hence we believe that they will prove to be solid trends along which the study of SN P systems will develop.

CONCLUSION

We have only briefly introduced the reader to a fast developing and much promising branch of membrane computing, inspired from the way the neurons communicate by means of electrical impulses. Although the model is rather reductionistic (we may say, minimalistic), it proves to be both powerful and efficient from a computational point of view.

FUTURE RESEARCH TOPICS

Many problems were already mentioned above, many others can be found in the papers listed below, and further problems are given in (Păun, 2007). Large research directions (trends) were also mentioned in a previous section. That is why we list here only a few more technical questions. Investigate the way the axon not only transmit impulses, but also amplifies them; consider not only "positive" spikes, but also inhibitory impulses; define a notion of *memory* in this framework, which can be read without being destroyed; provide ways for generating an exponential working space (by splitting neurons? by enlarging the number of synapses?), in such a way to trade space for time and provide polynomial solutions to computationally hard problems; investigate further the feature used in (Leporati et al., 2007a), (Leporati et al., 2007b), i.e., the fact that checking the applicability of a spiking rule is done in one clock tick, irrespective of the complexity of the regular expression which control the rule; define systems with a dynamical synaptic structure; compare the SN P systems as generator/acceptor/transducers of infinite sequences with other devices handling such sequences; investigate further systems with exhaustive and other parallel ways of using the rules, as well as systems working in a non-synchronized way; find classes of (accepting) SN P systems for which there is a difference between deterministic and non-deterministic systems; find classes which characterize levels of computability different from those corresponding to finite automata (semilinear sets of numbers or regular languages) or to Turing machines (recursively enumerable sets of numbers or languages). Several of these research topics were preliminarily investigated – see the additional reading provided below – but serious research efforts are still needed.

REFERENCES

Cavaliere, M., Egecioglu, O., Ibarra, O.H., Ionescu, M., Păun, Gh., &. Woodworth, S. (2007). Unsynchronized spiking neural P systems: decidability and undecidability. *Proceedings of 13rd DNA Based Computers Conference, Memphis, USA.*

Chen, H., Freund, R., Ionescu, M., Păun, Gh., & Perez-Jimenez, M.J. (2007). On string languages generated by spiking neural P systems. *Fundamenta Informaticae*, *75*(1-4), 141–162.

Chen, H., Ishdorj, T.-O., Păun, Gh., & Perez-Jimenez, M.J. (2006a). Spiking neural P systems with extended rules. In (Gutierrez-Naranjo et al., eds., (2006)), vol. I, 241–265.

Chen, H., Ishdorj, T.-O., Păun, Gh., Perez-Jimenez, M.J. (2006b). Handling languages with spiking neural P systems with extended rules. *Romanian Journal of Information Sciience and Technology, 9*(3), 151–162.

Chen, H., Ionescu, M., & Ishdorj, T.-O. (2006c). On the efficiency of spiking neural P systems. *Proceedings of the 8th International Conference on Electronics, Information, and Communication*, Ulanbator, Mongolia, June 2006, 49–52.

Ciobanu, G., Păun, Gh., &. Perez-Jimenez, M.J. (Eds.) (2006). *Applications of Membrane Computing.* Berlin: Springer.

Gerstner, W., & Kistler, W. (2002). *Spiking Neuron Models. Single Neurons, Populations, Plasticity.* Cambridge Univ. Press.

Gutierrez-Naranjo, M.A. et al., eds. (2006). *Proceedings of Fourth Brainstorming Week on Membrane Computing.* Sevilla: Fenix Editora.

Gutierrez-Naranjo, M.A. et al., eds. (2007). *Proceedings of Fifth Brainstorming Week on Membrane Computing.* Sevilla: Fenix Editora.

Ibarra, O.H., Păun, A., Păun, Gh., Rodriguez-Paton, A., Sosik, P., & Woodworth, S. (2007). Normal forms for spiking neural P systems. *Theoretical Computer Science, 372*(2-3), 196–217.

Ionescu, M., Păun, Gh., & Yokomori, T. (2996). Spiking neural P systems. *Fundamenta Informaticae, 71*(2-3), 279–308.

Ionescu, M., Păun, Gh., & Yokomori, T. (2007). Spiking neural P systems with exhaustive use of rules. *International Journal of Unconventional Computing*, in press.

Korec, I. (1996). Small universal register machines. *Theoretical Computer Science*, 168, 267–301.

Maass, W., &. Bishop, C., eds. (1999). *Pulsed Neural Networks.* MIT Press.

Păun, Gh. (2000). Computing with membranes. *Journal of Computer and System Sciences, 61*(1), 108–143, and Turku Centre for Computer Science (TUCS) Report 208, November 1998.

Păun, Gh. (2002). *Membrane Computing. An Introduction.* Berlin: Springer.

Păun, A., & Păun, Gh. (2007). Small universal spiking neural P systems. *BioSystems, 90*(1), 48–60.

Păun, Gh. (2007). Twenty six research topics about spiking neural P systems. In (Gutierrez-Naranjo et al., eds., (2007)), 263–280.

Păun, Gh., Perez-Jimenez, M.J., & Rozenberg, G. (2006a). Spike trains in spiking neural P systems. *International Journal of Foundations of Computer Science, 17*(4), 975–1002.

Păun, Gh., Perez-Jimenez, M.J., & Rozenberg, G. (2006b). Infinite spike trains in spiking neural P systems. Submitted.

ADDITIONAL READING

Alhazov, A., Freund, R., Oswald, M, &. Slavkovik, M. (2006). Extended variants of spiking neural P systems generating strings and vectors of non-negative integers. In (Hoogeboom et al., eds, 2007), 123–134.

Binder, A., Freund., R., Oswald., M., & Vock, L. (2007). Extended spiking neural P systems with excitatory and inhibitory astrocytes. In (Gutierrez-Naranjo et al., eds., (2007)), 63–72.

Chen, H., Ionescu, M., Păun, A., Păun, Gh., & Popa, B. (2006). On trace languages generated by spiking neural P systems. In *Proceedings of Descriptional Complexity of Formal Systems Conference,* Las Cruces, NM, June 2006.

Chen, H., Ishdorj, T.-O., & Păun, Gh. (2007). Computing along the axon. *Progress in Natural Science, 17*(4), 417–423.

Garcia-Arnau., M., Perez., D., Rodriguez-Paton, A., & Sosik, P. (2007). Spiking neural P systems: Stronger normal forms. In (Gutierrez-Naranjo et al., eds., (2007)), 157–178.

Hoogeboom, H.J., Păun, Gh., Rozenberg, G., & Salomaa, A., eds. (2007). *Membrane Computing, International Workshop, WMC7, Leiden, The Netherlands, 2006, Selected and Invited Papers*, LNCS 4361, Berlin: Springer.

Ibarra, O.H. & Woodworth, S. (2007). Characterizations of some restricted spiking neural P systems. In (Hoogeboom et al., eds, 2007), 424–442.

Ibarra, O.H., Woodworth, S., Yu, F., & Păun, A. (2006). On spiking neural P systems and partially blind counter machines. In *Proceedings of Fifth Unconventional Computation Conference, UC2006*, York, UK, September 2006.

Ionescu, M., & Sburlan, D. (2007). Some applications of spiking neural P systems. In (Gutierrez-Naranjo et al., eds., (2007)), 213–227.

Păun, Gh., Perez-Jimenez, M.J., & Rozenberg, G. (2007). Computing morphisms by spiking neural P systems. *International Journal of Foundations of Computer Science*, to appear.

Ramirez-Martinez, D., & Gutierrez-Naranjo, M.A. (2007). A software tool for dealing with spiking neural P systems. In (Gutierrez-Naranjo et al., eds., (2007)), 299–314.

Chapter V
Simulation of the Action Potential in the Neuron's Membrane in Artificial Neural Networks

Juan Ramón Rabuñal Dopico
University of Coruña, Spain

Javier Pereira Loureiro
University of Coruña, Spain

Mónica Miguélez Rico
University of Coruña, Spain

ABSTRACT

In this chapter, we state an evolution of the Recurrent ANN (RANN) to enforce the persistence of activations within the neurons to create activation contexts that generate correct outputs through time. In this new focus we want to file more information in the neuron's connections. To do this, the connection's representation goes from the unique values up to a function that generates the neuron's output. The training process to this type of ANN has to calculate the gradient that identifies the function. To train this RANN we developed a GA based system that finds the best gradient set to solve each problem.

INTRODUCTION

Due to the limitation of the classical ANN models (Freeman, 1993) to manage time problems, over the year 1985 began the development of recurrent models (Pearlmutter, 1990) capable to solve efficiently this kind of problems. But this situation didn't change until the arrival of the Recurrent Backpropagation algorithm. Before this moment, the more wide used RANN were Hopfield networks and Boltzman machines that weren't effective to treat dynamic problems. The powerful of this new type of RANN is based on the increment of the number of connections and the whole recursivity of the network. These characteristics, however, increment the complexity of the training algorithms and the time to finish the convergence process. These problems have slow down the use of the RANN to solve static and dynamic problems.

However, the chances of RANN are very big compared to the powerful of feedforward ANN. For the dynamic or static pattern matching, the RANN developed until now offer a better performance and a better learning skill.

Most of the studies that have already been done about RANN, have been center in the development of new architectures (partial recurrent or with context layers, whole recurrent, etc.) and to optimize the learning algorithms to achieve reasonable computer times. All of these studies don't reflect changes in the architecture of the process elements (PE) or artificial neurons, that continue having an input function, an activation function and an output function.

The PE architecture has been modified, basing our study in biological evidences, to increment the RANN powerful. These modifications try to emulate the biological neuron activation that is generated by the action potential.

The aim of this work is to develop a PE model with activation output much more similar to the biological neurons one.

BACKGROUND

Artificial Neural Networks

An Artificial Neural Network (ANN) (Lippmann, 1987; Haykin, 1999) is an information-processing system that is based on generalizations of human cognition or neural biology and they are electronic or computational models based on the neural structure of the brain. The brain basically learns from experience. An Artificial Neural Network consists on various layers of parallel procesing elements or neurons. One or more hidden layers may exist between the input and the output layer. The neurons in the hidden layer(s) are connected to the neurons of a neighboring layer by weighting factor that can be adjusted during the training process. The ANN's are organized according to training methods for specific applications.

There are two types of ANN's, the first one with only feed forward connections is called feed forward ANN, and the second one with arbitrary connections without any direction, are often called Recurrent ANN (RANN). The most common type of ANN consists on different layers, with some neurons on each of them and connected with feed-forward connections and trained with the back propagation algorithm (Johansson et al., 1992).

The numbers of neurons contained in the input and output layers are determined by the number of input and output variables of a given problem. The number of neurons of a hidden layer is an important consideration when solving problems usign multilayer feed-fordward networks. If there are fewer neurons within a hidden layer, there may not be enough opportunity for the neural network capture the intricate relationships between the inputs and the computed output values. Too many hidden layer neurons not only require a large computational time for accurate training, may also result in overtraining situation (Brion et al., 1999). A neural network is said to be "overtrained" when the ANN focuses on the characteristics of individual data points rather than just capturing the general patterns in the entire training set. The optimal number of neurons in a hidden layer can be estimated as two-thirds of the sum of the number of input and output neurons.

An ANN has a remarkable ability to derive meaning from complicated or imprecise data. The ANN can be used to extract patterns and detect trends that are too complex to be noticed by either humans or other computer techniques. "Training" of an ANN model is a procedure by which the ANN repeatedly processes a set of test data (input-output data pairs), changing the values of its weights according to a predetermined algorithm in order to improve its performance. Backpropagation is the most popular algorithm for training feed-forward ANN's (Lippman,

1987). A trained ANN can be considered as an "expert" in the category of information it has received to analyse. The two main advantages of the ANN are:

- **Adaptive learning:** An ability to learn how to do tasks based on the data given for training.
- **Fault tolerance:** Partial destruction of the network leads to the corresponding degradation of performance.However, some network capabilities may be retained even with major network damage.

The ANN's have shown to be a powerful tool in many different applications. If significant variables are known, but not their exact relationships, an ANN is able to perform a kind of function fitting by using multiple parameters on the existing information and predict the possible relationships for the coming future.

In recent years, the need to develop more powerful systems which are capable of solving time problems. Recurrent ANN is the best option in the ANN field (Williams et al., 1989) for this kind of problems. This kind of network is more complex than traditional ones, so that the problems of network development are more acute. Their structure has a much higher number of connections, which complicates both the training process (Hee et al., 1997) and the architecture adjustment.

Genetic Algorithm

A GA (Holland, 1975) is a searching method developed by John Holland and based in the emulation of the natural evolution. This method is contained in the Evolutionary Computation techniques.

A GA (Goldberg, 1989) is a search technique inspired in the world of biology. More specifically, the Evolution theory by Charles Darwin (Darwin, 1859) is taken as a basis for its working. GAs are used to solve optimization problems by copying the evolutionary behaviour of species: from an initial random population of solutions, this population is evolved mainly by means of selection, mutation and crossover genetic operators, taken from natural evolution. By applying this set of operations, the population goes through an iterative process in which it reaches different states, each one is called generation. As a result of this process, the population is expected to reach a generation in which it contains a good solution to the problem. In GAs the solutions of the problem are codified as a string of bits or real numbers.

While there are many different types of selection operation, the most common type is roulette whell selection. In roulette wheel selection, individuals are given a probability of being selected that is directly proportionate to their fitness. Two individuals are then chosen randomly based on these probabilities and produce offspring. There are many different kinds of crossover operation, but the most common type is a single point crossover. In single point crossover, you choose a point at which you swap one part of gens from a parent to the other.

Genetic Algorithms are very effective way of quickly finding a reasonable solution to a complex problem. They do an excellent job of searching trough a large and complex search spaces. Genetic Algorithms are most effective in a great space for which little is known and we didn't any more about the problem that we want to solve. You may know exactly what you want a solution to do but have no idea how you want it to go about doing it. This is where genetic algorithms thrive. They produce solutions that solve the problem in ways you may never have even considered.

Biological Neuron Dynamics

The activation output of a neuron cell is explained from the electrical characteristics of the cell's membrane (Taylor et al., 2003). This membrane is more permeable to the potassium ions (K+) than to the sodium ions. The potassium chemical gradient gets to take the potassium ions out of the cell by diffusion, but the strong attraction of the organic ions trend to maintain the potassium inside.

The result of these opposite forces is an equilibrium where there are more sodium and chlorine ions (Cl-) outside and more organic and potassium ions inside.

This equilibrium makes a potential difference through the cell's membrane of 70 to 100 mV (millivolts), being negative the intracell fluid. This potential is called cell's repose potential.

The influences of the excitatory inputs that arrive to one cell from another are added in the axonal join and cause a successive depolarization of the membrane. It is then that the potential gets inverted up to +35 mv in only 0.1 milliseconds. The resultant depolarization alters the permeability of the cellular membrane relating to sodium ions. As a result, there is an incoming flux of positive sodium ions that enter the cell influencing even more the depolarization.

This self-generated effect causes the action potential (Figure 1). After an action potential, the potential of the membrane becomes more negative than the repose potential for some milliseconds, up to get the ion equilibrium. To begin with the action potential, it is required a sudden increment of approximately 15-30 mv, considering -65 mv the stimulation threshold. Within an excitable fiber it can't be got a second action potential while the membrane is depolarized.

The period during which it can't be launched another action potential, even with a very powerful stimulus, it is called "absolute refractory period" and it lasts about 0.5 milliseconds. Then, it is followed by a "relative refractory period" that lasts from 1 to 1.5 milliseconds, while it is needed a bigger stimulus than usually to get the action potential, because the ionic equilibrium of the repose potential hasn't been recovered yet.

This point can't be excited in approximately 1 millisecond, the time that it takes to recover the repose potential. This refractory period limits the transmission frequency of the nervous impulse to about 1000 per second.

TIME DECREASED ACTIVATION

Obviously, the behaviour of the PE in the ANN is not very similar, about the activation phase, to what we have just described. In the traditional

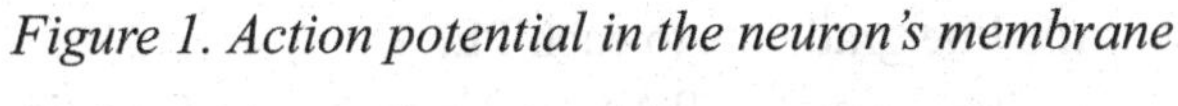

Figure 1. Action potential in the neuron's membrane

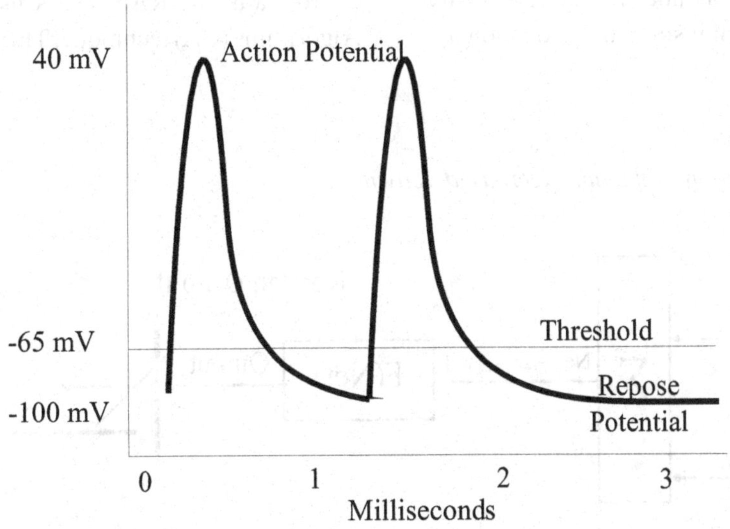

process elements, over the input function, which ponders the weights with the activation of the previous layer, it is applied an activation function, being the hyperbolic and the sigmoid more used. If the output of this function is positive and the activation occurs, in that instant, the neurons that get the activation will have as input the product of the weight of the connection and the value of the activation.

This way of propagating the activation of the neuron works in a very punctual way, without giving the idea of continuity that can be seen in the natural model, where the activation flows from a maximum point (action potential) to get the "zero" (repose potential).

In this work it has been added to the classical architecture of process element, the function of the action potential, approximating it by a line that goes from the output level of the activation function to the zero level.

This model of ANN changes the conception of the connections between neurons that go from being a number (an only numerical value) to be represented by a function, a line with negative gradient that is characterized by the gradient value (Figure 2).

Now, the neuron functioning has to be explained as a time function. That is, the activation of a neuron is the join of a series of inputs in a point of the time. This not only affects to the output neurons in that instant, the activation affects to the following neurons in n next instants too (Figure 3).

The value of n depends on the gradient of that connection.

In the Figure 4 it can be observed the line formula that emulates the time decreased activation of the natural neurons. In this formula y represents the value of the neuron's output that is received by the following neurons. The letter b represents the output of the neuron's activation function. The letter x represents the time from the activation and the letter **m** is the line gradient that, as already commented, is always negative.

At the moment when the activation occurs, the time is 0 and the output is $y = b$, from this moment and depending on the gradient, the output y decreases until 0 through a time $x = t$.

Training Process

When we have defined the new architecture, to apply this type of ANN to solve real problems, it's necessary a training algorithm that adjusts the weights and, in this case, the gradients of the line formula used for the time decreased activation for each Neuron in the ANN. To this new kind of PE the learning algorithms used until now aren't valid, because we need to calculate the line gradients to each PE.

To train this RANN, it's used the Genetic Algorithms (GA) technique. This technology has

Figure 2. Representation of the time decreased activation

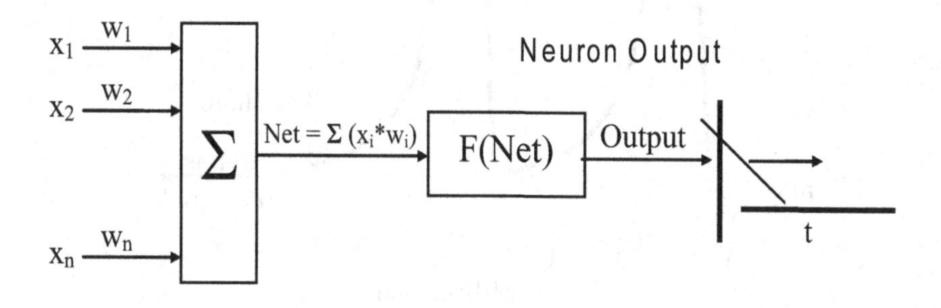

Figure 3. Modelling of the neuron's activation

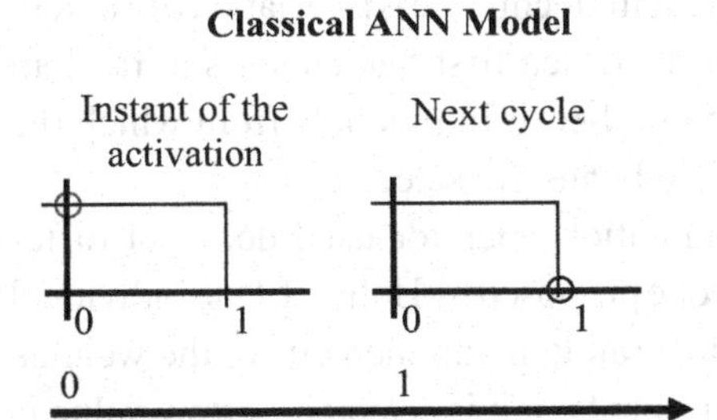

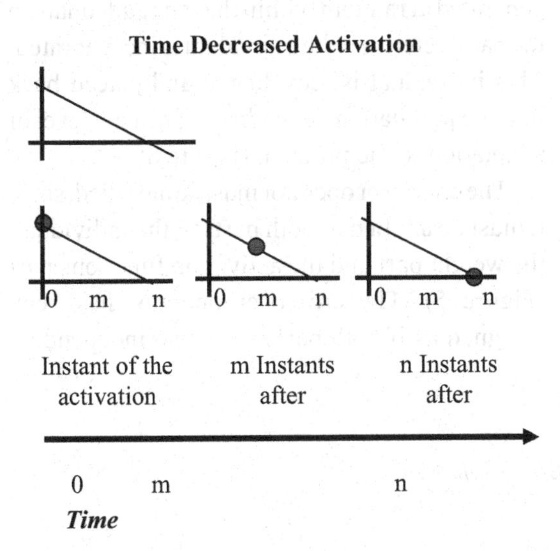

been recently applied to the ANN field (Angeline et al., 1994) (Ku et al., 1999) but it has just done a good results to the training of different types of ANN (calculating weights) and to the design of the network architectures (number of layers, number of neurons per layer and activation functions).

SYSTEM DESCRIPTION

The first thing to use a GA is to codify the possible solutions as an array of n elements. Each of these solutions will be an individual of the population. In this case, we can codify the ANN as an array of gradients or weights (depending on the ANN if it is recurrent or not) from the input to the output layer. In order to use a GA, first one must codify the possible solutions as an array with n elements or cells, which is going to represent a possible solution, that is, an individual in the population that represents a configuration of ANN.

In this case, each ANN has been codified with two arrays. A weight array goes from the initial layer to the output (Figure 5). The network connection weights, can be seen as a $n*n$ matrix, where n stands for the total number of neurons in the network, so that if there is a connection between i and j neurons, the weight of that connection will be stored in the matrix intersection of row *i* with column *j*. If it is a feedforward ANN, just a limited number of weights are stored in the top part of the matrix, while if the network is a recurrent one, the matrix will be fully occupied. To represent the time decreased activation of the neurons we need another $n*n$ matrix. This matrix is similar to the previous one, but in this case, the values represents the line gradients (*m*) of the formula of the Figure 4 (in this case the values will be always negative) .The other array contains the activation functions for each neuron in the network (Figure 6). This array has two positions for each neuron: the first one with the type of activation function (linear, threshold, sigmoid, or tangent hyperbolic) and the second one with the configuration parameters of the chosen function for that neuron.

The ANN training process allows to choose whether the network is recurrent or not, the parameters common to all the neurons in the network, such as the value limit of the weights and the recurrent network parameters, such as stage or continuous training and the number of internal iterations of the network before the output.

The GA to be applied uses an elitist strategy of protection of the best individual against mutation and substitution, and the codification of the individuals is made through real numbers. Once the ANN and GA general parameters have been selected, the second phase is the creation of a random population of individuals to start

Figure 4. Line formula used by the time decreased activation

$$y = mx + b$$

applying the genetic operators (Figure 7). Once the population has been initialized, the genetic operators of selection, crossover and mutation of individuals start to be applied.

In order to measure the fitness of each individual, as the classical training algorithms, it is used the mean square error (MSE) between the outputs of each ANN and the optimal outputs.

The selection of individuals to which the crossover operator is going to be applied is made first. Three values common in GA have been selected for this operator: random selection within a population of individuals, the famous Montecarlo Technique, in which an individual's chance to be selected is directly proportional to its adjustment value ruleta, and a tournament among individuals technique. This technique is a variation of random selection, in which first one chooses at random a certain number of individuals from which the best-adapted ones are selected.

The mutation operator used does not differ from the one proposed by Holland. One individual is chosen at random and then one of the weights or line gradients in it is selected. A new value is generated at random within the range adequate to the parameter, and the individual is incorporated. This individual is reevaluated and placed back into the population depending on its new level of adaptation to the problem (Figure 8).

The crossover operator must be modified, since it must be applied to both parts of the individual: the weight part and the activation functions part (Figure 5). The crossover operator has been designed as if both parts were two independent

Figure 5. Codification of the ANN. (a) Weights (b) line gradients

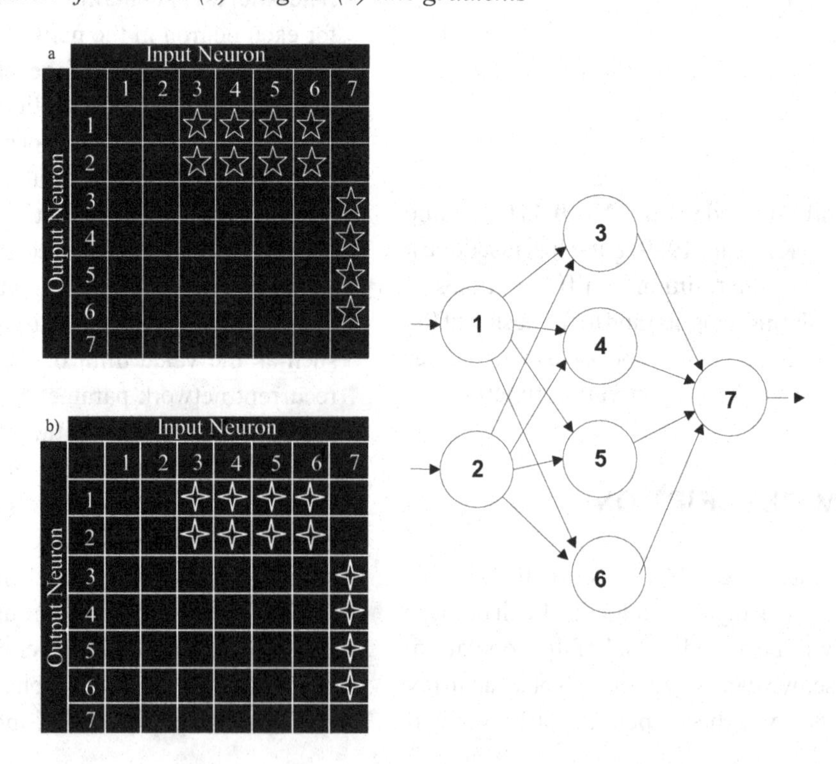

Figure 6. Codification of the activation functions. (a) Hyperbolic (b) sigmoid

Activation F.	Parameter
Hyperbolic	
Sigmoid	
Threshold	1.0
Lineal	0.85
Hyperbolic	
Sigmoid	
Threshold	1.5

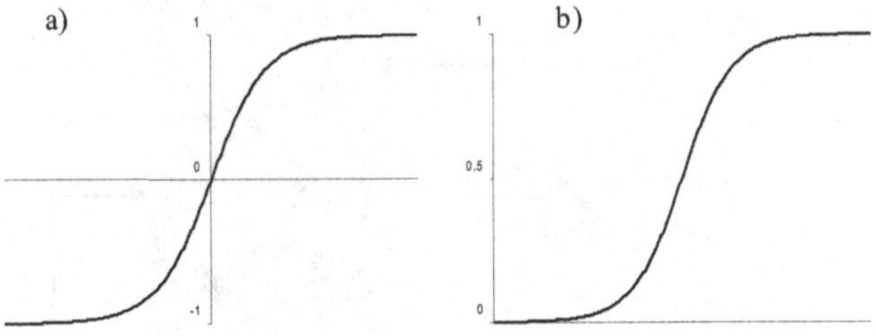

Figure 7. Genetic Algorithm used

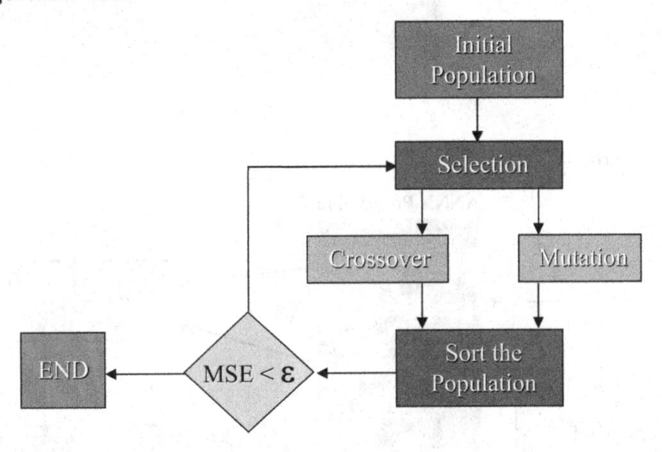

individuals and two crossover operators were applied. In each of these two operations, the crossover operator behaves in the usual way: a divide point is chosen at random and the new individuals are generated. The result of these two crossover operations are two new individuals with a weight part and a line gradient part from each parent. Then the new individuals are evaluated in order to apply the individual substitution strategy (Figure 9).

Apart from the usual crossover operator with one crossover point that has just been described, tests have been carried out with a two-point crossover operator and with the uniform operator which selects genes from one of the parents in each position according to a uniform distribution.

Once the new individuals have been obtained, it only remains to choose which individuals in the population are going to leave their place to the new ones. Apart from the usual and more Darwinist technique of individual substitution, which eliminates the worse adapted ones, two additional techniques have been tested in order to avoid homogenization. The first one is about substituting parents, i.e. if the offspring adapts better than the parents, these are substituted, in the opposite case, the offspring is eliminated. The second technique is based on a substitution according to similarity of error level. Those individuals with an equal or similar error level to the new offspring are found among the population, and one of them is chosen at random to be substituted. In the Figure 10 we can see all the parameters for the implemented Genetic Algorithm.

Also, it's codified if the ANN is recurrent or not and the generic parameters of the PE, as the

Figure 8. Mutation operation

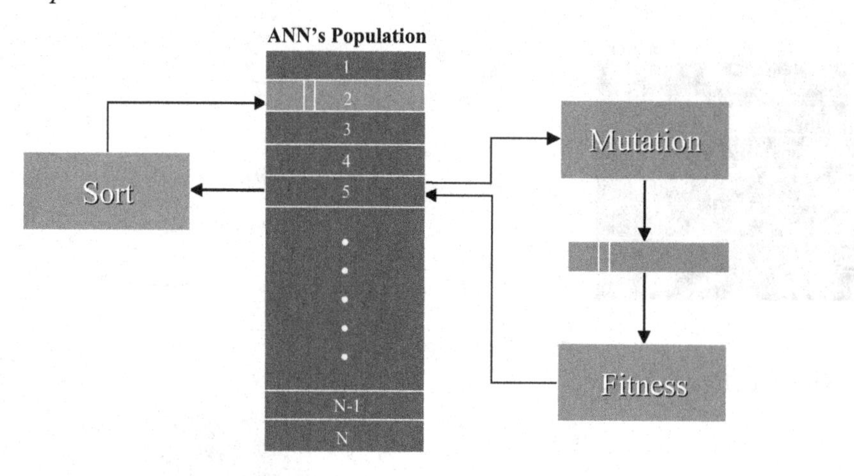

Figure 9. Crossover operation

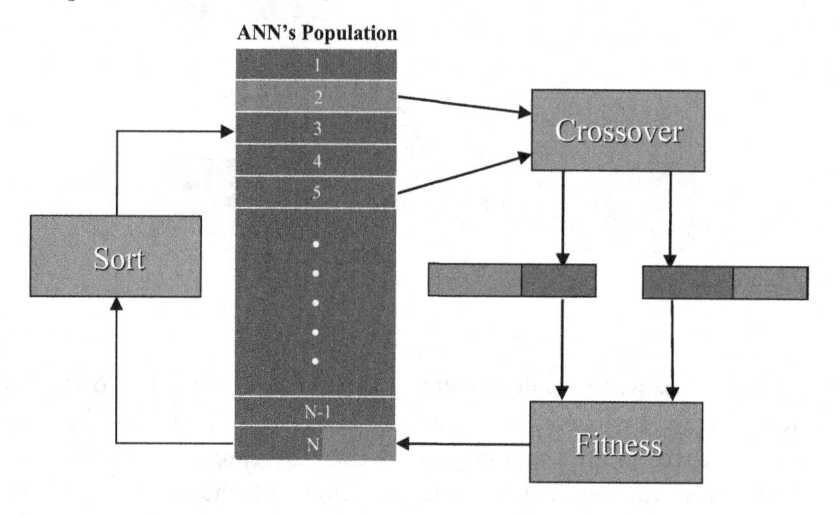

high limit of weights and gradients, the type of activation functions (lineal, threshold, hyperbolic tangent and sigmoid). To the RANN the parameters codified are: continuous or epochs training and the number of internal iterations to evaluate an input. In the Figure 11 we can see all the parameters for the ANN architecture.

Results

It is necessary to use Recurrent ANN architectures for the prediction of time series and for modelling this type of problems. To the training and validation of this new RANN architecture, 2 classical time series in the field of the statistics have been used (William, 1990) (Box, 1976), the number of sunspots and the tobacco production in USA.

Figure 10. Genetic Algorithm parameters

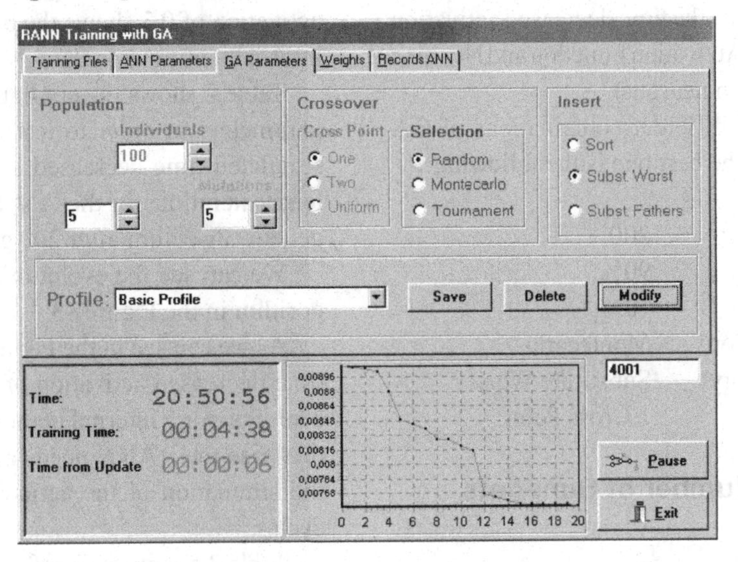

Figure 11. Parameters of the ANN Architecture

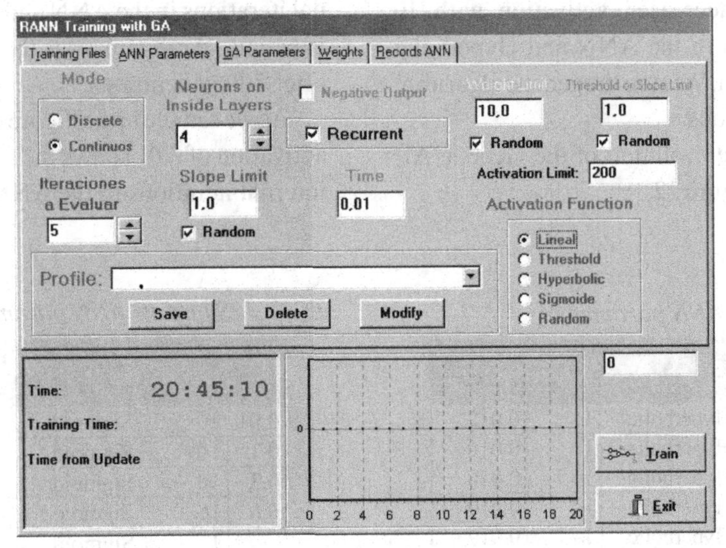

Another problem to be solved will be the prediction of a classical laboratory random time series: the Mackey-Glass series (Mackey, 1977).

Most of these time series cases are structures with one single input and one single output. The input corresponds to a number value of the t moment, while the system's output is the prediction of the number value at t+1 (short term predictions).

First, it's showed the parameters used in the ANN developed to solve the first two problems (sunspots and tobacco). These parameters mean: Time (the scale of the line gradient), Iterations (number of internal cycles of the ANN before it produces the output. Only with a value of iterations greater than 1 the time decreased activation option is active), Activation Function and Neurons (number of hidden neurons).

We test with different combinations of GA parameters and the best one is the following:

Individuals:	200
Crossover Rate:	90%
Mutation Rate:	10%
Selection operation:	Montecarlo
Substitute Strategy:	Substitute Worst
Cross Points:	1 Cross Point

Results for Number of Sunspots Time Series

Table 1 shows the configuration of the ANN parameters in order to test the system and the simulated time decreased activation with 10 internal iterations in the ANN and Hyperbolic activation function. We stop the genetic algorithm after 200 generations.

We can see the evolution of the Genetic Algorithm in the Figure 12.

As we can see in the Figure 12, a time decreased activation of 0.3 shows the best evolution for 10 internal iterations in the ANN.

Table 2 shows the configuration of the ANN parameters in order to test the system and the simulated time decreased activation with 5 internal iterations in the ANN. We also stop the genetic algorithm after 200 generations.

We can see the evolution of the Genetic Algorithm in the Figure 13.

As we can see in the Figure 13, in this case, a time decreased activation of 0.6 shows the best evolution for 5 internal iterations in the ANN. In this case the RANN needs more speed to make the simulation of the action potential between cycles.

Table 3 shows the configuration of the ANN parameters in order to test the system and the simulated time decreased activation with 10 internal iterations in the ANN and Sigmoid activation function. We again stop the genetic algorithm after 200 generations.

As we can see in the Figure 14, a time decreased activation of 0.01 shows the best evolution for 10 internal iterations in the ANN.

Table 1. Sunspots ANN parameters

Time	Iterations	Act. Funct.	Neurons	Mean Square Error
0.01	10	Hyperbolic	1	0.012
0.1	10	Hyperbolic	1	0.010
0.3	10	Hyperbolic	1	0.010
0.6	10	Hyperbolic	1	0.010
--	1	Hyperbolic	1	0.019

Table 2. Sunspots ANN parameters

Time	Iterations	Act. Funct.	Neurons	Mean Square Error
0.01	5	Hyperbolic	1	0.012
0.1	5	Hyperbolic	1	0.013
0.3	5	Hyperbolic	1	0.010
0.6	5	Hyperbolic	1	0.010
--	1	Hyperbolic	1	0.019

Table 3. Sunspots ANN parameters

Time	Iterations	Act. Funct.	Neurons	Mean Square Error
0.01	10	Sigmoid	1	0.013
0.1	10	Sigmoid	1	0.017
0.3	10	Sigmoid	1	0.018
0.6	10	Sigmoid	1	0.016
--	1	Sigmoid	1	0.024

Table 4. Sunspots ANN parameters

Time	Iterations	Act. Funct.	Neurons	Mean Square Error
0.01	5	Sigmoid	1	0.019
0.1	5	Sigmoid	1	0.017
0.3	5	Sigmoid	1	0.020
0.6	5	Sigmoid	1	0.018
--	1	Sigmoid	1	0.024

Figure 12. Evolution of the GA in the first 200 generations (10 internal iterations in the hyperbolic ANN)

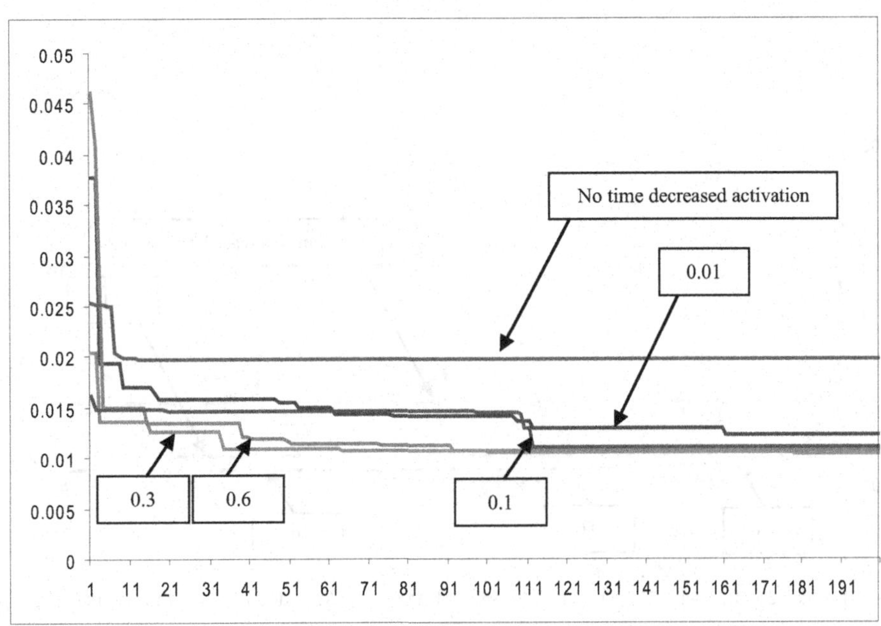

Table 4 shows the configuration of the ANN parameters in order to test the system and the simulated time decreased activation with 5 internal iterations in the ANN. We also stop the genetic algorithm after 200 generations.

As we can see in the Figure 15, in this case, a time decreased activation of 0.1 shows the best evolution for 5 internal iterations in the ANN. In this case (as the hyperbolic activation function tests) the RANN needs more speed to make the simulation of the action potential between cycles.

In these tests we can see the hyperbolic activation function with time decreased of 0.3 and 10 internal iterations in the ANN shows the best evolution. Then we make another test without limitation in the number of generations of the GA and with different number of hidden neurons.

As we can see in the Table 5, with 3 neurons in the hidden layer we can obtain the lower MSE, and the best solution for the prediction of this time series.

Results for Tobacco Production in USA Time Series

With the same operation of the previous point and with different values for the simulated time decreased activation, in the Table 6 we can observe the results for different combinations.

As we can see in Table 7, all the MSE are very similar, but the hyperbolic activation function with only one hidden neuron produces the best value. In the following Figures we can see the structure of this RANN and the comparative of the predictions that it produces.

Figure 13. Evolution of the GA in the first 200 generations (5 internal iterations in the hyperbolic ANN)

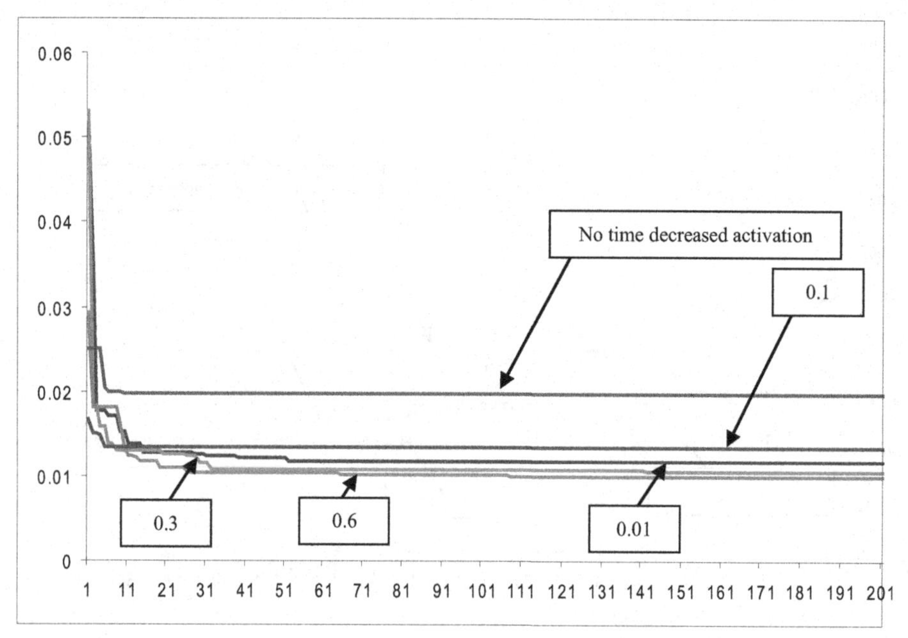

Figure 14. Evolution of the GA in the first 200 generations (10 internal iterations in the sigmoid ANN)

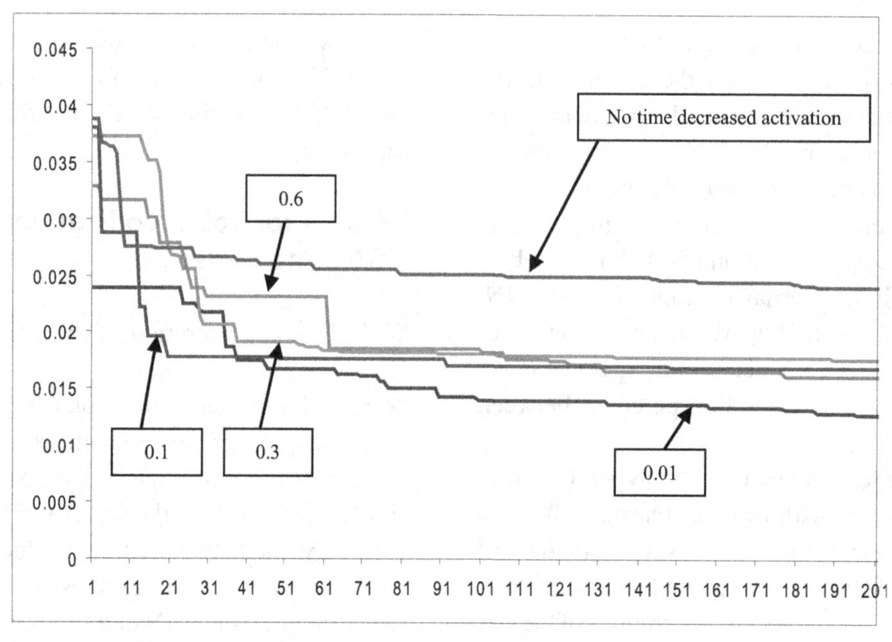

Figure 15. Evolution of the GA in the first 200 generations (5 internal iterations in the sigmoid ANN)

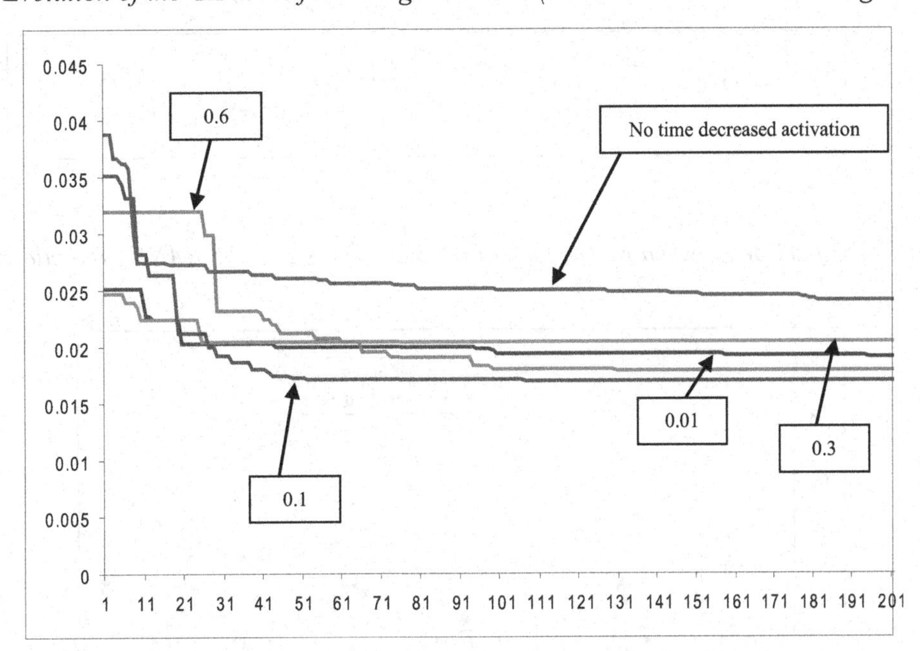

Table 5. Sunspots ANN parameters

Act. Funct.	Neurons	Mean Square Error
Hyperbolic	1	0.0069
Hyperbolic	2	0.0055
Hyperbolic	3	0.0039
Hyperbolic	4	0.0041
Hyperbolic	5	0.0040
Sigmoid	3	0.0047

Results for Mackey-Glass Time Series

The Mackey-Glass equation (Mackey-Glass, 1977) is an ordinary differential delay equation:

$$\frac{dx}{dt} = \frac{ax(t-\tau)}{1+x^c(t-\tau)} - bx(t)$$

Choosing $\tau = 30$, the equation becomes chaotic, and only short-term predictions are feasible. Integrating the equation in the rank $[t, t+\delta t]$ we obtain Equation 1.

With the same operation of the previous point and with different values for the simulated time decreased activation, in the Table 9 we can observe the results for different combinations.

As we can see in Table 9 the hyperbolic activation function with 3 hidden neurons produces the best value. In the following Figures we can see the structure of this RANN and the comparative of the predictions that it produces.

FUTURE TRENDS

Now, when we have just reached the results showed above, these investigations are going to continue in the same way. We will try to develop an ANN model as similar as possible to the natural neuron function, especially, in the activation phase.

Equation 1.

$$x(t+\Delta t)=\frac{2-b\Delta t}{2+b\Delta t}x(t)+\frac{\alpha\Delta t}{2+b\Delta t}\left[\frac{x(t+\Delta t-\tau)}{1+x^{c}(t+\Delta t-\tau)}+\frac{x(t-\tau)}{1+x^{c}(t-\tau)}\right]$$

Figure 16. Comparison between real values of the time series and the RANN predictions (3 hidden neurons)

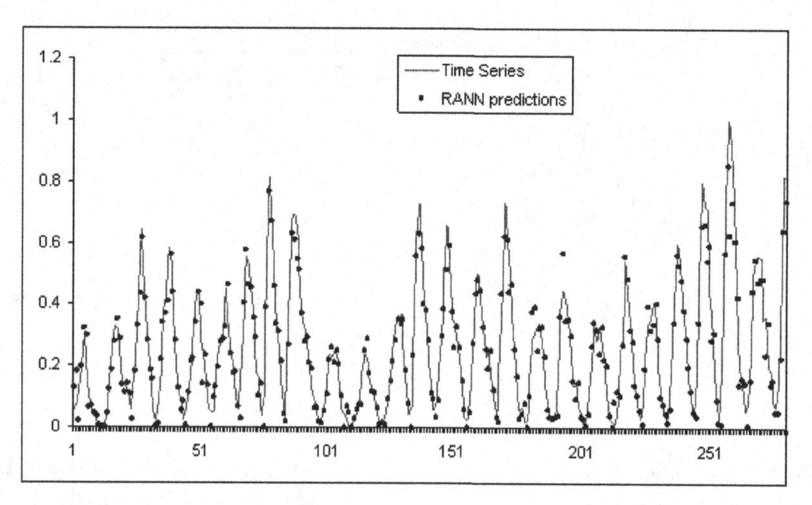

Table 6. Values of the connection weights of the RANN and the line gradients of the time decreased activation

NETWORK CONNECTION WEIGHTS					
Neuron	**1**	**2**	**3**	**4**	**5**
1	**-1.682**	**0.992**	**-1.101**	**1.606**	5.965
2	**2.866**	**1.429**	**-4.339**	**0.044**	9.932
3	**-7.999**	**4.963**	**5.398**	**-2.304**	9.843
4	**0.765**	**2.160**	**7.411**	**-9.971**	9.991
5	0.572	-1.069	0.011	-9.989	-0.564

LINE GRADIENTS					
Neuron	**1**	**2**	**3**	**4**	**5**
1	**-0.347**	**-1.000**	**-0.146**	**-0.089**	-0.197
2	**-0.776**	**-0.096**	**-0.103**	**-0.597**	-0.031
3	**-0.999**	**-0.154**	**-0.929**	**-0.979**	-0.985
4	**-0.614**	**-0.472**	**-0.446**	**-0.236**	-0.156
5	-0.096	-0.132	-1.000	-0.451	-1.000

Table 7. Tobacco production ANN parameters

Time	Iterations	Act. Funct.	Neurons	Mean Square Error
0.1	10	Hyperbolic	1	0.0071
0.1	10	Sigmoid	1	0.0075
0.1	10	Hyperbolic	2	0.0073
0.1	10	Sigmoid	2	0.0075
0.1	10	Hyperbolic	3	0.0074
0.1	10	Sigmoid	3	0.0074

With the studies done, over the physiology and the functioning of the natural neuron, we think that the process of action potential unchaining is the more essential phase.

Within this phase it can be highlighted two clearly distinct aspects. In the one hand, the shape of the curve that represents the depolarization of the cell's membrane.

Although in this work we approximate the curve with a line with negative gradient, we are working to substitute with a logarithm function. In the other hand we want to simulate the response of the natural neuron to consecutive activations.

As we have commented in a prior point, in the natural neuron, just after the activation phase there exists a period where the neuron isn't able to unchain other activation. This period is called absolute refractory period and, after this, there is another period where only a stronger input can reach an activation response of the neuron. The configuration of these periods's duration is other parameter that will include in the new type of RANN.

Table 9. Mackey-Glass ANN parameters

Time	Iterations	Act. Funct.	Neurons	Mean Square Error
0.3	5	Hyperbolic	1	0.0010
0.3	5	Sigmoid	1	0.0010
0.3	5	Hyperbolic	2	0.0004
0.3	5	Sigmoid	2	0.0010
0.3	5	Hyperbolic	3	0.0003
0.3	5	Sigmoid	3	0.0010

To study the response of these new characteristics, we are going to use the time series because of it's a field much well studied in the mathematics area, but it demands quite experience to catalogue the series types. Then, it's very interesting to prove that the RANN can reach better results than the time series theory, because a better functioning of the RANN techniques.

CONCLUSION

With the development of this work, it's been tried to near the classical PE architecture to the natural neurons model. Also, it's been tried to improve the internal ANN functioning, increasing their information storage capacity. In this sense, the first advance was the development of the RANN because they had a bigger number of neuron's connection. In this new type of RANN the level of complexity is increased because of the time decreased activation used.

Table 8. Values of the connection weights of the RANN and the line gradients of the time decreased activation

NETWORK CONNECTION WEIGHTS			
Neuron	**1**	**2**	**3**
1	**1.800**	**1.178**	8.404
2	**-9.343**	**2.646**	-2.719
3	-1.015	-0.216	2.041

LINE GRADIENTS			
Neuron	**1**	**2**	**3**
1	**-0.201**	**-0.579**	-0.214
2	**-0.385**	**-0.166**	-0.670
3	-0.855	-0.855	-0.574

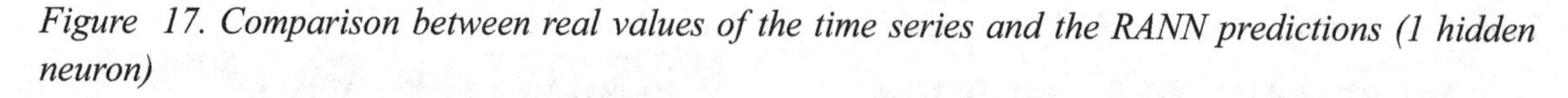

Figure 17. Comparison between real values of the time series and the RANN predictions (1 hidden neuron)

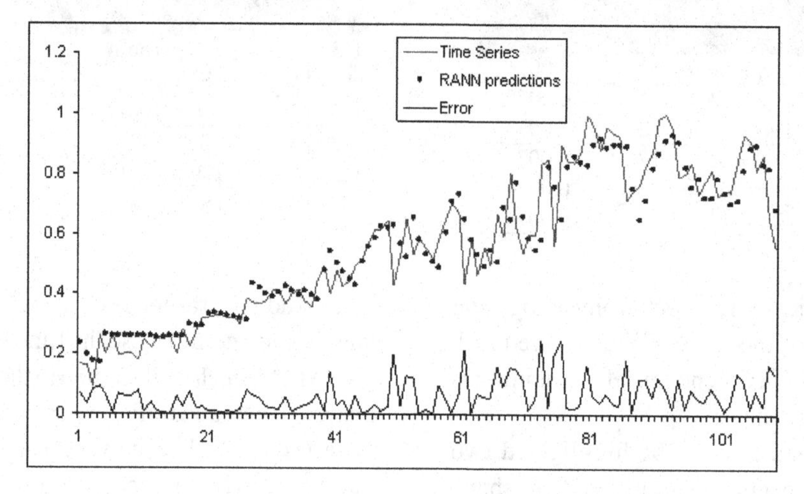

Figure 18. RANN architecture (1 hidden neuron)

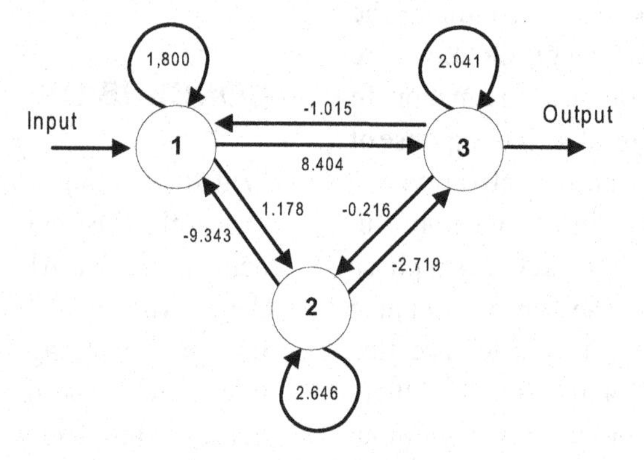

With regard to the internal network functioning, as it can be studied in the result tables, in the time series forecast problems the use of time decreased activation improve the training and the mean squared error level. This is caused by the persistence of the internal activation states of the ANN. Then, these ANN are able to adapt themselves in a more flexible way to the resolution of this kind of problems.

FUTURE RESEARCH DIRECTIONS

In this chapter a real life performance is simulated into an ANN and training it through the evolutionary computation techniques: the activation of a neuron by means of the generation of an action potential. This idea of modelling the action potential for emulating real life performances (functioning of the biological neuron) could be used either for simulating the functioning of other

Figure 19. Comparison between real values of the time series and the RANN predictions (3 hidden neurons)

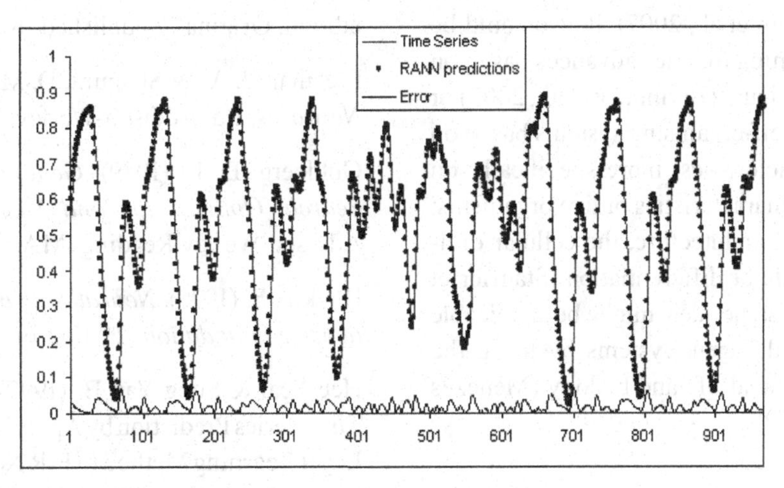

Table 10. Values of the connection weights of the RANN and the line gradients of the time decreased activation

NETWORK CONNECTION WEIGHTS					
Neuron	1	2	3	4	5
1	2.761	0.344	-1.442	-8.568	9.363
2	-8.207	0.124	8.488	0.606	2.259
3	-9.308	-6.459	5.006	-4.964	-9.484
4	-0.689	9.857	-5.133	-2.419	-6.343
5	-0.375	0.089	-5.364	-0.550	-0.578

LINE GRADIENTS					
Neuron	1	2	3	4	5
1	-0.715	-1.000	-0.537	-0.911	-0.621
2	-0.868	-1.000	-1.000	-1.000	-0.458
3	-0.746	-1.000	-0.252	-1.000	-0.854
4	-0.045	-0.816	-0.673	-1.000	-1.000
5	-1.000	-0.763	-0.851	-1.000	-1.000

neurons or for developing methodologies that enable the estimation of the complexity of protein structures (Balbín et al., 2007). It also could be used for performing robotics advances based on biological behaviours (Bermudez et al., 2007) or for developing self-organising systems based on biological characteristics, more specifically on the three main characteristics of live organisms: the multicellular architecture, the cellular division and the cellular differentiation (Stauffer et al., 2007). The same idea might be applicable to design of multi-agent systems by using the concepts of artificial life and biology (Menezes et al., 2007)

REFERENCES

Almeida, F., Mateus, L., Costa, E., Harvey, I. & Coutinho, A. (2007). *Advances in Artificial Life. LNAI 4648*. Springer, Berlin.

Angeline, P.J., Saunders, G.M. & Pollack, J.B. (1994). An evolutionary algorithm that constructs recurrent neural networks. *IEEE Transactions on Neural Networks*, 5(1), 54–65.

Balbín, A. & Andrade, E. (2007). Designing a Methodology to Estimate Complexity of Protein Structures. *Advances in Artificial Life. LNAI 4648*. Springer, Berlin.

Bermudez, E. & Seth A.K. (2007). Simulations of Simulations in Evolutionary Robotics. *Advances in Artificial Life. LNAI 4648*. Springer, Berlin.

Brion, G.M. and Lingireddy, S. (1999). A neural network approach to identifying nonpoint sources of microbial contamination. Water Res.,33(14), 3099-3106.

Box, G. E. & Jenkins, G. M. (1976). *Time Series Analysis Forecasting and Control.* Ed. Holden Day.

Darwin, C. (1864). *On the Origin of Species by means of Natural Selection or the Preservation of Favoured Races in the Struggle for Life*. Cambridge University Press, Cambridge, UK, sixth edition. Originally published in 1859.

Freeman, J. A.. & Skapura, D. M. (1993). *Neural Networks*. Ed. Addison-Wesley, USA.

Goldberg, D. E. (1989). *Genetic Algorithms in Search, Optimization and Machine Learning.* Addison-Wesley Reading, MA.

Haykin, S. (1999). *Neural Networks: a Comprehensive Foundation.* 2nd Edition, PrenticeHall.

Hee Yeal & Sung Yan B. (1997). An Improved Time Series Prediction by Applying the Layer-by-Layer Learning Method to FIR Neural Networks. *Neural Networks*. 10(9), 1717-1729.

Holland, J. H. (1975). *Adaptation in Natural and Artificial Systems.* University of Michigan Press.

Johansson, E. M., Dowla, F. U. & Goodman, D. M. (1992). Backpropagation learning for multi-layer feed-forward neural networks using the conjugate gradient method. *International Journal of Neural Systems*, 2 (4), 291-301.

Ku, K.W.C., Man, W.M. & Wan, C.S. (1999). Adding learning to cellular genetic algorithms for training recurrent neural networks. *IEEE Transactions on Neural Networks*, 10(2), 239–252.

Lippmann, R. P. (1987). An Introduction to Computing with Neural Nets, *IEEE ASSP Magazine*.

Mackey M. & Glass L. (1977). Oscillation and chaos in physiological control systems. *Science*, 197-287.

Menezes, T. & Costa, E. (2007). Designing for Surprise. *Advances in Artificial Life. LNAI 4648*. Springer, Berlin.

Pearlmutter, B. A. (1990). Dynamic Recurrent Neural Networks. *Technical Report CMU-CS*, School of Computer Science, Carnegie Mellon University, 88-191.

Stauffer, A., Mange, D. & Rossier J. (2007). Self-organizing Systems Based on Bio-inspired Properties. *Advances in Artificial Life. LNAI 4648.* Springer, Berlin.

Taylor, D.J., Green, N. P. & Stout, G.W. (2003). *Biological Sciences*, 3rd ed. United Kingdom, Cambridge University Press.

Williams, R.J. & Zipser, D. (1989). A Learning Algorithm for Continually Running Fully Recurrent Neural Networks. *Neural Computation* 1, 270-280.

William W. S. (1990). *Time Series Analysis. Univariate and Multivariate Methods.* Ed. Addison Wesley, USA.

ADDITIONAL READING

Adaptive Behavior. ISSN: 1059-7123. Sage Publications LTD.

Network–Computation in Neural Systems. ISSN: 0954-898X. IOP Publishing LTD.

Neural Computation. ISSN: 0899-7667. MIT Press.

Neural Computing & Applications. ISSN: 0941-0643. Springer.

Neural Networks. ISSN: 0893-6080. Pergamon-Elsevier Science LTD.

Neural Processing Letters. ISSN: 1370-4621. Springer.

Neurocomputing. ISSN: 0925-2312. Elsevier Science BV.

Porto, A., Araque, A., Rabuñal, J., Dorado, J. & Pazos, A. (2007). A New Hybrid Evolutionary Mechanism Based on Unsupervised Learning for Connectionist Systems. *Neurocomputing.* 70(16-18), 2799-2808.

Rabuñal J.R. & Dorado, J. (2006). *Artificial Neural Networks in Real-Life Applications.* Idea Group, Hershey.

Chapter VI
Recent Methodology in Connectionist Systems

Ana B. Porto Pazos
University of A Coruña, Spain

Alberto Alvarellos González
University of A Coruña, Spain

Alejandro Pazos Sierra
University of A Coruña, Spain

ABSTRACT

The Artificial NeuroGlial Networks, which try to imitate the neuroglial brain networks, appeared in order to process the information by means of artificial systems based on biological phenomena. They are not only made of artificial neurons, like the artificial neural networks, but also they are made of elements which try to imitate glial cells. An important glial role related with the processing of the brain information has been recently discovered but, as the functioning of the biological neuroglial networks is not exactly known, it is necessary to test several and different possibilities for creating Artificial NeuroGlial Networks. This chapter shows the functioning methodology of the Artificial NeuroGlial Networks and the application of a possible implementation of artificial glia to classification problems.

INTRODUCTION

Recently Connectionist Systems (CSs) which pretend to imitate the neuroglial nets of the brain were built. These systems were named Artificial NeuroGlial Networks (ANGNs) (Porto, 2004). These ANGNs are not only made of neuron, but also from elements which imitate the astrocytes of the Glial System. During the last decades Neuroscience has advanced remarkably, and increasingly complex neural circuits, as well as the Glial System, are being observed closely. New discoveries are now unveiling that the glia is intimately linked to the active control of neural activity and takes part in the regulation of synaptic neurotransmission (Perea & Araque, 2005). In that case, it may be useful to integrate into the artificial models other elements that are not neurons.

Despite the latest achievements about the astrocytes, there is not exact information about their influence on brain processing or about the regulation of the astrocyte-neuron communication and its consequences.

Artificial models of neuroglial circuits have been developed as an attempt of studying the effects of the neuroglial activity that have been observed until now at the Neuroscience laboratories (Pasti et al., 1997; Araque et al., 1999; Araque et al., 2001; Perea & Araque, 2002; Perea & Araque, 2005, Martin & Araque, 2006), as well as of giving expression to the hypothesis and phenomena of these activity. In that way, the initially performed works have involved the elaboration of a ANGNs building methodology (Porto, 2004). The essence of this methodology, its application to classification problems and the latest results achieved after mimicking different functioning options of the Artificial NeuroGlial Networks are shown in the present chapter. The implementation of the mentioned options enables the study of these networks from the Artificial Intelligence viewpoint and the application of the ANGNs for solving real problems.

BACKGROUND

The specific association of processes with synapses and the discovery of two-way astrocyte-neuron communication (Perea & Araque, 2005) have demonstrated the inadequacy of the previously held view regarding the purely supportive role for these glial cells. Instead, future progress requires rethinking how the dynamics of the coupled neuron-glial network can store, recall, and process information.

It is a novel research field that is here covered from the Artificial Intelligence viewpoint; no CS considering the glial system had been developed ever before.

In this regard, the RNASA (Artificial Neural Networks and Adaptive Systems) laboratory, our group from the University of A Coruña performs two interrelated research works from the viewpoint of Computer Science. One of these types involves the elaboration of "biological" computational models to reach a better understanding of the structure and behaviour of both neurons (LeRay et al., 2004, Fernández et al., 2007), and astrocytes (Porto, 2004). The second type of works considers behaviour observed in the brain circuits and the studied biological phenomena in order to create CSs; these systems should test if the presence of such phenomena provides advantages for information processing (Porto et al., 2005; Porto et al. 2006, Porto et al. 2007). No actualised publications of other research groups have been found regarding the later works, although some first attempts of incorporating the astrocyte functions are appearing, as the work of Xi Shen & Philippe De Wilde (2006). These authors model the increase of the blood flow within the brain capillary vessels according the neuronal activity and the neuron-astrocyte pairing; however, it is only a mathematical modelling that does not use the knowledge about astrocytes on the implementation of CSs.

Also Nadkarni et al. (2007) elaborated a mathematical framework, apart from the connectionist models, for modelling the synaptic interactions between the neurons and the astrocytes, bearing in mind the tripartite synapse concept (Araque et al., 1999). This model also includes the quantitative description of the experimental conclusions related to the synaptic boosting and to the increase of spontaneous postsynaptic currents that occur when astrocyte activity is involved (Liu et al., 2004) (Fiacco et al., 2004). This model tries to provide a conceptual basis for more complex experimental protocols, as it quantifies an adaptive synapses that changes its reliability depending on the astrocyte behaviour. The intervention of the astrocytes can modulate a neuron network and its subsequent activity by providing other capabilities and an additional plasticity that it would not exist in absence of the glial cells.

Regarding to our works, it should be mentioned that an exhaustive analyses of the current CSs was necessary before designing the ANGNs. It is known that the CSs offer an alternative to classic computation for problems of the real world that use natural knowledge (which may be uncertain, imprecise, inconsistent and incomplete) and for which the development of a conventional programme that covers all the possibilities and eventualities is unthinkable or at least very laborious and expensive. In (Pazos, 1991) we find several examples of successful applications of CSs: image and voice processing, pattern recognition, adaptive interfaces for man/machine systems, prediction, control and optimisation, signals filtering, etc.

After the analyses of the CSs it was observed that all the design possibilities, for their architecture as well as for their training process, are basically oriented towards minimising the error level or reducing the system's learning time. As such, it is in the optimisation process of a mechanism, in case the ANN, that we must find the solution for the many parameters of the elements and the connections between them. Conflicts exist between the functioning of several types of CSs and brain networks, due to the use of methods that do not reflect reality. For instance, in the case of a multilayer perceptron, which is a simple CS, the synaptic connections between the process elements have weights that can be excitatory or inhibitory, whereas in the nervous system, are the neurons that seem to represent these functions, not the connections; recent research (Perea & Araque, 2005; Perea & Araque, 2007) indicates that the cells of the Glial System, more concretely the astrocytes, also play an important role.

Another example that not reflect reality concerns the learning algorithm known as Backpropagation, which implies that the change of the connections value requires the backwards transmission of the error signal in the ANN. It was traditionally assumed that this behaviour was impossible in a natural neuron, which, according to the "dynamic polarisation" theory of Ramón y Cajal (1911), is unable to efficiently transmit information inversely through the axon until reaching the cellular soma; new research however has discovered that neurons can send information to presynaptic neurons under certain conditions, either by means of existing mechanisms in the dendrites or else through various interventions of glial cells such as astrocytes.

Lastly, it should be also mentioned that the use of a supervised learning method by the CS implies the existence of an "instructor", which in the context of the brain means a set of neurons that behave differently from the rest in order to guide the process. At present, the existence of this type of neurons is biologically indemonstrable, but the Glial System seems to be strongly implied in this orientation and may be the element that configures an instructor that until now had not been considered.

Due to this, and bearing in mind the latest discoveries in Neuroscience, it was analysed how CSs should be designed for integrating the observed brain phenomena in order to contribute to the advance in the optimisation of the information processing. By means of these systems, whose structure and functioning are described later, not only it is intended to benefit the Artificial Intelligence, but it is also pretended to improve the Neuroscience with the study of brain circuits since other point of view, incorporating artificial astrocytes.

ARTIFICIAL NEUROGLIAL NETWORKS

ANGNs include both artificial neurons and processing control elements that represent the astrocytes (Porto, 2004).

The design of the ANGNs has been oriented towards classification problems that are solved by means of simple networks, i.e. multilayer networks, although future research may lead to the design of models in more complex networks. It seems a logical approach to start the design of these new models with simple CSs, and to orientate the latest discoveries on astrocytes in information processing towards their use in classification networks, since the control of the reinforcement or weakening of the connections in the brain is related to the adaptation or plasticity of the connections, which lead to the generation of activation ways. This process can therefore improve the classification of the patterns and their recognition by the ANGN (Porto et al., 2007).

Searching for a Learning Method

In order to design the integration of the astrocytes into the Artificial Neural Networks (ANNs) and elaborate a learning method for the resulting ANGNs that allows us to check whether there is an improvement in these systems, we have analysed the main existing training methods. We have analysed Non-Supervised and Supervised Training methods, and other methods that use or combine some of their characteristics. After considering the existing methods for training multilayer ANNs, none of them seemed adequate to be exclusively used. We have found that Backpropagation and other based-gradient methods have difficulties to train ANNs with more complex information processing elements than those traditionally used (Dorado, 1999; Dorado et al., 2000; Rabuñal et al., 2004). Therefore, these methods could not be used in order to achieve our essential aim, i.e. to analyze the effect of including astrocytes in the ANNs. Process elements's modifications in ANNs have been made using Evolutionary Algorithms (EAs) (Whitley, 1995) for the training and we thought could be a good possibility to be taken into account when it comes to training our CSs. It's known EAs are particularly useful for dealing with certain kind of problems. They are less likely to be trapped in local minima than traditional gradient-based search algorithms, although they are rather inefficient in fine-tuned local search (Yao, 1999).

Moreover, numerous authors have considered improving the efficiency of evolutionary training by incorporating a local search procedure into the evolution, i.e., combining EA's global search ability with local search's ability to fine tune. For instance, the hybrid Genetic Algorithms/Backpropagation approach, used by Lee (1996) used Genetic Algorithms (GAs) to search for a near-optimal set of initial connection weights and then used Backpropagation to perform local search from these initial weights. Hybrid training has been used successfully in many application areas (Kinnebrock, 1994; Taha and Hanna, 1995; Yan et al, 1997; Yang et al., 1996; Zhang et al., 1995).

Taking into account the above mentioned, we designed a new hybrid learning method for training ANGNs. Such method looks for optimal connection weights in two phases. In one phase, unsupervised learning, it modifies weights values following rules based on the behaviour of brain circuits studied. In the other phase, supervised learning, the weights are searched by means of GAs. The results (Porto et al., 2007) demonstrated that the combination of these two techniques was found to be efficient because GAs are good at global search (Whitley, 1995; Yao, 1999) and the use of this EA has enabled us to easily incorporate to CSs the artificial astrocytes which have facilitated a fine-tuned local search. The basic differences with regard to other hybrid methods (Erkmen et al., 1997; Lee, 1996; Yan et al, 1997; Zhang et al., 1995) are that we did not use the GAs to search for initial connection weights and the local search technique used is based on astrocytes behaviour, which had never been implemented until the present time. Details of this new hybrid learning method can be seen further on.

The Functioning of the ANGNs

The functioning of the ANGNs follows the steps that were successfully applied in the construction and use of CSs: design, training, testing and execution.

The design is based on feed-forward multilayer architectures which are totally connected, without back propagation or lateral connections, and oriented towards the classification and recognition of patterns.

ANGNs hybrid training method combines non-supervised learning (first phase) with the supervised learning that uses the GAs evolutionary technique (second phase).

The first phase is based on the behaviour of astrocytes (Perea & Araque, 2005). This behaviour has been incorporated to the ANNs as control elements (CE). The ANGNs have one CE per layer which is responsible for monitoring the activity of every neuron in that layer (Figure 1).

Since the GAs requires individuals, the first phase creates a set of individuals to work with. Each individual of the GA consists of as many values as the connection weights existing in the CS, and each arbitrary set of values of all the weights constitutes a different individual. We study the CS functioning with all the individuals. Every individual is modified as each training pattern passes on to the network, according to how the activity of the neurons has been during the passage of that pattern. For each individual, every pattern or input example of the training set is presented to the network during a given number of times or iterations. These iterations allow the modification of the individual by means of the application of the rules based on brain circuits behaviour, and these iterations constitute a pattern cycle. The number of iterations can be changed and can be established for any cycle. As regards these modifications over the connection weights, we selected a set from all the different possibilities analyzed in our computational models (Porto, 2004; LeRay et al., 2004) that will be explained in the following section.

Going on with the training, when a pattern cycle has finished, the error for that pattern is calculated and stored. It will be the difference between the CS output obtained and the one desired. Later on, when all the training patterns have been passed on to the CS, the mean square error (MSE) of that individual is calculated. We have opted for the MSE because it gives a relative measure to the patterns that are fed to the network to compare the errors between different architectures and training sets. Also, the square in the numerator favours the cases of individuals for which the output of the network is close to the optimal values for all the examples.

The process is the same for all the individuals. This phase constitutes a non-supervised training, because the modifications of the connections

Figure 1. CS with artificial neurons (AN) and control elements (CE)

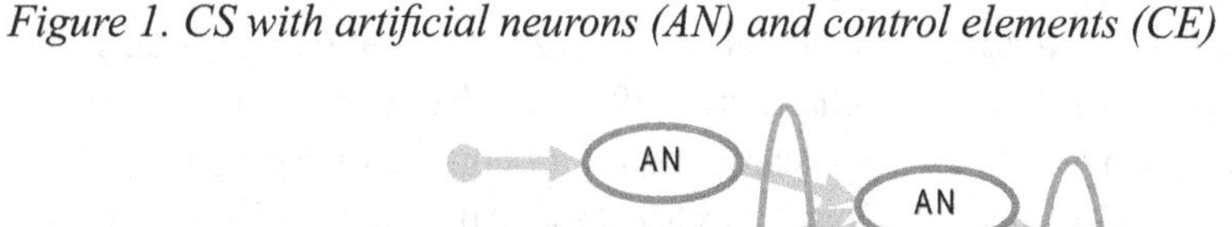

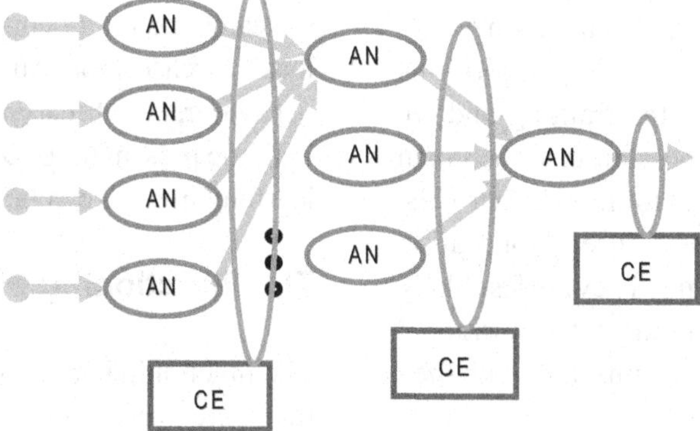

weights do not consider the error of the output, but take place at any time according to the activation frequency of each neuron, simulating reinforcements and inhibitions in the brain circuits.

The second and last learning phase consists of applying GAs to the individuals according to their MSE, which was stored in the first phase. The GAs phase carries out the corresponding crossovers and mutations and selects the new individuals with which the first and second phases will be repeated until the least possible error, or preferably no error, is obtained. The second phase is considered a supervised training because the GAs takes into account the error made by the network to select the individuals that will be mutated and crossed-over, i.e. it makes the changes in the weights according to that error.

During the testing phase (once training has finished) it is necessary to test the ANGN generalization capacity. If this capability is correct, ANGN will be ready for its subsequent use. We want to emphasize that at this phase, and in the course of all the subsequent runs, the brain behaviour introduced in the non-supervised learning phase will always be applied, since it is a part of the ANGN in all its stages and it participates directly in the information processing. The input patterns will be presented to the net during the iterations that were determined in the training phase, which allow the ANGN to carry out their activity.

LATEST RESULTS

The analysis of the cerebral activities has opened various ways to convert ANNs into ANGNs and provide them with a potential that improves their contribution to the information processing. We established possibilities of conversion that were classified by us according to what happens with connections between neurons, the activation value of the neurons, and combinations of both (Porto & Pazos, 2006). By analysing these options it can be noticed that, in most of them, the artificial astrocytes must induce network modifications based on neuronal activity. With our work we have tested the first proposed options, related to the connections between neurons. Such options were implemented, as it has been mentioned, at the non-supervised stage. The efficacy of the non-supervised stage was shown by means of comparing the ANGNs with the corresponding ANNs trained only by means of GAs. We made said comparison at solving two classification problems of various complexities. The tests started with a very simple problem: MUltipleXor (Porto et al., 2005). Later, in a more complicated domain, initial tests were performed with the IRIS flower problem (Porto et al., 2007). Such tests have been recently completed with multiple simulations whose summary results are following shown.

IRIS Flower Problem

This problem has served to test the ANGNs functioning when dealing with multiple classification tasks. IRIS flower problem uses continuous input values, different activation functions in artificial neurons of different layers, and twice as many training patterns. This example has been carefully analysed in the field of ANNs since A. Fisher first documented it in 1936 (Fisher, 2006). It consists on identifying a plant's species: Iris setosa, Iris versicolor and Iris virginica. This case has 150 examples with four continuous inputs which stand for 4 features about the flower's shape. The four input values represent measurements in millimeters of the: petal width, petal length, sepal width, sepal length. We have selected one hundred examples (33,3% of each class) for the training set and fifty examples for the test set, with the purpose of achieving a great generalization capability.

The learning patterns have been found to have 4 inputs and 3 outputs. The three outputs are boolean ones, representing each Iris species. By doing it in this manner (three boolean outputs instead of a multiple one) additional information can be provided about whether the system's outputs are reliable or not. That is, due to the outputs' intrinsic features, only one of them must possess the true value, standing for the type of flower it has classified, while the rest have a false value. Therefore, if two or more outputs have true values, or if all of them are false, we may conclude that the value classified by the system is an error and the system can not classify that case. The values corresponding to the four input variables have been normalized in the interval (0-1) so that they are dealt with by the CSs.

We have started from optimal architecture and parameters which were obtained by our research group in previous works (Rabuñal et al., 2004): 5 hidden neurons, tangent hyperbolic activation functions and threshold function (0.5) in the output neurons. By using ANN with these features and trained exclusively by means of GAs, J. Rabuñal et al. (2004) reaching an adjustment better than the previous best example of work for solving IRIS flower with ANN, in which A. Martinez & J. Goddard (2001) used BP for the training and a hidden neuron more than J. Rabuñal et al. (2004). These good results demonstrated the GA efficacy for simplifying and solving this problem. We have compared our ANGNs with these ANNs trained exclusively by means of GAs taking in account a maximum value for the weights of "1".

The ANN architecture may be observed in Figure 2.

The tests were carried out by keeping the same ten populations of individuals and the same random seed which originates the selection of the individuals for crossover and mutation.

In order to draw an adequate comparison we established the same GAs parameters for all the tests: Population size of 100 individuals; Montecarlo technique in order to select the individuals; Darwinian substitution method; Crossover rate 90% and mutation rate 10%; and a single cut point for crossover.

We want point out that these GAs options are not the ones which provide good results. Our purpose was that said parameters coincide in the CSs to be compared.

In this problem, we establish four thousand generations for training the ANGNs in all the simulations.

Table 1 shows the best results obtained for the ten populations using the ANN training only with GAs.

Table 2 shows the best results achieved with the ANGNs for the same ten populations.

If we compare the two tables, ANGNs reached minor MSE and major percent accuracy in the

Figure 2. ANN architecture for IRIS flower problem

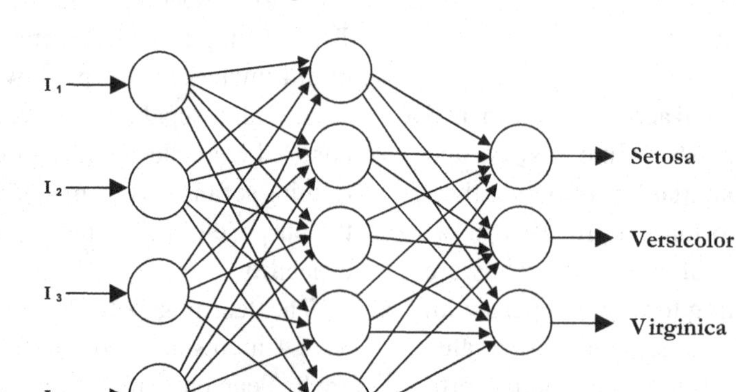

Table 1. IRIS: Results of ANN training only with GA

		MSE	% Test	Seconds	Generation
POPULATIONS	1	0,42	50	0:00:01	11
	2	0,39	58	0:00:01	21
	3	0,34	48	0:00:06	117
	4	0,37	60	0:00:00	7
	5	0,25	54	0:01:17	1455
	6	0,33	56	0:00:01	4
	7	0,39	54	0:00:01	13
	8	0,38	58	0:00:01	19
	9	0,4	62	0:00:01	11
	10	0,44	60	0:00:00	8

Table 2. IRIS: ANGN results of the ten populations

		MSE	% Test	Seconds	Generation
POPULATIONS	1	0,09	82	0:00:12	44
	2	0,04	76	0:03:28	780
	3	0,06	80	0:04:31	1045
	4	0,02	84	0:06:15	1001
	5	0,07	82	0:01:17	297
	6	0,04	80	0:02:33	404
	7	0,09	82	0:06:32	1516
	8	0,08	80	0:01:28	316
	9	0,07	64	0:00:13	51
	10	0,09	72	0:06:25	1485

test phase. Although training time was shorter in the case of ANN training only with GAs, it is important to emphasize that the simulations continued until 4000 generations in order to unify the results; also the percent accuracy of the test with ANNs did not reach more than the indicated value during the 4000 generations.

Considering the results obtained so far with all the simulations, it was checked the effectiveness of the chosen options of artificial astrocyte behaviour for the non-supervised stage. The reason why these satisfactory results were achieved with the ANGNs is attributed to what was expected when the hybrid method was designed. The individuals, once evaluated, are presented to the GAs arranged by their MSE. The changes made to the weights during the unsupervised learning phase cause this resulting order to be different from a potential order obtained in case this phase did not exist. This is so because their MSE varies according to this new methodology. The process of searching for the global minimum carried out by the GAs adds to the local adjustment process implemented thanks to this non-supervised stage.

FUTURE TRENDS

The research in this novel field intends to continue working, by means of Artificial Intelligence systems, on the reproduction of the observed biological processes. More specifically, the present, as well as future, the main goal of the intended research works is to increase and to take advantage of the current knowledge regarding nervous information processing. On the one hand, this will improve the computer information processing by means of CSs and, on the other hand, it will enable the consolidation of the hypothesis about the complex functioning of the biological neuroglial networks.

It has already been mentioned that, as the way in which the astrocytes modulate the physiology of a neural network depending on neural activity has not been quantified yet, there are many possibilities of influence over connection weights (Porto & Pazos, 2006). We have started to study what happens modyfied the connection weights according to the neural activity. We keep on analysing other synaptic modification possibilities based on brain behaviour to apply them to ANGNs which can solve real problems.

CONCLUSION

Considering the behaviour recently observed in glial cells (Perea & Araque, 2005; Perea & Araque, 2007) we have decided to create CSs that integrate some phenomena which are believed to be based on this behaviour. We have incorporated to the CSs new elements which will participate together with the artificial neurons in the information processing. The initial essential objective was to check if these new CSs were more efficient at solving problems currently solved by means of ANNs.

We have oriented the design of these ANGNs toward classification problems that are usually solved by means of feed-forward multilayer ANNs, fully connected, without back propagation and without lateral connections. We have thought it would be more adequate to start using the latest research about information processing in CSs for classification or recognition, since the control of the connections in the brain lead to the generation of activation ways.

We believe that the enhancement achieved with our results is firmly based on what occurs in brain circuits, given that it coincides with the most recent observations (Perea and Araque, 2007). Based on these new findings, glia is now considered as an active partner of the synapse, dynamically regulating synaptic information transfer as well as neural information processing. These results achieved show that the synaptic modifications introduced in the CSs, and based on the modelled brain processes (Porto, 2004, LeRay et al., 2004), enhance the training of the analysed multilayer architectures. The unsupervised mechanism permits a local adjustment in the search of the optimal solution. Moreover, GAs were found to be very efficient in the training because they helped to adapt the CSs to the global optimal solution.

In our opinion, CSs are still in a development phase, possibly even the initial one. Their real potential has not been reached, or even hinted at yet. We must remember that the innovation of the existing ANNs models towards the development of new architectures is conditioned by the need to integrate the new parameters into the learning algorithms so that they can adjust their values. New parameters, that provide the process element models of the ANNs with new functionalities, are harder to come by than optimizations of the most frequently used algorithms that increase the calculations and basically work on the computational side of the algorithm. The ANGNs integrate new elements and thanks to a hybrid method this approach did not complicate the training process. Thus, we believe that the research with these ANGNs directly benefits Artificial Intelligence because it can improve information processing capabilities which would allow us to deal with a wider range of problems. Moreover, this has indirectly benefited Neuroscience since experiments with computational models that simulate brain circuits pave the way for difficult experiments carried out in laboratories, as well as providing new ideas for research.

FUTURE RESEARCH DIRECTIONS

Some algorithms that might cover all the astrocyte functioning items based on empirical data from physiological experimentation should be defined; this definition will enable the continuation of the research in this field, especially when we face with the way in which a given astrocyte modifies the connections of the artificial neurons.

The presence of such a large amount of possibilities (Porto, 2004; Porto & Pazos, 2006) is an advantage that allows the experimentation of various possibilities when facing a real problem, but it presents at least one disadvantage. How do we know which is the best option to solve a determined problem? Mathematically speaking, it is impossible to know that the final choice is indeed the best. The use of Evolutionary Computation techniques is proposed in order to avoid such disadvantage. It will enable the acquisition of the most suitable artificial glia parameters and ANGNs-fitted architectures. An experimentation that might gradually use the different proposed options is therefore expected by means of the definition of optimisation protocols and the establishment of artificial glia behaviour protocols.

Moreover, given that it has been proved that the glia acts upon complex brain circuits, and that the more an individual's brain has developed, the more glia he has in his nervous system (Ramón y Cajal, 1911), we want to apply the observed brain behaviour to more complex network architectures. Particularly after having checked that a more complex network architecture achieved better results in the problem presented here.

For the same reason, we intend to analyse how the new CSs solve complex problems, for instance time processing ones where totally or partially recurrent networks would play a role. These networks could combine their functioning with this new behaviour.

ACKNOWLEDGMENT

This work was partially supported by grants from the Spanish Ministry of Education and Culture (TIN2006-13274), and grants from A Coruña University, Spain (Proyectos para equipos en formación-UDC2005 and UDC2006).

REFERENCES

Araque, A., Púrpura, V., Sanzgiri, R., & Haydon, P. G. (1999). Tripartite synapses: glia, the unacknowledged partner. *Trends in Neuroscience*, 22(5).

Araque, A., Carmignoto, G., & Haydon, P. G. (2001). Dynamic Signaling Between Astrocytes and Neurons. *Annu. Rev. Physiol*, 63, 795-813.

Dorado, J. (1999). *Modelo de un Sistema para la Selección Automática en Dominios Complejos, con una Estrategia Cooperativa, de Conjuntos de Entrenamiento y Arquitecturas Ideales de Redes de Neuronas Artificiales Utilizando Algoritmos Genéticos*. Phd Thesis. Facultad de Informática. University of A Coruña, Spain.

Dorado, J., Santos, A., Pazos, A., Rabuñal, J.R., & Pedreira, N. (2000). Automatic selection of the training set with Genetic Algorithm for training Artificial Neural Networks. In: *Proc. Genetic and Evolutionary Computation Conference GECCO'2000*, (pp. 64-67) Las Vegas, USA.

Erkmen, I., & Ozdogan, A. (1997) Short term load forecasting using genetically optimized neural network cascaded with a modified Kohonen clustering process, in: *Proc. IEEE Int. Symp. Intelligent Control* (pp. 107–112).

Fernández, D., Fuenzalida, M., Porto, A., & Buño, W. (2007). Selective shunting of NMDA EPSP component by the slow after hyperpolarization regulates synaptic integration at apical dendrites of CA1 pyramidal neurons. *Journal of Neurophysiology*, 97, 3242-3255.

Fiacco, T.A., & McCarthy, K.D. (2004). Intracellular astrocyte calcium waves in situ increase the frequency of spontaneous ampa receptor currents in Ca1 pyramidal neurons. *Journal of Neuroscience*, 24, 722-732.

Fisher, R.A. (1936). The use of multiple measurements in taxonomic problems, *Annals of Eugenics*, 7, 179-188.

Kinnebrock, W. (1994). Accelerating the standard backpropagation method using a genetic approach. *Neurocomputing* 6, 583–588.

Lee, S.W. (1996). Off-line recognition of totally unconstrained handwritten numerals using multilayer cluster neural network, *IEEE Trans. Pattern Anal. Machine Intell.* 18, 648–652.

LeRay, D., Fernández, D., Porto, A., Fuenzalida, M., & Buño, W. (2004). Heterosynaptic Metaplastic Regulation of Synaptic Efficacy in CA1 Pyramidal Neurons of Rat Hippocampus. *Hippocampus*, 14, 1011-1025.

Liu, Q-S, Xu, Q., Arcuino, G., Kang, J., & Nedergaard, M. (2004). Astrocytes mdiated activation of neural kinate receptors. *Proceedings of the National Academy of Sciences,* Vol. 101 (pp. 3172-3177). USA

Martín, E.D., & Araque, A. (2006). Astrocytes and The Biological Neural Networks. In J. Rabuñal (Ed.), *Artificial Neural Networks in Real-Life Applications* (pp. 22-45) Hershey PA, USA: Idea Group Inc.

Martínez, A. & Goddard, J. (2001). Definición de una red neuronal para clasificación por medio de un programa evolutivo. *Revista Mexicana de Ingeniería Biomédica* 22 nº 1, 4-11.

Nadkarni, S., & Jung, P. (2007). Modeling synaptic transmission of the tripartite synapse. *Physical Biology*, 4, 1-9.

Pasti, L., Volterra, A., Pozzan, R., & Carmignoto, G. (1997). Intracellular calcium oscillations in astrocytes: a highly plastic, bidirectional form of communication between neurons and astrocytes in situ. *Journal of Neuroscience*, 17, 7817-7830.

Perea, G., & Araque, A. (2002). Communication between astrocytes and neurons: a complex language. *Journal of Physiology*. Paris: Elsevier Science.

Perea, G., & Araque, A. (2005). Properties of synaptically evoked astrocyte calcium signal reveal synaptic information processing by astrocytes. *The Journal of Neuroscience*, 25(9), 2192-2203.

Perea, G., & Araque, A. (2007). Astrocytes Potentiate Transmitter Release at Single Hippocampal Synapses. *Science*, 317(5841), 1083 – 1086.

Porto, A. (2004). *Computational Models for optimizing the Learning and the Information Processing in Adaptive Systems*, Ph.D. Thesis, Faculty of Computer Science, University of A Coruña, Spain.

Porto, A., Araque, A., & Pazos, A. (2005). Artificial Neural Networks based on Brain Circuits Behaviour and Genetic Algorithms. *LNCS*. 3512, 99-106.

Porto, A., & Pazos, A. (2006). Neuroglial behaviour in computer science. In J. Rabuñal (Ed.): *Artificial Neural Networks in Real-Life Applications* (pp. 1-21). Hershey, PA: Idea Group Inc.

Porto, A., Araque, A., Rabuñal, J., Dorado, J., & Pazos, A. (2007). A New Hybrid Evolutionary Mechanism Based on Unsupervised Learning for Connectionist Systems. *Neurocomputing*, 70(16-18), 2799-2808.

Rabuñal, J. (1998). *Entrenamiento de Redes de Neuronas Artificiales con Algoritmos Genéticos.* Tesis de Licenciatura. University of A Coruña. Spain.

Rabuñal, J., Dorado, J., Pazos, A., Pereira, J. & Rivero, D. (2004). A New Approach to the Extraction of Rules from ANN and their Generalization Capacity through GP, Neural Computation, 16(7), 1483-1523.

Ramón y Cajal, S. (Ed.). (1911). *Histologie du sisteme nerveux de l'homme et des vertebres.* Maloine, Paris.

Shen, X., & De Wilde, F. (2006) Long-term neuronal behaviour caused by two synaptic modification mechanisms. Department of Computer Science, Heriot-Watt University, United Kingdom.

Taha M.A. & Hanna A. S. (1995). Evolutionary neural network model for the selection of pavement maintenance strategy. *Transportation Res. Rec.* 1497, 70–76.

Whitley, D. (1995). Genetic algorithms and neural networks. In *Genetic Algorithms in Engineering and Computer Science* (vol. 11, pp. 1-15). John Wiley & Sons Ltd.

Yan, W., Zhu, Z. & Hu R. (1997). Hybrid genetic/BP algorithm and its application for radar target classification. In: Proc. *IEEE National Aerospace and Electronics Conf. NAECON.* Part 2 of 2. 981–984.

Yang, J.M., Kao, C.Y. & Horng, J.T. (1996). Evolving neural induction regular language using combined evolutionary algorithms. In: *Proc. 1st Joint Conf. Intelligent Systems*/ISAI/IFIS 162–169.

Yao X. (1999). Evolving Artificial Neural Networks. In: *Proc. IEEE,* Vol. 87 nº9 1423-1447.

Zhang, P., Sankai, Y. & Ohta, M. (1995). Hybrid adaptive learning control of nonlinear system. In: *Proc. American Control Conf.* Part 4 of 6, 2744–2748.

ADDITIONAL READING

Haydon, P. G. & Araque, A. (2002). Astrocytes as modulators of synaptic transmission. In Volterra A, Magistretti P, Haydon (eds.) *Tripartite Synapses: Synaptic transmission with glia.* PG. Oxford University Press. pp: 185-198.

Haydon, P. G. (2001). Glia: listening and talking to the synapse. Nat. *Rev. Neurosci., 2,* 185-93.

Hebb, D. O. (1949). *The Organization of Behaviour.* J. Wiley. N.Y.

Hines, M. (1994). The NEURON simulation program. In J. Skrzypek. Norwell (ed.) *Neural Network Simulation Environments,* MA: Kluwer, p. 147-163.

Holland, J.H. (1975). *Adaptation in Natural and Artificial Systems.* University of Michigan Presss. Ann Arbor, MI, USA.

Hopfield, J. (1989). *Neural networks: algorithms and microchardware.* Institute of Electrical and Electronics Engineers. Ed. Piscataway. N.J.

Kimelberg, H.K. (1983). Primary Astrocyte Culture. A Key to Astrocyte Function. *Cellular and Mollecular Neurobiology, 3,* 3 pag.1-16.

Kimelberg, H.K. (1989). Neuroglia. *Rev. Investigación y Ciencia, 153.* pág. 44-55. Ed. Prensa Científica. Barcelona.

Kuwada, J.Y. (1986). Cell recognition by neuronal growth cones in a simple vertebrate embryo. *Science, 233.* 740-746.

Largo, C., Cuevas, P., Somjen, G.G., Martin del Rio, R. & Herreras, O. (1996). The effect of depressing glial function in rat brain in situ on ion homeostasis, synaptic transmission, and neuron survival. *J. Neurosci., 16,* 1219-1229.

McCulloch, W.S., & Pitts, W. (1943). A Logical Calculus of Ideas Immanent in Nervous Activity. *Bulletin of Mathematical Biophysics,* 5, 115-133.

Mennerick, S. & Zorumski, C.F. (1994). Glial contribution to excitatory neurotransmission in cultured hippocampal cells. *Nature, 368* 59-62.

Noremberg, M.D., Hertz, L. & Schousboe, A. (1988). *Biochemical Pathology of Astrocytes.* Alan R. Liss Inc.

Pazos, A. & Col. (1991). *Estructura, dinámica y aplicaciones de las Redes de Neuronas Artificiales.* Centro de Estudios Ramón Areces, S.A. España.

Pfrieger, F.W. & Barres, B.A. (1997). Synaptic efficacy enhanced by glial cells in vitro. *Science, 277,* 1684-1687.

Ramón y Cajal, S. (1904). *Textura del Sistema Nervioso del Hombre y los Vertebrados. Tomo II.* Madrid.

Von Neumann, J. (1956). *The Computer and the Brain.* Yale University Press.

Wiener, N. (Ed.). (1985). *Cibernética.* Tusqets editores.

Chapter VII
A Biologically Inspired Autonomous Robot Control Based on Behavioural Coordination in Evolutionary Robotics

José A. Fernández-León
University of Sussex, UK & CONICET, Argentina

Gerardo G. Acosta
Univ. Nac. del Centro de la Prov. de Buenos Aires & CONICET, Argentina

Miguel A. Mayosky
Univ. Nac. de La Plata & CICPBA, Argentina

Oscar C. Ibáñez
Universitat de les Illes Balears, Palma de Mallorca, Spain

ABSTRACT

This work is intended to give an overview of technologies, developed from an artificial intelligence standpoint, devised to face the different planning and control problems involved in trajectory generation for mobile robots. The purpose of this analysis is to give a current context to present the Evolutionary Robotics approach to the problem, which is now being considered as a feasible methodology to develop mobile robots for solving real life problems. This chapter also show the authors' experiences on related case studies, which are briefly described (a fuzzy logic based path planner for a terrestrial mobile robot, and a knowledge-based system for desired trajectory generation in the Geosub underwater autonomous vehicle). The development of different behaviours within a path generator, built with Evolutionary Robotics concepts, is tested in a Khepera© robot and analyzed in detail. Finally, behaviour coordination based on the artificial immune system metaphor is evaluated for the same application.

INTRODUCTION

The main goal of this chapter is to describe the authors' experiences in developing trajectory generation systems for autonomous robots, using artificial intelligence (AI) and computational intelligence (CI) methodologies. During this engineering work, some questions have arisen that motivated the exploration of new techniques like ER and the behaviour coordination with artificial immune systems. This maturing process as well as new questions and hypotheses for the suitability of each technique are presented in the following paragraphs.

In order to provide paths to an autonomous mobile robot, being it terrestrial, aerial or aquatic, there are some basic building blocks that must be necessary present. One essential feature needed consists on on-board sensory systems to have perception of the world and the robot's presence in the environment. This will be called the navigation system. Another necessary feature is the low-level trajectory generation from the next target position and the robot's current position, referred as the guidance system. Finally, the lowest level feedback loops allowing the robot to describe a trajectory as close as possible to the proposed path (Fossen, 2002), (Meystel, 1991), named the control system.

A top hierarchy module is responsible of generating the next target positions for the robot, and then, the whole trajectory or path. This module is called the mission planner and varies according to the mobile robot application domain. The mission plan can be given beforehand (static planning) or it can be changed on-line as the robot movement progresses in the real world (dynamic planning or replanning). Mission replanning is the robot's response to the changing environment (obstacle avoidance, changes in mission objectives priorities, and others).

The navigation, the guidance and the mission planner systems providing trajectories for the mobile robot, may be considered as a supervisory control layer giving appropriate set points to the lower level controllers, in a clear hierarchical structured control (Acosta et. al, 2001). Consequently, every control layer could be approached with many different and heterogeneous techniques. Nevertheless, to better focus within the scope of this book, current technology on mobile robot path generation with AI and CI techniques will be analyzed in more detail in next sections.

BACKGROUND

The guidance system is usually designed to navigate joining two points, routing over intermediate points between actual robot position and the target position. These points, called waypoints, vary in number from one approach to another, but in every case, they represent intermediate goals to fulfil when going towards the final one. Even further, a complete mission is split in a great number of intermediate trajectories; each conformed by more than one waypoints. The navigation system comprises the data fusion necessary to locate precisely in 3D the robot rigid body (considering it as holonomic or non-holonomic) and the target position. The components within the navigation system can be geostationary positional system (GPS), an inertial navigations system (INS), a compass, a depth sensor, sonars, lasers, video images, and others. Thus, the navigation system provides the mission planner system, the guidance system and the control system with accurate data to achieve their objectives. Although this problem is faced with several techniques, AI is employed to organize the information in such a way that sensor readings are transformed into a kind of perception. Several approaches use artificial vision or pattern recognition with sonars, also considered as machine intelligence (Antonelli, Chiaverini, Finotello & Schiavon, 2001), (Conte, Zanoli, Perdon & Radicioni 1994), (Borenstein & Koren, 1991), (Warren, 1990), (Yoerger, Bradley, Walden, Singh & Bachmayer, 1996), (Hyland &

Taylor, 1993), (Panait & Luke, 2005). Regarding the controller, it is well known that nonlinearities and uncertainties are very well managed by computational intelligence approaches like fuzzy logic, artificial neural networks or neurofuzzy controllers. Literature is profuse and exceeds the scope of this chapter. The interested reader is referred to (Harris, 1993) or (Antsaklis, 1993). The mission planner, sometimes referred as task and path planner in a task decomposition perspective, has to face a complex problem: the trajectory generation for the mobile robots. Indeed, autonomous operation of mobile robots in real environments presents serious difficulties to classical planning and control methods. Usually, the environment is poorly known, sensor readings are noisy and vehicle dynamics is fairly complex and non-linear. Since this represents a difficult problem to solve due to the great amount of data required to take a decision in an effective time, it has attracted the AI community. Three main lines of activity have arisen within this AI researcher's community: (1) planning-based systems, (2) behaviour-based systems, and (3) hybrid systems. Starting with the intention of emulate partially some features of human intelligence, AI techniques application to control were gradually shifting to CI based control, which is more descriptive of the problems approached. The first two application examples in the following section belong to this stage of maturing of the knowledge in path planning.

Recently, streams of research and technological implementations are going towards the use of bio-inspired techniques. Particularly in robotics, the research community is trying to take advantage not only of human beings' problem solving paradigms as part of the nature, but also from other natural systems. A great deal of activity currently growing is coming from evolutionary systems and immune systems. In effect, several researchers and practitioners addressed the question of how biological systems adapt and interact with the environment (Abbot & Regehr, 2004), (Forde, Thompson & Bohannan, 2004). In particular they are interested in understanding and synthesizing complex behaviours in the context of autonomous robotics. Artificial Evolution was proposed as a suitable technique for the study of physical and biological adaptation and their harnessing. In (Nolfi & Floreano, 2000), dynamic modular hierarchies of neurocontrollers are proposed to face stability and scalability issues, in what they called *Evolutionary Robotics* (ER). ER is a sub-field of Behaviour-Based Robotics (BBR) (Arkin, 1998). It is related to the use of evolutionary computing methods in the area of autonomous robotic control. One of the central goals of ER is the development of automated methods to evolve complex behaviour-based control strategies (Nolfi & Floreano, 2000), (Nelson, Grant, Galeotti & Rhody, 2004). Another central goal of ER is to design morphological models. Some examples of these works comprises evolutionary design (Bentley, P., 1999), the evolution of robot morphologies (Lipson & Pollack, 2000), and complete structures from simple elements (Hornby & Pollack, 2001), (Reiffel & Pollack, 2005). Scaling up in ER relates to the creation of complex behaviours from simple ones. The sequential and hierarchical organization of a complex activity in systems with many behaviours, such as robot's path generation (Fernández-León, Tosini & Acosta, 2004), (Maaref & Barref, 2002), (Nelson, Grant, Galeotti & Rhody, 2004), remains one of the main issues for biologically inspired robotics and systems neuroscience. It is expected that understanding such systems in a robotic context will also shed some light into their natural counterparts, giving for instance a better understanding of human motor capabilities, their neural and embodied underpinnings, and their adaptive and recovery properties. It is being a statement that stability and scalability in robot's behaviour can be addressed by mean of the explicit incorporation of ideas derived from neuroscience and the study of natural behaviour (e.g., theories of evolutionary learning, skill acquisition, and behaviour emergence (Dopazo, Gordon, Perazzo & Risau-Gusman, 2003). ER concepts can be used to obtain controllers that adapt robot's behaviour according to its sensory input, focusing on scaling up during the path generation task in an unknown environment. A frequently used strategy to implement the scaling up of behaviours is the Layered Evolution (LE) (Togelius, 2003). It provides satisfying results in cases where some of traditional methodologies failed. Based on a comparison between Incremental Evolution (Urzelai & Floreano, 1999) and Modularized Evolution (Calabretta, Nolfi & Parisi, 2000), (Nolfi, 1997), LE uses the concept of subsumption architecture in the evolutionary process, (Brooks,1986), (Téllez & Angulo, 2004). Key concepts in LE are Modularity (Nolfi & Floreano, 2000), Adaptability (Di Paolo, 2000-2003) and Multiplicity. LE proposes an evolution sequence from bottom layers (basic behaviours) to upper levels (complex behaviours). A different fitness function is created for each level. Only when evolution on a given level reaches a suitable fitness score, a new (upper) level is created and the lower level gets frozen. Evolution then continues on the new hierarchical level. A *Coordination Structure* is a key point, necessary to coordinate emergent behaviours. It can be generated in several ways (e.g., expert user knowledge, as a part of an evolution process, or by mean of an artificial immune system).

These ideas will be illustrated with an experiment using artificial evolution in autonomous path generation devised for the Khepera© micro-robot in the third case study presented in the sequel. In this way, the next case studies show the progress from traditional AI and CI approaches to bio-inspired ones, which seem the most promising.

APPLICATION EXAMPLES

A traditional approach with a central world model to achieve path planning is analyzed firstly. The practical problem consists on the autonomous guidance and control of a terrestrial vehicle in a completely unknown environment. The *autonomous guided vehicle* (AGV) of Figure 1 was used as the non-holonomic experimental platform.

The method proposed combines optimum path planning techniques with fuzzy logic to avoid obstacles and to determine the shortest

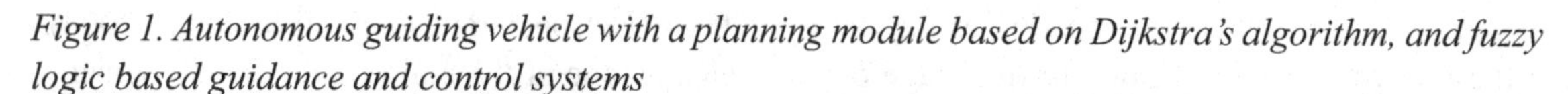

Figure 1. Autonomous guiding vehicle with a planning module based on Dijkstra's algorithm, and fuzzy logic based guidance and control systems

path towards its goal. The technique computes the potential surface using Dijkstra's algorithm in a moving window, updating the cost map as it moves with the information of obstacles and target obtained by the ultrasonic sensor, responsible of the navigation module. A Fuzzy Logic Controller (FLC) controls the wheels of a differential drive robot to the angle of minimum potential. This ensures a smooth trajectory towards the objective. A second FLC controls the average speed of the platform.

To minimize calculations, a 20 x 20 cm grid was defined and the cost and potential functions were computed only considering points on the grid. A window was defined covering a few cells on the grid in the neighbourhood of the vehicle. To compute the Dijkstra's algorithm, the local goal was established in the intersection of the line drawn from the actual position to the global goal, with the border of the active window. The case for a 13x8 window and a 20x20 grid was analyzed with the mobile position fixed with respect to the window. Sometimes the objective may fall in an obstacle but when the windows moves, this local objective will change. If the final objective is really occupied by an obstacle, a trap situation will occur. Previsions in the algorithm should be taken to cope with those situations, as well as wrong trajectories generation due to the application of such a sub-optimal method.

Later, cost map is placed on the active window based on the information obtained by the sensors. Then, with this partial cost map Dijkstra's algorithm is computed in the window. First, the actual position of the vehicle is taken as the ground node and later, the local destination is used as the reference node. Potentials are added to obtain single potential surfaces that hold the optimum path. A new iteration follows: the window is moved one-step in the grid and the previous procedure is repeated until the local objective matches the global one. More details can be found in (Calvo, Rodríguez & Picos, 1999).

The configuration of the objects was changed with good results in most of the cases. A comparison was also made with the Virtual Force Field technique (VFF) (Borenstein & Koren, 1991) giving shortest and smoothest trajectories but with longer computing times. With this traditional path planning, the robot performs good enough to face autonomous displacement in real world. However, the mission planner was static and the robot's ability to move avoiding obstacles is constraint to the previous user programming skills. The navigation scenario was glimpsed beforehand with no possibility of smart reaction to a new moving obstacle or a change in the target point. Also, the robot had no ability to learn from its own experience, which is a very important feature in ER.

A second example is the desired trajectory generation of an *Autonomous Underwater Vehicle* (AUV) devoted to pipeline inspections. In this case, the path planner is an instance of a dynamic mission planner (DMP) built from the experience of remote operated vehicles (ROV) users compiled as a production system of forward chained rules. The experimental vehicle was the Geosub constructed by Subsea7, shown in Figure 2. It is robust enough to allow surveys at depths of several thousands of meters. A main goal of this research and development was to evaluate experimentally if the current technology (by the year 2004) was able to face autonomous inspections in deep water, practically with minimum human intervention. The expert system called *EN4AUV* (Expert Navigator for Autonomous Underwater Vehicle) (Acosta, Curti, Calvo, & Mochnacs,2003-2005), showed the possibility of incremental growth of the knowledge base as more experience is gained on AUV navigation and pipeline inspection, although consistency within it must always be considered in these cases.

When compared to ROV based inspections, AUV ones allow a smoother navigation and then a more reliable data acquisition, because there is not any umbilical cable to any ship or platform. In

Figure 2. Autonomous underwater vehicle with a planning module based on an expert system (EN4AUV) and classical approaches for guidance, control and navigation systems

addition, missions can be done in a cheaper way due to less support infrastructure required. These contrasts become more evident as the mission depth increases. Hence, within the Framework Program Fifth, the EU funded AUTOTRACKER Project was thought to show that the technology was mature enough to face this autonomous underwater pipelines and cables inspections. The main features of the full AUTOTRACKER architecture, focussing in the dynamic mission planner to generate the AUV trajectories, are described in (Acosta, Curti, Calvo, Evans & Lane, 2005-in press).

The navigation system consisted of a geopositional system (GPS) giving an absolute position in global coordinates (latitude, longitude), a depth sensor, and an inertial navigation system based on gyroscopes that gives a relative position when the AUV is submerged. For the inspection task, data was acquired with a multi-beam echo sounder (MBE). Raw data from this MBE is processed by the MBE Tracer module, which can also generate a predicted target trajectory, using image-processing techniques (Tena Ruiz, Petillot & Lane, 2003). Position and direction of the target under inspection is estimated this tracer, together with a priori information about the lay down of the pipeline or cable, called legacy data. They are combined through a sensor fusion module (SFM). Using the information from the SFM and a priori information, the dynamic mission planning module, decided the path to follow on the different situations: search, track, reacquire, and skip from current position. The obstacle avoidance system (OAS) took the trajectories proposed by the *EN4AUV* expert system and validated or modify them using the information coming from the forward looking sonar and the exclusion zones in legacy data. Then both, the *EN4AUV* and the OAS are the main components of the DMP.

When doing the underwater inspections, complex situations might appear (like the sudden appearance of a fishing net, a complex pattern of more than one pipeline over the seabed, or a detour due to obstacle detection, and many others). These situations were coded as possible scenarios, in about fifty rules. As the knowledge about different situations increases, the knowledge base (KB) describing new scenarios may be simply and naturally completed and updated, yielding an incremental KB growth. Each scenario triggers different searching or tracking pattern of trajectory (*search, back-to-start, skip,* and *track*). Scenarios are based mainly on two ideas: the survey types and the AUV status as regards as the inspected target. The type of survey is defined a priori in the mission settings, to establish the number of pipelines/cables to be tracked, the navigation depth, and other mission features. The AUV status changes when the SFM updates its sensors and classify the situation as target seen, target lost, target seen intermittently, and avoiding obstacles. The KB conceptualization is presented in Figure 3.

The scenarios developed for the sea trials during 2004 were diverse (14 in total). Some examples are: 1st Scenario: The AUV is tracking an exposed pipeline, navigating on top, at a fixed offset smaller or equal than 5 meters. Both the MBE and the MAG detect it. 2nd Scenario: The AUV is tracking a buried pipeline on top, at a fixed offset smaller or equal than 5 meters. The MBE may not be able to detect it, but the MAG can track it anyway. 3rd Scenario: The AUV is tracking an intermittent (intermittently exposed and buried) pipeline at a fixed offset. This is a sequence of alternative appearance of scenarios number one and two. The preliminary sea trials performed in the North Sea near Scotland in August/September 2004 and November 2005 showed promising results and that with traditional AI approaches, the problem of autonomous underwater inspection was possible (Acosta, Curti, Calvo, Evans & Lane, 2005-in press).

Previous examples, as well as the prior and contemporary work of several other researchers

Figure 3. Knowledge base conceptualization to build the dynamic mission planner

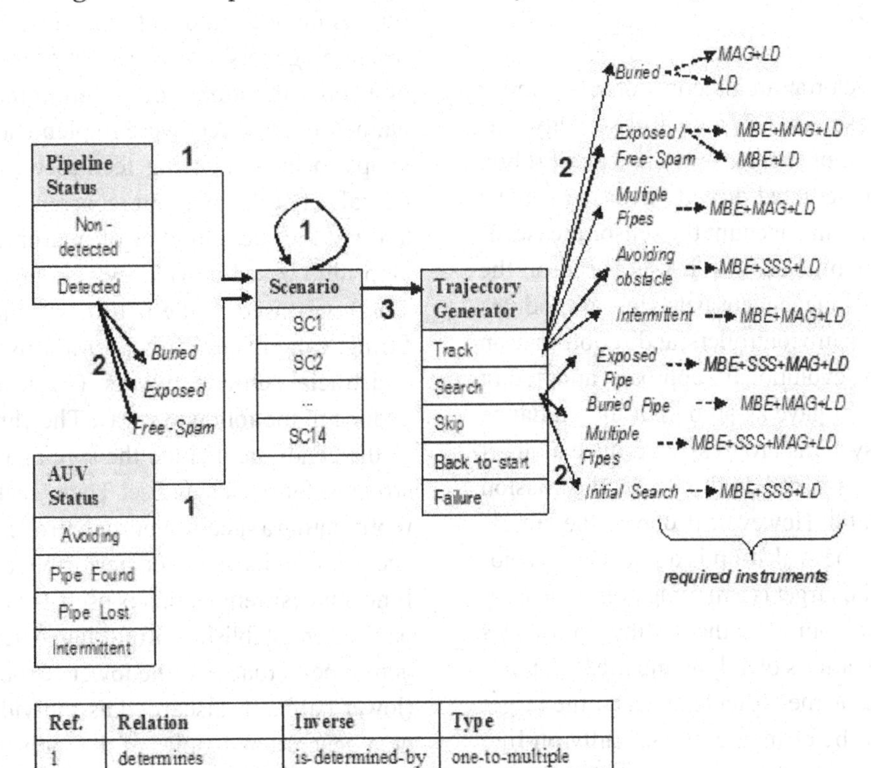

Ref.	Relation	Inverse	Type
1	determines	is-determined-by	one-to-multiple
2	generalization-for	type-of	one-to-one
3	selects	is-selected-by	one-to-multiple

and practitioner showed that robots were able to operate autonomously with minimum human intervention. They could also carry out complex tasks successfully. However, they are still unable to *learn autonomously* from their own experience. Next generation of autonomous robots surely should have this skill. In our opinion, this is an inflection point in robot's development. The most promising methodologies to achieve this robot ability seem to be the ones provided by ER. They try to emulate natural or bio-inspired behaviours. Furthermore, they have the additional attractive feature of facilitating robots interaction, enabling multiple robots problem solving approaches. In what follows, the remaining sections of this chapter are devoted to present our work within this research area.

EVOLUTIONARY ROBOTICS APPROACH

In order to develop an architecture for autonomous navigation based on ER, a control strategy was designed and implemented over a Khepera© robot. This strategy actuated directly over the robot's wheels, taking into account the sensor measures and the mission objectives. It was based on the generation of independent behavioural modules supported by neurocontrollers, and a coordination structure. The evolutionary approach adopted in this experiment gave as an output the guidance and control systems progressively, once a target position is determined. In this sense, the mission planner is static. However, if one of the simple behaviours to be scaled-up is an obstacle avoidance one or the target is a moving point, the final complex behaviour after the scaling-up process will exhibit features of a dynamic mission planner. In effect, as the obstacle appears, the target positions will be changing dynamically on-line as the robot moves.

Working Assumptions for the Autonomous Robot Application

In this case study, the following working conditions were assumed: (a) the robot moves on a flat ground; (b) inertial effects, as well as non-holonomic characteristics of the mobile robot, are not taken into account; (c) the robot moves without slipping; (d) the environment is structured but unknown, with some elements fixed (e.g., walls, corridors, passages, doors, etc.) while others, like goal position references (light sources) and obstacles, can be modified for each run; (e) environment variable conditions (e.g., light influence from other sources) are not managed directly but considered as perturbations to the controller.

Construction of the Neurocontrollers

An initial set of neurocontrollers, called population, is mapped into individuals with their representing genotypes. These genotypes are made of a constant number of chromosomes. Individual neurocontrollers were implemented for each simple behaviour using feed-forward recurrent neural networks, without recurrence in any level and with a fixed number of neurons. A genetic algorithm based on Harvey's proposal (Harvey, 1992) was used in the neurocontroller's weight fitting, called learning algorithm in the context of artificial neural networks. This learning stage is done in the following way. The chromosomes of the genotype include the sign and the weight strength for each synapse. Then, each genotype representing a specific neurocontroller is awarded according to its observed performance through a fitness measurement that is used as a comparison parameter, establishing a ranking. After that, those genotypes situated at the lower part of this scale (lower half) are discarded as individuals in the next generation. Copies of the individuals at the upper part replace these individuals.

From Basic Behaviours to Path Generation

The selected simple behaviours to be scaled-up to obtain the more complex path generation behaviour were:

Phototaxis: The robot's ability to reach a light source as a target position point. The reference for this part of the work was (Nolfi & Floreano, 2000). The fitness function was composed of two variables:

$$\Phi_1 = k \cdot (1 - i) \quad (1)$$

where k is proportional to the average value of the frontal sensors measurement[1] in Figure 4 ($0 \leq k \leq \gamma$, with γ a defaulted value); i is the absolute value of the difference among the two node-motor activities, representing the deviation angle expressed in radians.

Therefore, a deviation is valid if it does not exceed 1 radian. The first component of equation (1) is maximized according to the proximity to a light source, while the second component is maximized when direct movements to the goal are generated.

Obstacle avoidance: The robot's skill to avoid obstacles, when going towards a particular point. The fitness function adopted Φ_2 is (Nolfi & Floreano, 2000):

$$\Phi_2 = z \cdot \left(1 - \sqrt{\Delta z}\right) \cdot (1 - j) \quad (2)$$

where z is the difference between the output value of m_1 and m_2. z expresses the deviation angle of the robot's trajectory in radians ($-2 \leq z \leq 2$); Δz is the absolute value of the algebraic difference between the node-motor activations m_1 and m_2, maximizing ($1-\sqrt{\Delta z}$) when both activations are equal. The term $\sqrt{\Delta z}$ enhances small differences of the motor-nodes; ($1-j$) is the difference between the maximum activation (value 1) and the maximum sensed value j of the proximity infrared sensors.

Wall following: Robot's abilities to follow a wall. These are intended to complement other reactive behaviours in narrow spaces and corridors. The

Figure 4. Down view of sensor deployment in the robot used in the experiments

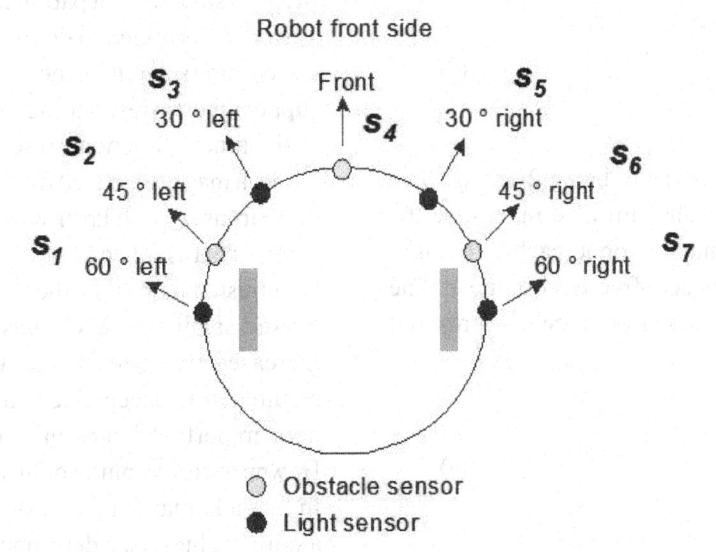

basic objective of wall seeking (wall-following generalization track) is the generation of a trajectory parallel to a wall, or eventually an obstacle. The fitness score used was:

$$\Phi_3 = \begin{cases} \left(1-|\Theta - V_8|\right)+\left(1-|\Theta - V_2|\right)\big/3 & if \quad left \\ \left(1-|\Theta - V_7|\right)+\left(1-|\Theta - V_5|\right)\big/3 & if \quad right \\ 0 & otherwise \end{cases} \quad (3)$$

where Θ is the minimum distance allowable between each sensor and a wall/obstacle. In all simulations a value of $\Theta=0.3$ was adopted.

Learning: This behaviour consists of the robot's approach to one of two possible light sources (targets) (Nolfi & Floreano, 2000), (Togelius, 2004). In half of the tests (learning stage), the objective to be reached varies without any predefined pattern. The robot does not know a-priori which light source should reach at the beginning of each test. Therefore, it is expected that the robot learns to discriminate its objective in a trial and error paradigm. Reinforcement learning (φ) is accomplished using the following score:

$$\varphi = \begin{cases} \delta & \text{if the goal reached is right} \\ -\delta & \text{if the goal reached is wrong} \\ 0 & \text{else} \end{cases} \quad (4)$$

where δ is a default value, being 2 for the proposed experiment. The aim is to maximize the number of times that the robot reaches the right objective in an obstacle-free environment. The fitness function is based on Togelius' proposal (Togelius, 2003):

$$\Phi_4 = \begin{cases} \sum_{i=0}^{n=200} \varphi_i & \textbf{if } \sum_{i=0}^{n=200} \varphi_i \geq 1 \\ \max(S_j) & \textbf{else} \end{cases} \quad (5)$$

where S_j refers to the sensed value for the jth sensor $(1 \leq j \leq 7)$.

Some fitness functions were normalized according to the generation number (Φ_1 and Φ_2 in [0; 1]), and Φ_3 range was in $[-\infty,\infty]$.

The above mentioned simple behaviours were combined for the implementation of the more complex behaviour of path generation to analyze whether the robot approaches to a certain source in a small closed environment avoiding obstacles (Nolfi & Floreano, 2000) and (Togelius, 2004).

Coordination of Behavioural Levels

The coordination among behaviours was done in a first approach using another FFNN taking the outputs of the behavioural modules and sensors as inputs (Figure 5). Outputs of the coordination module directly control the actuators. The fitness score adopted was:

$$\Phi_{coord} = \Theta_A\left(1-\Theta_B\right) \quad (6)$$

where Θ_A is the maximum of all light sensors and Θ_B is the maximum of all proximity sensors.

Experimental Results

In a first step, several parameters (shown in Table 1) were determined to do the experiments. After the learning stage, those neurocontroller that better support the corresponding behaviour according to the fitness function, were selected.

As it may be deduced from Table 1, the number of neurons in each layer were considered fixed a priori. That is, instead of letting network size to be adjusted as part of the evolutionary process, a fixed small size ANN was adopted. This fact increased the speed in the learning phase and resulted in an acceptable real-time performance, both important issues in practical prototyping. However, from a philosophical standpoint, this is in fact a human intervention that may constrain a pure evolutionary development.

Figure 5. Behaviours coordination to yield a more complex one in the process known as scaling-up

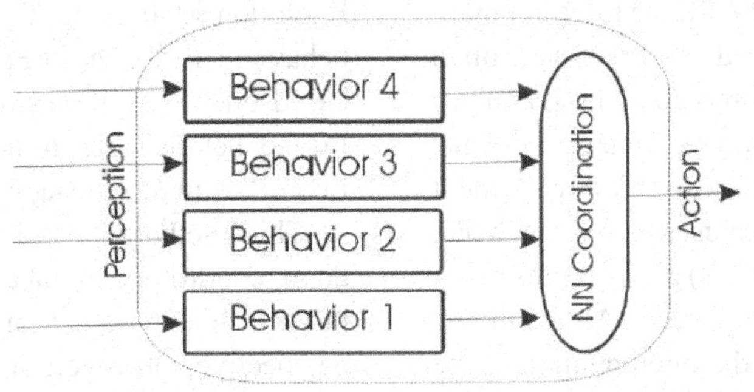

Table 1. Specifications parameters for the experimental tests of the evolutionary approach to path generation in an autonomous robot

Parameter	Value
Number of input neurons	8 plus 1 additional input
Number of hidden neurons	6
Number of output neurons	2
Number of runs	10
Number of iterations per run	200
Number of generations	300
Number of objects in the environment	0 (phototaxis|learning) and 10 (obstacle avoidance)
Initial population	random
Synaptic weight range	[-1; 1]
Initial range of synaptic weight	[0; 0.01]
Selection percentage	50%
Mutation rate	5%
Light sensors range	[0; 1]
Sensors proximity range	[0; 1]
Type of activation function	sigmoid
Type of mutation	Uniform
Type of selection	Ranking

To implement the neurocontrollers we used a simulator of the Khepera© robot called YAKS. The tests consisted of the robot moving on a labyrinth-like environment. Since in reactive path generation security is essential, it is possible to distinguish two situations to analyze: (a) if an obstacle is detected in the robot proximity (sides or front), the obstacle avoidance behaviour has priority over the attraction to the goal behaviour; and (b) the actuators' action arises as a combination between the obstacle avoidance and the attraction toward the objective. Therefore, two possible situations are necessary to study: (1) convex obstacle avoidance; (2) concave obstacle avoidance. Robustness of the final controller with regard to sensor imperfections (e.g., robustness to environmental variations and morphologic variations) was also analyzed. The obtained results showed that the controller was capable of overcoming environmental variations, but no morphologic variations in most tests.

Once behaviours were implemented, tests were carried out on a real robot Khepera©. The environment consisted of a 105x105 cm. arena delimited by walls, with different internal walls of white plastic and a compact green rubber floor. A light source equipped with a bulb of 75 Watts was placed perpendicularly to the floor of the arena at approximately 3 cm. high. Please refer to Figure 6. The test room presented other non-controlled light sources, such as fluorescent lights (zenithal) and natural light from windows. Some experiments reported in the literature use a dark environment only lighted by the source used as reference like in (Nolfi & Floreano, 2000). This represents a fully-controlled test environment. Although this alternative can simplify the interpretation of the final results, the experiment proposed in this approach was aimed to analyzing the controller's behaviour in partially uncontrolled environment, which is closer to a real situation. The prototype performs very well avoiding concave and convex obstacles, phototaxis, and wall seeking behaviours and hence emerging a sophisticated path generation behaviour learned from its own experience. Evolution performance for the obstacle avoidance behaviour is shown in Figure 7. Comparison is based on fitness scores and convergence time. Further details of results are given in (Fernández-León, Tosini & Acosta, 2004-2005).

This methodology then demonstrated very good conditions to face real-world problems in an efficient way. From a pure evolutionary perspective, however, it must be quoted that this methodology is still too much dependent on the human engineer previous knowledge on the problem to solve. Evolution was done from a rigid and prescribed framework. Subdivision in atomic tasks, individual fitness functions and coordination rules are strongly user dependant, leaving small chance for self-organization and feature

Figure 6. Evolutionary path generation tested in a small robot a) in simulation environment and b) in a real environment

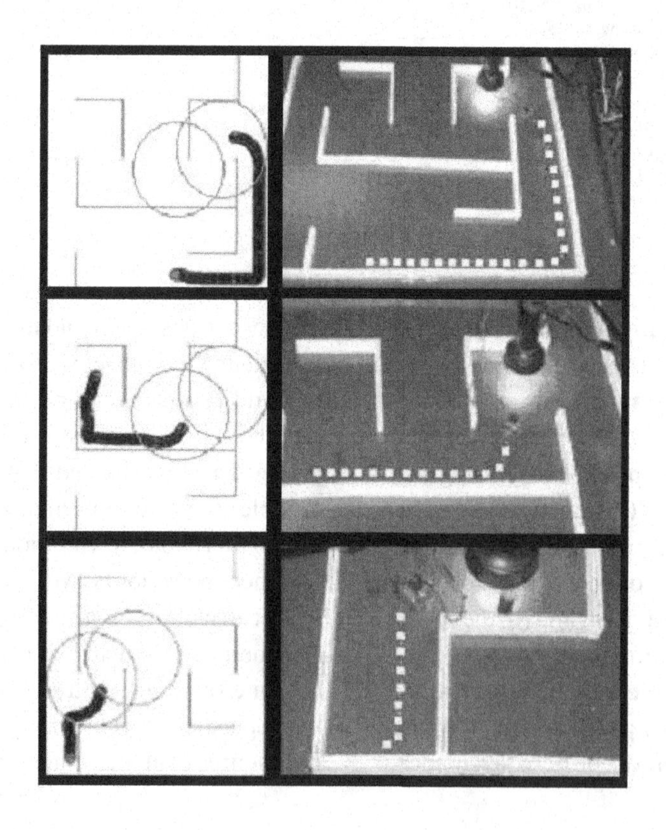

discovering. A generalization of scalability to closely emulate biological systems, in the sense of self-organization, is still an open subject.

BEHAVIOUR COORDINATION BASED ON THE ARTIFICIAL IMMUNE SYSTEM

Studies on self-organizing systems have emerged as a major part of the biological-inspired approach in response to challenges introduced by control systems design complexity in real problems (Nolfi & Floreano, 2000). Due to its distributed characteristic, the main ideas of self-organising design are the use of principles in nature to develop and control complex adaptive systems. This organization tends to be robust enough to reject perturbations (Jakobi, 1997).

The basic building blocks of these complex adaptive systems are *agents*. Agents are semi-autonomous, or autonomous, units that seek to maximize some merit figure or fitness by its evolution over time. These agents interact and connect each other in unpredictable and unplanned ways. However, from the set of interactions, regularities emerge to form patterns that feed-back on the system and inform the interactions of the agents (Tay & Jhavar, 2005). Examples of complex adaptive systems are the immune system (Timmis, Knight, de Castro & Hart, 2004) (Roitt, 1997), the self-organized artificial neural system (Haykin, 1999), and the artificial evolution (Nolfi & Floreano, 2000). In particular, *coordination* and *behaviour integration* are also central topics in the study of self-organizing collective behaviours in nature (Baldassarre, Nolfi & Parisi, 2002). As stated in (Nolfi, 2005), the interaction among simple behavioural rules followed by each system component might lead to rather complex collective behaviours

A central issue is then to achieve behaviour coordination able of self-adaptation. Then the objective of this section is to describe a feasible technique for behaviours coordination based on the artificial immune system metaphor, which seems to meet these requirements.

Figure 7. Performance of the best, the mean, and the worst neurocontroller over generations. Measure carried out according to the fitness function on a population of FFNN controllers in obstacle avoidance when reaching the goal. Behaviour development is obtained in approximately 40 generations.

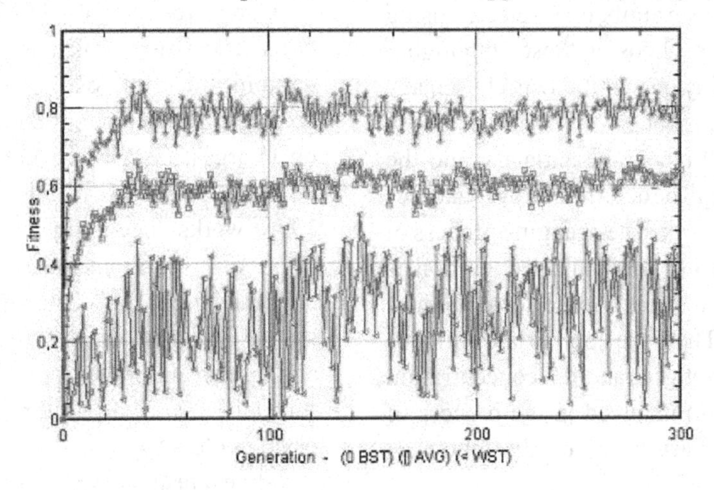

Immune System Inspired-Algorithms as Behaviour Coordination

The proposed behaviour coordination has a dynamic process of decision making defining a network structure of conditions-action-restrictions items. The incoming information provided by each sensor is considered as an "antigen", meanwhile the robot as an "organism". The robot (organism) contains "antibodies" in a network organization to recognize the antigens and to perform immune responses. Thus, the condition-action behaviour of the robot could be related with the antigen-antibody response of the immune system. The antigen represents the current state of the environment (Vargas, de Castro, Michelan & Von Zuben, 2003).

In our recent experiments, we develop an AIS-based system for behaviour coordination to solve the problem of tracking a target (i.e, a black line in the floor or a pipeline in the seabed). In them, the immune network dynamics was modelled through Farmer's model (Farmer, Packard & Perelson, 1986), selecting the most appropriate antibody to deal with external information. The concentration of an antibody may stimulate or suppress other antibodies of the network, and it may depend on the concentration of other antibodies through network links. Based on the resulting concentration level, an antibody is selected using a ranking algorithm, choosing those with higher concentrations. Only one antibody is chosen at a time to control the robot.

Antibodies will recognize antigens, and the action to be taken will be determined by dynamics of the immune network. The changing process of antibodies involved in antigenic recognition is planned to be implemented to perform autonomous navigation. Then, the appropriate antibodies recognize the antigen, and the concentration of each antibody (*a*) involved in the process of recognition could vary based on the following equations.

$$\frac{da_i^t}{dt} = \left(\sum_{j=1}^{N} m_{ij} a_j(t) - \sum_{k=1}^{N} m_{ik} a_k(t) + m_i - k_i \right) \cdot a_i(t) \tag{7}$$

$$a_i(t) = \frac{1}{1 + \exp(0.5 - a_i(t))} \tag{8}$$

The terms involved in equations (7)-(8) for the calculation of the concentration of the *i*-th antibody at the instant *t*, are: *N*, number of antibodies that compose the network; m_i, affinity between antibody *i* and a given antigen; m_{ji}, affinity between antibody *j* and antibody *i*, in other words the degree of stimulation; m_{ik}, affinity between antibodies *k* and *i*, in other words, the degree of suppression; k_i, natural death coefficient of antibody *i*. Equation (8) is a squashing function used to impose boundaries on the concentration level of each antibody.

As stated, the Khepera© robot has light and proximity sensors (Figure 4). The antigens refer to the direction of light sources, direction of the obstacles, and the proximity of them respect to the robot body. In this new approach, the sensory system was divided in the zones showed in Figure 8.

During robot evaluation, the network of antibodies tries to recognise antigens, and the network dynamics determines the action (behaviour) to be taken. The behaviour coordination was based on (Vargas, de Castro, Michelan, Von Zuben, 2003) and (Ishiguro, Kondo, Shirai & Uchikawa, 1996) works. The antigens coding corresponds to possible situations that activate under certain conditions (e.g. an obstacle is near, or a light source is placed on the far-right-front side). The actions that antibodies represent are, for example, turn left, turn right, go forward, avoid obstacle, or do phototaxis.

The experiments carried out contrast with Vargas et al.'s work in that an evolutionary adjustment mechanism was not used, in which the immune "idiotopes" (restrictions to antibody execution) are evolved using genetic algorithms. This difference was done on purpose just to find out if it was possible to evolve using life-time adaptation without human intervention. Another difference with Vargas et al.'s work was that behaviours could relate with evolved neural networks, instead of simple rules for controlling robots. The use of evolved neural networks as behaviour enhances the robot robustness when performing tasks like path generation in real world, due to its intrinsic robust characteristics (Nolfi & Floreano, 2000).

Figure 8. Definition of zones for antigen recognition based on waypoints, pipe segments, zones, and obstacles recognition

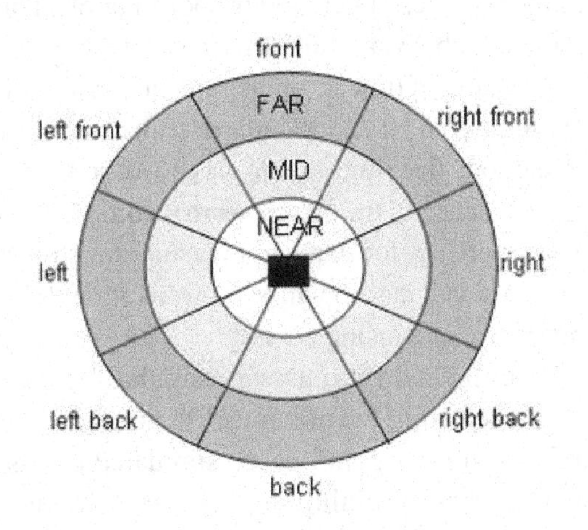

In order to evaluate the AIS-based coordination, a computer simulated environment for the Khepera© robots, the previously introduced YAKS, was used to cope with the target tracking problem with obstacles in the environment. In this case, the target was a black line representing a pipeline deployment. These preliminary experiments demonstrated the feasibility of this approach, even when dealing with perturbations. The following tests were performed: *test 1*: control case; *test 2*: sensors light gain ±5%; *test 3*: motor gain ±5%; *test 4*: random noise in affinities (m) among antibodies ±50%; *test 5*: random noise in antibody concentrations (a) ±50%. In Figure 9 the final trajectory is depicted. Results show that the AIS-based coordination could deal with the expected robot performance. The robot was capable to develop all its dedicated tasks, without risk of collision.

Figure 9. Example of tracking trajectory generation using AIS-based behaviour coordination

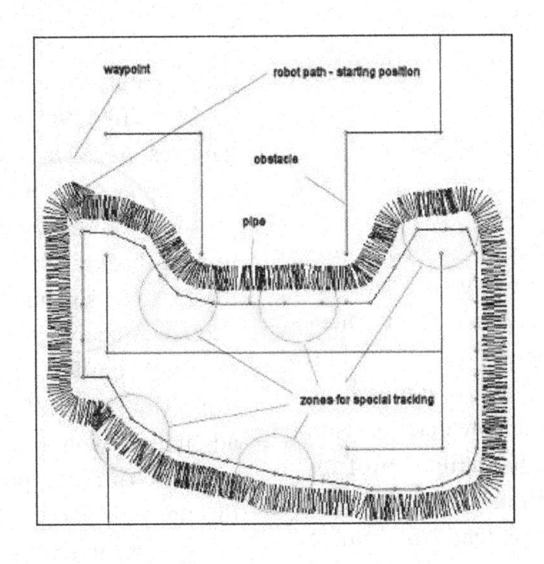

Table 2. Unsuccessful outcomes for the simulator. Code expresses the level of importance of the unsuccessful outcome (the highest, the worst)

Code	Unsuccessful outcome
1	The goal was not reached but the robot did appropriate gate passing.
2	The goal was reached but the robot did not pass through gates AB.
3	Passed through the gate and stopped (stuck) near the gate.
4	Became trapped and could not escape before pass the gate.

Figure 10. Experimental settings for comparison between layered evolution and AIS based behaviour coordination

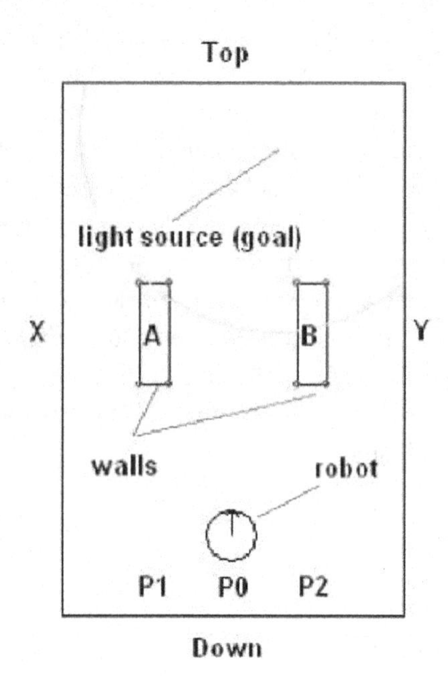

A Comparative Analysis

A conceptual comparison among different behaviour coordination approaches is always necessary to analyze their scopes, possibilities and applicability ranges. However, in engineering areas it is also need an exhaustive comparison among techniques within a specific domain of application. Although this section is far from being such exhaustive analysis, it pretends to evaluate in computer simulations the most promising techniques for behaviour coordination, within ER. They are the layered evolution coordination and the AIS-based behaviour coordination. The simulated robot was allocated to explore during a fixed amount of time its environment (maximum 300 iterations). If it was successful in its task completion, the time taken was recorded. If it was unsuccessful the causes were recorded. Four causes were set for experiments and are shown in Table 2. All significant tests were at the 95% confidence level using a t-test.

Experimental settings were similar than the ones presented in the previous ER experimental section, with Khepera©-based simulated robots. The robot has the ability to sense obstacles (e.g. walls and objects) and to sense the proximity to the target (light source). Similar experiment settings are proposed also in Whitbrook (2005), Vargas et al. (2003), and Ishiguro et al. (1996). The task consisted on passing through a gate (AB) in a small simulated arena (Figure 10).

In the first aim, the robot is required to navigate reactively in a safe way around the arena and stop near the light source (goal) once it had passed through the gate AB once. It was restricted to pass through gaps XA or BY at any time due their stretch separation with walls. It was given a fixed number of iterations to complete the task, starting from P0, P1, or P2 (initial positions) 10 times (independent tests). Note that a long-term adaptation would require that the robot goes as many times as possible to the goal, passing through the gate in any direction. This was not carried out here for short. As it may be seen, the

Table 3. Frequency of failure of simulated robot performance considering all experiments from different starting positions

Coordination approach	Unsuccessful code	Causes of failures (Freq.)	% All failed trials	% Causes of failure
Layered	1	2	6.7%	14%
(ruled-based)	2	0	0%	0%
	3	7	23.3%	50%
	4	5	16.7%	36%
AIS	1	0	0%	0%
	2	10	33.3%	100%
	3	0	0%	0%
	4	0	0%	0%

Table 4. Summary of statistics for time (CPU cycles) to reach the goal using the simulator considering only successful cases (goal reached)

Co-ordination approach	Mean			Standard deviation			95% confidence interval		
	P0	P1	P2	P0	P1	P2	P0	P1	P2
Layered	185.43	233.7	249.7	27.6	21.14	15.1	[212.01;287.3]	[159.8; 211]	[200.1;267.2]
AIS	98.9 1	19.8 6	9.6	56.4 4	1.4	13.38	[58,5;139.3] [	90.2;149.4]	[60;79.2]

problem is difficult because the dimensions of the environment requires a high degree of precision to steer towards the centre of the small gate and the robot needs to be able to reach the goal in a safe way.

The following Tables 3 and 4 summarize experimental results. In them, it was found that the performance of the behaviour coordination depends on the correct choice of the following parameters:

- ***D***: the distance between the robot an obstacles (when it is small, there are more frequent collisions with walls and obstacles and then frequent trapped situations tend to appear);
- ***S***: the scope of sensor readings, mainly those related with obstacle avoidance;
- ***B***: the simple behaviours implemented to emerge a more complex behaviour

A trial and error procedure was used to determine these parameters in order to obtain acceptable results in previous experiments, because there is not a straightforward method for determining them. Although adequate parameter selection was not relevant to this research it is an important issue in mobile robotics. In Krautmacher & Dilger (2004) found that AIS code was dependent on the choice of free parameter used.

Table 4 shows that the performance rate was good over all experiments with AIS. In average, the robot takes 96 CPU cycles to approach to the target, failing in none of 30 trials. The worst performance rate obtained for layered coordination may be found in the following fact. Layered coordination depended on the sensing performance of the obstacle avoidance simple behaviour. Due to the shape of the lowest part of internal walls, this obstacle avoidance module could not be able to sense the obstacles effectively. This coordination approach could not surpass this inconvenience giving the control to another behavioural module. In other words, layered coordination is done over static modules in a prescribed (by rules) way, meanwhile AIS coordination is based on self-adjustment during task performance. Then, changes in simple behaviour to be coordinated (like an adaptation of simple behaviours) will cause changes in the coordination method itself. However, the observed problems for the AIS coordination were related to reaching the target through the narrow corridors (XA and BY ones). This difficulty could be solved with slower movements.

Comparisons between both approaches of ER behaviour coordination showed no significant differences. Further experiments are planned to test the long-term tracking problem where adaptation is essential. In these cases, the rule-based coordination seems to be less adequate than the AIS one. In addition, an improvement in the emergent behaviour may also appear if simple behaviours modules are constructed by learning in a noisy and uncertain environment just before they are coordinated.

FUTURE TRENDS

ER is a key tool to obtain self-organized adaptive systems, like the ones required in autonomous robots interacting with the real world. Layered Evolution and AIS approaches to behaviour coordination share many similarities. In the authors' opinion, the challenge to obtain a bio-inspired robot control within ER is to deepen in the way the emergent behaviour is achieved.

In the previously described case studies of ER, it was shown that genetics algorithms obtained "fittest" behaviours for each task (e.g. phototaxis, obstacle avoidance, and others). In this way, this means that evolutionary algorithms are suitable for optimizing the solution space in proposed problems, which besides is not new. The immediate question is about the suitability of AIS as optimization methods. According to de Castro & Von Zuben (2002) and Aickelin & Dasgupta (2005), AIS are probably more suited as an optimizer where multiple solutions are of benefit. AIS can be made into more focused optimizers by adding hill-climbing or other functions that exploit local or problem-specific knowledge. AIS can exploit its potential either when some concept of "matching" is needed. In some sense, AIS works as algorithms that adapt their solution space during test time meaning that they are suitable for problems that change over time and need to be several times with some degree of variability.

When AIS is compared with hand-design ruled-based and fittest evolutionary coordination systems, it has the potential benefit of an adaptive pool of antibodies that can produce adaptive and robust coordination. Therefore, the benefits of this computation can be used to tackle the problem of dealing with changing or noisy environments. In this robot control context AIS-based behaviour coordination is useful because it does not optimize a path, i.e. in finding the best match for a planned path in a priori unknown environment. Instead, AIS require a set of antibodies that are a close match, but which are at the same time distinct from each other for successful environmental situation recognition. This is the main interest around most of AIS implementations in literature and from our own experience, because the designer can implement different autonomous methods for path generation. AIS only require positive examples. In addition, the patterns that it has learnt can be explicitly examined (see de Castro & Von Zuben (2002)). Besides, AIS, as well as the majority of other heuristics that require parameters to operate, has the drawback of parameter settings that are domain dependent. Universal values for them are not available.

The introduction of fuzziness within the AIS properties (see de Castro & Timmis, 2003) in antigenic recognition, suggests that fuzzy logic might be appropriate to model several aspects and mechanisms of the immune system, particularly to combine the multiple responses of simultaneous immune network outcome.

In self-organized adaptive systems, some interesting questions still remain open, like how a bio-inspired system reacts to discover new optima solutions, or how hybrid solutions can enhance behavioural coordination by mean of adding other computational intelligence techniques. For example, AIS could be combined with GA for discovering more suitable immune networks for a specific problem. This could correspond to the primary response of the AIS and to the convergence phase with an immune network based on GAs. Therefore, this combination could exploit further adaptations that are precisely what GAs is lacking (Gaspar & Collard, 1999).

From a strictly theoretical point of view, the adaptive nature of behaviours has several consequences that are not well understood yet. For instance, motor actions partially determine sensor patterns in an also autonomous sensory-motor loop. Therefore, on-line coordination between sensors and actuators can enhance adaptively the robot ability to achieve its goal, as suggested in (Nolfi, 2005). This fact resembles the natural adaptation of animals to interact with their surroundings. This is a cutting-edge research topic in mobile robots.

CONCLUSION

The problem of path generation in autonomous robots was addressed from the stand point of artificial intelligence techniques. Traditional approaches, in which authors have been working, were exemplified as case studies. This was a comparison framework for the new trends in autonomous path and task planning approaches like the ones in ER. A considerable amount of research has been done in the last decade in this area, and a special emphasis was put into autonomous robot path generation with environmental adaptation behaviour, in which authors are currently researching.

From an engineering viewpoint, a deterministic behaviour is still difficult to be designed straightforward because it requires a special ability from the designer to infer the rules governing the interactions between the robot and the environment that will precisely determine the desired emergent behaviour. In addition, such a design methodology is opposite to the inherent philosophy of free evolution. In fact, evolution include a necessary learning process in which the rules governing the robot/environment interactions are progressively assigned, modified through a process of random variation, and refined during the lifetime as a part of the adaptive process. This process allows discovering and retaining useful properties emerging from such interactions without the need to identify the relation between rules governing the interactions and the resulting behaviours. Difficulties arise when trying to optimize a merit figure, engineers' obsession, like minimum energy consumption, shortest path and others. In such cases, the proper selection of a fitness function may be more an art than a prescribed method. Another perceptible drawback of ER is the need of a learning period previous to the task development with a certain degree of success. This training can be carried out off-line or on-line. The on-line learning as in the case of humans, has involved the risk of non-stabilities and the impossibility to assure a complete convergent algorithm.

In spite of these apparent disadvantages of ER in contrast to other more traditional techniques, it seems to be the future research in robotics, for it is being demonstrated that it not only can contribute to solve real engineering problems, but also in spreading light to cognitive and bio-inspired sciences like Biology and Neurosciences in the understanding of open problems.

ACKNOWLEDGMENT

This work was partially supported by the Programme Alban, the European Union Programme of High Level Scholarships for Latin America, scholarship No. E05D059829AR, by The Peter Carpenter CCNR Award, University of Sussex, UK, and also recognised by the National Council of Scientific and Technological Research (CONICET), Argentina, Res. 38 Type-I from 08/01/2004. In addition, this work was partially supported by a Marie-Curie International Fellowship within the 6th European Community Framework Programme (Project 03027 – AUVI).

REFERENCES

Abbott, L. F. & Regehr W.G. (2004). Synaptic Computation. *Nature*, Vol. 431, pp. 796-803.

Acosta, G. G., Alonso, C., & Pulido, B. (2001). Basic Tasks for Knowledge Based Supervision in Process Control. *Engineering Applications of Artificial Intelligence*, Vol. 14, N° 4, Elsevier Science Ltd/IFAC, August 2001, pp. 441-455.

Acosta, G.G., Curti, H.J., Calvo, O.A, & Mochnacs, J. (2003). An Expert Navigator for an Autonomous Underwater Vehicle. *Proceedings of SADIO/Argentine Symposium on Artificial Intelligence ASAI '03*, Buenos Aires, Argentina, Sept. 3-5 2003, in CD.

Acosta, G.G., Curti, H.J., & Calvo, O.A. (2005). *Autonomous Underwater Pipeline Inspection in AUTOTRACKER PROJECT: the Navigation Module*. Proc. of the IEEE/Oceans'05 Europe Conference, Brest, France, June 21-23, 2005, pp. 389-394.

Acosta, G.G., Curti, H.J., Calvo, Evans, J., & Lane, D. (2007). *Path Planning in an autonomous vehicle for deep-water pipeline inspection,* submitted to the Journal of Field Robotics.

Antonelli, G., Chiaverini, S., Finotello, R., & Schiavon, R. (2001). Real-Time Path Planning and Obstacle Avoidance for RAIS: an Autonomous Underwater Vehicle. April 2001, *IEEE Journal of Oceanic Engineering*, Vol. 26, N°2, pp. 216-227.

Antsaklis P.J., & Passino K.M. (1993). *An Introduction to Intelligent and Autonomous Control.* Boston, Dordrecht, London, 1993, Kluwer Academic Pub.

Arkin, R.C. (1998). *Behaviour-Based Robotics.* The MIT Press, Cambridge, Massachusetts, 1998.

Baldassarre, G.; Nolfi, S., & Parisi, D. (2002). *Evolving mobile robots able to display collective behaviour.* In Helmelrijk C.K. (ed.). International Workshop on Self-Organization and Evolution of Social Behaviour, pp. 11-22. Zurich: Swiss Federal Institute of Technology.

Borenstein, J., & Koren, Y. (1991). The vector field histogram-fast obstacle avoidance for mobile robots. June 1991, *IEEE Transactions on Robotics and Automation*, Vol. 7, Issue 3, pp. 278 – 288.

Brooks, R.A. (1986). A Robust Layered Control System for a Mobile Robot. March 1986, *IEEE Journal of Robotics and Automation*, Vol. RA-2, N° 1, pp. 14-23.

Calabretta, R, Nolfi, S., Parisi, D., & Wagner, G. P. (2000). *Duplication of modules facilitates functional specialization.* Artificial Life, 6(1), 69-84.

Calvo, O., Rodríguez, G., & Picos, R. (1999). Real Time Navigation Of Autonomous Robot With Fuzzy Controller And ultrasonic Sensors. Palma de Mallorca, Spain, September 1999, *Proc. of the European Society of Fuzzy Logic and Technology Symposium EUSFLAT-ESTYLF99*, pp 95-98.

Conte, G., Zanoli, S., Perdon, A., & Radicioni, A. (1994). *A system for the automatic survey of underwater structures*. Brest, France, 13-16 Sept. 1994, Proc. of OCEANS '94, Vol. 2, pp. 92-95.

Di Paolo, E. (2000). Homeostatic adaptation to inversion of the visual field sensorimotor disruptions. *Proceedings of SAB 2000*, MIT Press.

Di Paolo, E. (2003). Evolving spike-timing dependent plasticity for single-trial learning in robots. *Philosophical Transactions of the Royal Society*, A 361, pp. 2299 - 2319.

Dopazo, H., Gordon, M.B., Perazzo, R.S., & Risau-Gusman, S.A. (2003). Model for the Emergence of Adaptive Subsystems. *Bulletin of Mathematical Biology*, 65, Elsevier, pp. 27–56.

Farmer, J., Packard, N. & Perelson, A. (1986). *The Immune System, Adaptation and Machine Learning*. Physica D. 22. 1986. pp. 187-204.

Fernández León, J.A., Tosini, M., & Acosta G.G, (2004). Evolutionary Reactive Behaviour for Mobile Robots Navigation. Singapore, December 1-3, 2004, *Proc. of the 2004 IEEE Conference on Cybernetics and Intelligent Systems (CIS04)*, pp. 532-537.

Fernández León, J.A, (2005). *Estudio de Neuro-Controladores Evolutivos para Navegación de Robots Autónomos*. Master in Systems Engineering. UNCPBA, Argentina.

Fernández-León, J.A., & Di Paolo, E. A. (2007). Neural uncertainty and sensorimotor robustness. *Proceedings of the 9th European Conference on Artificial life ECAL 2007*. Springer-Verlag. 2007.

Forde, S.E., Thompson, J., & Bohannan, B.J. (2004). M. *Adaptation varies through space and time in a coevolving host-parasitoid interaction*. Nature, vol. 431.

Fossen, T. (2002). *Marine Control Systems*. Trondheim, Norway, Marine Cybernetics Ed. 2002.

Harris, C.J., Moore, C.G., & M. Brown. (1993). *Intelligent Control: Aspects of Fuzzy Logic and Neural Nets*, World Scientific, 1993.

Harvey, I. (1992). *The SAGA cross. The mechanics of recombination for species with variable length genotypes*. Amsterdam: North-Holland. R. Manner and B. Manderick (Eds.), Parallel Problem Solving from Nature 2.

Haykin, S. (1999). *Neural Networks: A comprehensive Foundation*. Second Edition. Prentice Hall.

Hyland, J. & Taylor, F. (1993). Mine avoidance techniques for underwater vehicles. *IEEE J. Oceanic Engineering*, Vol. 18, 1993, pp.340-350.

Ishiguro, A., Kondo, T., Shirai, Y., & Uchikawa, Y. (1996). Immunoid: An Architecture for Behaviour Arbitration Based on the Immune Networks. *Proceedings of the 1996 IEEE/RSJ International Conference on Intelligent Robots and Systems*, volume 3, pp. 1730 - 1738.

Jacobi, N. (1997). Evolutionary robotics and the radical envelope of noise hypothesis. *Journal of Adaptive Behaviour*. 1997.

Krautmacher, M., & Dilger, W. (2004). AIS based robot navigation in a rescue scenario. *Lecture Notes Computer Science*, 3239, pp. 106-118.

Maaref, H. & Barref, C. (2002). Sensor-based navigation of a mobile robot in an indoor environment. *Robotics and Autonomous Systems*, Elsevier, Vol. 38, pp. 1-18.

Meystel, A. (1991). *Autonomous Mobile Robots: Vehicles With Cognitive Control*, World Scientific, 1991.

Nelson, A.L., Grant, E., Galeotti, J.M., & Rhody, S. (2004). Maze exploration behaviours using an integrated evolutionary robotic environment. *Robotic and Autonomous Systems*, 46, pp. 159-173.

Nolfi, S. (1997). *Using emergent modularity to develop control systems for mobile robots*. Adaptive Behaviour (5) 3-4: 343-364.

Nolfi, S. & Floreano, D. (2000). *Evolutionary Robotics: The Biology, Intelligence, and Technology of Self-Organizing Machines*. MA: MIT Press/Bradford Books. Ref.: Phototaxis: pp. 121-152; Obstacle avoidance/Navigation: pp. 69-92; Learning: pp.153-188.

Nolfi S. (2006). *Behaviour as a complex adaptive system: On the role of self-organization in the development of individual and collective behaviour*. Unpublished yet, internal communication with author.

Panait, L. & Luke, S. (2005). *Cooperative Multi-Agent Learning: The State of the Art. Autonomous Agents and Multi-Agent Systems*. 11(3) : 387–434. Springer 2005.

Roitt, I. (1997). Essential Immunology. Ninth Edition. *Pub. Blackwell Science*.

Seth, A.K. & Edelman, G. (2004). M. Theoretical neuroanatomy: Analyzing the structure and dynamics of networks in the brain. *Complex Networks*. E. Ben-Naim, H. Fraunfelder, & Z. Toroczkai (eds). p. 487-518, Springer-Verlag, Berlin. 2004.

Tay, J. & Jhavar, A. (2005). CAFISS: A complex adaptive framework for immune system simulation. *2005 ACM Symposium on Applied Computing (SAC'05)*, pp. 158-164.

Téllez, R. & Angulo, C. (2004). Evolving cooperation of simple agents for the control of an autonomous robot. *Proceedings of the 5th IFAC Symposium on Intelligent Autonomous Vehicles (IAV04)*, Lisbon, Portugal.

Tena Ruiz, I., Petillot, Y., & Lane, D. (2003). Improved AUV navigation using side-scan sonar, *Proceedings of IEEE OCEANS 2003*, 3, 22-26 Sept. 2003, pp. 1261 – 1268.

Timmis, J., Knight, T., de Castro, L., & Hart, E. (2004). *An overview of artificial immune systems*. In Paton et al., editors, Computation in Cells and Tissues: Perspective and Tools for Thought. Natural Computation Series, pp. 51-86. Ed. Springer.

Togelius, J. (2003). *Evolution of the Layers in a Subsumption Architecture Robot Controller*. Master of Science in Evolutionary and Adaptive Systems. University of Sussex, UK.

Togelius, J. (2004). Evolution of a Subsumption Architecture Neurocontroller. *Journal of Intelligent & Fuzzy Systems*, 15:1, pp. 15-21.

Urzelai, J. & Floreano, D. (1999). *Incremental Evolution with Minimal Resources*. Proceedings of IKW99. 796-803. Computation. Nature, Vol. 431. 1991.

Vargas, P., de Castro, L., Michelan, R., & Von Zuben, F. (2003). *Implementation of an immunogenetic network on a real Khepera II robot*. 2003 IEEE Congress on Evolutionary Computation, Volume 1, 8-12 Dec. 2003, pp. 420 – 426.

Warren, C. (1999). A technique for autonomous underwater vehicle route planning. *IEEE J. Oceanic Engineering*, Vol. 15, 1990, pp. 199-204.

Whitbrook, A. (2005). *An idiotypic immune network for mobile robot control*. Msc Thesis, University of Nottingham.

Yoerger, D., Bradley, A., Walden, B., Singh, H., & Bachmayer, R. (1996). *Surveying a subsea lava flow using the autonomous bentic explorer (ABE)*. Proc. 6th Int. Advanced Robotics Program, 1996, pp. 1-21.

ADDITIONAL READING

Several publications on topics discussed in this chapter can be proposed to be consulted at your leisure. A list of references on these topics is given as follows. Evolutionary Robotics: (Nolfi & Floreano, 2000), (Harvey, Di Paolo, Wood, Quinn & Tuci, 2005),(Fernández-León & Di Paolo, 2007), and (Fernández-León, Tosini & Acosta, 2005); AIS: (de Castro & Timmis, 2002); IA applied to path planning: (Acosta, Curti, Calvo, Evans & Lane, 2007).

ENDNOTE

1 The Khepera© robot's sensor S_3 and S_4.

Chapter VIII
An Approach to Artificial Concept Learning Based on Human Concept Learning by Using Artificial Neural Networks

Enrique Mérida-Casermeiro
University of Málaga, Spain

Domingo López-Rodríguez
University of Málaga, Spain

J.M. Ortiz-de-Lazcano-Lobato
University of Málaga, Spain

ABSTRACT

In this chapter, two important issues concerning associative memory by neural networks are studied: a new model of hebbian learning, as well as the effect of the network capacity when retrieving patterns and performing clustering tasks. Particularly, an explanation of the energy function when the capacity is exceeded: the limitation in pattern storage implies that similar patterns are going to be identified by the network, therefore forming different clusters. This ability can be translated as an unsupervised learning of pattern clusters, with one major advantage over most clustering algorithms: the number of data classes is automatically learned, as confirmed by the experiments. Two methods to reinforce learning are proposed to improve the quality of the clustering, by enhancing the learning of patterns relationships. As a related issue, a study on the net capacity, depending on the number of neurons and possible outputs, is presented, and some interesting conclusions are commented.

INTRODUCTION

Hebb (1949) introduced a physiological learning method based on the reinforcement of the interconnection strength between neurons. It was explained in the following terms:

When an axon of cell A is near enough to excite a cell B and repeatedly or persistently takes part in firing it, some growth process or metabolic change takes place in one or both cells such that A's efficiency, as one of the cells firing B, is increased.

This kind of learning method has been widely applied to recurrent networks in order to store and retrieve patterns in terms of their similarity. Models that used this learning rule were the bipolar model (BH) presented by J. J. Hopfield in 1982 (Hopfield, 1982) representing a powerful neural model for content addressable memory, or its analogical version, among others. These networks, although successful in solving many combinatorial optimization problems, present two main problems when used as content-addressable memory: their low capacity and the apparition of spurious patterns.

The capacity parameter α is usually defined as the quotient between the maximum number of patterns to load into the network and the number of used neurons that obtains an acceptable error probability (usually p_{error}=0.05 or 0.01). It has been shown that this constant is approximately α=0.15 for BH.

This value means that, in order to load K patterns, more than K/α neurons will be needed to achieve an error probability lower than or equal to p_{error}. Or equivalently, if the net is formed by N neurons, the maximum number of patterns that can be loaded in the net (with that error constraint) is $K<\alpha N$.

Since patterns are associated to states of the network with minimal energy, we wonder about what happens with these states if the network capacity is exceeded.

The main idea of this chapter holds that when patterns are very close each other, or if the net capacity is exceeded, then local minima corresponding to similar patterns tend to be combined, forming one unique local minimum. So, although considered as a limitation of the net as associative memory, this fact can explain the way in which the human brain form concepts: several patterns, all of them similar to a common typical representative, are associated and form a group in which particular features are not distinguishable.

Obviously, enough samples are needed to generalize and not to distinguish their particular features in both cases: artificial and natural (human) concept learning. If there are few samples from some class, they will still be retrieved by the net individually, that is, as an associative memory.

NEURAL BACKGROUND

Associative memory has received much attention for the last two decades. Though numerous models have been developed and investigated, the most influential is Hopfield's Associative Memory, based on his bipolar model (Hopfield, 1982). This kind of memory arises as a result of his studies on collective computation in neural networks.

Hopfield's model consists in a fully-interconnected series of bi-valued neurons (outputs are either -1 or +1). Neural connection strength is expressed in terms of weight matrix $W=(w_{i,j})$, where $w_{i,j}$ represents the synaptic connection between neurons i and j. This matrix is determined in the learning phase by applying Hebb's postulate of learning Hebb, and no further synaptic modification is considered later.

Two main problems arise in this model: the apparition of spurious patterns and its low capacity.

Spurious patterns are stable states, that is, local minima of the corresponding energy function of the network, not associated to any stored (input) pattern. The simplest, but not the least important, case of apparition of spurious patterns is the fact of storing, given a pattern, its opposite, i.e. both X and $-X$ are stable states for the net, but only one of them has been introduced as an input pattern.

The problem of spurious patterns is very fundamental for cognitive modelers as well as practical users of neural networks. Many solutions have been suggested in the literature. Some of them (Parisi, 1986) (Hertz et al., 1987) are based on introducing asymmetry in synaptic connections. However, it has been demonstrated that synaptic asymmetry does not provide by itself a satisfactory solution to the problem of spurious patterns, see (Treves et al., 1988) (Singh et al., 1995). Athitan et al. (1997) provided a solution based on neural self-interactions with a suitably chosen magnitude, if Hebb's learning rule is used, leading to the near (but not) total suppression of spurious patterns.

Crick (1983) suggested the idea of unlearning the spurious patterns as a biologically plausible solution to suppress them. With a physiological explanation, they suggest that spurious patterns are unlearned randomly by human brain during sleep, by means of a process that is the reverse of Hebb's learning rule. This may result in the suppression of many spurious patterns with large basins of attraction. Experiments have shown that their idea leads to an enlargening of the basins for correct patterns along with the elimination of a significant fraction of spurious patterns (van Hemmen et al., 1991). However, a great number of spurious patterns with small basins of attraction do survive. Also, in the process of undiscriminate reverse learning, there is a finite probability of unlearning correct patterns, what makes this strategy unacceptable.

On the other hand, the capacity parameter α is usually defined as the quotient between the maximum number of patterns to load into the network, and the number of used neurons that achieve an acceptable error probability in the retrieving phase, usually $p_e = 0.01$ or $p_e = 0.05$. It was empirically shown that this constant is approximately $\alpha = 0.15$ for BH (very close to its actual value, $\alpha = 0.1847$, see (Hertz et al., 1991)). The meaning of this capacity parameter is that, if the net is formed by N neurons, a maximum of $K \leq \alpha N$ patterns can be stored and retrieved with little error probability.

McElliece et al. (1987) showed that an upper bound for the asymptotic capacity of the network is $\frac{1}{2\log N}$, if most of the input (prototype) patterns are to remain as fixed points. This capacity decreases to $\frac{1}{4\log N}$ if every pattern must be a fixed point of the net.

By using Markov chains to study capacity and the recall error probability, Ho et al. (1992) showed results very similar to those obtained by McEliece, since for them it is $\alpha = 0.12$ for small values of N, and the asymptotical capacity is given by $\frac{1}{4\log N}$.

Kuh et al. (1989) manifested roughly similar estimations by making use of normal approximation theory and the theorems about exchangeables random variables.

Hopfield's Model

Hopfield's bipolar model consists in a network formed by N neurons, whose outputs (states) belong to the set {-1,1}. Thus, the state of the net at time t is completely defined by a N-dimensional state vector $\mathbf{V}(t) = (V_1(t), V_2(t), \ldots, V_N(t)) \in \{-1,1\}^N$.

Associated to every state vector there is an energy function, expressed in the following terms:

$$E(\mathbf{V}) = -\frac{1}{2}\sum_{i=1}^{N}\sum_{j=1}^{N} w_{i,j} V_i V_j + \sum_{i=1}^{N} \theta_i V_i \qquad (1.1)$$

where $w_{i,j}$ is the connection weight between neurons i and j, and θ_i is the threshold corresponding to i-th neuron (since thresholds are not used in the case of associative memory, from now on all of them will be considered 0). This energy function determines the behavior of the net.

Hopfield's Associative Memory

Let us consider $\{X^{(k)} : k = 1,\ldots,K\}$, a set of bipolar patterns to be loaded into the network. In order to store these patterns, weight matrix W must be determined. This is achieved by applying Hebb's classical rule for learning. So, the change of the weights, when pattern $X = (X_i)$ is introduced into the network, is given by $\Delta w_{i,j} = X_i X_j$. Thus, the final expression for the weights is:

$$w_{i,j} = \sum_{k=1}^{K} X_i^{(k)} X_j^{(k)} \quad (1.2)$$

In this case, the energy function that is minimized by the network can be expressed in the following terms:

$$E(\vec{V}) = -\frac{1}{2}\sum_{i=1}^{N}\sum_{j=1}^{N}\sum_{k=1}^{K} X_i^{(k)} X_j^{(k)} V_i V_j \quad (1.3)$$

In order to retrieve a pattern, once the learning phase has finished, the net is initialized with the known part of the pattern (called probe). Then, the dynamics makes the network converge to a stable state (due to the decrease of the energy function), corresponding to a local minimum. Usually this stable state is close to the initial probe.

If all input patterns form an orthogonal set, then they are correctly retrieved, otherwise some errors may happen in the recall procedure. But, as the dimension of the pattern space is N, there is no orthogonal set with cardinality greater than N. This implies that if the number of patterns exceed the number of neurons, errors may occur. Thus, capacity can not be greater than 1 in this model.

MREM Model with Semi-Parallel Dynamics

The Multivalued REcurrent Model (MREM) consists of a recurrent neural network formed by N neurons, where the state of each neuron i is defined by its output V_i (i=1,…,N), taking values in any finite set $\mathcal{M} = \{m_1, m_2, \ldots, m_L\}$. This set does not need to be numerical.

The state of the network, at time t, is given by a N-dimensional vector, $\mathbf{V}(t) = (V_1(t), V_2(t), \ldots, V_N(t)) \in \mathcal{M}^N$. Associated to every state vector, an energy function, characterizing the behaviour of the net, is defined:

$$E(V) = -\frac{1}{2}\sum_{i=1}^{N}\sum_{j=1}^{N} w_{i,j} f(V_i, V_j) + \sum_{i=1}^{N}\theta_i(V_i) \quad (2.1)$$

where $w_{i,j}$ is the weight of the connection from the j-th neuron to the i-th neuron, and $f : \mathcal{M} \times \mathcal{M} \to \mathbb{R}$ can be considered as a measure of similarity between the outputs of two neurons, usually verifying the following similarity conditions:

1. For all $x \in \mathcal{M}$, $f(x,x) = c \in \mathbb{R}$.
2. f is a symmetric function: for every $x,y \in \mathcal{M}$, $f(x,y) = f(y,x)$.
3. If $x \neq y$, then $f(x,y) \leq c$.

and $\theta_i : \mathcal{M} \to \mathbb{R}$ are the threshold functions. Since thresholds will not be used for content addressable memory, henceforth we will consider θ_i be the zero function for all $i \in I$.

The introduction of this similarity function provides, to the network, of a wide range of possibilities to represent different problems (Mérida et al., 2001) (Mérida et al., 2002). So, it leads to a better representation than other multi-valued models, like SOAR and MAREN (Erdem et al., 1996) (Ozturk et al., 1997), since in those models most of the information enclosed in the multi-valued representation is lost by the use of the sign function that only produces values in {-1,0,1}.

It is clear that MREM, using bipolar ($\mathcal{M}$ = {1,1}) or bi-valued ($\mathcal{M}$ = {0,1}) neurons, along with the similarity function given by $f(x, y) = xy$ and constant bias functions, reduces to Hopfield's model. So, this model can be considered a powerful generalization of Hopfield's model.

In every instant, the net evolves to reach a state of lower energy than the current one.

It has been proved that the MREM model with its associated dynamics always converges to a minimal state. This result is particularly important when dealing with combinatorial optimization problems, where the application of MREM has been very fruitful (López-Rodríguez et al, 2006) (Mérida-Casermeiro et al., 2001a) (Mérida-Casermeiro et al., 2001b) (Mérida-Casermeiro et al., 2002a) (Mérida-Casermeiro et al., 2002b) (Mérida-Casermeiro et al., 2003) (Mérida-Casermeiro et al., 2004) (Mérida-Casermeiro et al., 2005).

MREM as Auto-Associative Memory

Now, let $\{X^{(k)} : k = 1,\ldots,K\}$ be a set of patterns to be loaded into the neural network. Then, in order to store a pattern, $\mathbf{X}=(X_1,X_2,\ldots,X_N)$, components of the W matrix must be modified in order to make $\mathbf{X}$ the state of the network with minimal energy.

Since energy function is defined, we modify the components of matrix W in order to reduce the energy of state $V=\mathbf{X}$ by the rule:

$$\Delta w_{i,j} = -2\frac{\partial E}{\partial w_{i,j}} = f(X_i, X_j).$$

The coefficient 2 does not produce any effect on the storage of the patterns, and it is here chosen for simplicity. Considering that, at first, W=0, that is, all the states of the network have the same energy and adding over all the patterns, the next expression is obtained:

$$w_{i,j} = \sum_{k=1}^{K} f(X_i^{(k)}, X_j^{(k)}) \tag{3.1}$$

Equation is a generalization of *Hebb's postulate of learning*, because the weight $w_{i,j}$ between neurons is increased in correspondence with their similarity.

It must be pointed out that, when bipolar neurons and the product function are used, $f(x,y)=xy$, the well-known learning rule of patterns in the Hopfield's network is obtained. In fact, this is equivalent to choose $f(x,y)=1$ if $x=y$, and otherwise $f(x,y)=0$.

In what follows, we will consider the similarity function given by: $f(x, y)=1$ if $x=y$ and -1, otherwise.

In order to recover a loaded pattern, the network is initialized with the known part of that pattern. The network dynamics will converge to a stable state (due to the decreasing of the energy function), that is, a minimum of the energy function, and it will be the answer of the network. Usually this stable state is next to the initial one.

How to Avoid Spurious States

When a pattern X is loaded into the network, by modifying weight matrix W, not only the energy corresponding to state $\mathbf{V} = X$ is decreased. This fact can be explained in terms of the so-called associated vectors.

Given a state $\mathbf{V}$, its associated matrix is defined as $G_{\mathbf{V}} = (g_{i,j})$ such that $g_{i,j} = f(V_i, V_j)$.

Its associated vector is $A_{\mathbf{V}} = (a_k)$, with $a_{j+N(i-1)} = g_{i,j}$, that is, it is built by expanding the associated matrix as a vector of N^2 components.

With this notation, the energy function can be expressed as:

$$E(\mathbf{V}) = -\frac{1}{2}\sum_{k=1}^{K} < A_{X^{(k)}}, A_{\mathbf{V}} > \tag{3.2}$$

where $<\cdot,\cdot>$ denotes the usual inner product.

Lemma 1. The increment of energy of a state $\mathbf{V}$ when pattern X is loaded into the network, by using Equation , is given by:

$$\Delta E(\mathbf{V}) = -\frac{1}{2} < A_X, A_{\mathbf{V}} > \tag{3.3}$$

Lemma 2. Given a state vector $\mathbf{v}$, we have $A_{\mathbf{V}} = A_{-\mathbf{V}}$. So $E(\mathbf{V}) = E(-\mathbf{V})$.

These two results explain why spurious patterns are loaded into the network. Let us see it with an example:

Suppose that we want to get pattern $X=(-1,1,-1,-1,1)$ loaded into a BH network. Then, its associated matrix will be:

$$G_X=\begin{pmatrix} 1 & -1 & 1 & 1 & -1 \\ -1 & 1 & -1 & -1 & 1 \\ 1 & -1 & 1 & 1 & -1 \\ 1 & -1 & 1 & 1 & -1 \\ -1 & 1 & -1 & -1 & 1 \end{pmatrix}$$

and therefore the associated vector will be:

$$A_X=(1,-1,1,1,-1,-1,1,-1,-1,1,1,\\ -1,1,1,-1,1,-1,1,1,-1,-1,1,-1,-1,1)$$

But, as can be easily verified, the vector $-X=(1,-1,1,1,-1)$ has the same associated matrix and vector than the original pattern X, that is, $A_{-X}=A_X$. By using the previous lemmas, we obtain that the increment of energy of both X and $-X$ is the same, so these two vectors are those whose energy decreases most:

$$\Delta E(X)=-\frac{1}{2}<A_X,A_X>=\\ -\frac{1}{2}<A_X,A_{-X}>=\Delta E(-X)$$

This fact also implies that the corresponding ΔW is the same for both vectors.

These results explain the well-known problem of loading the opposite pattern of Hopfield's associative memory.

When using MREM, spurious patterns are generated by the network in the same way. For example, when we load the pattern $X=(3,3,2,1,4,2)$, also the pattern $X_1=(4,4,3,2,1,3)$ is loaded, but also $X_2=(1,1,4,3,2,4)$, since all of them have the same associated vector, and produce the same decrease in the energy function. So, in MREM, the number of spurious patterns appearing after the load of a vector into the net is greater than the corresponding in BH.

Since all associated vectors are vectors of N^2 components taking value in $\{-1, 1\}$, their norms are equal, $\| A_{\mathbf{V}} \|_E = N$ for all **v**. This result implies that what is actually stored in the network is the orientation of the vectors associated to loaded patterns.

From the above expression for the increment of energy, and using that components of associated vectors are either -1 or 1, the following expression for the decrease of energy when a pattern is loaded is obtained:

$$-\Delta E(\mathbf{V})=\frac{1}{2}(N-2d_H(\mathbf{V},X))^2 \tag{3.4}$$

where $d_H(\mathbf{V},X)$ is the Hamming distance between vectors **v** and X.

After this explanation, we propose a solution for this problem:

The augmented pattern $\hat{X}$, associated to X, is defined by appending to X the possible values of its components, that is, if $\mathcal{M}=\{m_1,\ldots,m_L\}$, then $\hat{X}=(X_1,\ldots,X_N,m_1,\ldots,m_L\}$.

Particularly:

- In case of bipolar outputs, $\mathcal{M}=\{-1,1\}$, and consequently it is $\hat{X}=(X_1,\ldots,X_N,-1,1)$.
- If $\mathcal{M}=\{1,\ldots,L\}$, then $\hat{X}=(X_1,\ldots,X_N,1,2,\ldots,L)$.

By making use of augmented patterns, the problem of spurious patterns is solved, as stated in the next result:

Lemma 3. The function Ψ that associates an augmented pattern to its corresponding associated vector is injective.

It can be shown that if augmented patterns are used, the state **V** whose energy decreases most when pattern X is introduced in the net, is $\mathbf{V} = X$. This result is deduced from Equation , applied to the case of $N + L$ components:

$$-\Delta E(\hat{\mathbf{V}}) = \frac{1}{2}(N + L - 2d_H(\hat{\mathbf{V}}, \hat{X}))^2$$

So, if $\mathbf{V} \neq X$ then $\hat{\mathbf{V}} \neq \hat{X}$ and $1 \leq d_H(\mathbf{V}, X) = d_H(\hat{\mathbf{V}}, \hat{X}) \leq N$, and the next inequality holds:

$$L - N = N + L - 2N \leq N + L$$
$$-2d_H(\hat{\mathbf{V}}, \hat{X}) \leq N + L - 2$$

Therefore

$$-2\Delta E(\hat{\mathbf{V}}) = (N + L - 2d_H(\hat{\mathbf{V}}, \hat{X}))^2 \leq \max\{(N-L)^2, (N+L-2)^2\} =$$
$$= (N + L - 2)^2 < (N + L)^2 = -2\Delta E(\hat{X})$$

which demonstrates our statement.

Then, in order to load a pattern X, it will suffice to load its augmented version, which will be the unique state maximizing the decrease of energy.

In this example, we can see how this method works. Consider, at first, that W=0, that is $E(V)$=0 for all state vector V. Then, in the original model MREM, the pattern X=(3,3,2,1,4,2) is loaded, and matrix W is updated. Note that, in this case, N=6 and L=4. Then, if Y=(4,4,3,2,1,3), we can compute:

$$E(X) = -\frac{1}{2}(N - 2d_H(X, X))^2 = -\frac{1}{2}6^2 = -18$$
$$E(Y) = -\frac{1}{2}(N - 2d_H(Y, X))^2 = -\frac{1}{2}(6 - 2 \cdot 6)^2 = -\frac{1}{2}(-6)^2 = -18$$

Therefore, Y has been also loaded into the network, since X is one global minimum of the energy function, and $E(Y)$=$E(X)$. With the original model, Y is a spurious pattern. Let us apply the technique of augmented patterns to solve this problem. In this case, the augmented patterns are: $\widehat{X} = (3,3,2,1,4,2,1,2,3,4)$ and $\widehat{Y} = (4,4,3,2,1,3,1,2,3,4)$. We can now compute the energy value associated to X and Y:

$$E(X) = -\frac{1}{2}(N + L - 2d_H(\widehat{X}, \widehat{X}))^2 = -\frac{1}{2}(6 + 4 - 0)^2 = -50$$
$$E(Y) = -\frac{1}{2}(N + L - 2d_H(\widehat{Y}, \widehat{X}))^2 = -\frac{1}{2}(6 + 4 - 2 \cdot 6)^2 = -2$$

This result ($E(X)$<$E(Y)$) implies that Y is not stored in the net, since it is not a minimum of the energy function. So, this technique is able to avoid the apparition of spurious patterns.

It must be noted that it will only be necessary to consider N neurons, their weights, and the weights corresponding to the last L neurons, that remain fixed, and do not need to be implemented.

SOME REMARKS ON THE CAPACITY OF THE NET

In Mérida et al. (2002), authors find an expression for the capacity parameter α for MREM model in terms of the number of neurons N and the number of possible states for each neuron, L, for the case in which N is big enough to apply the Central Limit Theorem ($N \geq 30$):

$$\alpha(N, L) \approx \frac{1}{N} + \frac{\frac{A^2}{z_\alpha^2} - B}{NC} \qquad (4.1)$$

where

$$A = N + 3 + (N-1)\frac{4-L}{L}, \quad B = 8(N-1)\frac{L-2}{L^2}, \quad C = \frac{8N}{L}$$

and z_α is obtained by imposing the condition that the maximum allowed error probability in retrieving patterns is p_{error}. For p_{error}=0.01, we get $z_\alpha \approx 2.326$.

Some facts can be extracted from the above expression.

For a fixed number of neurons, capacity is not bounded above:

Suppose N fixed. Equation (4.2) can be rewritten in the following form:

$$\alpha(N,L) \approx \frac{1}{N^2 z_\alpha^2}\left(2L + 4(N-1) + z_\alpha^2 + 2(N-1)\frac{N-1+z_\alpha^2}{L}\right) \quad (4.2)$$

If we make L tend to ∞, we get $\lim_{L\to\infty} \alpha(N,L) = \infty$, since the coefficient of L in this expression is positive.

What actually happens is that $\alpha(N,\cdot)$, as a function of L, has a minimum at the point $L_0(N) = \sqrt{(N-1)(N-1+z_\alpha^2)} \approx N$ for z_α=2.326. It is a decreasing function for $L<L_0(N)$ and increasing for $L\geq L_0(N)$.

One consequence of this result is that, for appropriate choice of N and L, the capacity of the net can be $\alpha(N,L)>1$.

This fact can be interpreted as a adequate representation of the multi-valued information, because, to represent the same patterns as MREM with N and L fixed, BH needs NL binary neurons and therefore the maximum number of stored patterns may be greater than N. So it is not a strange thing that the capacity can reach values greater than 1, if the patterns are multi-valued, MREM needs much less neurons to represent the pattern than BH.

For a fixed number of possible outputs, capacity is bounded below by a positive constant:

Suppose L is fixed. Equation can be rewritten as follows:

$$\alpha(N,L) \approx \frac{1}{L z_\alpha^2}\left(2 + \frac{4(L-1)+2z_\alpha^2}{N} + \frac{2(L-1)^2 + (L-2)z_\alpha^2}{N^2}\right) \quad (4.3)$$

It can be easily seen that this expression represents a function whose value decreases as N grows. So, a net with more neurons than other, and the same possible states, will present less capacity than the second one.

Thus, a minimum positive capacity can be computed for each possible value of L, verifying $\alpha_{min}(L) = \lim_{N\to\infty} \alpha(N,L) = \frac{2}{L z_\alpha^2} > 0$.

$\alpha_{min}(L)$ coincides with the asymptotic capacity for the net with L possible neuron outputs. For example, if L=2 (as in BH), an asymptotic capacity of $\alpha_{min}(2)$=0.1847 is obtained, exactly the capacity for BH provided in other works (Hertz et al., 1991).

Figure 1. Many individual patterns are loaded into the net, forming a group of local minima of the energy function (left). When the number of patterns to be loaded is greater than the capacity of the network, the corresponding local minima are merged (right). The formation of these new local optima can be interpreted as the apparition and learning of the associated concepts.

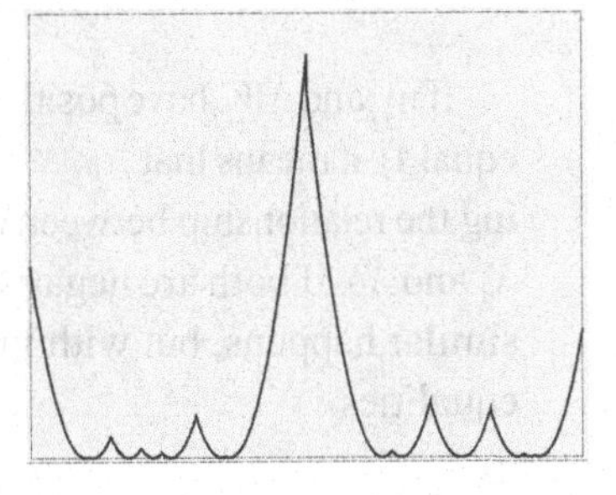

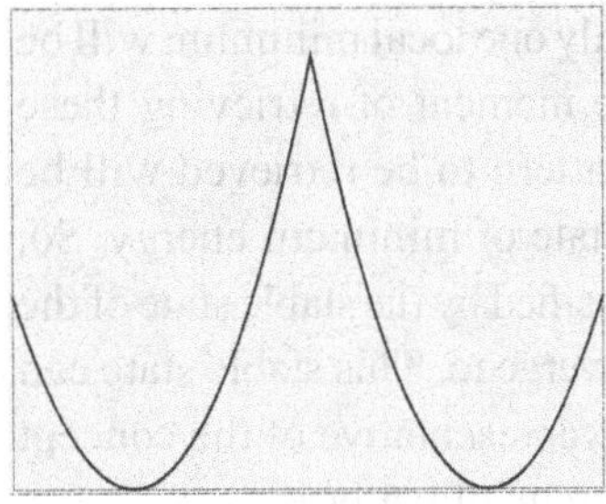

WHEN CAPACITY IS EXCEEDED

This work tries to explain what may happen psychologically in the human brain. When a reduced number of patterns has to be memorized, the brain is able to remember all of them when necessary. Similarly, when the capacity of the net is not exceeded, the net is able to retrieve exactly the same patterns that were loaded into it. But when the brain receives a great amount of data to be recognized or classified, it distinguishes between some groups of data (in an unsupervised way) and thus forming concepts. This kind of behaviour is also simulated by neural networks, as we will show next.

Then, learning rules as Hebb's (or the more general given by equation (3.1), where connection between neurons is reinforced by the similarity of their expected outputs), may produce classifiers that discover some knowledge from the input patterns, like the actual number of groups in which the data are divided. Then, an unsupervised clustering of the input pattern space is automatically performed. This unsupervised clustering generates the concept space, formed by the equivalence classes found for the input patterns.

If a pattern, say X, is to be loaded in the net, by applying equation (3.1), a local minimum of the energy function E is created at $V=X$. If another pattern X' is apart from X, its load will create another local minimum. But, if X and X' are close each other, these two local minima created by the learning rule will be merged, forming one local minima instead.

Then, if a group of patterns is loaded into the net (overflowing its capacity), and all of them are close each other, only one local minimum will be formed, and at the moment of retrieving these data, the unique pattern to be retrieved will be associated to the state of minimum energy. So, patterns can be classified by the stable state of the net which they converge to. This stable state can be considered as a representative of the concept associated to that group of patterns.

This way, the learning of new concepts corresponds to the saturation of individuals, that is, the only way to learn a general concept is by presenting to the learner (the network, or the human brain, both cases have the same behaviour) a great number of individuals satisfying that concept.

For example, if a boy has to learn the concept 'chair', he will be presented a series of chairs, of many styles and sizes, until he is able to recognize a chair, and distinguish it from a table or a bed.

RELATIONSHIPS LEARNING

Equation for the learning rule, generalization of Hebb's one, shows that the only thing taking part in updating weight matrix is the pattern to be loaded into the net at that time. So, it represents a very 'local' information, and does not take account of the possible relationships that pattern could have with the already stored ones. So, it is convenient to introduce an additional mechanism in the learning phase, such that the information concerning to relationships between patterns is incorporated in the update of the weight matrix. In what follows, we will consider that the similarity function is $f(x,y) = 2\delta_{x,y} - 1$, that is, its value is 1 if $x = y$ and -1 otherwise.

Relationships learning method: Suppose that we have the (augmented) pattern X_1 stored in the net. So, we have the weight matrix $W = (w_{i,j})$. If pattern X_2 is to be loaded into the network, by applying equation (3.1), components of matrix ΔW are obtained.

If $w_{i,j}$ and $\Delta W_{i,j}$ have positive signum (both values equal 1), it means that $X_{1i} = X_{1j}$ and $X_{2i} = X_{2j}$, indicating the relationship between components i and j of X_1 and X_2. If both are negative valued, something similar happens, but with inequalities instead of equalities.

Figure 2. Scheme of the relationship between RL and overflowing network capacity

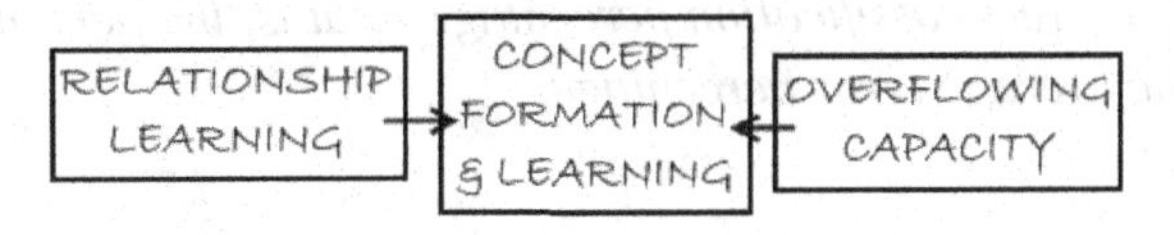

Figure 3. Scheme for the formation and learning of concepts

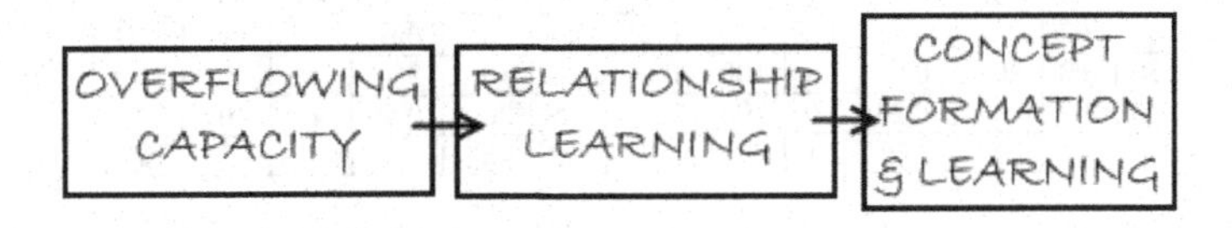

So, the fact of $w_{i,j}$ and $\Delta W_{i,j}$ having the same signum is a clue of a relationship that is repeated between components i and j of patterns X_1 and X_2. In order to reinforce the learning of this relationship, we propose a novel technique, presenting also another kind of desirable behavior: The model proposed before, given by equation (3.1), is totally independent of the order in which patterns are presented to the net. This fact does not actually happen in the human brain, since every new information is analyzed and compared to data and concepts previously learned and stored.

So, to simulate this kind of learning, a method named RL is presented:

Let us multiply by a constant, $\beta > 1$, the components of matrices W and ΔW where the equality of signum is verified, i.e., the components verifying $w_{i,j} \cdot \Delta W_{i,j} > 0$. Hence the weight matrix learned by the network is, after loading pattern X_2:

$$w'_{i,j} = \begin{cases} w_{i,j} + \Delta W_{i,j} & \text{if } w_{i,j} \cdot \Delta W_{i,j} < 0 \\ \beta[w_{i,j} + \Delta W_{i,j}] & \text{if } w_{i,j} \cdot \Delta W_{i,j} > 0 \end{cases} \tag{5.1}$$

Similarly, if there are some patterns $\{X_1, X_2, \ldots, X_R\}$ already stored in the network, in terms of matrix W, and pattern X_{R+1} is to be loaded, matrix ΔW (corresponding to X_{R+1}) is computed and then the new learning rule given by equation (5.1) is applied.

It must be noted that this method satisfy Hebb's postulate of learning quoted before.

This learning reinforcement technique, RL, has the advantage that it is also possible to learn patterns one by one or by blocks, by analyzing at a time a whole set of patterns, and comparing the resulting ΔW to the already stored in the net. Then, for instance, if $\{X_1, \ldots, X_R\}$ has already been loaded into the net in terms of matrix W, we can load a whole set $\{Y_1, \ldots, Y_M\}$ by computing $\Delta W = (\sum_{k=1}^{M} f(y_{ki}, y_{kj}))_{i,j}$ and then applying equation (5.1).

As shown in Figure 2, RL and the capacity of the network are related via their ability to make the learner (the neural network, or the human brain, whatever the case) form and learn concepts.

With this idea in mind, one can think that what the capacity overflow actually induces is the learning of the existing relationships between patterns: as explained in Figure 1, several local minima, representing stored patterns, are merged when network capacity is overflowed. This implies a better recognition of the relationship between similar patterns, which usually represent the same

Table 1. Average clustering results on 10 runs of these algorithms, where n_a indicates the number of obtained clusters, p_{cc} is the correct cassification percentage (that is, the percentage of simulations in which $n_a = n$) and Err. is the average error percentage.

K	n=3						n=4					
	RL			IRL			RL			IRL		
	n_a	p_{cc}	Err.	n_a	p_{cc}	Err.	n_a	p_{cc}	Err.	n_a	p_{cc}	Err.
75	3.1	90	0.53	3.1	90	0.53	4.7	60	3.46	4.1	90	0.26
150	3.3	70	0.33	3.1	90	0.13	4.8	70	1.06	4.4	80	0.73
300	3.7	70	0.60	3.3	80	0.43	5.5	30	3.06	4.6	60	4.80
450	3.5	70	0.37	3.1	90	0.17	4.9	60	0.55	4.4	80	0.31
600	3.2	80	0.25	3.2	80	0.25	5.4	50	0.61	4.6	50	0.31
750	3.3	80	0.06	3.0	100	0.00	5.4	30	3.49	4.6	60	3.12
900	3.2	90	3.23	3.0	100	0.00	5.5	20	0.56	4.8	40	0.42
Av.	3.3	78	0.76	3.1	90	0.21	5.2	46	1.82	4.5	66	1.42

concept. This concept is thus formed and learned. So, the process of formation and learning concepts can be seen in Figure 3.

Iterative relationships learning method: RL can be improved in many ways. In this sense, an iterative approach, IRL, to enhance the solution given by RL, is presented.

Suppose that, by using equation (5.1) of RL, matrix W_X related to pattern set $X = \{X^{(k)} : k \in \mathcal{K}\}$ has been learned and denote by $Y^{(k)}$ the stable state reached by the network (with weight matrix W_X) when beginning from the initial state given by $\mathbf{V} = X^{(k)}$.

Then, the cardinal of $\{Y^{(k)} : k \in \mathcal{K}\}$ is (no multiplicities) the number of classes that RL finds, n_a.

$Y := \{Y^{(k)} : k \in \mathcal{K}\}$ can be considered (with all multiplicities included) as a new pattern set, formed by a noiseless version of patterns in X. So, if applying a second time RL to Y, by using equation (5.1) to build a new matrix W_Y, better results are expected than in the first iteration, since the algorithm is working with a more refined pattern set.

This whole process can be repeated iteratively until a given stop criterion is satisfied. For example, when two consecutive classifications assign each pattern to the same cluster.

SIMULATIONS

a. In order to show the ability of RL and IRL to perform a clustering task, as mentioned above, several simulations have been made whose purpose is the clustering of discrete data.

Several datasets have been created, each of them formed by K 50-dimensional patterns randomly generated around n centroids, whose components were integers in the interval [1,10]. That is, the n centroids were first generated and input patterns were formed from them by introducing some random noise modifying one component of the centroid with probability 0.018. So, the Hamming distance between input patterns and the corresponding centroids is a binomial distribution B(50,0.018). Patterns are equally distributed among the n clusters. It must be noted that patterns may have Hamming distance even 5 or 6 from their respective centroid, and new clusters can be formed by this kind of patterns.

So, a network with $N = 50$ neurons taking value in the set $\mathcal{M} = \{1,\ldots,10\}$ has been considered. The parameter of learning reinforcement has been chosen $\beta = 1.5$. It has been observed that similar results are obtained for a wide range of values of β.

The results obtained in our experiments are shown in Table 1. It can be observed not only the

Figure 4. The cameraman image

Figure 5. The resulting image after performing the unsupervised clustering

low classification error (from 0% to 4.80% on average), but in addition these new techniques get the exact, or very approximate, number of groups in which the pattern set is actually divided in almost every simulation. In fact, whenever the number n_a of discovered clusters equals n, an error percentage of 0% is obtained, retrieving in those cases the initial centroids. It can also be verified that IRL clearly outperforms RL in most cases, getting a more accurate classification and improving the estimation of the number of clusters, as seen in the last row of the table.

b. A more practical example is given below. In this case, it has been studied how, by overflowing the network capacity, the net is able to extract relationships between patterns and form clusters or groups representing concepts.

Here, the problem is to cluster an image, attending to its contents. It's a rather simple example, but it has many applications in image segmentation and understanding. For more details on image processing techniques, see Egmont-Petersen et al. (2002) and references therein.

In our case, the well-known cameraman image (Figure 4) has been used as benchmark.

This image (256x256 pixels) was divided in windows of size 8x8 pixels. Each window represents a pattern, rearranging its values columnwise from top to bottom. So, there were 1024 patterns of dimension 64.

These patterns were loaded into a MREM network with 64 neurons. Then, the net was initialized with every pattern, and the corresponding stable state (a new window) was used to substitute the original window. The resulting clustering can be viewed in Figure 5. It should be noted that there were 14 different clusters, detected automatically.

The stable states achieved by the net correspond to concepts. Since pattern components were gray values of pixels, the network has learnt concepts related to these graylevels.

The interesting point is that all this work has been done in an unsupervised way. Other clustering methods need to adjust the number of clusters before the training, and are based on Euclidean distance, which implies that patterns are embedded in the Euclidean space. However, discrete patterns like the ones described in this work can be non-numerical (qualitative) and may have a different topology from the usual one in the Euclidean space.

FUTURE TRENDS

Recently, concept learning has received much attention from the Artificial Intelligence community. There are sectors of this community interested in using techniques such as fuzzy and classical logics to study this process.

Regarding associative memories and neural networks, the current trend is to adapt or develop learning rules in order to increase the capacity of the network.

CONCLUSION

In this work, we have explained that the limitation in capacity of storing patterns in a recurrent network has not to be considered as determinant, but it can be used for the unsupervised classification of discrete patterns, acting as a concept learning system.

The neural model MREM, based on multivalued neurons, has been developed, as a generalization of the discrete Hopfield model and as an auto-associative memory, improving some of the undesirable aspects of the original Hopfield model: some methods and results for the net not to store spurious patterns in the learning phase have been shown, reducing so the number of local minima of the energy function not associated to an input pattern.

By applying a slight modification to Hebb's learning rule, a mechanism to reinforce the learning of the relationships between different patterns has been introduced to the first part of the process, incorporating knowledge corresponding to several patterns simultaneously. This mechanism can be repeated iteratively to enhance the relationship learning procedure.

One of the main facts expressed in this work is that network capacity, which can be viewed as a restriction for the network as associative memory, becomes a powerful ally when forming and learning concepts (groups of similar patterns), since it implies the learning of relationships between patterns mentioned above.

In addition, simulations confirm the idea expressed in this work, since, by overflowing the capacity of the net, we can get optimal results of classification in many cases. This technique is therefore very useful when tackling classification and learning problems, with the advantage of being unsupervised.

FUTURE RESEARCH DIRECTIONS

This chapter presents a new open research line, from the fact that some new modifications of the learning rule, reinforcing other aspects of the relationships among patterns, may be developed.

An interesting point to be studied is the possible generalization of Hebb's learning rule such that the capacity of a neural network increases, since it is crucial for many applications.

REFERENCES

Athithan, G., & Dasgupta, C. (1997). *On the Problem of Spurious Patterns in Neural Associative Models*, IEEE Transactions on Neural Networks, vol. 8, no. 6, 1483-1491.

Crick, F., & Mitchinson, G. (1983). *The Function of Dream Sleep*, Nature, vol. 304, 111-114.

Egmont-Petersen, M., de Ridder, D., & Handels, H. (2002). *Image processing with neural networks—a review*, Pattern Recognition, vol. 35, 2279-2301.

Erdem, M. H., & Ozturk, Y. (1996). *A New family of Multivalued Networks*, Neural Networks 9,6, 979-89.

Hebb, D. O. (1949). *The Organization of Behavior.* New York, Wiley.

Hemmen, J. L. van, & Kuhn, R. (1991). *Collective Phenomena in Neural Networks*, E. Domany, J. L. van Hemmen and K. Shulten, eds. Berlin: Springer-Verlag, 1-105.

Hertz, J., Krogh, A., & Palmer, R. G. (1991). *Introduction to the Theory of Neural Computation.* Lecture Notes, Volume 1, Addison Wesley.

Hertz, J. A., Grinstein, G., & Solla, S. A. (1987). *Heidelberg Colloquium on Glassy Dynamics*, J. L. van Hemmen and I. Morgenstern, eds. Berlin: Springer-Verlag, 538-546.

Ho, C. Y., Sasase, I. and Mori, S. (1992). On *the Capacity of the Hopfield Associative Memory.* In Proceedings of IJCNN 1992, II196-II201.

Hopfield, J. J. (1982). *Neural Networks and Physical Systems with Emergent Collective Computational Abilities.* In Proceedings of the National Academy of Science, USA, 79, 2554-2558.

Hopfield, J. J. (1984). *Neurons with graded response have collective computational properties like those of two-state neurons*, Proceedings of the National Academy of Sciences USA, 81, 3088-3092.

Kuh, A., & Dickinson, B. W. (1989). *Information Capacity of Associative Memory.* IEEE Transactions on Information Theory, vol. IT-35, 59-68.

López-Rodríguez, D., Mérida-Casermeiro, E., Ortiz-de-Lazcano-Lobato, J. M., & López-Rubio, E. (2006). *Image Compression by Vector Quantization with Recurrent Discrete Networks.* Lecture Notes in Computer Science, 4132, 595-605.

McEliece, R. J., Posner, E. C., Rodemich, E. R., & Venkatesh, S. S. (1990). *The Capacity of the Hopfield Associative Memory.* IEEE Transactions on Information Theory, vol. IT-33, no. 4, 461-482.

Mérida-Casermeiro, E., Muñoz-Pérez, J., & Benítez-Rochel, R. (2001). *A recurrent multivalued neural network for the N-queens problem,* Lecture Notes in Computer Science 2084, 522-529.

Mérida-Casermeiro, E., Galán-Marín, G., & Muñoz-Pérez, J. (2001). *An Efficient Multivalued Hopfield Network for the Travelling Salesman Problem.* Neural Processing Letters, 14:203-216.

Mérida-Casermeiro, E., & Muñoz-Pérez, J. (2002). *MREM: An Associative Autonomous Recurrent Network.* Journal of Intelligent and Fuzzy Systems, 12 (3-4), 163-173.

Mérida Casermeiro, E., Muñoz-Pérez, J. & García-Bernal, M.A. (2002). *An Associative Multivalued Recurrent Network*, IBERAMIA 2002, 509-518.

Mérida-Casermeiro, E., Muñoz-Pérez, J., & Domínguez-Merino, E. (2003). *An N-parallel Multivalued Network: Applications to the Travelling Salesman Problem.* Computational Methods in Neural Modelling, Lecture Notes in Computer Science, 2686, 406-413.

Mérida-Casermeiro, E., & López-Rodríguez, D. (2004). *Multivalued Neural Network for Graph MaxCut Problem*, ICCMSE, 1, 375-378.

Mérida-Casermeiro, E., & López-Rodríguez, D. (2005). *Graph Partitioning via Recurrent Multivalued Neural Networks.* Lecture Notes in Computer Science, 3512:1149-1156.

Ozturk, Y., & Abut, H. (1997). *System of associative relationships (SOAR)*, In Proceedings of ASILOMAR.

Parisi, G. (1986). *Asymmetric Neural Networks and the Process of Learning*, J. Phys. A: Math. and Gen., vol 19, L675-L680.

Singh, M. P., Chengxiang, Z., & Dasgupta, C. (1995). *Analytic Study of the Effects of Synaptic Asymmetry*, Phys. Rev. E, vol. 52, 5261-5272.

Treves, A. & Amit, D. J. (1988). *Metastable States in Asymmetrically Diluted Hopfield Networks*, J. Phys A: Math. and Gen., vol. 21, 3155-3169.

ADDITIONAL READING

Abu-Mustafa, Y.S. & St. Jacques, J-M. (1985). *Information capacity of the Hopfield model*, IEEE Trans. on Information Theory, Vol. IT-31, No. 4, 461-464.

Amari, S. (1977). *Neural theory of association and concept formation*, Biological Cybernetics, vol. 26, pp. 175-185.

Amit, D. J. (1989). *Modeling Brain Functions. Cambridge*, U.K.: Cambridge Univ. Press.

Amit, D.J., Gutfreund, H., & Sompolinsky, H. (1987). *Statistical mechanics of neural networks near saturation*, Annals Physics, New York, vol. 173, 30–67.

Athithan, G. (1995). *Associative storage of complex sequences in recurrent neural networks*, in Proceedings IEEE International Conference on Neural Networks, vol. 4, 1971–1976.

Athithan, G. (1995). *A comparative study of two learning rules*, Pramana Journal Physics, vol. 45, no. 6, pp. 569–582.

Athithan, G. (1995). *Neural-network models for spatio-temporal associative memory*, Current Science, 68(9), 917–929.

Bowsher, D. (1979). *Introduction to the Anatomy and Physiology of the Nervous System*, 4th ed. Oxford, U.K.: Blackwell, 31.

Cernuschi-Was, B.(1989). *Partial Simultaneous Updating in Hopfield Memories*, IEEE Trans. on Systems, Man and Cybernetics, Vol. 19, No. 4, 887-888.

Chuan Kuo, I. (1994). *Capacity of Associative Memory*, Ph.D. Thesis, University of Southern California.

Eccles, J. G. (1953). *The Neurophysiological Basis of Mind*. Oxford: Clarendon.

Forrest, B. M., & Wallace, D. J. (1991). *Storage capacity of learning in Ising-spin neural networks*, in Models of Neural Networks, E. Domany, J. L. van Hemmen, and K. Shulten, Eds. Berlin: Springer-Verlag, 121–148.

Gardner, E. (1988). *The space of interactions in neural-network models*, Journal Physics A: Math. and Gen., vol. 21, pp. 257–270.

Gross, D. J., & Mezard, M. (1984). *The simplest spin glass*, Nuclear Phys., vol. B 240 IFS121. DD. 431-452.

Grossberg, S. (1982). *Studies of Mind and Brain*. Boston: Reidel.

Grossberg, S. (ed.), (1986). *The Adaptive Brain*; Vol. I: Cognition Learning, Reinforcement, and Rhythm: Vol. II: Vision. Speech, Language, and Motor Control. Amsterdam, The Netherlands, North-Holland.

Guyon, I., Personnaz, L., Nadal, J.P., & Dreyfus, G. (1988). *Storage and retrieval of complex sequences in neural networks*, Physics Review A, vol. 38, p. 6365.

Hassoun, M.H., & Watta, P.B. (1995). *Alternatives to energy function-based analysis of recurrent neural networks*, in Computational Intelligence: A Dynamical System Perspective, M. Palaniswami et al., Eds. New York: IEEE Press, pp. 46–67.

Hertz, J., Krogh, A., & Palmer, R. G. (1991). *Introduction to the Theory of Neural Computation*. Reading, MA: Addison-Wesley, 40.

Herz, A. V. M., Li, Z., & van Hemmen, J. L. (1991). *Statistical mechanics of temporal association in neural networks with transmission delays*, Physics Review Letters, vol. 66, no. 10, p. 1370.

Hinton, G. E., & Anderson, J. A. (eds.), (1981). *Parallel Models of Associative Memory*, Hillsdale, NJ: Erlbaum.

Hopfield, J. J., & Tank, D. W. (1985). *Neural computation of decisions in optimization problems*, Biological Cybernetics, vol. 52, 141-152.

Kohonen, T. (1977). *Associative Memory: A System-Theoretic Approach*. Berlin: Springer-Verlag.

Krauth, W., & Mezard, M., (1987). *Learning algorithms with optimal stability in neural networks*, Journal Physics A: Math. and Gen., vol. 20, L745–L752.

Lee, B. W., & Sheu, B. J. (1991). *Modified Hopfield neural networks for retrieving the optimal solution*, IEEE Transactions on Neural Networks, vol. 2, p. 137.

Little, W. A., & Shaw, G. L. (1978). *Analytic study of the memory storage capacity of a neural network*, Math. Biosci., vol. 39, 281-290.

Little, W. A. (1974). *The existence of persistent states in the brain*, Muth. Biosci., vol. 19, pp. 101-120.

Loève, M. (1977). *Probability Theory*, Vol. I, Springer-Verlag.

Marr, D. (1969). *A theory of cerebellar cortex*, Journal Physiology, vol. 202, p. 437.

McEliece, R. J. (1977). *The Theory of Information and Coding*, vol. 3 of Encyclopedia of Mathematics and Its Application. Reading, MA: Addison-Wesley.

McEliece, R.J., & Posner, E. C. (1985). *The number of stable points of an infinite-range spin glass memory*, Telecommunications and Data Acquisition Progress Report, vol. 42-83, Jet Propulsion Lab., California Inst. Technol. Pasadena, 209-215.

Nakano, K. (1972). *Associatron-A model of associative memory*, IEEE Transactions on Systems, Man and Cybernetics, vol. SMC-2, 380-388.

Noback, C. R.. & Demarest, R. J. (1975). *The Human Nervous System: Basic Principles of Neurobiology*. New York: McGraw-Hill, 87.

Palm, G. (1980). *On associative memory*, Biol. Cybern., vol. 36, 19-31.

Parisi, G. (1986). *Asymmetric neural networks and the process of learning*, Journal Physics A: Math. and Gen., vol. 19, pp. L675–L680.

Psaltis, D., Hoon Park, C., & Hong, J. (1988). *High order memories and their optical implementation*, Neural Networks, vol. 1, 149-163.

Smetanin, Y. (1994). *A Las Vegas method of region-of-attraction enlargement in neural networks*, in Proceedings 5th International Workshop Image Processing Computer Opt. (DIP-94), SPIE, vol. 2363, 77–81.

Tank, D. W., & Hopfield, J. J. (1987). *Collective computation in neuron like circuits*, Scientific American, pp. 62–70.

Tank, D. W., & Hopfield, J. J. (1986). *Simple optimization networks: An A/D converter and a linear programming circuit*, IEEE Transactions on Circuits Systems. vol. CAS-33. DD. 533-541.

Venkataraman, G., & Athithan, G. (1991). *Spin glass, the travelling salesman problem, neural networks, and all that*, Pramana Journal of Physics, vol. 36, no. 1, 1–77.

Verleysen, M., Sirletti, B., Vandemeulebroecke, A., & Jespers, P. G. A. (1989). *A high-storage capacity content-addressable memory and its learning algorithm*, IEEE Transactions on Circuits Systems, vol. 36, p. 762.

Wozencraft, J. M., & Jacobs, I. M. (1965). *Principles of Communication Engineering*. New York: Wiley.

Chapter IX
Artificial Cell Systems Based in Gene Expression Protein Effects

Enrique Fernández-Blanco
University of A Coruña, Spain

Julian Dorado
University of A Coruña, Spain

Nieves Pedreira
University of A Coruña, Spain

ABSTRACT

The artificial embryogeny term overlaps all the models that try to adapt cellular properties into artificial models. This chapter presents a new model for artificial embryogeny that mimics the behaviour of biological cells, whose characteristics can be applied to solution of computational problems. The paper contains the theoretical development of the model and some test executed in an implementation of that model. The presented tests apply the model to simple structure generation and provide promising results with regard to its behaviour and applicability to more complex problems. The objective of the chapters is to be an introduction of the artificial embryogeny and shows an example of a model of these techniques.

INTRODUCTION

Use biology as inspiration for the creation of computational models is not a new idea: science has already been the basis for the famous artificial neuron models (Hassoun, 1995), the genetic algorithms (Holland, 1975), etc. The cells of a biological organism are able to compose very complex structures from a unique cell, the zygote, with no need for centralized control (Watson & Crick, 1953). The cells can perform such process thanks to the existence of a general plan, encoded in the DNA for the development and functioning of the system. Another interesting characteristic of natural cells is that they form systems that are tolerant to partial failures: small errors do not induce a global collapse of the system. Finally, the tissues that are composed by biological cells present parallel information processing for the coordination of tissue functioning in each and every cell that composes this tissue. All these characteristics are very interesting from a computational viewpoint.

Another point of view is to study the biological model as a design model. At present human designs use a top-down view, this methodology has served well. However thinking on the construction of software and hardware systems with a high number of elements, the design crisis is served. Verify formally the systems when interactions and possible states grows, becomes near impossible due the combinatorial explosion of configuration using a traditional way. Living systems suggest interesting solutions for these problems, such as that the information defining the organism is contained within each part. Consequently, if the designers want to increase the complexity of the systems, one way is to study the biological model trying to mimic its solutions.

This paper presents the development of a model that tries to emulate the biological cells and to take advantage of some of their characteristics by trying to adapt them to artificial cells. The model is based on a set of techniques known as *Artificial Embryogeny* (Stanley & Miikkulainen, 2003) or *Computational Embryology* (Kumar, 2004).

BACKGROUND

The Evoluationary Computation (EC) field has given rise to a set of models that are grouped under the name of Artifial Embryogeny (AE), first introduced by Stanley and Mikkulainnen (Stanley & Miikkulainen, 2003). This group refers to all the models that try to apply certain characteristics of biological embryonic cells to computer problem solving, i.c. self-organisation, failure tolerance, and parallel information processing.

The work on AE has two points of view. On the one hand could be found the grammatical models based on L-systems (Lindenmayer, 1968) which do a top-down approach to the problem. On the other hand could be found the chemical models based on the Turing's ideas (Turing, 1952) which do a down-top approach.

On the last one, the starting point of this field could be found in the modelling of gene regulatory networks, performed by Kauffmann in 1969 (Kauffman, 1969). After that work, several develops were carried out on subjects such as the generation of complex behaviour by the differential expression of certain genes. This behaviour causes a cascade influence on the expressions of others genes (Mjolsness, Sharp & Reinitz, 1995).

The work performed by the scientific community can be divided into two main branches. The more theoretical branch uses the emulation of cell capabilities such as cellular differentiation and metabolism (Kaneko 2006; Kitano et al., 2005) to create a model that functions as a natural cell. The purpose of this work is to do an in-depth study of the biological model.

The more practical branch mainly focuses on the development of a cell inspired-model that might be applicable to other problems (Bentley, 2002; Kumar & Bentley (eds), 2003; Stanley & Miikkulainen, 2003). According to this model, every cell would not only have genetic information that encodes the general performance of the system, it would also act as a processor that communicates with the other cells. This model is mainly applied to the solution of simple 3D spatial problems, robot control, generative encoding for the construction of artificial organisms in simulated physical environments and real robots, or to the development of the evolutionary design of hardware and circuits (Endo, Maeno & Kitano, 2003; Tufte & Haddow, 2005).

The most relevant models are the following: the Kumar and Bentley model (Kumar & Bentley (eds), 2003), which uses the theory of fractal proteins (Bentley, 2002) for the calculation of protein concentration; the Eggenberger model (Eggenberger, 1996), which uses the concepts of cellular differentiation and cellular movement to determine cell connections; and the work of Dellaert and Beer (Dellaert & Beer, 1996), who propose a model that incorporates the idea of biological operons to control the model expression, where the function assumes the mathematical meaning of a Boolean function.

All these models can be regarded as special cellular automats. In cellular automats, a starting cell set in a certain state will turn into a different set of cells in different states when the same transition function is applied to all the cells during a determined lapse of time in order to control the message concurrence among them. The best known example of cellular automats is Conway's "Game of Life" (Conway, 1971), where this behaviour can be observed perfectly. Whereas the classical conception specifies the behaviour rules, the evolutionary models establish the rules by searching for a specific behaviour. This difference comes from the mathematical origin of the cellular automats, whereas the here presented models are based on biology and embryology.

These models should not be confused with other concepts that might seem similar, such as Gene Expression Programming (GEP) (Ferreira, 2006). Although GEP codifies the solution in a string, similarly as how it is done in the present work, the solution program is developed in a tree shape, as in classical genetic programming (Koza et. al., 1999) which has little or nothing in common with the presented models.

BIOLOGICAL INSPIRATION

A biological cellular system can be categorized as a complex system following the identification characteristics of a complex system stated by Nagl (Nagl, Parish, Paton & Warner, 1998). The cells of a biological system are mainly determined by the DNA strand, the genes, and the proteins contained by the cytoplasm. The DNA is the structure that holds the gene-encoded information that is needed for the development of the system. The genes are activated or transcribed thanks to the protein shaped-information that exists in the cytoplasm, and consist of two main parts: the sequence, which identifies the protein that will be generated if the gene is transcribed, and the promoter, which identifies the proteins that are needed for gene transcription.

Another remarkable aspect of biological genes is the difference between constitutive genes and regulating genes. The latter are transcribed only when the proteins identified in the promoter part are present. The constitutive genes are always transcribed, unless inhibited by the presence of the proteins identified in the promoter part, acting then as gene oppressors.

The present work has tried to partially model this structure with the aim of fitting some of its abilities into a computational model; in this way, the system would have a structure similar that is similar to the above and will be detailed in section 4.

PROPOSED MODEL

Various model variants were developed on the basis of biological concepts. The present work uses for its tests an implemented version of our model used in (Fernandez-Blanco, Dorado, Rabuñal, Gestal & Pedreira, 2007). The proposed artificial cellular system is based on the interaction of artificial cells by means of messages that are called proteins. These cells can divide themselves, die, or generate proteins that will act as messages for themselves as well as for neighbour cells.

The system is supposed to express a global behaviour towards the generation of structures in 2D. Such behaviour would emerge from the information encoded in a set of variables of the cell that, in analogy with the biological cells, will be named genes.

One promising application, in which we are working, could be the compact encoding of adaptive shapes, similar to the functioning of fractal growth or the fractal image compression.

Model Structure

The central element of our model is the artificial cell. Every cell has a binary string-encoded information for the regulation of its functioning. Following the biological analogy, this string will be called DNA. The cell also has a structure for the storage and management of the proteins generated by the own cell and those received from neighbourhood cells; following the biological model, this structure is called cytoplasm.

DNA

The DNA of the artificial cell consists of functional units that are called genes. Each gene encodes a protein or message (produced by the gene).

The structure of a gene has four parts (see Figure 1):

- **Sequence:** The binary string that corresponds to the protein that encodes the gene
- **Promoters:** The gene area that indicates the proteins that are needed for the gene's transcription.
- **Constituent:** This bit identifies if the gene is constituent or regulating
- **Activation percentage (binary value):** The percentage of minimal concentration of promoters proteins inside the cell that causes the transcription of the gene.

Cytoplasm

The other fundamental element for keeping and managing the proteins that are received or produced by the artificial cell is the cytoplasm. The stored proteins have a certain life time before they are erased. The cytoplasm checks which and how many proteins are needed for the cell to activate the DNA genes, and as such responds to all the cellular requirements for the concentration of a given type of protein. The cytoplasm also extracts the proteins from the structure in case they are needed for a gene transcription.

Figure 1. Structure of a system gene

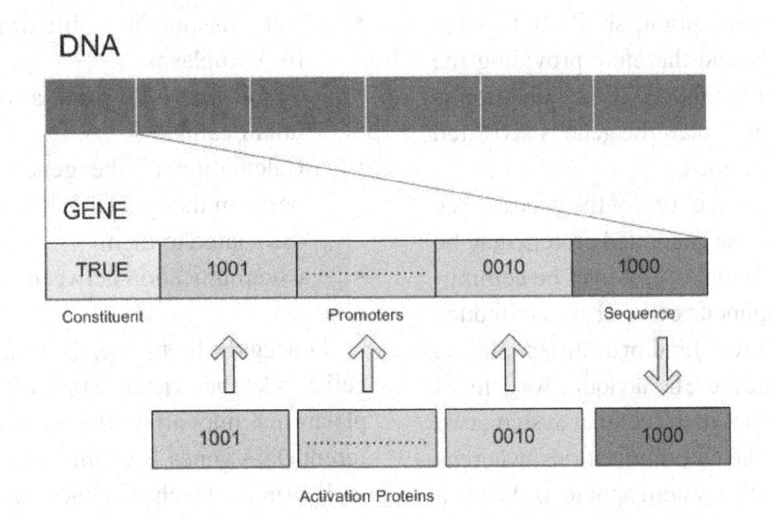

Model Functioning

The functioning of genes is determined by their type, which can be constituent or regulating. The transcription of the encoded protein occurs when the promoters of the non-constituent genes appear in a certain rate at the cellular cytoplasm. On the other hand, the constituent genes are expressed during all the "cycles" until such expression is inhibited by the present rate of the promoter genes.

$$\text{Protein Concentration Percent} >= (\text{Distance} + 1) * \text{Activation Percent} \quad (1)$$

The activation of the regulating genes or the inhibition of the constituent genes is achieved if the condition expressed by equation (1) is fulfilled, where *Protein Concentration Percentage* represents the cytoplasm concentration of the protein that is being considered; *Distance* stands for the Hamming distance between one promoter and the considered protein; and *Activation Percentage* is the minimal percentage needed for the gene activation that is encoded in the gene. This equation is tested on each promoter and each protein. If the condition is fulfilled for all the promoters, that gene is transcribed. According to this, if gene-like promoters exist in a concentration higher than the encoded concentration, they can also induce its transcription, similarly to what happens in biology and therefore providing the model with higher flexibility. If the condition is fulfilled for each promoter, the gene is activated and therefore transcribed.

After the activation of one of the genes, three things can happen: the generated protein may be stored in the cell cytoplasm, it may be communicated to the neighbour cells, or it may induce cellular division (mitosis) and/or death (apoptosis). In order to modulate these behaviours, we defined three threshold values in the cellular system, two of which manage the communications, whereas the third regulates the system apoptosis. When a cellular gene produces a protein, the concentration of the protein is checked to be below the value of the lower communication threshold. If such is the case, the protein is kept inside the cell cytoplasm. If the concentration is higher than the lower communication threshold value but lower than the upper communication threshold, the cell sends that protein to one of its randomly selected neighbour proteins. If the concentration exceeds the higher communication threshold, the protein is sent to all the neighbour cells. The third implemented threshold is the apoptosis threshold; it determines the protein concentration that is needed to induce cellular death. Finally, the mitosis and apoptosis procedures are executed when the cell generates certain types of proteins that were identified as "special". Therefore, if a protein whose sequence was identified as leftwards mitosis is produced and the cell has not been divided during that "cellular cycle", the action will be executed. Five special proteins were identified in the system: apoptosis, upwards mitosis, downwards mitosis, leftwards mitosis, and rightwards mitosis.

The different events of a tissue are managed in the cellular model by means of "cellular cycles". Such "cycles" will contain all the actions that can be carried out by the cells, restricting sometimes their occurrence. The "cellular cycles" can be described as follows:

- Actualisation of the life time of proteins in the cytoplasm
- Verification of the life status of the cell (cellular death)
- Calculation of the genes that react and perform the special behaviour that may be associated to them
- Communication between proteins

During the first stage, the functioning of each cell checks the existence of proteins in the cytoplasm that may affect the promoters of the different DNA genes. This information is then used to determine which are genes will be transcribed

Box A.

```
FOR every Protein at the cytoplasm do
  TTL_j = TTL_j - 1;
  IF (TTL_j = 0)
    P_i = P_i - 1;   protein j ∈ Type i
  ENDIF
ENDFOR

WHILE no New Cycle DO

  FOR each genes promoter
    Return %P_a = (P_a / ∑ P_b) * 100
     a, b ∈ 1,N
  ENDFOR

  IF Requirement a type protein AND P_a > 0
    P_a = P_a - 1;
  ENDIF
ENDDO
```

after the recovery of the necessary proteins from the cytoplasm and the generation of the encoded proteins in the selected genes. Once the genes are transcribed with the generated proteins, it is determined which cellular reactions accompany them: every protein can either be communicated to a neighbour cell, stored in the cytoplasm, or carry out one of the special actions of a mitosis or an apoptosis. The behaviour of the cytoplasm in each cycle of the system can be seen in the pseudo-code in Box A, where P_x identifies the number of x type-proteins and TTL_y represents the lifetime of the y type-protein.

According to this pseudo-code, the cytoplasm starts by updating the lifetime of its stored proteins and then returns the information on the concentrations of the system's N proteins. Importantly, the concentration is used to calculate the genes activation. The cytoplasm concentration of proteins decreases when they are required for genes transcription, which may coincide with specific actions. When all the possible actions of the cellular cycle are terminated, the process restarts by updating the cytoplasm with the proteins identified for storage and those received through communications with neighbour cells during the previous cycle.

Genome Evaluation

A classical approach of EC proposes the use of Genetic Algorithms (GA) (Fogel, Owens & Walsh, 1966; Goldberg, 1989; Holland, 1975) for the optimisation, in this case, of the values of the DNA genes (binary strands). Each individual of the GA population will represent a possible DNA strand for problem solving.

In order to calculate the fitness value for every individual in the GA or the DNA, the strand is introduced into an initial cell or zygote. After simulating during a certain number of cycles, the contained information is expressed and the characteristics of the resulting tissue are evaluated by means of various criteria, according to the goal that is to be achieved.

Figure 2. GA gene structure

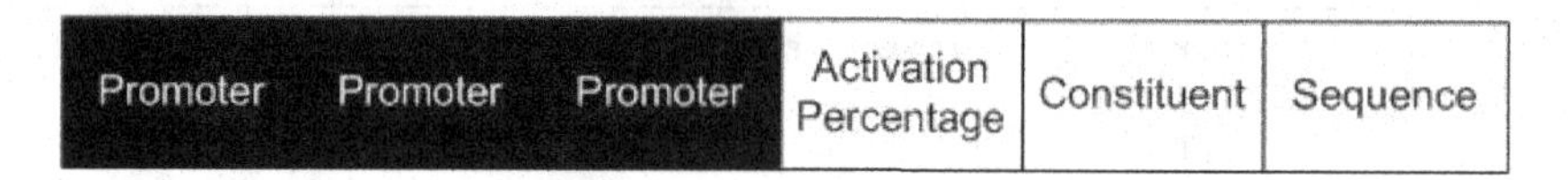

Figure 3. Example of GA genes for the encoding of cellular genes

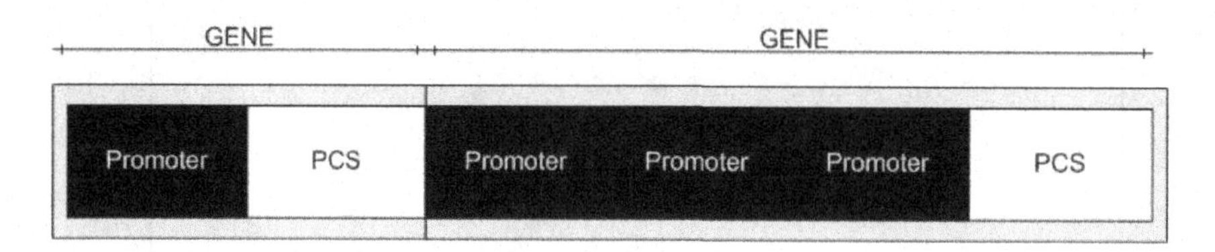

The encoding of the individual genes follows a structure that is similar to the one described in Figure 2, where the number of promoters of each gene may, vary but the white and indivisible section "Activation Percentage – Constituent – Sequence" (PCS) must always be present. The PCS sections determine the genes of the individual, and the promoter sections are associated to the PCS sections, as shown in Figure 3.

The search of a set of structures similar to those shown in Figure 3 required the adaptation of the crossover and mutation GA operations to this specific problem.

Since the length of the individuals is variable, the crossover had to be performed according to these lengths. When an individual is selected, a random percentage is generated to determine the crossover point of that individual; for instance, in Figure 4, the point that was selected as crossover point corresponds to 60% or the length. After selecting the section in that position, a crossover point is chosen for the section selected in the other parent. Once this has been done, the crossover point selection process is repeated in the second selected parent in the same position as in the previous individual; on the figure this point corresponds to 25%. From this stage on, the descendants are composed in the traditional way, since they are two strings of bits. We could execute a normal bit strings crossover, but the previously mentioned steps guarantee that the descendants are valid solutions for the DNA strands transformation.

It should be highlighted that, by using this type of variable length crossover, the number of gene promoters, together with the change in sections types, may induce the genes behaviour to vary drastically when the new individuals are evaluated. Another effect of this type of crossover is the absence of promoters for a given gene: if this occurs, the gene is either never expressed, or, if it is constituent, never inhibited.

With regards to mutation, it should be mentioned that the types of the promoter or PCS sections are identified according to the value of the first string bit. Bearing that in mind, together with the variable length of individuals, the mutation operation had to be adapted so that it could modify not only the number of these sections, but also the value of a given section.

The probability of executing the mutation is usually low, but this time it even had to be divided into the three possible mutation operations that the system contemplates. Various tests proved that the most suitable values for the distribution of the different mutation operations, after the selection of a position for mutation, were the following: for 20% of the opportunities, a section (either a

Figure 4. Crossover example

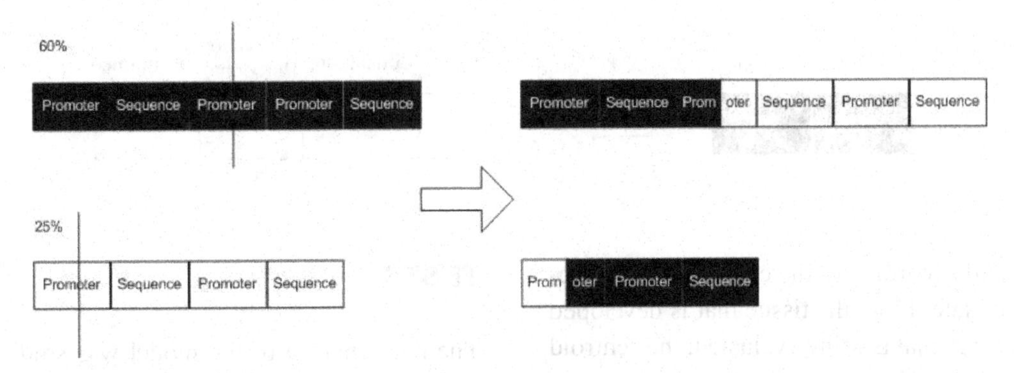

Figure 5. Types of mutations

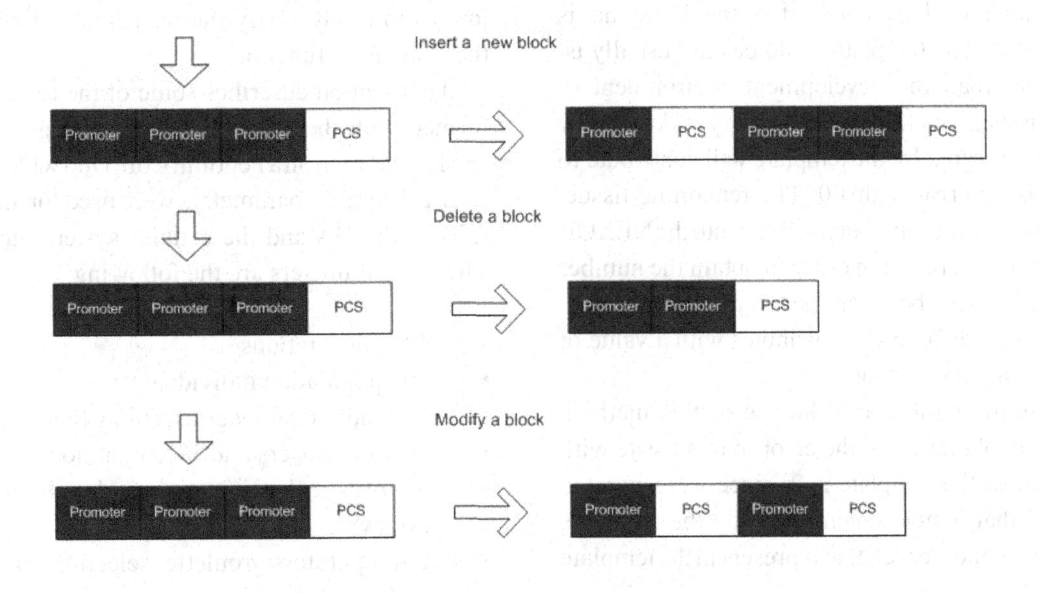

promoter or a PCS) is added, similarly to the first case of Figure 5; for another 20%, the existing section is removed, similarly to the second case of Figure 5; and finally, for the remaining 60% of the opportunities, the value of one of the bits of the section is randomly changed. The latter may provoke not only the change of one of the values, but also the change of the section type: if the bit that identifies the section is changed, the information of that section varies. For instance, if a promoter section turns into a PCS section, the promoter sequence turns into the gene sequence, as can be seen in the third case of Figure 5, and constitutive and activation percentage values are generated as in the case of Figure 6.

Changes of the Genome Evaluation

After reaching this development level and presenting the test set in (Fernandez-Blanco, Dorado, Rabuñal, Gestal & Pedreira, 2007), the bottleneck of the model turned out to be the development of the evaluation functions, since in every new figure the development of the function was time-consuming and not reusable. In order to solve this problem, the evaluation function was

Figure 6. Example of a type change

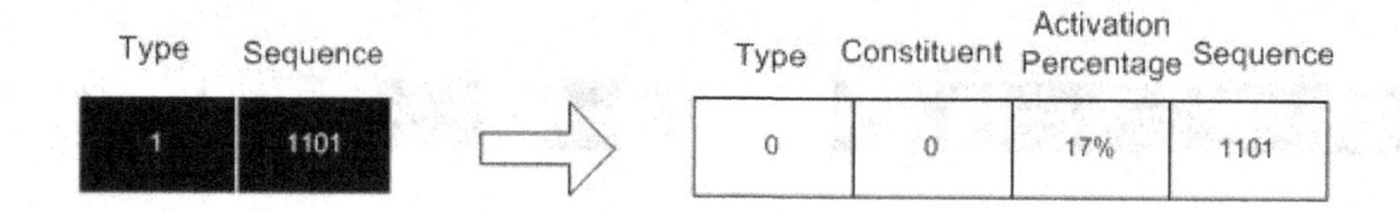

developed according to the concept of a correction template. From the tissue that is developed by the DNA that is being evaluated, the centroid is calculated. This point would be the center of the solution template, which is merely a matrix of Boolean values representing the figure that is aimed at. The template could be (and usually is) smaller than the development environment of the tissue, which means that every cell that may not be covered by the template will contribute to the tissue error with 1.0. The remaining tissue, covered by the template, will execute the NEXOR Boolean operation in order to obtain the number of differences between the template and the tissue. Each difference contributes with a value of 1.0 to the tissue error.

Figure 7 illustrates the use of this method. We can observe that the error of this tissue with regard to the template is 2, since we generated a cell that is not contemplated by the template, whereas another cell that is present in the template is really missing.

TESTS

The performance of the model was studied by applying it to simple 2D spatial problems. We selected this type of problems because they allow us to easily verify the results and simplify the evaluation functions.

This section describes some of the tests performed with the proposed model, using a Java implementation on a Pentium Core Duo with 1GB RAM. The same parameters were used for all the tests in the GA and the cellular system model. The GA parameters are the following:

- 1000 generations
- 500 population individuals
- 600 individual length sections (maximum)
- 90% crossovers and 10% mutations
- 50 pre-evaluation tissue development cycles
- GA operators: roulette selection, parent replacement, and one point crossover

Figure 7. Tissue + Template. Example of Template use

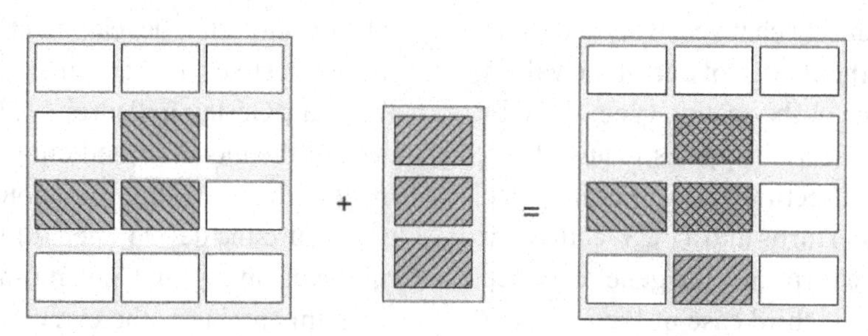

The cellular system parameters were fixed as follows:

- Protein lifetime: 3 cycles
- 20x20 development environment
- Division-inducing proteins: 1000 (north), 1100 (south), 1010 (west), and 1110 (east).
- Apoptosis-inducing protein: 0000 string
- 5% and 10% communication limit values and 10% apoptosis limit value

We also carried out experiments to select the sequences that were identified as division as well as apoptosis sequences. We tested randomly selected sequences and sequences with the biggest possible Hamming distance between them. The result of the tests showed that the specification of the sequences did not have any influence on the development of the tests.

Growing Tests

We carried out several tests to reach a better understanding of how the proposed model functions and to extract some conclusions. The tests involve the search of DNA strands whose associated performance might be the growth of simple structures with a minimum number of genes.

Cellular bars growth with different lengths and orientations is the structure that most simplifies the complexity, the visual study of the structure, and the analytical study of the genes. The choice of this simple structure accelerates the search for results, since it simplifies the evaluation of the structures, making it quicker and more efficient, as we could see in (Goldberg 1989).

The following tables show the genes of every test, every row representing a gene. The abbreviations of the headings have the following meaning:

- **N** refers to the order number within the gene's DNA
- **Pmtr** are the promoter sequences of the gene
- **Act** is the gene activation percentage
- **Cons** indicates the constitutive genes
- **Seq** is the protein that encodes the gene
- **Special** indicates if the gene has any associated behaviour

Length 5 Vertical Bar

The first test sought to minimize the number of genes needed to build a 5 cells vertical bar located at the center of the environment. To achieve this, we established a template that considered the number of sections and the distance of the figure centroid to the center as penalisations. The evaluation function is shown in equation (2).

We found an individual with three genes (see Table 1) that gave rise to a 5 sections structure.

The first gene works as the clock signal of the system, because it is constitutive and does not have any inhibiting promoters. The second gene induces the structure to grow upwards; it will only be activated during the first development stages of the structure because it has a very high activation percentage. Finally, the third gene is easily activated and acts as a regulator, because it will not allow the activation of the gene that induces the growth.

Figure 8 shows the final stage of the tissue. After the changes of the cycles 2, 4, 6 and 8, the tissue has become stable. The error of this solution is 0.0205 and mainly due to the fact that the structure is not at the center and only grows upwards. Between the options of minimizing the number of genes and obtaining a more center-close solution, the system opted for a DNA with only 3 genes, of which only one has a special action grown N.

Equation (2).

$$\text{Template-related differences} + 0.01 * |\text{center} - \text{centroid}| + 10^{-4} * \text{Number of sections}$$

Length 7 Vertical Bar

The second test consisted in developing a 7 cells vertical bar from the GA population of the 5 cells vertical bar.

We obtained the same genes, but this time the percentages and the gene order were changed, as can be seen in Table 2. This test seems to indicate that the resulting genes function as the previous ones, although they are not located in the same position.

Figure 9 shows the development of the structure and the final status with an error of 0.0305. The changes, besides the structure growth from 5 to 7 until its stabilisation for the remaining cycles, occurred during the cycles 10 and 12. The changes match their shape and cycle with those of the 5 length structure and occur at the even cycles, so the structure seems to have learned this growing pattern until reaching the expected size.

Table 1. Genes for the growth of a 5 elements vertical bar

N	Pmtr	Act	Cons	Seq	Special
1	1	.9771E-7 t	rue	0100	
2	0000	45.2033	false	1010	Grown N
3	1000	3.8838 f	alse	0101	

Figure 8. Final stable structure of the tissue

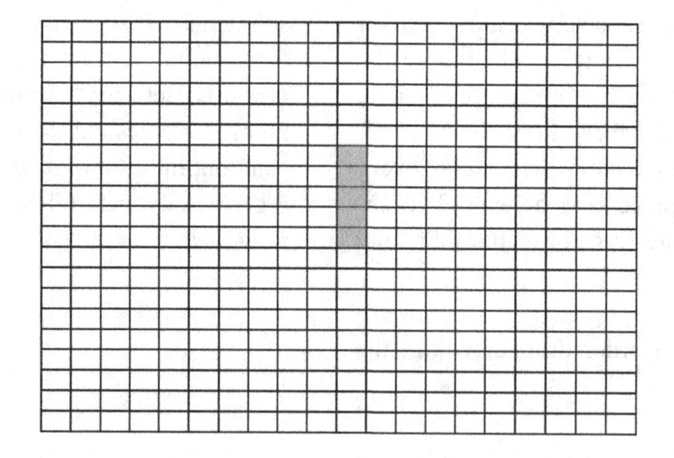

Length 15 Vertical Bar

Finally, we tried to grow a 15 elements vertical bar from the GA population of the 5 elements vertical bar. Although the result was correct, the solution has certain peculiarities: the growth of this structure created a lateral side that was supported by the superior edge and completed the remaining bar with some elements that are downwards generated. In this case, the genes that played the role of the work regulating clock signal were mixed with the growth generating genes. This test concluded that genes explanation is not always easy: at the same time that they express their own behaviour (see gene 2 at Table 3), they can be linked with the expression of other genes. The genes that provide the solution are those of Table 3, where the error committed is 0.0305.

Table 2. Genes for the growth of a 7 elements vertical bar

N	Pmtr	Act	Cons	Seq	Special
1	0000	45.2926	false	1010	Grown N
2	1110	3.645254 f	alse	0010	
3	9	.2864E-20 t	rue	0001	

Besides being the gene that induces upwards structure growth, it is also the one that later serves as a clock signal for other genes. The resulting tissue can be seen in Figure 10.

Length 3 Horizontal Bar

After the previous tests, the orientation change for structure growth still remained to be done. For horizontal growth, we used the population obtained after the vertical growth of a length 3 bar together with a consequently modified template. The resulting configuration is shown in Table 4.

Although the data are different, the three roles that were identified in the previous tests were also detected in the present test. The growth of the subsequent DNA results in a 3 cells structure that grows leftwards.

The structure undergoes changes during the cycles 3 and 7. These cycles match the ones where the starting length 3 vertical structure also underwent its changes. The final development of the structure can be observed in Figure 11.

3x3 Square

The last of these simple tests was the construction of a 3x3 square from the population of the length 3 horizontal bar of the previous section.

The solution has 11 genes and the clock signal role is played by the 7 first genes, which are almost identical (they vary between 2 sequences, they are constitutive, and they do not have promoters, therefore the activation percentage is not important). These 7 genes are summed up in the first row of the Table 5. Although the exact reason is not known, nature also tends to complicate the solution with redundant information, as explained in (Kaneko, 2006).

The remaining genes generate the shape and stabilize it. A group of genes would work as system clock by periodically introducing the 0100 and 0101 proteins. Genes 8 and 10 are expressed due to their low activation percentages. In the long run, they could induce the system to distribute the concentration among several proteins and then genes 9 and 11 would not be expressed. These two genes induce the growth and, due to their high activation percentage, their expression would only be possible during the first cycles,

Figure 9. Final status of the length 7 structure

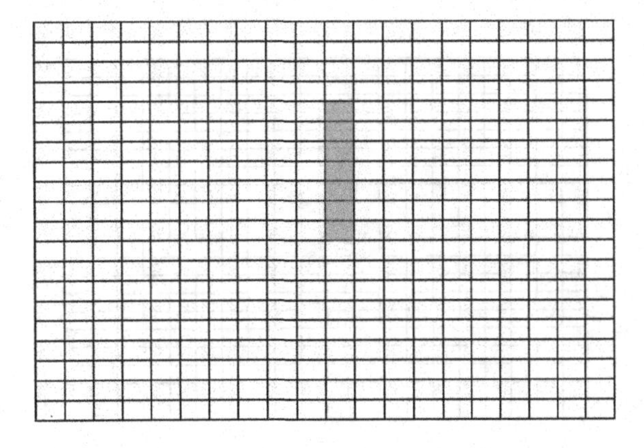

Table 3. Genes for the growth of a 15 elements vertical bar

N	Pmtr	Act	Cons	Seq	Special
1	1010	5.1819E-17	true	0110	
2		8.1266E15	true	1010	Grown N
3	1000	8.5467E-15	false	1000	Grown S

Table 4. Growth of a 3 elements horizontal bar

N	Pmtr	Act	Cons	Seq	Special
1		6.924E13	true	0100	
2	1001	1.2171E-26	false	1101	
3	0101	47.4337	false	1100	Grown W

Figure 10. Final status of the 15 length structure

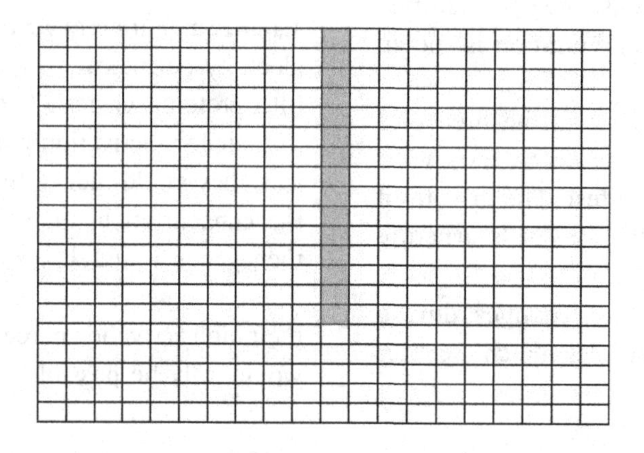

Figure 11. Final development of the length 3 horizontal bar

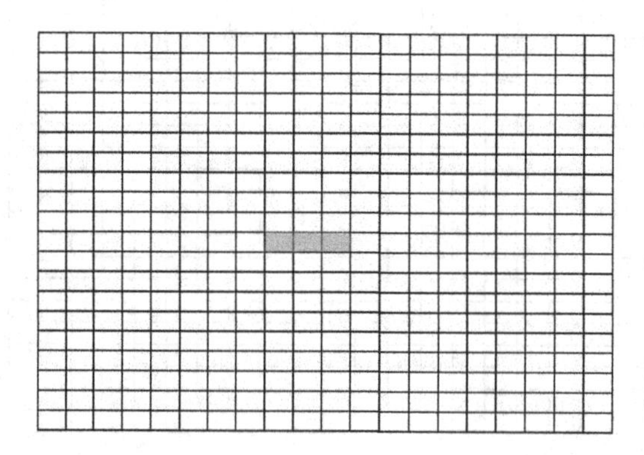

Table 5. Role-based summary of genes at the 3x3 square

N	Pmtr	Act	Cons	Seq	Special
1-7		*	true	0100-1010	
8	0000-0011	2.0101E-13	false	0001	
9	0000	45.125	false	1100	Grown W
10	0010	1.4626E-29	false	0011	
11	0000	41.1894	false	1000	Grown S

when the percentage has not been distributed yet. The development of the structure can be seen in Figure 12.

In this case, the tissue reaches its stable stage very early, in cycle 8. This solution has a 0.0216 error.

Considerations

Other tests were carried out to explore the generation capabilities of the model. Some further testing examples are: leaving only the edges of an initial square shape, generating a cross shape from a unique cell, or trying to generate a square structure in an environment with non-valid positions. Some of these examples could see on Figure 13.

Starting from a satisfactory 5x5 square test result, other tests are performed in order to evaluate the flexibility of the DNA solution that was found. The dark points of the image at the right of Figure 14 are positions that can not be occupied by the cells; this fact induces a higher error rate, but the used DNA was originated from an environment without such obstacles. The results are shown in Figure 14.

As it can be observed in Figure 14, in both types of blockades, the tissue development is affected by the obstacles placed in the environment.

The cells try to solve the problem and tend to build the square trying to avoid the obstacles. The two obtained solutions take up an approximately 5x5 box square area, although they do not use all the possible boxes of such area and they only use 14 cells. During the GA DNA search no obstacles were used to reach the DNA fitness so this proves that flexibility is inherent in the model used.

After all these tests, we can conclude that the system searches for certain functions/roles instead of concrete values in order to make the DNA strands work.

It was observed that the system can be configured in a simple way so as to change the range of shapes or their orientation. Our current work is focused on combining these compact genomes (which perform simple growing tasks), scaling them, and combining them in order to develop, together with other compact genomes, more complexes figures. One of the means for carrying out this process would involve a set of initial genes that generates a general shape such as the one show in Figure 12; from this shape, another set of genes would elaborate other different structures that could be combined with the previously achieved ones.

FUTURE TRENDS

The model showed here, have been applied to the development of simple forms like: squares, crosses, etc., but development of complex forms and the reutilization of genes to archive it, is still an open question. The way to archive these complex forms is not clearly, so a lot of different attempts may be tried until the best way is found.

Figure 12. Final development of the 3x3 square structure

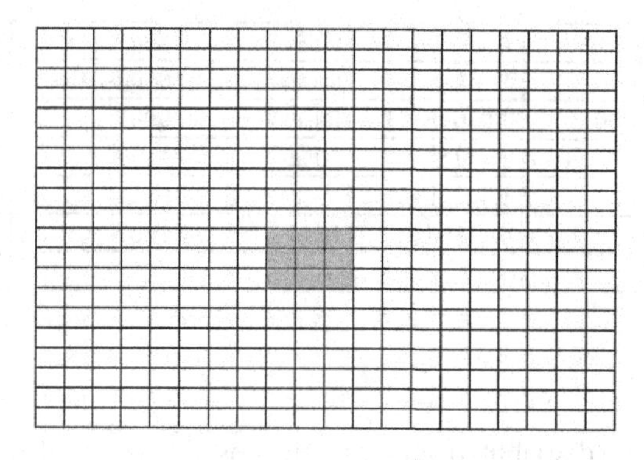

Figure 13. Test examples

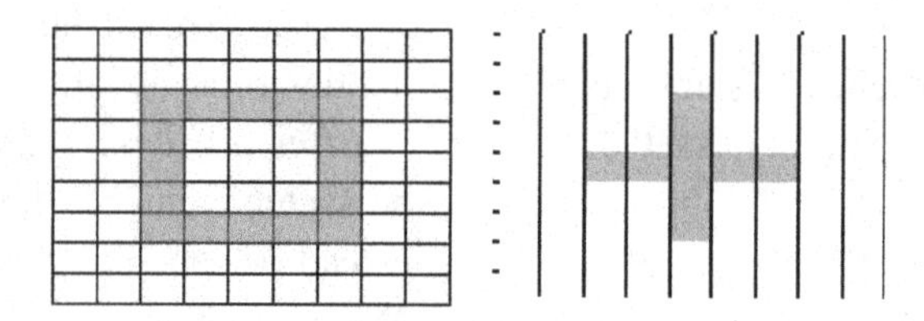

Figure 14. Failure tolerance example

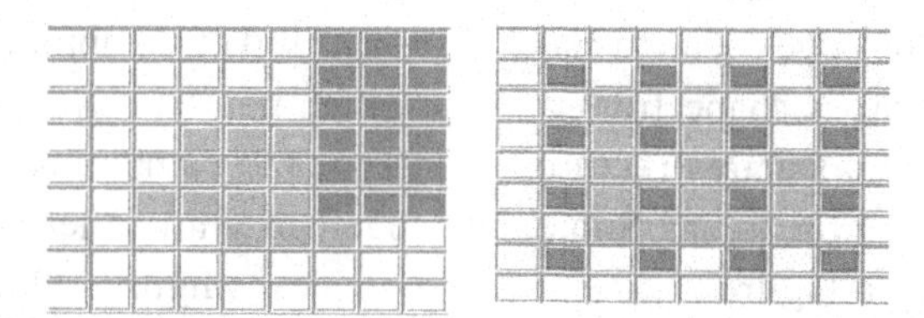

This group is currently working in one of the possible applications of this model: its use for image compression similarly as fractal compression works. The fractal compression searches the parameters of a fractal formula that encodes itself the starting image. The present model searches the gene sequence that might result in the starting image. In this way, the method based on template that has been presented in this paper can be used for performing that search, using the starting image as template.

CONCLUSION

Taking into account the model, we can say that the use of certain properties of biological cellular systems is feasible for the creation of artificial structures that might be used in order to solve certain computational problems.

Some behaviours of the biological model have been also observed in the artificial model: information redundancy in DNA, stability after achieving the desired shape, or variability in gene behaviour. These characteristics give to the model interesting qualities to be applied in different contest and problems.

FUTURE RESEARCH DIRECTIONS

On the last years the artificial embryogeny has had a very quickly development. As it was mentioned, models which are sorted as artificial embryogeny kind could be framed into the grammatical approach or into the chemical approach. On the last years, a lot of grammatical approach models have been adapted to generate artificial neural networks. This work is an interesting point of investigation, because it has application on evolvable artificial neural networks, such as archive an evolvable design of a network for a particular problem.

On the chemical approach models, as the one presented in this paper, could also include new characteristics such as the displacement of cells around their environment, or a specialisation operator that blocks pieces of DNA during the expression of its descendants, as happens in the natural model. These incorporations may induce new behaviours that make it applicable to new problems.

One of the questions is to develop models which could combine both approaches. The result will be a hybrid model which will has characteristics from both approaches.

Explore other possible applications of the artificial embryogeny models could be an interesting works in areas like the evolutionary robot controllers design, the evolutionary hardware, design of generative encoding for the construction of artificial organisms in simulated physical environments, etc.

Another problem that could be easy detected is to find a strategy search to work with the model. Both approaches, chemical and grammatical, have to search into enormous search spaces. Many models have failed into their applicability because it was difficult to search the correct configuration for the rules of the system. Find a general way to develop that search may be an interesting work.

Finally, remarks that even the model showed here have been used for the development of forms; it could be used to develop other task like the processing information. This point can be development using the genetic regulatory networks theory so rules can be defined as the combination of different genes. This theory has a Boolean algebra like structure as it could be seen in (Kauffman, 1969).

REFERENCES

Bentley, P.J. (2002) *Digital Biology*. Simon and Schuster, New York.

Conway J.H. (1971) *Regular Algebra and Finite Machines*. Chapman and Hall, Ltd., London.

Dellaert F., Beer R.D. (1996) *A Developmental Model for the Evolution of Complete Autonomous Agent* In From animals to animats: Proceedings of the Forth International Conference on Simulation of Adaptive Behaviour, Massachusetts, 9-13 September 1996, pp. 394-401, MIT Press.

Eggenberger P. (1996) *Cell Interactions as a Control Tool of Developmental Processes for Evolutionary Robotics* In From animals to animats: Proceedings of the Forth International Conference on Simulation of Adaptive Behaviour, Massachusetts, 9-13 September 1996, pp. 440-448, MIT Press.

Endo K., Maeno T., Kitano H. (2003) *Co-evolution of morphology and walking pattern of biped humanoid robot using evolutionary computation -designing the real robot.* ICRA 2003: pp. 1362-1367

Fernandez-Blanco E., Dorado J., Rabuñal J.R., Gestal M., Pedreira N.(2007) *A New Evolutionary Computation Technique for 2D Morphogenesis and Information Processing.* WSEAS Transactions on Information Science & Applications Volume 4 pp. 600-607. April 2007. WSEAS Press.

Ferreira C. (2006) *Gene Expression Programming: Mathematical Modelling by an Artificial Intelligence* Springer, Berlin.

Fogel, L.J., Owens A. J. and Walsh, M.A. (1966) *Artificial Intelligence through Simulated Evolution*. Wiley, New York.

Goldberg, D.E. (1989) *Genetics Algorithms in Search, Optimization and Machine Learning.* Addison-Wesley.

Hassoun M.H. (1995) *Fundamentals of Artificial Neural Networks*. University of Michigan Press, MA, USA

Holland J.H. (1975) *Adaptation in natural and artificial systems*. University of Michigan Press, Ann Arbor, MA, USA.

Kaneko K. (2006) *Life: An Introduction to Complex Systems Biology*. Springer Complexity: Understanding Complex Systems, Springer Press.

Kauffman S.A. (1969) *Metabolic stability and epigenesis in randomly constructed genetic nets*. Journal of Theoretical Biology 22 pp. 437-467.

Kitano H. et al. (2005) *Using process diagrams for the graphical representation of biological networks*, Nature Biotechnology 23(8) pp. 961 - 966.

Koza J. et. al. (1999) *Genetic Programming III: Darwin Invention and Problem Solving.* MIT Press, Cambridge, MA.

Kumar S. and Bentley P.J. (editors). (2003) *On Growth, Form and Computers*. Elsevier Academic Press. London UK.

Kumar S. (2004). *Investigating Computational Models of Development for the Construction of Shape and Form*. PhD Thesis. Department of Computer Science, University Collage London.

Lindenmayer, A. (1968) *Mathematical models for cellular interaction in development: Part I and II*. Journal of Theorical Biology. Vol. 18 pp. 280-299, pp. 300-315.

Mjolsness E., Sharp D.H. and Reinitz J. (1995) *A Connectionist Model f Development*. Journal of Theoretical Biology 176 pp. 291-300.

Nagl S.B., Parish J.H., Paton R.C. and Warner G.J. (1998) *Macromolecules, Genomes and Ourselves*. Chapter in *Computation in cells and tissues.Perspective and tools of thought*. Paton R., Bolouri H., Holcombe M., Parish J.H. and Tateson R. Springer Press

Stanley, K. & Miikkulainen, R. (2003) *A Taxonomy for Artificial Embryogeny*. In Proceedings Artificial Life 9, pp. 93-130. MIT Press.

Tufte G. and Haddow P. C. (2005) *Towards Development on a Silicon-based Cellular Computing Machine*. Natural Computing 4(4): 387-416.

Turing, A.(1952) *The chemical basis of morphogenesis*. Philosofical Transactions of the Royal Society B, vol.237, pp. 37-72

Watson J.D. & Crick. F. H. (1953) *Molecular structure of Nucleic Acids*. Nature vol. 171, pp. 737–738.

ADDITIONAL READING

Bentley, P. J. & Kumar, S. (1999) *The ways to grow designs: Acomparison of embryogenies for an evolutionary design problem*. In Proceedings of the Genetic and Evolutionary Computation Conference (GECCO-1999) pp. 35-43) San Francisco, CA. Morgan Kaufmann

Bongard, J.C. & Paul, C. (2000) *Investigating morphological symmetry and locomotive efficiency using virtual embodied evolution*. In Proceedings of yje Sixth International Conference on Simulation of Adaptative Behavior.(pp. 420-429) Cambridge, MA, MIT Press.

Bongard, J.C. (2002) *Evolving modular genetic regulatory networks.* In Proceedings of the 2002 Congress on Evolutionary Computation Piscataway, NJ. IEEE Press.

Bongard, J.C. & Pfeifer R. (2001) *Repeated structure and dissociation of genotypic abd phenotypic complexity in artificial ontogeny.* In L. Spensor, E.D. Goodman, A. Wu, W.B. Langdon, H.M. Voight, M. Gen, S. Sen, M. Dorigo, S. Pezeshk, M.H. Garzon & E. Burke (Eds). Proceedings of the Genetic and Evolutionary Computation Conference. (pp. 829-836). San Francisco, CA. Morgan Kaufmann

Dellaert, F. (1995). *Toward a biologically defensible model of development.* Master's thesis, Case Western Reserve University, Cleveland, OH.

Dellaert, F. & Beer, R. D. (1994). *Co-evolving body and brain in autonomous agents using a developmental model* (Tech. Rep. CES-94-16). Cleveland, OH: Dept. of ComputerEngineering and Science, Case Western Reserve University

Dellaert, F., & Beer, R. D. (1994). To*ward an evolvable model of development for autonomous agent synthesis.* In R. A. Brooks & P. Maes (Eds.), *Proceedings of the Fourth International Workshop on the Synthesis and Simulation of Living Systems (Artificial Life IV).* Cambridge, MA: MIT Press.

Fleischer, K., & Barr, A. H. (1993). *A simulation testbed for the study of multicellular development: The multiple mechanisms of morphogenesis.* In C. G. Langton (Ed.),*Artificial life III* (pp. 389–416). Reading, MA: Addison-Wesley.

Gruau F. (1994) *Neural networks synthesis using cellular encoding and the genetic algorithm.* Doctoral dissertation, Ecole Normale Superiere de Lyon, France.

Gruau, F., Whitley, D., & Pyeatt, L. (1996). A *comparison between cellular encoding and direct encoding for genetic neural networks.* In J. R. Koza, D. E. Goldberg, D. B. Fogel, & R. L. Riolo (Eds.), Genetic Programming *1996:* Proceedings of the First Annual Conference (pp. 81–89). Cambridge, MA: MIT Press

Hart, W. E., Kammeyer, T. E., & Belew, R. K. (1994). *The role of development in geneticalgorithms* (Tech. Rep. CS94-394). San Diego, CA: University of California

Hornby, G. S. & Pollack, J. B. (2001) *The advantages of generative grammatical encodings for physical design.* In Proceedins of the 2002 Congress on Evolutionary Computation. Piscataway, NJ. IEEE Press.

Hornby, G. S. & Pollack, J. B. (2002) *Creating high-level components with a generative representation for body-brain evolution.* Artificial Life 8(3).

Jakobi, N. (1995) *Harnessing morphogenesis.* In Proceedings of Information Processing in Cells and Tissues (pp. 29-41). Liverpool, UK: University of Liverpool.

Lindermayer, A. (1974) *Adding continuous components to L-systems.* In G.

Kitano, H. (1990) *Design neural networks using genetic algorithms with graph generation system.* Complex systems, 4 pp. 461-476.

Kauffman, S. A. (1993). *The origins of order.* New York: Oxford University Press.

Kumar, S. and Bentley P.J. (editors) (2003). *On Growth, Form and Computers.* Academic Press. London UK.

Kumar, S. and Bentley P.J. (2003) *Computational Embryology: Past, Present and Future.* Invited Chapter in Ghosh and Tsutsui (Editors). Theory and Application of Evolutionary Computation: Recent Trends. Springer Velak, UK.

Kumar, S. (2004). *A Developmental Biology Inspired Approach to Robot Control.* In Artificial

Life 9 (ALIFE), Proceedings of the Ninth Internacional Conference on the Simulation and Síntesis of Living Systems. MIT Press, Cambridge, MA

Miller, J. (2003) *Evolving Developmental Programs for Adaptation Morphogenesis, and Self-Repair.* 7th European Conference on Artifial Life 2003.

Otter, T. (2004) *Toward a New Theoretical Framework.* In Proceedings of the Genetic and Evolutionary Computation Congress (GECCO-2004). San Francisco, CA. Morgan Kaufmann

Prusinkiewicz, P., & Lindenmayer, A. (1990). *The algorithmic beauty of plants.* Heidelberg, Germany: Springer-Verlag

Stanley, K. & Miikkulainen, R. (2002) *Continual coevolution through complexification.* In Proceedings of the Genetic and Evolutionary Computation Congress (GECCO-2002). San Francisco, CA. Morgan Kaufmann.

Stanley, K. & Miikkulainen, R. (2003) *Evolving neural networkthrough augmenting topologies.* Evolutionary Computation, 10(2), pp. 99-127.

Chapter X
Computing vs. Genetics

José M. Barreiro
Universidad Politécnica de Madrid, Spain

Juan Pazos
Universidad Politécnica de Madrid, Spain

ABSTRACT

This chapter first presents the interrelations between computing and genetics, which both are based on information and, particularly, self-reproducing artificial systems. It goes on to examine genetic code from a computational viewpoint. This raises a number of important questions about genetic code. These questions are stated in the form of an as yet unpublished working hypothesis. This hypothesis suggests that many genetic alterations are caused by the last base of certain codons. If this conclusive hypothesis were to be confirmed through experiementation if would be a significant advance for treating many genetic diseases.

INTRODUCTION

The mutual, two-way relationships between genetics and computing (see Table 1) go back a long way and are more wide-ranging, closer and deeper than what they might appear to be at first sight. The best-known contribution of genetics to computing is perhaps evolutionary computation. Evolutionary computation's most noteworthy representatives are genetic algorithms and genetic programs as search strategies. The most outstanding inputs from computing to genetics are reproductive automata and genetic code deciphering. Therefore, section 2 will deal with von Neumann reproductive automata. Section 3 will discuss genetic code. Section 4 will introduce the well-know χ^2 test because of this importance in establishing the working hypothesis. Later, section, will address genome deciphering. And finally section 6 will establish the conjecture or working hypothesis, which is the central conclusion of the paper, and define the future research lines.

SELF-REPRODUCING AUTOMATA

The most spectacular contribution of computing to genetics was unquestionably John von Neumann's premonitory theory of self-reproducing automata, i.e. the construction of formal models of automata capable of self-reproduction. Von Neumann gave a conference in 1948 titled "The General and Logical Theory of Automata" (Von Neumann, 1951, 1963) establishing the principles of how a machine could self-reproduce. The procedure von Neumann suggested was at first considered an interesting logical and mathematical speculation more than anything else. However, von Neumann's view of how living beings reproduced (abstractedly simpler than what it might appear) was acclaimed five years later, when it was confirmed, after James D. Watson and Francis Harry C. Crick (1953(a)) discovered the model of DNA.

It was as of 1950s that Information Theory (IT) exercised a remarkable influence on biology, as it did, incidentally, on many other fields removed from the strictly mathematical domain. It was precisely as of then that many of the life sciences started to adopt concepts proper to IT. All the information required for the genesis and development of the life of organisms is actually located in the sequence of the bases of long DNA chains. Their instructions are coded according to a four-letter alphabet A, T, C and G. A text composed of the words written with these four letters constitutes the genetic information of each living being. The Nobel prize-winning physicist Erwin Schrödinger (1944) conjectured the existence of genetic code, which was demonstrated nine years later by Watson and Crick (1953(a), (b)), both awarded the Nobel prize for this discovery. It was in the interim, in 1948, when von Neumann established how a machine could self-reproduce.

Table 1. Computing vs. genetics

From genetics to computing	From computing to genetics
Natural Computation (NC) ≡ Evolutionary Computation (EC) [Genetics Algorithms (GA) + Evolution Strategies (ES) + Evolutionary Programming (EP)] + Neural Networks (NN) + Genetic Programming	1940 Claude Elwood Shannon (1940) defended his PhD thesis titled "An Algebra for Theoretical Genetics".
1966 Fogel, Owens and Walsh (1966) establish how finite state automata can be evolved by means of unit transformations and two genetic operators: selection and mutation.	1944 Erwin Schrödinger (1983) conjectured that genetic code existed.
1973 Rechemberg (1973) defined the evolutionary strategies of finite state machine populations.	1948 John Von Neumann (1966) established the principles underlying a self-reproducing machine.
1974 Holland (1975) and disciples defined genetic algorithms.	1953 Crick (Watson, 1953) luckily but mistakenly named the small dictionary that shows the relationship between the four DNA bases and the 20 amino acids that are the letters of protein language genetic code.
1992 Koza (1992) proposed the use of the evolutionary computation technique to find the best procedure for solving problems, which was the root of genetic programming.	1955 John G. Kemeny (1955) defined the characteristics of machine reproduction and how it could take place.
1994 Michalewitz (1992) established evolutionary programs as a way of naturally representing genetic algorithms and context-sensitive genetic operators.	1975 Roger and Lionel S. Penrose (Penrose, 1974) tackled the mechanical problems of self-reproduction based on Homer Jacobson's and Kemeny's work.
	1982 Tipler (1982) justified the use of self-reproducing automata.

Self-Reproduction

In his utopian novel "Erewhon", which is the mirror image of "nowhere", Samuel Butler (1982) explores the possibility of machines using men as intermediaries for building new machines. There have been many examples of machines built by other machines in the last century. Steam engines were used to build other steam engines, and machine tools made all sorts of devices. Long before the computer era, there were mechanical and electrical machines that formed metal to build engines. The Industrial Revolution was largely possible thanks to machine tools: machines that were conceived exclusively to build other machines. However, devices that can be used to make other devices were not exactly the type of machine tools that the English Prime Minister, Benjamin Disraeli, had in mind when he said, "The mystery of mysteries is to view machines making machines", i.e. machines building other machines without human involvement. In other words, machines that self-reproduce or, if you prefer a less controversial term, are self-replicate.

What is the meaning of the word reproduction? As John G. Kemeny (1955) pointed out, if reproduction is understood as the creation of an object that is identical to the original one from nothing, it is evident that a machine cannot self-reproduce, but neither could a human being. For reproduction not to violate the principle of energy conservation, some raw material is required. What characterises the reproduction of life is that the living organism is capable of creating new, similar organisms from the inert matter in its surroundings. If we accept that machines are not alive and we insist on the fact that the creation of life is a fundamental characteristic of reproduction, the problem is settled: a machine is incapable of self-reproduction. The problem is therefore reformulated so as not to logically rule out the reproduction of machines. To do this, we need to omit the word "living". So, we will stipulate that the machine should be capable of creating a similar new organism from simple components existing in its surroundings.

Scientists have been dreaming about creating machines programmed to produce replicas of themselves since 1948. These replicas would produce other replicas and so on, without any limit whatsoever. The theory that established the principles of how such a feat could be achieved was first formulated in 1948. This theory has two aspects, which could be termed logical and mechanical. The mechanical question was addressed, among others by Lionel S. and Roger Penrose (1974) and will not be considered here. The logical part, which is our concern here, was first researched by von Neumann at the Institute for Advanced Study in Princeton. It was there that von Neumann suggested the possibility of building a device that had the property of self-reproduction. The method involved building another describable machine from which it followed logically that this machine would carry a sort of tail that would include the code describing how to reproduce the body of the machine and how to reproduce the actual code. According to Kemeny (1955), a colleague of von Neumann, the basic body of the machine would be composed of a box containing the constituent parts, to which a *tail* would be added that stored the units of information. From the mechanical viewpoint, it was considered that the elementary parts from which the machine would have to be built would be rolls of tape, pencils, rubbers, empty tubes, quadrants, photoelectric cells, motors, batteries and other similar devices. The machine would assemble these parts from the surrounding raw material, which it would organise and transform into a reproduction of itself. As von Neumann's aim was to solve the logical conditions of the problem, the incredible material complications of the problem were left aside for the time being.

Von Neumann's proposal for building machines that have the reproductive capability of the living organisms was originally considered as an interesting mathematical speculation more than anything else, especially taking into account that computers back then were 30 or more tonne giants and were little more than devices for rapidly performing mathematical operations. How could we get a machine to produce a copy of itself? A command from a human programmer to "Reproduce!" would be out of the question, as the machine could only respond "I cannot self-reproduce because I don't know who I am". This approach would be as absurd as if a man gave his partner a series of bottles and glass flasks and told her to have a child. In von Neumann's opinion, any human programmer proposing to create a dynasty of machines would have to take the following three simple actions:

1. Give the machine a full description of itself.
2. Give the machine a second description of itself, which would be a machine that has already received the first description.
3. Finally, order the machine to create another machine that is an exact copy of the machine in the second description and order the first machine to copy and pass on this final order to the second machine.

The most remarkable thing about this logical procedure is that, apart from being simpler than it may appear, it was von Neumann's view of how living creatures reproduce. A few years after his conference, his ideas were confirmed when the biologists Crick and Watson (1953 (a) and (b)) found the key to genetic code and discovered the secret of organic reproduction. It was essentially the same as the sequence for machine reproduction that von Neumann had proposed. In living beings, deoxyribonucleic acid (DNA) plays the role of the first machine. The DNA gives instructions to ribonucleic acid (RNA) to build proteins; RNA is like DNA's "assistant". Whereas the RNA performs the boring task of building proteins for its parent organisms and offspring, the DNA plays the brilliant and imaginative role of programming its genes, which, in the case of a human baby, will decide whether it has blonde or brown hair and whether it will be of an excitable or calm temperament. In short, DNA and RNA together carry out all the tasks that the first von Neumann machine has to perform to create the second machine of the dynasty. And, therefore, if we decide to build self-reproductive machines, there is important biological evidence that von Neumann came across the right procedure to do so a long time ago.

But, one might wonder, why would anyone want to build computers that make copies of themselves? The procedure could at best be bothersome. Suppose that someone went to bed after having spent the evening working at his computer and, when he woke up the next day found that there were two computers instead of one. What would these regenerating computers be useful for? The answer is that they will be used at remote sites to perform difficult and dangerous tasks that people cannot do easily. Consequently, we have to consider at length the possible location of such places. What is it that is holding back human biological development? Why, over thirty-five years after man first set foot on the moon, is there still no permanent lunar colony? Over three quarters of a century have passed since man first managed to fly and most human beings are still obliged to live on the surface of the Earth. Why? The astronomer Tipler (1980), from the University of California in Berkeley, answered this question very clearly when stated that it was the delay in computer not rocket technology that was preventing the human race from exploring the Galaxy. It is in space, not on Earth, where the super intelligent self-reproducing machines will pay off, and it is in space where the long-term future of humanity lies. It is fascinating to consider how Tipler and others who have studied the far-off future consider how von Neumann machines will make it possible, firstly to colonise the solar system of planets and then the Milky Way with over 150,000 million suns.

Of course, none of the machines mentioned here have, as far as we know, been built, but there is nothing to stop them from being built. Scientifically, nothing stands in the way of their construction. Whether it is technologically feasible to make replicas is another question, and a tricky one at that.

GENETIC CODE

Code vs. Cipher

Technically, a code is defined as a substitution at the level of words or sentences, whereas a cipher is defined as a substitution at the level of letters. Ciphering means concealing a message using a cipher, whereas coding means concealing a message using a code. Similarly, the terms decipher is applied to the discovery of a ciphered message, i.e., in cipher, whereas the term decode is applied to the discovery of a coded message. As we can see, the terms code and decode are more general, and are related to both codes and ciphers. Therefore, these two terms should not be confused through misuse. For example, Morse code, which translates the letters of the English alphabet and some punctuation marks to dots and dashes, is not a code in the sense of a form of cryptography because it does not conceal a message. The dots and dashes are simply a more convenient form of representing the letters for telegraphy. Really, Morse code is nothing other than an alternative alphabet.

Code is formally defined as follows. Let A* be a free monoid (Hu, 1965) engendered by the set A, i.e. A* is the set of finite-length words formed by means of concatenation (the associative law of internal composition) with the symbols of A and with a neutral element, namely, the empty word. A code $C = \{c_1,...,c_i\}$, then, is a subset of A*, where the elements of c_i are the words of the code and n_i denotes the size of the word c_i.

A code is said to be binary, ternary or, generally, n-ary when A is formed by, respectively, two, three or, generally, n symbols. If all the words are of the same length, C is said to be a fixed-length or block code. Otherwise, C is said to be a variable-length code.

Let σ be the alphabet of a source of information, which is defined as the source that emits symbols of an effect $\sigma\{s_1,...,s_5\}$, whose probability of appearance is given by $\pi_i (1 \le i \le r)$. Then a coding is an application φ of σ in C, which is extended to an application φ* of σ* in C*. And, of course, decoding is an application ψ of C* in σ*. A code has only one decoding, i.e. φ* is injective, and ψ or ψ* is the identity.

Genetic Code

The expression genetic code is now used to mean two very different things. The lay public uses it often to refer to the full genetic message of the organism. Molecular biologists use it to allude to the small dictionary that shows how to relate the four-letter language of the nucleic acids to the twenty-letter language of the proteins in the same way as Morse code relates the dots-and-dashes language to the twenty-six letters of the alphabet. Here we will use the term in this sense. However, the technical term for such a rule of translation is not, strictly speaking, *code* but cipher, just as Morse code should be called Morse cipher. This Crick did not know at the time, which was a stroke of luck, as *genetic code* sounds much better than *genetic cipher*. In actual fact, Crick correctly referred to the set of bases contained in the chromosomes as ciphered text or key, but added that the term *key* or ciphered text was too limited. The chromosomal structures are at the same time the instruments that develop what they foresee. They represent both the legal text and the executive power or, to use another comparison, they are both the architect's plans and the builder's workforce.

Code is the core of molecular biology just like the periodic table of elements is at the heart of chemistry, but there is a profound and transcendental difference. The periodic table is almost certainly valid and true all over the universe. However, if there is life in other worlds, and if this life is based on nucleic acids and proteins, of which there is no guarantee, it is very likely that the code there would be substantially different. There are even small code variations in some terrestrial organisms. Genetic code, like life itself, is not an aspect of the eternal nature of things, but, at least partly, product of an accident.

The central lemma of genetic code is the relationship between the sequence of the bases of DNA or of its transcribed m-RNA and the sequence of protein amino acids. This means that the sequence of bases of a gene is collinear with the sequence of amino acids of its product polypeptide. What makes this code, which is the same in all living beings, so marvellous is its simplicity. A set of three bases or codon specifies an amino acid. The t-RNA molecules, which work as protein synthesis adaptors, read the codons sequentially.

Crick et al.'s 1961 experiments (Crick et al., 1961) established that genetic code had the following characteristics:

1. **The alphabet:** A1 = {A,G,C,T}.
2. **Coding relationship:** A group of three bases codes an amino acid. This group of three bases is called, as mentioned above, codon or triplet.
3. **Non-optimality:** The fact that there is a code is such is likely to be due to structural, electrochemical and physical criteria applied to the molecules involved in the described processes. The optimal base is actually 3 (Pazos, 2000).
4. **Non-overlapping:** In a code of non-overlapping triplets, each codon specifies only one amino acid, whereas in a code of overlapping triplets, ABC specifies the first amino acid, BCD the second, CDE the third and so on. Studies of the amino acid sequence in the protein cover of mutants of the tobacco mosaic virus indicated that normally only one amino acid altered, leading to the conclusion that genetic code does not overlap.
5. **Codon reading sequentiality:** The sequence of bases is read sequentially from a fixed starting point. There are no commas, i.e. genetic code does not require any punctuation or signal whatsoever to indicate the end of a codon or the start of the next one. There is only an "initiation signal" in the RNA, which indicates where the reading should start. This signal is the AUG codon that codes the amino acid methionine.
6. **Code inefficiency or degeneracy:** There is more than one *word* or coding codon for most amino acids. Mathematically, this means that there is a *superjective* application between the codons and the amino acids plus the chain initiation and termination signals, which is transcendental for the subject of this paper. Table 2 shows genetic code. Only tryptophan and methionine are encoded by a single triplet. Two or more triplets encode the other eighteen. Leucine, arginine and serine are specified by six codons each. However, under normal physiological conditions, the code is not ambiguous: each codon designates a single amino acid. From this table, the respective amino acid can be located, given the position of the bases in a codon, as follows. Suppose we have the m-RNA codon 5'AUG 3', then we start at A in Table 2, then go to U and then to G and we find methionine.

Figure 1 shows a finite-state automaton that recognises the DNA alphabet and translates the codons into amino acids.

7. **Importance of the bases in the codon:** The codons that specify the same amino acid are called synonyms, e.g. CAU and CAC are synonyms for histidine. Note that the synonyms are not distributed arbitrarily

Figure 1. A finite-state automaton that translates genetic code to amino acids

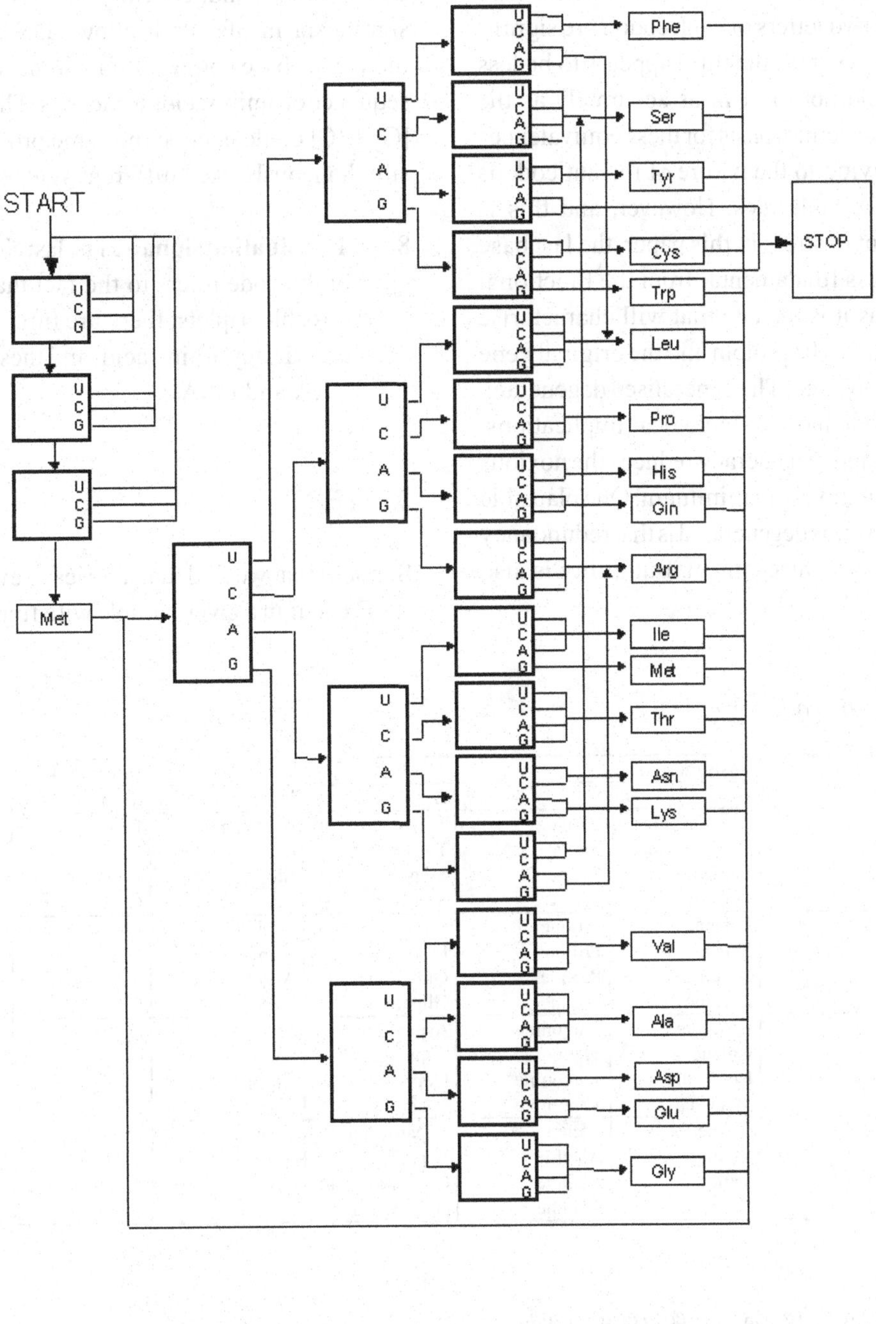

in Table 2. An amino acid specified by two or more synonyms occupies only one cell, unless there are over four synonyms. The amino acids in each cell are specified by codons whose first two bases are the same but differ as to the third, e.g. GUU, GUC, GUA and GUG. Most synonyms differ only as to the last base of the triplet.

Looking at the code, we find that XYC and XYU always code the same amino acid, whereas XYG and XYA almost always do. It is clear then that the first two letters of each codon are significant factors, whereas the third appears to be less important and not to fit in as accurately as the others. The structural basis for these equivalences is evident owing to the nature of the anticodons of the t-RNA molecules. However, and this is what we want to stress in this paper, the last base of the codon is fundamental from the functional viewpoint, as it is the one that will characterise the behaviour of the protein and the original gene in which it is located. The generalised degeneracy of genetic code has two biological implications. On the one hand, degeneracy reduces the noxious effects of mutations to a minimum. One plausible reason why code is degenerated is that redundancy provides a safety mechanism. But utility is also a possible ground: it is quite plausible that the silent mutations may have long-term benefits. On the other hand, code degeneracy can also be significant insofar as it allows DNA to largely modify its base composition without altering the sequence of amino acids it encodes. The content of [G] + [C] could encode the same proteins if they simultaneously used different synonyms.

8. **Termination signals:** The last characteristic of the code refers to the fact that of the 64 possible triplets, there are three that do not encode any amino acid, and these are UAG, UAA and UGA.

χ^2 TEST

Suppose that we find that a series of events E_1, E_2, …, E_j occur in a given sample with frequencies o_1,

Table 2. Genetic code

Position 1 (5' end)	Position 2				Position 3 (3' end)
	U	C	A	G	
U	Phenylalanine	Serine	Tyrosine	Cysteine	U
	Phenylalanine	Serine	Tyrosine	Cysteine	C
	Leucine	Serine	Stop	Stop	A
	Leucine	Serine	Stop	Tryptophan	G
C	Leucine	Proline	Histidine	Arginine	U
	Leucine	Proline	Histidine	Arginine	C
	Leucine	Proline	Glutamine	Arginine	A
	Leucine	Proline	Glutamine	Arginine	G
A	Isoleucine	Threonine	Asparagine	Ser	U
	Isoleucine	Threonine	Asparagine	Ser	C
	Isoleucine	Threonine	Lysine	Arginine	A
	Met (Start)	Threonine	Lysine	Arginine	G
G	Valine	Alanine	Aspartic acid	Glycine	U
	Valine	Alanine	Aspartic acid	Glycine	C
	Valine	Alanine	Glutamic acid	Glycine	A
	Valine	Alanine	Glutamic acid	Glycine	G

Table 3. Sample with associated frequencies

Event	E1	E2	…	Ej
Observed frequency	o1	o2	…	oj
Expected frequency	e1	e2	…	ej

$o_2, \ldots, o_j$, called "observed frequencies", and that, according to the rules of probability, they would be expected to occur with frequencies $e_1, e_2, \ldots, e_j$, called theoretical or expected frequencies, as shown in Table 3.

A measure of the discrepancy between the observed and expected frequencies is given by the χ^2 statistic, as

$$\chi^2 = \frac{(o_1 - e_1)^2}{e_1} + \frac{(o_2 - e_2)^2}{e_2} + \ldots\ldots\ldots + \frac{(o_j - e_j)^2}{e_j} = \sum_{i=1}^{j} \frac{(o_j - e_j)^2}{e_j}$$

where, if the total frequency is N, $\Sigma o_i = \Sigma e_j = N$.

If $\chi^2 = 0$, the observed and expected frequencies are exactly equal, whereas if $\chi^2 > 0$, they are not. The greater the value of χ^2, the greater the discrepancies between the two frequencies are.

The χ^2 sample distribution is very closely approximated to the chi-square distribution given by

$$Y = Y_0 (\chi^2)^{1/2(v-2)} e^{-1/2 \chi^2} = Y_0 \chi^{v-2} e^{1/2 \chi^2}$$

where v is the number of degrees of freedom given by:

a. v = k-1, if the expected frequencies can be calculated without having to estimate population parameters from the sample statistics.
b. v = k-1-m, if the expected frequencies can only be calculated by estimating m parameters of the population from the sample statistics.

In practice, the expected frequencies are calculated according to a null hypothesis H_0. If, according to this hypothesis, the calculated value of χ^2 is greater than any critical value, such as $\chi^2_{0.95}$ or $\chi^2_{0.99}$, which are the critical values at the significance levels of 0.05 and 0.01, respectively, it is deduced that the observed frequencies differ significantly from the expected frequencies and the H_0 is rejected at the respective significance level. Otherwise, the H_0 will be accepted or, at least, not rejected.

Now, looking at Table 4 taken from Jack Lester King and Thomas H. Jukes (1969), we ran the χ^2 test on these data and found that the value of χ^2 is 477.809. As this value is much greater than the expected $\chi^2_{0.95}$, which, for the 19 degrees of freedom in respect of the twenty amino acids, is 38.6, we find that the observed frequencies differ very significantly from the expected values, thereby rejecting the H_0. Accordingly, it is to be expected that the bases are not associated as triplets at random and, therefore, an explanation needs to be sought.

DECIPHERING THE GENOME

As already mentioned, genes are usually located in the chromosomes: cellular structures whose main component is deoxyribonucleic acid, abbreviated to DNA. DNA is formed by complementary chains, made up of long sequences of nucleotide units. Each nucleotide contains one of the four possible nitrogenised bases: adenine (A), cytosine (C), thymine (T) and guanine (G), which only associate in two possible ways with each other: A with T and C with G. Usually, some portions of DNA form genes and others do not. In the case of human beings, the portions that are genes make up only approximately 10% of total DNA. The remainder appears to have nothing to do with protein synthesis; it is, until it finds a functionality, genetic trash so to speak. In any case, the reading and interpretation of this set of symbols that make up DNA can be compared to deciphering the hieroglyphics of life. If Jean-François Champollion's deciphering of hieroglyphic script from the Rosetta Stone was arduous and difficult, imagine deciphering 3×10^9 symbols from a four-letter alphabet. To give an idea of the magnitude of the endeavour, suffice it to say, for example, that the sequence of DNA nucleotides written would take up a space equivalent to 150

Table 4. Amino acids and triplets associated with their frequencies according to King and Jukes (1969)

AMINO ACIDS	Triplets	Number of appearances	Observed frequency (%)	Expected number	Expected frequency (%)	$\frac{(O_i - O_i)^2}{e_i}$
Serine	UCU.UCA.UCC UCG.AGU.AGC	443	8.0	472	8.6	1.782
Leucine	CUU.CUA.CUC CUG.UUA.UUG	417	7.6	434	7.9	0.666
Arginine	CGU.CGA.CGC CGG.AGA.AGG	230	4.1	582	10.6	212.9
Glycine	GGU.GGA.GGC GCG	406	7.4	390	7.1	0.656
Alanine	GCU.GCA.GCC GCG	406	7.4	395	7.2	0.306
Valine	GUU.GUA.GUC CUG	373	6.8	330	6.0	5.603
Threonine	ACU.ACA.ACC ACG	340	6.2	373	6.8	2.919
Proline	CCU.CCA.CCC CCG	275	5.0	275	5.0	0
Isoleucine	AUU.AUA.AUC	209	3.8	280	5.1	18.0
Lysine	AAA.AAG	395	7.2	302	5.5	28.639
Glutamic acid	GAA.GAG	318	5.8	258	4.7	13.953
Aspartic acid	GAU.GAC	324	5.9	192	3.5	90.750
Phenylalanine	UUU.UUC	220	4.0	121	2.2	81
Asparagine	AAU.AAC	242	4.4	225	4.1	1.284
Glutamine	CAA.CAG	203	3.7	210	3.8	0.233
Tyrosine	UAU.UAC	181	3.3	165	3.0	1.551
Cysteine	UGU.UGC	181	3.3	137	2.5	14.131
Histidine	CAU.CAC	159	2.9	164	3.0	0.152
Methionine	AUG	99	1.8	99	1.8	0
Tryptophan	UGG	71	1.3	88	1.6	3.284
TOTAL		**5,492**	**100.0**	**5,492**	**100.0**	$\chi^2 = 477.809$

volumes similar to the telephone book of a city like Madrid, with four million inhabitants. Or, to establish a more illustrative comparison, if a virus gene that has 3,000 pairs of bases takes up one page of a book composed of 3,000 letters and a gene of a bacterium, which contains three million pairs of bases, would be equivalent to a 1,000-page book, the human genome, composed of three thousand million bases, would take up a library of a thousand books. Perhaps one form of deciphering the hieroglyphics would be what Eric Steven Lander of the MIT proposed when he said that just as the organisation of the chemical elements in the periodic table lent coherence to a mass of previously unrelated data, we will end up discovering that the tens of thousands of genes of existing organisms are made of combinations of a much smaller number of simpler genetic modules or elements or, so to speak, primordial genes. The important point is that human genes should not be considered as completely different from each other. Rather, they should be seen as families that live all sorts of lives. Having completed the table, structural genetics will leave the field open to functional genetics or the practical use of the table. For example, the difference between two people boils down to no more than 1% of the bases. Most genes have only two, three or four variants, that is, some 300,000 principal variants.

Going back now to deciphering the genome, sequencing the genome refers to solely stating the bases A, T, C and G of the genome, i.e. reading without understanding. Therefore, it is to spell out or, at most, babble the genome. For the time being, genome sequencing simply involves determining how the thousands of millions of bases of which it is composed are chained. As Daniel Cohen (1994) said, who we follow in this section, it is not hard to realise how arid the task is, as it involves examining a text that more or less reads as follows:

TCATCGTCGGCTAGCTCATTCGACCATCG-TATGCATCACTATTACTGATCTTG...,

and goes on for millions of lines and thousands of pages. Of course, it was to be expected that a language that has a four-letter alphabet would be much more monotonous than contemporary languages, whose Latin, Cyrillic alphabets, etc., are composed of over twenty letters.

But sequencing the genome is not the last stop, as it has to be deciphered, i.e. its meaning has to be understood like learning to read letters and converting them into ideas. And this is the tricky thing. There is, in actual fact, no way of foreseeing when, how and to what extent it will be possible to decipher the sequenced genome. It would be marvellous, but not very realistic, to suppose that learning to read the genome would unexpectedly lead to its understanding just as children learn the letters and suddenly cross some mysterious threshold and start to understand what they are reading. Reality is, however, much tougher, and the deciphering of the genome looks like a very hard and thankless task. This is due not so much to the poverty of the alphabet but to the ignorance of the language.

Just to give an illustrative example of what we have just said, try to read, albeit for no more than twenty minutes, and aloud (no cheating!), a novel written in a language you don't know (Turkish, Serbo-Croatian, etc.) transliterated to the twenty-six letter Latin alphabet and see what a headache you get. Suppose now that you have no choice, because you are on a desert island and the only book you have is written in one of these languages and you do not have dictionary on hand. Therefore, you would have to make do with what little you know and be patient and perseverant. If you do this, you will end up becoming acquainted with some features of the unknown language. For example, you will identify recurrent patterns, establish analogies, discover some rules, other meanings, interesting similarities, etc., etc., etc. The first thing we find is that the genetic language has a peculiarity, which, at least in principle, is disconcerting: it does not just consist of sequences furnished with a precise meaning, the sequences of interest which are called genes. It all includes jumbled paragraphs situated both between gene and gene, intergenes, and inside the genes, intragenes, which divide the meaningful sequence. To date, no one has been able to find out what all this filling is for. And, what is even more exasperating is that these extravagant series of letters make up over ninety per cent of the genome, at least, the human genome, which is an interminable list of genes and intergenes with non-coding intragenes situated within the very genes.

Now consider a volume of poetry by Quevedo (1995), but written in an unknown language and according to the following genomic style, although for the readers' comfort, the text has been translated into Spanish:

kvlseimkmifdsqmoieurniñaedpvaeighmlkehimdogcoleizoapglkchdjthzkeivauozierdmofmoimthaaoekkkkkkkghdmorsleidjtldifltsfifesithgmelimlkajchwchmqurozqdaverirlmeoarusndorkejmtsimeormtlehoekdmzriglalmethoslerthrosazlckthmekromlsdigquemelthnlslrejtestalkhlrjjsletehrejrozthiolalelyrletuolgdhnartlgldrtlmalkjfsdñanaioemzlekthldiimsekjrmlkthomsdlkgldheozkelldesgyrureiotpleghkdssseruieormdkfjhjkddddgoghfjdsenbvcxxwqsd ...

Biologists are now beginning to distinguish the characteristic sequences that initiate and terminate the words of the genome and can better identify genes from what are not genes. Therefore, with a bit of training, geneticists will manage to pick out, from the above mess, the following sentence:

Quitar codicia no añadir dinero hace a los hombres ricos Casimiro

But, even after identifying the words of the poem, they would still be a long way away from imagining what Casimiro, to whom the poem is addressed, was like.

Returning to the analogy with the human language, the task facing those who are sequencing the genome is to make an inventory of the thousands of words of the biological dictionary of humanity. Only the inventory. The explanation of the words will come afterwards. To give an idea of the endeavour, this dictionary, at a rate of thirty genes and approximately a million characters per page, will have three thousand pages, given that the genome has three thousand million bases. The equivalent, in pages and weight, of the two-volume Diccionario de la Real Academia de la Lengua Española.

But, unlike usual dictionaries, the words will be arranged not in alphabetical but in chromosomal order. Instead of being grouped in the respective sections A to Z, the words of the human genomic dictionary would be arranged in 23 chapters, one for each pair of chromosomes duly numbered from 1 to 23 in the conventional order attributed to them by biologists. The genes of the first pair are the longest, whereas those of the twenty-first and twenty-second are the shortest. These are, as mentioned above, the chromosomes termed "autosomes", i.e. non sexual, from the Greek *soma*, meaning body. The twenty-third pair is the sex chromosomes X and Y.

So, in the first chapter, a chromosome of *x* million bases would take up *x* pages at a rate of a million characters per page, in the second, *y*..., and in the twenty-first, *z*. "Beaconing" and mapping the genome is equivalent to paginating the dictionary; ordering the pages from 1, 2, 3, ... to three thousand, in 23 chapters, yet without filling the pages in. At this stage, we still do not know what there is on each of the pages, except for a thousand words, set out here and there. Having done the pagination or, if you prefer, map-making, the experts will be able to start on the systematic sequencing of the genes, i.e., fill the pages with words still without explanation.

Then, in a few years time, when we have the sequence of the whole human genome, we will see that page 1 of chapter 1 contains the words, or their genomic equivalent, not in alphabetical order (a, aal, aam, Aaronic, ..., ab, aba, abaca), but arranged as stipulated by nature through the chromosomes (e.g., grace, disgrace, graceless, followed by a word, like "everyday", which bears no resemblance to its foregoer, then gracecup, graced, followed by an "incongruous" plum and division and, then, disgraceful, aggrace, begrace, etc.). Looking at these words more carefully, we find that many are composed of the letters "grace" and we wonder what this common factor means. Further on, we find that the common factor "gen" appears in genetic and also, in another place in another chapter, in genesis, genius, ingenious, genial, etc. All the "grace" and all the "gen" can then be entered into a computer to discover how often they appear and fully examine the sequences in which they are accommodated. The whole thing can then be reclassified taking note of the significant statistical data. On other occasions, researchers may come across a series of the style: complain, complainant, complaint, plaint, plaintive, plaintiff, and may be fortunate enough to know the meaning of its common factor, in this case, "plaint", because it is associated with some hereditary particular or other that has already been studied at length. They will then be able to venture to try out new rules of construction, new combinations, taking the same common root. This is how the cartouches helped to decipher hieroglyphics. In other words, genes, like the words of the language can be grouped by families, the words of different human languages that revolve around the same concept share the same root. Likewise, it is to be supposed that the genes whose sequence is similar fulfil similar functions. As a matter of fact, it now appears that genes that resemble each other come from one and the same ancestral gene and form families of genes with similar functions. These are termed multigenic families, whose genes may be disseminated across several chromosomes. Often, neighbouring genes have a similar spelling even if they do not start with the same letters. And homonymic genes are often synonyms, genes that are written differently but have similar meanings. This will mean that they can be used like *puntales* (a fragment of plain text associated with a fragment of ciphered text) were used to decipher secret codes.

The analysis will be gradually refined and fine-tuned. Finally, we will know how to distinguish the equivalents of linguistic synonyms. By detecting the common roots, we will even learn to trace back the genealogy of certain genes, i.e. follow their evolutionary lineage. Complex biological functions like breathing, digestion or reproduction are assimilated to sentences whose words are inscribed on different pages of the genomic dictionary. Now, if there is a first book, the genetic dictionary written according to the topographical order of the chromosomes, evolution has written another thousand after learning this dictionary, on physiology, growth, ageing, immunity, etc., to the book of thought, which, doubtless, will never be finished. It is even conceivable that one and the same gene could acquire a different meaning depending on its position in one or other genetic sentence, just as a word in human language depends on the context. Additionally, biological functions have been invented as evolution has become more complex. It is likely that genetic combination was at the same time enriched by new rules integrating earlier levels in as many other Russian dolls of *algorithms*, placed inside each other in a subtle hierarchy.

In sum, the syntax and style of the genetic language has gradually been refined, and it remains for us to discover its semantics and pragmatics. The messages that determine eye colour, skin texture or muscular mass are doubtless not the same as those that induce the immune system, cellular differentiation or cerebral *wiring*. Obviously, many fundamental concepts of this language are unknown. Even after sequencing is complete, there will still be a lot to research to do and it will take perhaps centuries of work to get things straight, if we ever do. The question is that DNA is neither an open book nor a videotape.

FUTURE TRENDS

Now already numerous, wide-ranging and rewarding, the interrelations between genetics and computing will expand increasingly in the future. The major trends that are likely to be the most constructive are:

a. **Biological computation:** Under this label, DNA computing deserves a mention. While the early work on DNA computing dates back to Adelman, there is still a long way to go before we can build what is now being referred to as the *chemical universal Turing machine* (CUMT). Its construction would have an extraordinary effect on the understanding of genetics and computer science, as it would allow theoretical and experimental approaches and models in both fields to be combined holistically.

b. **DNA, the brain and computers have one thing in common:** All three process information. Consequently, it is evident that, at a given level of abstraction, their operational principles will be the same. This will lead, sooner or later, to the discovery of an information theory that accounts for the behaviour of the three processors and even more. In actual fact, the world has been considered so far as being formed by matter and energy, both of which are, since Einstein's famous formula of human destiny, $E=\pm mc^2$, equivalent. Now, to understand today's world (both at the macroscopic level, for which the general theory of relativity accounts, i.e. black holes, and the microscopic level accounted for by quantum physics, i.e. Wheeler's delayed choice experiment), information needs to be added to the matter-energy equation. Now, this theory would of course encompass Shannon's communication theory, but would go further than Shannon's premise does. It might perhaps only retain his notions of the information unit "bit" and negative entropy. One of the authors is already working in this field and expects to have some preliminary results to report in a few months' time.

c. **Genetics-computation hybridization:** The exchange of approaches between genetics and computation will provide a hybrid form of dealing with problems in both fields. This will improve problem solving in both domains. For example, geneticists will be able to routinely apply concepts commonly used in computing, like abstraction or recursiveness. This way they will acquire profound skills for solving complex problems. Additionally, the huge quantities of information that DNA employs to develop its full potential, as well as the complexity of its workings, will be excellent guides for dealing with problems in the world of computing.

CONCLUSION AN FUTURE RESEARCH LINES

The classical scientific dogma, which is or should be inculcated to any university student, is that first conjectures or working hypotheses are formulated and are then tested. But, of course, to formulate such conjectures or hypotheses, the facts, as such, need to be taken into account, and these facts are:

a. Proteins owe their function to their structure or folding, i.e. to their shape, which depends on the order of the amino acid sequence of which they are composed. And this order is again determined by the sequence of the DNA bases.
b. From Table 4, taken from King and Jukes (King, 1969), we calculated the χ^2 and found that the χ^2 test results offer no doubt as to the fact that the distribution of the triplets and their translation to amino acids is not due to chance, quite the opposite.
c. According to genetic code, we know that several triplets yield the same amino acid.

This leads us to formulate the proposed working hypothesis or conjecture:

An individual's *situation* will depend on what triplet and in what position it yields a particular amino acid.

Testing:

To test this conjecture, we have to, and this is what we are in the process of doing, take the following steps.

S1. Determine a genetic disease of unique aetiology

S2. Try to associate the possibilities of "triplets versus generated amino acids" relationships within the gene causing the disease and establish, if possible, a causal relationship and, if not, a correlation. For this purpose, we have to define the right sample size.

S3. If a causality or correlation greater than 0.8 is established, go to S1 with a new case. If the number of cases is greater than the proposed sample size and causality or correlation was established for all cases, accept the hypothesis, if not, reject it.

Of course, when genetic code has been completely deciphered, this type of hypotheses will make no sense, because the DNA will explain its message and its resultant consequences. But, in the meantime, it is a way of understanding why some diseases occur. This will, of course, lead to its prevention, cure or, at least, to the relief of its effects, through genetic engineering.

REFERENCES

Butler, S. (1982). *Erewhon.* Barcelona: Bruguera.

Cohen, D. (1994). *Los Genes de la Esperanza.* Barcelona: Seix Barral.

Crick, F.H.G.; Barnett, L.; Brenner, S. & Watts-Tobin, R.J. (1961) General Nature of the Genetic Code for Proteins. *Nature*, 192, 1277-1232.

Fogel, L. & Atmar, J.W (Eds.) (1992) *Proceedings First Annual Conference on Evolutionary Programming.*

Holland, J. H. (1975) *Adaptation in Natural and Artificial Systems.* Ann Arbor, Michigan: University Michigan Press.

Hu, S-T. (1965). *Elements of Modern Algebre.* San Francisco, Ca: Holden-Day, Inc.

Kemeny, J. G. (1955). Man Viewed as a Machine. *Scientific American,* 192 (4), *58-67.*

King, J. L. & Jukes, T. H. (1969). Non Darwinian Evolution. *Science*, 164, 788-798.

Koza, J.R. (1992). *Genetic Programming.* Reading, MA: The MIT Press.

Michalewitz, Z. (1992). *Genetic Algorithms + Data Structures = Evolutionary Programs.* New York: Springer Verlag.

Pazos, J. (2000). El Criptoanálisis en el Desciframiento del Código Genético Humano. In J. Dorado et al. (Eds.), *Protección y Seguridad de la Información* (pp. 267-351). Santiago de Compostela: Fundación Alfredo Brañas.

Penrose, L. J. (1974). Máquinas que se Autorreproducen. In R. Canap, et al., *Matematicas en las Ciencias del Comportamiento* (pp: 270-289). Madrid: Alianza Editorial, S.A.

Quevedo, F. de. (1995). *Poesía Completa I* (pp: 43-44). Madrid: Turner Libros, S.A..

Rechenberg, I. (1973). *Evolutionsstrategie: Optimierung Technischer Systeme nach Prinzipien der Biologischen Evolution.* Stuttgart, Germany: Fromman-Holzboog Verland.

Schrödinger, E. (1944). *What is Life?* Cambridge: Cambridge Universities Press. Spanish translation (1983). *¿Qué es la Vida?* (pp: 40-42). Barcelona: Tusquets Editores, S.A.

Tipler, F. J. (1980). Extraterrestrial Intelligent Beings Do Not Exist. *Quarterly Journal of the Royal Astronomical Society,* 21, 267-281.

Tipler, F. J. (1982). We Are Alone In Our Galaxy, *New Scientist*, 7 October, 33-35.

Von Neumann, J. (1951). La Teoría Lógica y General de los Autómatas. In L.A. Jeffres (Ed), *Cerebral Mechanisms in Behaviour* (pp. 1-41). New York: John Willey and Sons.. And in, J. von Neumann (1963). *Collected Works* (pp 288-328). Oxford: Pergamon.

Von Neumann, J. (1966). *Theory of Self-Reproducing Automata*. Urbana. Illinois: Illinois University Press.

Watson, J. D. & Crick, F. H. C. (1953 a). Molecular Structure of Nucleic Acid. A Structure for Deoxyribose Nucleic Acid. *Nature* 171, 737-738.

Watson, J. D. & Crick, F. H. C. (1953 b). Genetic Implications of the Structure of Deoxyribonucleic Acid. *Nature* 171, 964-967.

ADDITIONAL READING

Adleman L. M. (1998). Computing with DNA. *Scientific American Magazine* nº 279 pp. 54-61.

Adleman L. M. (1994). Molecular computation of solutions to combinatorial problems. *American Association for the Advancement of Science* Vol. 266, Issue 11 pp. 102-1024 Washington, DC, USA Univ. of Southern California, LA.

Bekenstein J. D. (2003) La información en el Universo Holográfico. *Investigación y ciencia*, nº 325, pp. 36-51.

Benenson Y, Gil B, Ben-Dor U, Adar R, Shapiro E. (2004). An Autonomous Molecular Computer for Logical Control of Gene Expression. *Nature*, 429:423–429.

Copeland, B. J. y Proudfoot, D. (1999). Un Alan Turing desconocido. Prensa Científica S.A. *Investigación y Ciencia* nº 273.

Feynman, R.P. (1960). There's Plenty of Room at the Bottom: An Invitation to enter a New World of Physics. *Engineering and Science* 10:23 (5) pp. 23-36.

Gifford D. K. (1994). On the Path to Computation with DNA. in Science, Vol. 266, pp. 993–994.

Gray J, Liu DT, Nieto-Santisteban M, Szalay AS, De Witt D, Heber G. (2005). Scientific Data Management in the Coming Decade. *Technical Report MSR-TR*-2005-10. Microsoft Research.

Griffiths, A.J.F.; Gelbart, W.M.; Miller, J.H. and Lewontin, R.C. (2002). *Genética Moderna* 1ª Edición en español. McGraw-Hill/Interamericana.

Gruska J. (1999). *Quantum Computing*. McGraw-Hill.

Kurzweil R. (2006). When Computers Take Over. Nature Publishing Group. *Nature*, Books and Arts, vol 437(440) pp. 421-422 23.

Lehn J-M. (2004). Supramolecular Chemistry: from Molecular Information Towards Selforganization and Complex Matter. *Reports on Progress In Physics* 67:249-265.

Lipton R. J. (1995). DNA solution of Hard Computational Problems. *Science*, 268: 542-545.

Lloyd S. (2002). Computational Capacity of the Universe. *Physical Review Letters* 10Volume 88, nº 23.

Lloyd S. (2000). Ultimate Physical Limits to Computation. *Nature* 406 Aug 31; 406(6799): 10 47-54.

Martin-Vide C.; Paun G.; Pazos J.; Rodríguez-Paton A. (2003) Tissue P systems. Elsevier. *Theoretical Computer Science*, Vol. 296, Number 2, 8, pp. 295-326(32).

Mount, D.W. (2004). *Bioinformatics: Sequence and Genome Análisis*. Edition: 2nd edition, Cold Spring Harbor Laboratory Press.

Nielsen M. A. and Chuang I. L. (2007). *Quantum Computation and Quantum Information*. Cambridge University Press. Cambridge UK.

Regev A., Shapiro E. (2002). Cellular Abstractions: Cells as Computation. *Nature*, 419: 343.

Russell, P.J. (2002). *iGenetics*. Benjamin Cummings. USA

Wing J. (2006). Computational Thinking. *Communications of the ACM*. Vol. 49, No. 3, pp. 33-35.

Wolfram S.A. (2002). *New Kind of Science*. Champaign, IL:Wolfram Media Inc.

Zauner K-P. (2005). Molecular Information Technology. *Critical Reviews in Solid State and Material Sciences*, Volume 30, nº 1, pp. 33-69(37).

Chapter XI
Artificial Mind for Virtual Characters

Iara Moema Oberg Vilela
Universidade Federal do Rio de Janeiro, Brazil

ABSTRACT

This chapter discusses guidelines and models of Mind from Cognitive Sciences in order to generate an integrated architecture for an artificial mind that allows various behavior aspects to be simulated in a coherent and harmonious way, showing believability and computational processing viability. Motivations are considered the quantitative, driving forces of the action selection mechanism that guides behavior. The proposed architecture is based on a multi-agent structure, where reactive agents represent motivations (Motivation Agents) or actions (Execution Agents), and cognitive agents (Cognition Agents) embody knowledge-based attention, goal-oriented perception and decision-making processes. Motivation Agents compete for priority, and only winners can activate their corresponding Cognition Agents, thus filtering knowledge processing. Active Cognition Agents negotiate with each other to trigger a specific Execution Agent, which then may change internal and external states, displaying the corresponding animation. If no motivation satisfaction occurs, frustration is expressed by a discharge procedure. Motivations intensities are then accordingly decreased.

INTRODUCTION

Convergence of artificial life, artificial intelligence and virtual environment techniques has given rise to intelligent virtual environments (Aylett and Luck, 2000; Aylett and Cavazza, 2001; Thalmann, 2003; Osório, Musse, Santos, Heinen, Braun and Silva, 2005). The simulation of inhabited virtual worlds where lifelike forms behave and interact may then provide more realistic visualizations of complex emergent scenarios. Applications are not restricted just to entertainment systems like game and virtual storytelling, as it may seem (Rist, André, and Baldes, 2003). Education (Antonio, Ramírez, Imbert, Méndez and Aguilar, 2005; Tortell and Morie, 2006; Chittaro, Ieronutti and Rigutti, 2005), training in dangerous situations involving people (Miao Hoppe and Pinkwart,

2006; Braga, 2006; Querrec, Buche, Maffre and Chevaillier, 2004), simulation of inhabited environment for security, adequacy analysis and evaluation or historical studies (Papagiannakis, Schertenleib, O'Kennedy, Arevalo-Poizat, Magnenat-Thalmann, Stoddart and Thalmann, 2005), product or service demonstration (Kopp, Jung, Lessmann and Wachsmuth, 2003), ergonomics (Colombo and Cugini , 2006; Xu, Sun and Pan, 2006), are some of many others possible uses for Intelligent Virtual Environments (IVE).

The key issue about virtual environments is *immersion*, or, as it is usually said, the suspension of disbelief. The system developer's first concern tends to be related to graphical and sound issues, since senses are very important to involvement in a virtual world. However, when related to dynamic and complex environments, the believability of the virtual world elements behavior becomes paramount, especially if it includes life simulation. There are many levels of activity to be simulated, and modeling depends heavily on which is the main concern of the application. For instance, if we are simulating a garden just for aesthetic appreciation, there is no point in simulating complex interactions occurring between plants, and plants and other organisms of the world. It is different, though, if the same world is being simulated for ecological analysis.

This problem becomes more complex when virtual humans or humanoids are to be simulated. To show believability, the behavior must express an internal life, some kind of goal-driven attitude, even when it is erratic as in drunk or crazy people. But on what basis is it possible to accomplish this?

Virtual environment inhabitants are usually developed as agents with some varying levels of autonomy, which encapsulate specific context-sensitive knowledge to accomplish their role in the system application. What may be considered "behavior" goes from just body movements to complex interactions with the environment, depending on the character role. It is expected that when the underlying mechanism of behavior production is more similar to the actual functioning in the real world, more the resulting virtual behavior will be believable. That tends to be particularly true when complex and rich virtual worlds are being created. If a VE is simple, it may suffice just to imitate the real behavior. But if the virtual world is complex, and relies on emergence to produce behavior, just to emulate it may be too risky.

The present chapter focuses on virtual character behavior as believable sequence of actions of humanlike virtual entities, and not on character computer animation, or virtual biomechanical body movements. Some approaches are discussed, and a motivation-driven architecture is proposed based on assumptions derived from cognitive science.

The first part of the chapter focuses on various forms of modeling the relationship between characters and their environment in order to produce believable behavior. The second part discusses different approaches for modeling character behavior, and suggests the convenience of integrating the underlying mechanisms in a single high-level structure, the artificial mind. The third part presents a proposed motivation-driven architecture for artificial minds. Finally, conclusions are presented and future trends are discussed.

CHARACTER INTELLIGENCE AND VIRTUAL ENVIRONMENT

As already pointed out, immersion is a key issue when virtual environments are considered. Of course, when they are populated, the "suspension of disbelief" requirement is extended to their inhabitants. Believable lifelike virtual characters have to behave as if internal forces are driving them. They have to show autonomy and some kind of environment awareness. How can this be accomplished? Two approaches are possible:

the first is to focus on producing behaviors similar to those seen in real life, no matter by what means; the second is to achieve believability by using plausible mechanisms to generate character behavior. While the former just tries to imitate real behavior, the latter goes further, trying to simulate scientific models of behavior, mind and/or brain. Between those approaches, a range of rigorousness in applying those models may exist, going from not taking them into account, to attempts to strictly implement them. Next, the two approaches are discussed regarding their use of Artificial Intelligence techniques.

Smart Environments

When underlying mechanisms of behavior are not relevant to the application, Artificial Intelligence techniques may be not restricted to the "living" entities of the virtual world. Intelligence may be conveniently distributed in the environment. This technique may increase the computational efficiency of the whole simulation, especially when there is a large population of virtual characters.

That is the case, for instance, of the popular game The Sims® (Schwab, 2004). From the player perspective, this game is about following and taking care of specific characters in their everyday life. Apparently, they are virtual humans with internal needs, aware of the environment elements that can fulfill them. However, they are relatively simple agents responding to messages sent by objects of the environment that are broadcasting what they have to offer to nearby characters. These objects encapsulate the knowledge of which needs they satisfy and what happens when a character interact with them (like the correspondent animations, sound effects and behaviors). This is called "smart terrain". The environment guides the sequence of character actions. If a character walks close to a refrigerator, the refrigerator sends a message advertising that it can satisfy hunger. When the character takes the food from the refrigerator, the food broadcasts that it needs cooking and the microwave oven broadcasts that it can cook food. So, the character next action is to put the food in the oven. The characters just have to walk around in their world and manage their needs priorities. They don't have anything like a "mind" or "brain" structure.

In PetroSim Project (Musse, Silva, Roth, Hardt, Barros, Tonietto, and Borba, 2003; Barros, Silva, and Musse, 2004), also, the characters' structure does not contain all the information they need to choose their next action. A module extracts environment information from the 3D model of the virtual world and stores it on a database. The automatic reading of the geometric model provides elements like streets, houses, etc., and then a hierarchical tree is built based on this information. User-defined attributes may be added to the tree, if they are considered relevant to the application (such as risk areas, for example). When characters need information about the world they search it on the database, instead of obtaining it directly from the environment. Even knowledge about the characters themselves, like psychological profile (leader, dependent, etc.), along with some information about the environment (such as daytime, possible accidents, etc.), is stored in the database. The purpose of this approach is to decrease the real-time processing, especially in the case of crowded virtual worlds.

The problem with this *smart environment* approach is that it doesn't leave much room for character idiosyncrasies. When action selection is conducted from outside the character, the resulting behavior tends to be somewhat standardized. After all, the purpose of this technique is mainly to reduce the complexity of the system. Also, the character autonomy is decreased, by definition. According to Merriam-Webster's Dictionary, autonomy is "the quality or state of being self-governing". Many inhabited virtual environment applications need a high degree of autonomy, and then a self-contained structure is necessary to model the characters. Usually the solution is to develop them as cognitive agents, as it is discussed next.

Virtual Characters as Cognitive Agents

The idea of developing a general structure for intelligent agents that remains constant across different knowledge domains is an interesting candidate to model autonomous characters that concentrate the reasoning necessary to their behavior. Two such architectures are mentioned here because of their frequent use in character modeling: SOAR (Laird, Newell, and Rosenbloom, 1987) and BDI (Rao and Georgeff, 1995).

SOAR stands for State, Operator, And Result, and follows the tradition of symbolic artificial intelligence paradigm by which problem solving is the core of human thinking, and may be described as searching for an a priori unknown procedure to go from an initial state of the world to a desired one. Inspired on Newell's General Problem Solver (Newell, 1990), this approach is based on a system of resolution of universal problems centered on heuristic research of spaces of problems.

BDI (Beliefs, Desires and Intentions) is another model that also tries to capture a general mechanism for human behavior. Agents' Beliefs are their knowledge. Desires are their set of long-time goals, and are considered to provide their motivations to act. They may be contradictory. Intentions can be viewed as a non-contradictory set of desires, or as a set of plans. The point is that intentions are what the agent is going to attempt to fulfill.

Both approaches treat goals and motivations as outside factors that guide behavior. But as was already said, autonomy is about self-governing, and a mind architecture has to control agents' motivations from inside itself. Next, some alternatives of these approaches are discussed.

VIRTUAL CHARACTER BEHAVIOR

Behavior as Body Movements

The visual appearance of the virtual world plays a major role in user immersion since it conveys a significant part of the information about elements and events occurring in the environment in an intuitive form. The same applies to virtual characters. Their visual image expresses not only many physical features like age, weight, strength, but also their inner state (joy, fatigue, excitation, etc.) and even personality and culture traits.

Many levels of computer animation are involved in the development of a virtual character appearance. Body physical features have to be combined with environmental elements such as wind force, ground type, lightening, etc. to produce visual images that reflect those interactions. For instance, hair (Volino and Magnenat-Thalmann, 2006) or cloth (Magnenat-Thalmann and Volino, 2005) movements may require an especially complex computation, depending on what level of detail is desired.

Another important aspect of a character appearance is emotional expression (Garcıa-Rojas, Vexo, Thalmann, Raouzaiou, Karpouzis, Kollias, Moccozet and Magnenat-Thalmann, 2006). Body language (including face movements) is an important way of communication. When natural language is not possible due to its great complexity, it can be partially replaced by an appropriate repertoire of face expression and/or body positions or movements. But some computation is still needed to match each character state to the corresponding expression. This may be troublesome if too many possibilities are available.

Besides the superficial appearance of physical movements, a character performs displacements in the environment that are supposedly subordinated to higher goals. Those displacements are supposed not only to reach an intended spot, but also to choose a good path avoiding both stationary and moving obstacles in the process. In computer games, the correspondent computational processing is known simply as *pathfinding*, and in some cases is the main AI implemented.

Frequently, when a character is trying to satisfy a goal, finding the path to a certain location is not the only choice to be made. For instance, maybe it should be better to run rather than simply to walk. Sometimes, to reach the goal it is necessary to jump or to climb something. Of course, it depends on which are the possible movements available to the character. Besides, it has to be reminded that each one of them has to match a different animation sequence. Usually based on graphs of positions and paths on the map, pathfinding may also include some knowledge about the environment that is necessary for choosing where to go to solve the current problem when the solution is not immediately available. Since this level of computation involves some kind of decision-making process, it may sometimes be rather complex. If the character faces a rich environment with many different possible solutions, some heuristics may be required.

Until this point, only simple and basic behaviors have been considered, those that can be described as *body movements*. They are executed through computer animation techniques controlled by basic AI algorithms of pathfinding and decision-making processing. Although far from complete, frequently they are responsible for all the character behavior implemented in many VEs, especially when computer games are considered. In fact, body movements are the minimum expected behavior performed by a character that has to look alive. Body expression and apparently intentional displacements are perhaps the minimum an observer needs to believe that there is a "living" entity present in a virtual environment. Despite their apparent simplicity, they may represent considerable real-time, complex computational processing, especially in large and very populated virtual environments. Perhaps for commercial reasons, the graphical aspects of character body movements are frequently the main concern regarding character believability. Altogether, these computational processes don't leave much resource for implementing more sophisticated behaviors. However, for many applications it may be important to have characters exhibiting more complex and adaptive behavior while still preserving an interesting appearance. Then, there has to be a character behavior structure including a flexible and context-sensitive action selection mechanism that doesn't consume too much computational resource. Such a process should then be responsible for the lower-level control of decision-making, path finding, and body expression and movements.

Behavior as Action Selection

Human observers evaluate characters believability not only by how they look. Their sequences of actions have to be consistent and reveal goal-oriented behaviors that look as if they are attempting to satisfy internal needs. Also, awareness of the possibilities and obstacles of the environment has to be expressed. Some kind of "internal life" has to be emulated. That is the way a character may seem intelligent and emotional.

Franklin (1995) has postulated the Action Selection Paradigm, by which the prevailing task of the mind is to bring about the next action. Actions serve internal drives and are controlled by the mind in situated autonomous agents. All cognitive functions (sensing, recognizing, reasoning, planning, etc.) are subordinated to mind control to produce a consistent action flow, coherent with the dynamic internal states of an agent and the current environment possibilities.

So, we can think of a character artificial mind as a central control that integrates and coordinates all cognitive functions in order to select actions that serve internal forces in its interaction with the environment. A central control can integrate perception, action and reasoning in a consistent way, since all those processes are subordinated to the same goals. This integration may also provide an interesting way of pruning the cognitive functions processing by using the internal drives as filters. Next, main properties necessary to build an efficient artificial mind structure are discussed.

Character Artificial Mind

Minsky (1986) proposed a theory of human cognition that he called the *Society of Mind.* The core of his theory is that mind is not the result of a single, intelligent and unified processing but, on the contrary, is produced by the work of thousands of specialized different sub-systems of the brain. Minsky uses the term *agent* to refer to the simplest units of the mind that can be connected to compose larger systems, or *societies.* What we know as mind functions are performed by these societies of agents, each one of them executing different tasks, with different, but specialized roles. Unlike Newell (1990) with his General Problem Solver, Minsky does not believe that a general thinking algorithm, method or procedure could describe human intelligence performance. Instead, he understands human thought as the result of the combined activity of more specialized cognitive processes. Each one of them has limited powers and does not have significant intelligence, being capable of interacting with only certain others. However, Minsky considers that human mind emerges from their interactions. Minsky does not distinguish "intellectual" and "affective" abilities, he considers both as mental abilities emerging from societies of agents organized in hierarchies. The Mental Society would then work as does a human organization, where there are, on the largest scale, gross divisions where each subspecialist performs smaller scale tasks.

According to Franklin (1995), cognition emerges as the result of interactions between relatively independent modules, and the criterion for evaluating such a mechanism is the fulfillment of the agent's needs in the current environment. Artificial mind architectures have to produce sequences of adaptive actions emerging from a non-monolithic structure of interactive modules with different functionalities serving internal drives. This structure specifies possibilities and restrictions of both perception and action in a known changing environment. Perception and action are not considered as if they were independent from one another, but rather they are seen as constantly giving mutual feedback to control their processing. That close relation must be part of an efficient modeling of both processes, as has already been stated in other studies (Vilela, 1998, 2000). Also, both processes depend on knowledge and on internal states. Transitory priorities may change the way of perceiving and acting in the current environment, in a dynamic interaction process.

Internal forces establish goals that orient action selection, and may be thought of as *motivations.* Their intensities determine goal priorities and their dynamics, allowing proactive behavior. Sevin and Thalmann (2005a, 2005b) consider motivations as essentially representing the *quantitative* side of decision-making. For them, proactivity includes the ability to behave in an opportunistic way by taking advantage of new situations, especially when they permit satisfaction of more than one motivation. Usually, there are many self-generated concurrently active motivations, even though one of them may be more prominent. So, there are many elements to be considered in the resolution process:

- **Priority:** Search for satisfaction of the most relevant motivation.
- **Opportunism:** Ability to change plans when new circumstances show interesting possibilities of motivation satisfaction.
- **Compromise actions:** Ability to change the course of action if it is possible to satisfy more than one motivation.
- **Quick response time:** Action selection must occur in real time.

In the model proposed by Sevin and Thalmann, motivations result from the evaluation of internal variables according to a threshold system and generate actions when combined with environmental information. There is one hierarchical decision loop per motivation, which generates sequences of locomotion actions (motivated behavior) that lead to locations where the motivation can be satisfied. All decision loops run in parallel and, at the end, the most activated action is executed. Besides locomotion actions, there are motivated actions, which can satisfy one or several motivations. Both types, when executed, alter internal variables related to their original motivations: locomotion actions increase them, and motivated actions decrease them. The threshold evaluation system represents a non-linear model of motivation changes, and it is specific to each motivation. Depending on two thresholds T1 and T2 and according to the value of its internal variable i, motivation M is evaluated in three possible regions—Comfort Zone, Tolerance Zone and Danger Zone—by the following equations, respectively:

$$M = T_1 e^{(i-T_2)^2} \quad if\ i < T_1$$

$$M = i \quad if\ T_1 \leq i \leq T_2$$

$$M = \frac{i}{(1-i)^2} \quad if\ i < T_2$$

If a Motivation lies in the Comfort Zone, it is not taken into account. Otherwise, its chance to be satisfied increases as its internal value increases. However, when it reaches the Danger Zone, its competitiveness is much more intense. The purpose of the action selection mechanism is to lead motivations to the Comfort Zone. A Motivation ***M*** is described as a "tendency to behave in particular ways" (Sevin & Thalmann, 2005b), and then it may be considered as a driving force of action selection. But this quantitative model does not account for an important human characteristic: the possibility of frustration. In normal situations, human motivations are not always satisfied. And when a person can not satisfy a motivation, she/he does not keep trying forever, and tends to express frustration in some emotional way. But by the model just described, when an agent reaches the danger zone, it can not abandon the corresponding motivation. The other motivations will continue to increase, and will also not be satisfied, because the agent is stuck in that former motivation. That is not a plausible human attitude. When someone becomes frustrated, some expression of this feeling may be seen by an observer. Besides, eventually she/he moves on, and tries to satisfy other motivations. That problem has to be tackled by a believable model.

Motivational Graphs architecture (Chiva, Devade, Donnart, and Maruéjouls, 2004) is another interesting approach to character motivation structure, developed in the context of computational games. It is based on activity propagation in a graph structure where nodes represent rule sets, instead of representing neurons like in connectionist systems. Each set is part of a plan decomposition, and propagates activity according to its internal rules. The final nodes correspond to actions that the character can perform. Environment input activates some nodes, and after energy propagation, the node with greater energy represents the action to be performed by the character. This architecture is meant to take advantage of both symbolic and connectionist approaches by hybridizing them in a single structure.

Another important aspect of mind structure that Franklin (1995) pointed out is the relation between Perception and Action. Since every interaction with the environment has to be guided by some knowledge, we may add its processing in the mutual perceiving-acting relationship. But instead of what may be suggested by the traditional sequence input- processing- output, these three activities are not thought of as independent processes. On the contrary, sensory processing is strongly determined by the current action, and motor activities greatly depend on sensory feedback. Previous works extending this point (Vilela, 1998 and 2000) argue that Perception, Action and Knowledge are integrated, goal-oriented cognitive processes. Indeed, we don't simply *look* at the environment, but we *search* for something we *know* that can be useful to what we *want to do*. Also, we try to do (always *checking* if it is really being done) what we *know* that can satisfy something we *need*, which is determined by our *motivations*. A motivation thus has an associated knowledge that has to answer questions like *how*, *where* and *what* is needed in order to satisfy it. This knowledge allows the interpretation of the current environment so as to guide the search for information and the choice of an adequate sequence of actions. So, we can think of a qualitative component (Minsky's agent or Franklin's module) that encapsulates Perception, Knowledge and Action processing related to its task.

Motivation can then be viewed as a propagating energy that assigns different strengths to character cognitive component. If it is assumed that motivation is a consequence of emotional states, this is coherent with a previous discussion (Vilela and Lima, 2001) of the activating role of emotion in cognitive processing. So, we can think of motivational components that define motivation, and are responsible for the propagation of the corresponding energy, as quantitative component.

As already pointed out, an artificial mind has to show unpredictable, flexible and dynamic choice of actions, compatible with human being behavior. The following characteristics summarize the above discussion about which characteristics are relevant to virtual character mind architecture:

- Behavior emerges from interaction between qualitative components (modules or agents).
- Qualitative components representing cognitive functions encapsulate Perception, Action and Knowledge processes for a specific task.
- Qualitative components representing Motivations propagate activation energy.
- Qualitative components have a quantitative element representing their activity level that codes motivation energy.
- Motivations have competing quantitative forces that establish priorities and drive behavior according to activity propagation among qualitative elements.
- Each motivation triggers only specific cognitive components by transferring their energy.
- Only those cognitive components with high activity level perform their tasks, and this functions as attentional and intentional knowledge filtering.
- Activity propagation among qualitative elements has to allow negotiations and competition between motivations according to environmental information.
- Satisfaction and frustration of a single motivation are specified as zones of the correspondent internal variable values, defined by threshold parameters.

ARTIFICIAL MIND AS MULTI-AGENT ARCHITECTURE

Although Minsky's term *agent* did not originally mean the same as it does in the context of Distributed Artificial Intelligence, they are not incompatible in principle. Usually, characters are developed as single agent structures, but it is possible to model them as a composition of closely interacting sub-agents. Vanputte, Osborn and Hiles (2002) developed the *composite agent architecture*, which combines related reactive and cognitive agents in a single structure. The purpose of this combination is to take advantage of both kinds of agents, improving the effectiveness of their correspondent functionalities at the higher-level agent. In the mentioned model, the composite agent has two agent sets and an internal environment. One set groups the Symbolic Constructor Agents (***SCAs***) and process sensory inputs from the outer environment (E_{outer}) in order to build the inner environment (E_{inner}). Each SCA is a cognitive agent designed to capture specific environment aspects, filtering and controlling their inputs so as not to overload the composite agent if the outer environment is rich and complex. The other set groups reactive agents (RAs), each one responsible for promoting a specific composite agent behavior, then coding what behaviors the agent can execute. The inputs to RAs come from E_{inner}

These Composite Agents use the "*sense-update-act*" model (Wooldridge, and Jennings, 1995). This means that they first sense the environment, then update their internal environmental representation, and finally decide what action to perform, before restarting the cycle. They can sense only those aspects perceived by the *active* SCAs, which are constantly updating an internal environment representation (E_{inner}). Combining E_{inner} information with goals and knowledge about the context, composite agents can generate appropriated actions.

Composite Agent Architecture is interesting for artificial minds design because it matches some of the characteristics enumerated earlier, like a multi-modular structure and emergent behavior. However, the "*sense-update-act*" model does not seem convenient to the intended Perception-Knowledge-Action relationship as previously discussed. Instead, a two-steps "*motivation→ knowledge-based perception/action*" cycle would be more appropriate. While in the first case the environment information filtering is made at the perception level, in the second, it is made at the motivation level. So, attention is determined from the outside in the composite agent architecture, and from the inside in the artificial mind structure. Although sometimes the environment can catch our attention (usually when extraordinary stimuli are present – like intense sounds or lights), most part of the time our senses are guided by our goals. Besides, information filtering is a potent way to decrease computational processing. If the outside determines it, the whole process may be slowed down in complex environments. But the inside has a constant structure that can be adequately designed.

The multi-agent structure here proposed is inspired in the composite agent architecture, but functionalities are differently distributed. First of all, there are three sets of sub-agents: Motivation agents, Cognition agents and Execution agents.

Motivation Agents code the character needs and desires, conveying the internal forces that drive the character behavior. They compete for priority and constitute the Inner Motivational Environment (IME), which represents the current emotional and affective state of the character. Only those Motivation Agents with higher-level of priority can act outside the IME by activating specific pre-defined Cognition Agents that have the knowledge to lower their activation level.

Cognition Agents are responsible for the "*knowledge based perception/action*" step of the iteration cycle. They store and represent the character knowledge, constituting the character cognitive structure. Like a motivational graph node described previously, each one embodies knowledge that can be activated by a certain quantitative internal force. They can be compared to Minsky's agents in the sense that they execute some specific and particular task, and they are related to each other in a hierarchical structure. In terms of implementation techniques, they are not necessarily restricted to rule sets. Their knowledge may be represented by any AI paradigm. The developer may choose whatever technique seems best suited to the task in question. A Cognition Agent uses the *sense-update-act* cycle but only in a very restricted scope. It searches for relevant information wherever it can find it (this is part of its knowledge), changes its own state accordingly, and then propagates the result to a specific agent, which is also determined by the processing of the encapsulated knowledge. It may be another Cognition Agent or an Execution one (see next). In the first case, the knowledge processing is refined, until it reaches the decision. In the latter case, an action is performed and the IME state is altered accordingly.

Execution Agents represent the result of this action selection mechanism. They are responsible for all the "behavior as body movement" that the character in question is capable of performing. This means changing the environment, presenting an animation sequence, and altering the inner state of the character according to the needs/desires satisfaction or frustration. Figure 1 shows the integration of the three agent sets and their interface with the outer environment.

In the next topics the proposed architecture is described in more details, and then a simple example and some words of conclusion are given about character design based on the proposed architecture.

Motivation Agents

Each Motivation Agent represents a specific character motivation. It is a *reactive agent*, the simplest class of agent (Mandiau and Strugeon, 2001), then having a relatively low computational complexity. A character motivation may be anything: from simple basic needs (hunger or thirst), to complex psychological and emotional whishes (fame, success, love, etc.). Anyway, it is coded with the same basic structure. Choosing what motivations are going to be represented in a character belongs to the art level of the system creation, being a developer's decision. From the point of view of the architecture, they are the driving forces of the behavior. As they code internal processes, they don't have direct contact with the environment. They affect and are affected only by other character agents.

The main attribute of a Motivation Agent is its **Intention**, which represents a tendency to act in certain way, possibly searching for some specific object to satisfy it. Like in Sevin and Thalmann's (2005) model, another important attribute is the motivation **Intensity** i, which is an internal variable that reflects how much the corresponding motivation is being significant at the current time. This variable increases autonomously in time in a rate determined by the motivation characteristics. Again like the mentioned model, there are two thresholds (T_{min} and T_{max}) that divide the variable value range in three zones, here named: comfort zone, tolerance zone and discharge zone. The latter is renamed because it is not considered as a *danger* zone, but instead it is just a zone that demands some kind of urgent procedure to decrease the Intensity value. Of course, motivation satisfaction is the more effective way, but sometimes this is not possible. In such cases, there must be an alternative way of doing it, which is called the **Discharge Procedure**, and has to

Figure 1. Multi-agent architecture for artificial mind

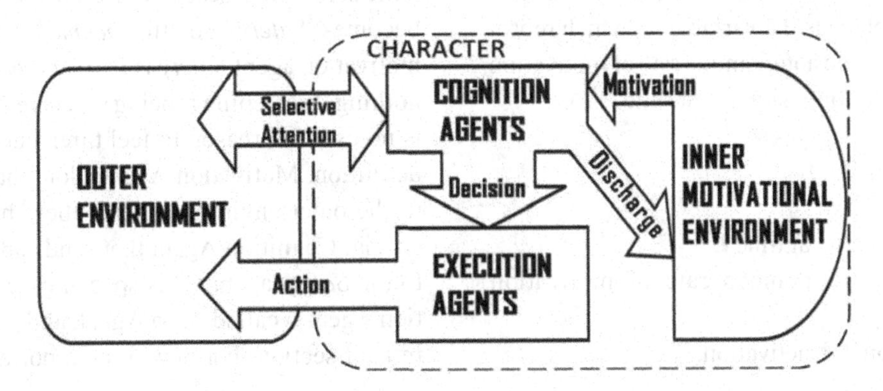

be included in the Cognition Agent activated by the Motivation Agent. It corresponds to the *Frustration Attitude* of the character toward that particular motivation. This procedure includes two routines: the activation of a special Execution Agent (the *Discharge Agent*), and the update of the Motivation Intensity value $\boldsymbol{i}$, by returning it to the $\boldsymbol{T_{max}}$ value. A Discharge Execution Agent may execute, for instance, a meaningless behavior, like walking in circles in the room expressing impatience. So, the Discharge Procedure may be as simple as this:

IF $\boldsymbol{i} > \boldsymbol{T_{max}}$
THEN
{Activate Discharge Execution Agent;
$\boldsymbol{i} = \boldsymbol{T_{max}}$ }

Updating $\boldsymbol{i}$ by returning it to the $\boldsymbol{T_{max}}$ value keeps the motivation in the competition for priority. So, it can be satisfied later, when the opportunity arrives. It is possible to make the character give up trying, if the procedure returns the Intensity value to the comfort zone after a certain limit of time, for example. The context may have to be analyzed to decide which discharges are possible, and which are not. Anyway, it is important to deal with the frustration, and this can be done with small variations of this simple procedure.

Since every Motivation Agent begins with $\boldsymbol{i} = 0$ (which means that no motivation is active), there has to be an internal procedure that changes this value to reflect the variation of each particular motivation with time. The attribute that allows this to take place is the variable $\boldsymbol{m}$, which represents the intrinsic autonomous rate of motivation growth. So, let Intensity $\boldsymbol{i}_{t+1}$ at time t+1 be:

$$i_{t+1} = i_t + m - d,$$

where:
$\boldsymbol{i_t}$: intensity value at time $\boldsymbol{t}$;
$\boldsymbol{m}$: intrinsic autonomous rate of motivation growth;
$\boldsymbol{d}$: motivational deactivation.

Motivation deactivation $\boldsymbol{d}$ occurs when specific actions take place, either satisfying the motivation or just discharging frustration. The Cognition Agent that was activated by the Motivation Agent sends the corresponding value.

Intrinsic autonomous rate growth $\boldsymbol{m}$ represents the self-generated component of a motivation. It codes the relative importance that the correspondent **Intention** has to the character in question. Basic natural needs (hunger, thirsty, etc.) probably have the same $\boldsymbol{m}$ for all characters of a single environment, except when there is a pathologic state associated with anyone of them (like anorexia, for instance). Complex psychological drives (fear, aggressivity, etc.) tend to vary between different characters, as they reflect their particular traits.

When a Motivation Agent wins the competition for priority, it triggers the activation of a chain of agents that eventually produce the Motivational deactivation value $\boldsymbol{d}$, which represents the amount of discharge obtained with the performed action. It may not be enough to take the intensity to the comfort zone. For instance, if a character is very hungry, but the only food obtainable in the environment is, say, an apple, it probably will not be sufficient to completely satisfy the hunger, although it may decrease it a little. In this case, $\boldsymbol{d}$ value is not high enough, and the *Hunger* agent will still compete for priority in the next cycle.

A special case is when a motivation can only be activated by an external stimulus. For instance, usually people feel fear only when confronted with something scary, and then the prior goal becomes *Safety*. So, the *normal* state of the motivation agent *Safety* is $\boldsymbol{i} = 0$ and $\boldsymbol{m} = 0$, when nothing threatening is being perceived, and there is no external reason to feel threatened. But, by definition, Motivation Agents don't have access to the outer environment. So, there has to be a special Cognition Agent that sends an alert sign to the *Safety* agent. This special case of Cognition Agent is called Alert Agent, and it is detailed in next section. For now, it is important to note that the value d sent by an Alert Agent has to be *negative*; otherwise, it is not able to increase the motivation intensity. So, it is not a deactivation factor, but, on the contrary, an activating component of the intensity equation. The i value may decrease either if the stimulus intensity decreases or the stimulus distance increases (by running away from the scary stimulus, for example), both causing the d absolute value to decrease.

Anyway, it is important to note that the determination of d value results from an evaluation made by a Cognition Agent, and not by the Motivation Agent. Mixed Motivation Agents can also exist, and may receive both negative and positive values of d, each sent by a different cognition agent. For instance, a hungry character may be stimulated by the vision of food. There is an internal component (in this example, a *Hunger* Agent) that increases autonomously, but is affected by environmental stimuli, which is perceived by an Alert Agent.

Inner Motivational Environment

The set of all Motivation Agents constitutes the Inner Motivational Environment, which has no direct contact with the Outer Environment (see above), and represents the motivational state of the character. Its purpose is to assure that only motivations with high priority can trigger their correspondent cognition agents. In other words, this means that it is responsible for the filtering process that decreases the computational complexity of the whole action selection process.

A Motivation Agent can be in one of three states: inactive, competitive and active. A motivation is considered *inactive* when it is in the comfort zone of intensity value. This means that it can't affect behavior because it is satisfied. From the moment the intensity value increases to the point of entering the tolerance zone, some action is needed to discharge it. So, its state becomes *competitive*, which means that it is now part of a *Competition* process. This process selects the prior motivations, which then pass to the *active* state. Only active Motivation Agents can trigger the cognitive processing by which Cognition Agents perform action selection.

Competition could be just a special kind of winner-take-all mechanism that does not restrict the result to just a single Motivation Agent, but selects a little set of them instead. Restricting the competition result to just one Agent would not allow a negotiation (as described in the next topic) between prior motivations in the cognitive structure. The importance of this negotiation is that it could permit opportunism or compromise actions, which are interesting behaviors, as already mentioned. Besides, a single winner mechanism overestimates little circumstantial differences between motivation intensities. So, the chosen mechanism establishes a proportional intensity threshold above which all motivations are activated. For instance, all agents included in the competition process can send each other a message like this for each competition cycle:

IF $\alpha i_k \geq i_n$
THEN
ReceiverState = Inactive
// Temporary change of receiver state ,

where:

$0 < \alpha < 1$, α being an arbitrary value chosen at design time.
i_k : sender intensity value;
i_n : receiver intensity value.

After sending all messages, the remaining agents that are still in the *competitive state* are set to *active state*, and are allowed to trigger their corresponding Cognition Agents. At this point, computation of intensities of all other Motivation Agents resets their states based on their motivation zone.

All those Motivation Agents that have intensity values above or equal αi_{max} (where i_{max} is the higher intensity) are going to trigger energy propagation in the cognitive structure as in the motivational graph discussed previously. This procedure has complexity ***O(n)*** on the number ***n*** of Motivations Agents, in the worst case. So, up to this point of the action selection process, reactive agents have not executed too complex processing, which is satisfactory from the point of view of computational requirements. The next step is cognitive processing, which usually is the more complex of any intelligent action selection mechanism. However, this previous competition is able to prune significantly this processing. Next, Cognition Agents and the cognitive negotiation performed by them are described.

Cognition Agents

A Cognition Agent integrates perception and decision-making in a same set of rules. Perception searches and analyses only environment information that is relevant to the agent task. Once the needed information is obtained, a decision-making process takes place and selects which agent to pass the intensity value. If it is possible that relevant environment elements could not be immediately available, additional knowledge and a decision-making process are necessary to choose where to look for them.

Alert Cognition Agents are special agents responsible only for tracking specific stimuli, objects or events in the environment, and evaluating them in terms of distance and/or intensity. Once this evaluation reaches predefined thresholds, its value is sent to a specific Motivation Agent, increasing its intensity value, as already described earlier.

The main purpose of the whole action selection mechanism is to reduce the total intensity of the Inner Motivational Environment by acting in the Outer Environment. When each Motivation Agent is designed, its corresponding way of satisfaction has also to be defined. This involves determining what environment elements are possibly relevant and must be searched, and what alternative actions may be performed to achieve the intended goal. Such knowledge can be either encapsulated in a single Cognition Agent, or distributed among many of them. This cognitive structure is similar to Minsky's society of agents, and to the motivational graph, both already previously discussed (see above). A hierarchical tree of Cognition agents is built, its root being the agent activated by the motivation in question, and its leafs being Cognition Agents that trigger specific Execution Agents. Different trees may share Cognition agents at any level, except the root. When the Motivation Agent activates the root, the intensity value ***i*** is passed to the Cognition Agent intensity. This intensity propagation continues in the cognitive tree, in a way similar to energy propagation in motivational graphs, allowing the negotiation mentioned previously. When intensities coming from more than one root converge in a shared node, the chance of selecting a leaf satisfying more than one motivation increases. As a consequence, a higher amount of discharging probably occurs in the Inner Motivational Environment.

Active leafs pass their intensity to specific Execution Agents and to the roots that have contributed to their activation. These roots, in turn, send this value (***d***) to their respective Motivation Agent.

At this point, if an active Motivation Agent in discharge zone does not pass at least to tolerance zone, its root Cognition Agent triggers the Discharge Procedure according to the context.

Next, a simple example is presented for the reader to feel a flavor of the architecture possible applications.

A Simple Example: Threatening Situations and Need for Safety

Threatening conditions are frequently included in virtual storytelling for creating thrill, but also in some simulations where human reactions to danger are to be evaluated. That is the case of some building analyses for safety conditions in case of accidents, like fire or explosions, for instance. Each virtual human in such simulations has to deal with high intensity stimuli (light, heat, noise, etc.), and the main motivation has to be related to the danger they represent. Since these are extreme conditions, we may usually skip minor motivations like hunger or thirst: adrenaline allows a human being to ignore them in such cases. So, we focus here on *safety* as a simple example of need. In more complex situations, other motivations like love and sympathy for others in danger can be easily included to the model by adding the correspondent Motivation and Cognition Agents.

A panic situation involves a variety of behaviors like: reflex (fright, shock, etc.); selection of course of action (running away, hiding, confronting, etc.); path finding on rapid changing environments, etc. All these elements generate internal conflicts that must be solved very quickly, using knowledge processing.

Modeling this kind of behavior using the proposed architecture involves, first of all, the description of a Motivation Agent we can call ***Safety*** (after its main goal). It can be specified as already described (see Figure 2).

The Motivational Drive $\boldsymbol{m}$ is the rate a motivation increases in time while not being satisfied. Since safety doesn't change in time, in normal persons (not paranoid, for instance) and in normal conditions (peace, everyday life), $\boldsymbol{m}$ is set to zero. The intensity of a *Safety* motivation then depends on stimuli coming from the Alert Cognition Agent *Danger*, which, when in presence of high intensity target stimuli at time $\boldsymbol{t}$, activates an alarm by sending the correspondent intensity d_t that is responsible for increasing the *Safety* Intensity. Tolerance threshold $\boldsymbol{T_{min}}$ and $\boldsymbol{T_{max}}$ are specified according to the character's personality. Calm people are more difficult to panic, while an anxious person tends to easily become frightened.

While the *Safety* intensity is between the two thresholds, a Cognition Agent here named *Protection* is activated as a root of a decision hierarchy that will select the better action to be executed. If the chosen action succeeds, a correspondent $\boldsymbol{p_t}$ component is sent back, at time $\boldsymbol{t}$, to decrease the Safety Intensity. Otherwise, from the moment the upper threshold is reached, behaviors expressing anxiety tend to emerge, and are executed by Discharge Execution Agents, like *Panic* (also specified according to the character personality). In this case, the intensity is correspondently decreased, keeping the intensity value in the upper threshold until a successful action is finally selected by *Protection* (or some catastrophe happens and the character dies).

Once the main Motivation Agent *Safety* is modeled, the *Danger* Alert Agent has to be specified taking into account the sensitivity of the characters to the stimuli signaling the danger situation. Likewise, the *Protection* graph depends on the knowledge the characters are expected to have about the environment and the danger involved. Both *Danger* and *Protection* Agents are Cognition Agents, which means that they can be embedded in a motivational graph (see above) that may include other factors (like, for instance, competition with or sympathy for others, already mentioned).

CONCLUSION AND FUTURE TRENDS

Virtual Characters are frequently related to computational games, but they can have many other applications, both in technology and research systems. Behavior mechanisms, though, are usually very specific of each application. But, as common expected properties of character behavior are believability and plausibility, it is important that action selection can be flexible, autonomous and adaptable to the environment to satisfy this criterion. A generic structure for artificial mind that can provide these qualities is interesting and useful. In complex environments such as those of Virtual Reality, this can only be attained through

Figure 2. Motivation agent safety

```
Motivation Agent Safety

Goal: Safety;
   Motivational Drive m = 0;               // default: there is no danger;
   Intensity i_{t+1} = i_t + d_t - p_t;    // d comes from an Alert Cognition Agent Danger;
   Tolerance Interval: T_min; T_max;       // depends on the character ´s personality;
   Target Cognition Agent: Protection;     // which sends back p;
   Discharge Procedure:
      IF (i_t > T_max)
      THEN
        { Call Execution Agent: Panic;
          i_t = T_max;
        }
```

emergent behavior resulting from a complex structure of modular functionalities for action selection in real time.

The architecture presented in this chapter assumes character motivations as the driving internal force of autonomous behavior. They are considered to be the quantitative component of the structure dynamics, and are responsible for filtering the attention-knowledge-decision interaction processing for action selection. This filtering may reduce the computational complexity of the whole process.

A character is considered here as a multi-agent structure including both reactive and cognitive agents. These agents are grouped according to three basic functionalities: Motivation agents, Cognition agents and Execution agents.

Motivation Agents are reactive and each one represents a single motivation. They compete with each other for priority based on their Intensity values. The set of all Motivation Agents constitutes the Inner Motivational Environment of the character. The Intensity increases autonomously based on a rate ***m*** defined at design time. Two threshold divide the intensity range in three zones: (a) comfort zone, where the agent is in an inactive state; (b) tolerance zone, where the agent goes to a competitive state, competing with other motivations for priority; and (c) discharge zone, where, if it does not win the competition and is set to active state, then frustration takes place and it has to discharge the excess of Intensity through some of a set of predefined discharge actions, chosen according to the context.

Higher intensity intentions win the competition and can activate their correspondent cognition agents. These agents integrate the necessary perception, knowledge and decision-making process for their intention satisfaction. They can be decomposed in hierarchical trees so as to allow sharing the common components between different intentions, in the same way as motivational graphs do (Chiva, Devade, Donnart, and Maruéjouls, 2004). As intensity propagates in these trees, they can negotiate the best way to simultaneously satisfy many active intentions.

Special kinds of Cognition Agents, the Alert Agents, are responsible for tracking the environment for specific stimuli, objects or events. They decide when a given environment element achieves certain levels of predefined parameters and must be considered by the character. In this case, they activate a correspondent motivation agent with a value associated to the urgency of the alert. This can change the course of action of the character in order to deal with the emergent situation. Execution Agents run actions, which means changing outer environment and inner motivations states, and they also exhibit the corresponding 3D animation sequence.

If motivation parameters, like thresholds or motivation growth rate, vary, it may be possible to obtain different emergent behaviors, simulating different personalities in an easy way. For instance, a calm character can be simulated with higher threshold than anxious ones, because it will take more time for it to achieve the discharge zone. Then, the probability of a frustration attitude decreases. Different ***m*** values may reflect the various degrees of importance for motivations between different characters. The choice of a discharge action is another way to express variation among personalities. Aggressive characters tend to express frustration by violent actions, while more inhibited ones tend to do it by subtle movements of impatience.

All this flexibility, though, is not computationally complex as it was expected to be. Indeed, the more complex part in a decision-making process is that involving environment information search and analysis, but the present architecture restricts this processing through motivation filtering. Since reactive agents (the simplest form of agents) are responsible for motivation processing, the pruning of the whole action selection computation is significant. The hierarchical cognitive structure that propagates activity through very specific rules rapidly reaches the final decision, making the processing even more efficient. Although motivational graphs use symbolic rules, it seems reasonable that they can be implemented by using other techniques, if they are more adequate to certain specific tasks. As a future work, motivation-oriented constructivist learning and cultural influences can be included in the cognitive structure.

FUTURE RESEARCH DIRECTIONS

There are three main areas where future perspectives can be discussed: (a) understanding of human behavior and cognition; (b) artificial mind models and architectures; (c) application fields.

Artificial Intelligence has always been about simulating human behavior and cognition through computational models, as it is well known. But until now only fragments of partial and isolated cognitive functions and properties have been actually implemented. But the development and convergence of many technologies, along with the progress in complex system formalism, cognitive modeling, and many other multidisciplinary subjects, is finally giving the opportunity to attempt to integrate these multiples partial models coming from many sources. Integration means building an artificial mind capable to coordinate and perform in a consistent and harmonious way behaviors so different as feeling, reasoning, dancing, eating, creating, etc. It is time to put together the work done by AI researchers till now. It is not an easy goal to pursue. But we have to remember that the whole does not function the same way as merely the sum of the parts. So, integrating would be the actual and effective way of testing partial models.

Models and architecture still have difficulties in incorporating dynamics to the system. Although psychological and pedagogical researches and practices have shown the superiority of constructivism over behaviorism, this has not been deeply approached by AI models, yet. Also, until now, it has not been possible to implement the way people influence each other from the point of view of their cognitive and knowledge structure and reasoning processing. Cultural issues are still to be approached by AI techniques.

Applications of Virtual Humans with an integrated Artificial Mind would expand the possibilities of complex simulations. There are areas, like serious games, that have already created education, training, cognitive ergonomics research, or management applications, but they have to unfold and cover much more aspects than what has been done. Maybe the path to be followed should be summarized in three words: integration, complexity, and dynamics.

ACKNOWLEDGMENT

This work was supported by CNPq and LAMCE, COPPE-UFRJ.

REFERENCES

Antonio, A., Ramírez, J. , Imbert, R., Méndez, G. & Aguilar, R. A. (2005). A Software Architecture for Intelligent Virtual Environments Applied to Education. *Revista Facultad de Ingeniería - Universidad de Tarapacá*, Arica – Chile, 13 (1), 47-55.

Aylett, R.& Cavazza, M. (2001). Intelligent Virtual Environments – A state of the art report. *Eurographics Conference*, Manchester, UK.

Aylett, R.& Luck, M. (2000). Applying Artificial Intelligence to Virtual Reality: Intelligent Virtual Environments. *Applied Artificial Intelligence*, 14(1), 3-32.

Barros, L., Silva, A. & Musse, S. (2004) "PetroSim: An Architecture to Manage Virtual Crowds in Panic Situations", in *CASA 2004 – Computer Animation and Social Agents*, Geneva, Switzerland.

Braga, L. A. F. (2006). *Escape Route Signage Simulation Using Multiagents and Virtual Reality.* D.Sc Thesis (in Portuguese), COPPE -Universidade Federal do Rio de Janeiro.

Colombo, G. & Cugini , U. (2006). Virtual humans and prototypes to evaluate ergonomics and safety. *Journal of Engineering Design*,.16(.2), 195-203.

Chittaro, L., Ieronutti, L. & Rigutti, S. (2005. Supporting Presentation Techniques based on Virtual Humans in Educational Virtual Worlds. *Proceedings of CW 2005: 4th International Conference on Cyberworlds*, (pp.245-252), IEEE Press, Los Alamitos, CA, US.

Chiva, E., Devade, J. Donnart, J.-Y. Maruéjouls, S. (2004). Motivational Graphs: A New Architecture for Complex Behavior Simulation. In Rabin, S. *AI Game Programming Wisdom 2*, 361-372 Charles River Media, Inc.

Franklin, S.(1995). *Artificial Minds*. MIT Press.

García-Rojas, A. Vexo, F. Thalmann, D. Raouzaiou, A. Karpouzis, K. Kollias, S. Moccozet L., Magnenat-Thalmann, N (2006). Emotional face expression profiles supported by virtual human ontology. *Computer Animation and Virtual Worlds*, 17(3-4) 259 – 269.

Kopp, S. , Jung, B. Lessmann N., Wachsmuth, I. (2003). Max - A Multimodal Assistant in Virtual Reality Construction. *KI-Küstliche Intelligenz* 4(03), 11-17, Bremen: arenDTap Verlag.

Laird, J. E., Newell, A., & Rosenbloom, P. S. (1987). Soar: An architecture for general intelligence. .*Artificial Intelligence, 33*(1): 1-64.

Magnenat-Thalmann, N. & Volino, P (2005). From early draping to haute couture models: 20 years of research. *Visual Computing*, 21:506–519,

Mandiau, R., Strugeon, G. (2001). Multi-agent Systems (in French). In Mandiau, R. and Strugeon, G. (Eds.), *Techniques De L'ingénieur* (Ed.), pp. 1-17, Paris.

Miao, Y., Hoppe, H. U. & Pinkwart, N. (2006). Naughty Agents Can Be Helpful: Training Drivers to Handle Dangerous Situations in Virtual Reality. In Kinshuk et al (Eds.), *Proceedings of the 6th IEEE International Conference on Advanced Learning Technologies* (p. 735-739).

Minsky, M. (1986). *The Society of Mind.* Simon and Schuster, New York.

Musse, S., Silva, A., Roth, B., Hardt, K., Barros, L., Tonietto, L. And Borba, M. (2003) "PetroSim:

A Framework to Simulate Crowd Behaviors in Panic Situations". In *MAS 2003 - Modeling & Applied Simulation*, Bergeggi, Italy.

Newell, A. (1990). *Unified Theories of Cognition*. Harvard Press: Cambridge, MA.

Osório, F. S.; Musse, S. R.; Santos, C. T.; Heinen, F.; Braun, A. & Silva, A. T. (2005). Intelligent Virtual Reality Environments (IVRE): Principles, Implementation, Interaction, Examples and Practical Applications. In: Fischer, Xavier. (Org.). *Virtual Concept (Proceedings - Tutorials)*. Biarritz, França, 1: 1-64.

Papagiannakis, G., Schertenleib, S., O'Kennedy, B., Arevalo-Poizat, M., Magnenat-Thalmann, N., Stoddart, A. & Thalmann, D. (2005). Mixing Virtual and Real Scenes in the site of ancient Pompeii. *Journal Of Computer Animation and Virtual Worlds*, 16(1)11 – 24. John Wiley and Sons Ltd. Chichester, UK.

Querrec, R., Buche, C., Maffre, E. & Chevaillier, P. (2004). .Multiagents systems for virtual environment for training: application to fire-fighting. *Special issue Advanced Technology for Learning of International Journal of Computers and Applications* 1: p. 25-34. ACTA Press.

Rao, S. A., & Georgeff. M. P. (1995) BDI Agents: From Theory to Practice. *Proc. of 1st international Conference on Multiple Agent System*.

Rist, T., André, E. & Baldes, S. (2003). A flexible platform for building applications with life-like characters. *Proceedings of the 8th international conference on Intelligent user interfaces*, p. 158-165. ACM Press New York, NY, USA

Schwab, B. (2004). *AI Game Engine Programming (Game Development Series)*. Charles River Media.

Sevin, E., Thalmann, D. (2005a). A Motivational Model of Action Selection for Virtual Humans, In *Computer Graphics International (CGI)*, IEEE Computer, SocietyPress, New York.

Sevin, E., Thalmann, D. (2005b). An Affective Model of Action Selection for Virtual Humans, In *Proceedings of Agents that Want and Like: Motivational and Emotional Roots of Cognition and Action symposium at the Artificial Intelligence and Social Behaviors Conference (AISB'05)*, University of Hertfordshire, Hatfield, England.

Thalmann, D. (2003). Concepts and Models for Inhabited Virtual Worlds. *Proceedings of the First International Workshop on Language Understanding and Agents for Real World Interaction*.

Tortell, R., Morie, J..F. (2006). Videogame play and the effectiveness of virtual environments for training. *Interservice/Industry Training, Simulation, and Education Conference (I/ITSEC)*. (Paper No. 3001, 1-9), Orlando , Florida, USA.

Vanputte, M., B. Osborn, J. Hiles, (2002). A Composite Agent Architecture for Multi-Agent Simulations. In: *Proceedings of the Eleventh Conference in Computer Generated Forces and Behavior Representation*. Orlando, Florida.

Vilela, I.M.O., P. M.V. Lima, 2001. Conjecturing the Cognitive Plausibility of an ANN Theorem-Prover. In: *Lecture Notes on Computer Science*, 2084 (1)- 822-829.

Vilela, I.M.O(1998). Integrated Approach of Visual Perception Computational Modeling (in Portuguese). MSc Dissertation, COPPE -Universidade Federal do Rio de Janeiro.

Vilela, I.M.O. (2000) An Integrated Approach of Visual Computational Modeling. *6th Brazilian Symposium on Neural Networks (SBRN)*, Rio de Janeiro, Brazil. IEEE Computer Society.

Volino, P, Magnenat-Thalmann, N. (2006) Real-Time Animation of Complex Hairstyles. - *IEEE Transactions on Visualization and Computer Graphics* 12(2) pp. 131-142.

Wooldridge, M., Jennings, N. (1995). Intelligent Agents: Theory and Practice. In: *Knowledge En-*

gineering Review, 10 (2), Cambridge University Press, U.K.

Xu, M., S.Sun,, Y. Pan, 2006. Virtual human model for ergonomics workplace design. *Zhongguo Jixie Gongcheng (China Mechanical Engineering)*,17(80),836-840.

ADDITIONAL READINGS

Brom, C. & Joanna Bryson (2006). *Action selection for Intelligent Systems*. Retrieved May, 2007 from www.eucognition.org/asm-whitepaper-final-060804.pdf

Cabral, J., Oliveira, L., Raimundo, G. & Paiva, A. (2006). *What voice do we expect from a synthetic character?* Paper presented at SPECOM'2006, St. Petersburg.

Champandard, A. J. (2003). *AI Game Development: Synthetic Creatures with Learning and Reactive Behaviors*, New Riders Games.

Conde, T. & Thalmann, D. (2006). An integrated perception for autonomous virtual agents: active and predictive perception. *Computer Animation and Virtual Worlds*, 17,457–468.

Damasio, A. (1994) *Descartes' Error: Emotion, Reason, and the Human Brain*, Avon Books.

Egges, A. & Magnenat-Thalmann, N.(2005). Emotional Communicative Body Animation for Multiple Characters. Paper presented at *V-Crowds'05*, Lausanne, Switzerland.

Egges, A. Papagiannakis, G. , & Magnenat-Thalmann, N. (2007). Presence and Interaction in Mixed Reality Environments. *The Visual Computer*, 23(5), 317-333.

Freud, S. (1895) Project for a scientific psychology, in *The Standard Edition of the Complete Psychological Works of Sigmund Freud*, 1991 - The Hogarth Press.

Funge, J. (1999). *AI for Computer Games and Animation: A Cognitive Modeling Approach*. AK Peters, Ltd, Wellesley, MA.

Gazzaniga, M. S. & LeDoux, J. E. (1978), *The Integrated Mind*, Plenum Press, New York.

Gazzaniga, M.S., Ivry, R., & Mangun, G.R. (2002). *Cognitive Neuroscience: The Biology of the Mind.* 2nd Edition, W.W. Norton.

Kurzban, R. & Aktipis, C.A. (2006). Modular Minds, Multiple Motives. in Schaller, M., Simpson, J. & Kenrick, D. (Eds.) *Evolution and Social Psychology*. Psychology Press, London.

Laird, J. E. & Duchi, J. C. (2000). Creating Human-like Synthetic Characters with Multiple Skill Levels: A Case Study using the Soar Quakebot. Paper presented at *American Association for Artificial Intelligence Fall Symposium Series on Simulating Human Agents.*

Laird, J.E. (2002). Research in Human-Level AI Using Computer Games. *Communications of the ACM* 45(1),32-35.

Langley, P., Laird, J. E. & Rogers, S. (2006). *Cognitive architectures: Research issues and challenges* (Technical Report). Computational Learning Laboratory, CSLI, Stanford University, CA.

LeDoux (1998). *The Emotional Brain: The Mysterious Underpinnings of Emotional Life.* Weidenfeld & Nicolson.

Luria, A. R. (1980). *Higher Cortical Functions in Man*, 2nd ed., Inc Publishers, New York.

Magnenat-Thalmann, N., Kim, H., Egges, A. & Garchery, S. (2005). Believability and Interaction in Virtual Worlds. *International Multi-Media Modeling Conference*, IEEE Computer Society Press.

Piaget, J. (1972). Affective unconscious and cognitive unconscious (in French). In: *Problèmes de psychologie génetique*. Éditions Denoel, Paris.

Rickel, J., Marsella, S., Gratch, J., Hill, R. Traum, D. & Swartout, W. (2002). Toward a New Generation of Virtual Humans for Interactive Experiences. *IEEE Intelligent Systems*, (July/August), 32-38.

Sevin, E. Thalmann, D. (2004). The Complexity of Testing a Motivational Model of Action Selection for Virtual Humans. *Computer Graphics International* (CGI), IEEE Computer Society-Press, Crete.

Sloman, A. (2001). Beyond Shallow Models of Emotion. *Cognitive Processing*, 2(1), 177-198

Sloman, A. (2002). Architecture-Based Conceptions of Mind, in *In the Scope of Logic, Methodology, and Philosophy of Science* (Vol II, 403–427), Synthese Library Vol. 316, Gärdenfors, P. , Wolenski, J. & Kijania-Placek, K.(eds.), Kluwer Academic Publishers.

Torres, J.A., Nedel, L. & Bordini, R.H. (2003). Using the BDI Architecture to Produce Autonomous Characters in Virtual Worlds. *Proceedings of the Fourth International Conference on Interactive Virtual Agents* (IVA 2003), Lecture Notes in Artificial Intelligence, Springer-Verlag, Berlin.

Tyrrell, T. (1993). *Computational Mechanisms for Action Selection.* Ph.D. Dissertation. Centre for Cognitive Science, University of Edinburgh.

Wooldridge, M. (2002). *An Introduction to MultiAgent Systems.* John Wiley & Sons.

Wray, R. Laird, J.E. Nuxoll, A. Stokes, D. & Kerfoot, A. (2004). Synthetic Adversaries for Urban Combat Training. *Proceedings of the Sixteenth Innovative Applications of Artificial Intelligence Conference*, 923-930.

Chapter XII
A General Rhythmic Pattern Generation Architecture for Legged Locomotion

Zhijun Yang
Stirling University, UK

Felipe M.G. França
Universidade Federal do Rio de Janeiro, Brazil

ABSTRACT

As an engine of almost all life phenomena, the motor information generated by the central nervous system (CNS) plays a critical role in the activities of all animals. After a brief review of some recent research results on locomotor central pattern generators (CPG), which is a concrete branch of studies on the CNS generating rhythmic patterns, this chapter presents a novel, macroscopic and model-independent approach to the retrieval of different patterns of coupled neural oscillations observed in biological CPGs during the control of legged locomotion. Based on scheduling by multiple edge reversal (SMER), a simple and discrete distributed synchroniser, various types of oscillatory building blocks (OBB) can be reconfigured for the production of complicated rhythmic patterns and a methodology is provided for the construction of a target artificial CPG architecture behaving as a SMER-like asymmetric Hopfield neural networks.

INTRODUCTION

Animal gait analysis is an ancient science. As early as two thousand years ago, Aristotle described the walk of a horse in his treatise (Peek & Forster, 1936) *De Incessu Animalium*: "The back legs move diagonally in relation to the front legs; for after the right fore leg animals move the left hind leg, then the left fore leg, and after it the right hind leg." However, he erroneously believes that the bound gait is impossible: "If they moved the fore legs at the same time and first, their progression would be interrupted or they would even stumble forward… For this reason, then, animals do not move separately with their front and back legs."

Following the legend, modern gait analysis also originated with a horse, namely, a bet concerning the animal's gait (Taft, 1955). In the 1870s, Leland Stanford, the former governor of the state of California, became involved in an argument with Frederick MacCrellish over the placement of the feet of a trotting horse. Stanford put 25,000 dollars behind his belief that at times during the trot, a horse had all of its feet off the ground. To settle the wager, a local photographer, Eadweard Muybridge, was asked to photograph the different phases of the gaits of a horse. As a matter of fact, Stanford was correct in his bold assertion.

Aristotle and Stanford's insights into horse gaits can be viewed as the classical representations of embryonic ideas which lead to the modern studies of rhythmic pattern formation. After the case of Stanford there followed about eighty years of silent time till the 1950s, when A. M. Turing (1952) analysed rings of cells as models of morphogenesis and proposed that isolated rings could account for the tentacles of hydra and whorls of leaves of certain plants. Meanwhile, A. L. Hodgkin and A. F. Huxley published their influential paper (1952) on circuit and mathematical models of the surface membrane potential and current of a giant nerve fibre. The history has never seen such a prosperous era in the development of science and technology during the recent fifty years. With the rapid development of computational methods and computer techniques, many great scientific interdisciplines such as neural networks have been born and grew astonishingly. Obviously, it is not exaggerative at all to say that Turing et al.'s pioneer works on pattern formation are the cradle of modern connectionism. It is also interesting to notice that the macro- and microscopic approaches have coexisted since the initial stage of the modern biological rhythmic pattern research, just as the two examples stated above.

Rhythmic Patterns in Artificial Neural Networks

It is widely believed that animal locomotion is generated and controlled, in part by central pattern generators (CPG), which are networks of neurons in the central nervous system (CNS) capable of producing the rhythmic outputs. Current neurophysiological techniques are unable to isolate such circuits from the intricate neural connections of complex animals, but the indirect experimental evidence for their existence is strong (Grillner, 1975, 1985; Stein, 1978; Pearson, 1993).

The locomotion patterns are the outputs of musculoskeletal systems driven by CPGs. The study of CPGs is an interdisciplinary branch of neural computing which involves mathematics, biology, neurophysiology and computer science. Although the CNS mechanism underlying CPGs is not quite clear to date, artificial neural networks (ANN) have been widely applied to map the possible functional organisation of the CPGs network into the muscular motor system for driving locomotion.

The constituents of the locomotory motor system are traditionally modelled by nonlinear coupled oscillators, representing the activation of flexor muscles and the activation of extensor muscles by, respectively, two neurophysiologically simplified motor neurons. Different types of neuro-oscillators can be chosen and organised in a designed coupled mode, and usually with appropriate topological shape to allow simulating the locomotion of relative animals (Bay & Hemami, 1987; Linkens et al., 1976; Tsutsumi & Matsumoto, 1984). All internal parameters and weights of coupled synaptic connections of the oscillator network are controlled by the environmental stimulations, CNS instructions and the network itself. The nature of the parallel and distributed processing (PDP) is the most prominent characteristic of this oscillatory circuit that can be canonically described by a group of ordinary differential equations (ODE), which may also be an autonomous system.

In other words, a complex biological pattern generator system such as the CPGs can be simplified and implemented in a phenomenological model which uses the concrete ANN network dynamics.

Implementing Artificial CPGs as Neighbourhood-Constrained Systems

From a philosophical point-of-view, one could see that the world is full of neighbourhood-constrained systems. Considering our case of CPGs consisting of purely inhibitory connections, which is essentially a neighbourhood-constrained system, the traditional research method is to investigate an ordinary differential equation (ODE) or partial differential equation (PDE) of the concerned variables over a time course for all neurons. Nevertheless, it may be difficult to construct an ODE/PDE group representing a complicated CPG architecture with various periodic solutions for various locomotion patterns. Therefore, qualitative dynamical analysis may be a simpler strategy than quantitative numerical approach.

However, because of the complexity of the neuronal locomotor system, accurate mathematical descriptions or even detailed qualitative analysis are usually impossible. Thus, one has to make use of simplifications to describe the observed phenomena. As an alternative, a series of novel PDP fundamentals and algorithms, namely scheduling by edge reversal (SER) and its generalisation, scheduling by multi-edge reversal (SMER) (Barbosa & Gafni, 1989; Barbosa, 1996; França, 1994), have been found to be especially efficient in treating topologically representable CPGs. By adopting a self-timing scheme which is a key technique underlying the SER approach, large-scale CPG systems can be constructed easily and naturally, with immunity of starvation and deadlock, or saying, these CPG models can operate without undesired problems.

These PDP algorithms give a potentially optimal solution to Edsger Dijkstra's paradigmatic dining philosophers problem (Dijkstra, 1971), which is a canonical resource-sharing problem. They were devised on the assumption that the target systems were under the heavy load and neighbourhood-constrained environment, i.e., processes are constantly demanding access to *all* shared-resources, and neighbouring processes in the system must alternate in their turns to operate. These scheduling mechanisms were proved having the potential to provide the greatest concurrency among scheduling schemes with neighbourhood-constrained feature, while it is also capable of avoiding traditional problems as deadlock and starvation.

Based on the aforementioned PDP algorithms, we present a novel structural approach to the modelling the complex behavioural dynamics with a new concept of oscillatory building blocks (OBB) (Yang & França, 2003; França & Yang, 2000). Through appropriate selection and organisation of appropriately configured OBB modules, different gait patterns can be achieved for producing complicated rhythmic outputs, retrieving realistic locomotion prototypes and facilitating the VLSI circuit synthesis in an efficient, uniform, and systematic framework. In addition to the formal introduction of the new concepts of OBB building blocks and OBB networks, we will also show their applications in simulating various gait patterns of hexapod animals and demonstrate the efficiency of using an OBB network to mimic some functionalities of the CPG models.

PREVIEW OF FUNDAMENTALS

After a brief introduction to the state of the art of neurolocomotion, we describe SER and SMER algorithms and show their potential as a theoretical background for our pattern generation strategy.

State-of-the-Art on Neurolocomotion

Legged animals usually adopt various gait patterns in their terrestrial locomotion for various reasons, e.g., avoiding dangers, adapting terrain, or just obeying the willingness of changing gaits. Although many biological experiments have shown that generation of animal's gait patterns is a result of interactions between CNS and feedback of external stimulations, which induces wide admission of the existence of CPGs, the neural mechanisms underlying CPGs are still not well understood. One of the crucial questions undetermined is whether a unique CPG is sufficient for governing switching among various gait patterns or different CPGs are required to generate different gaits in the real-life biological systems (Collins, 1995). So far many models have been suggested on CPGs mechanism of vertebrate and invertebrate animals, for instance, the biped (Bay & Hemami, 1987; Taga et al., 1991; Taga, 1995), quadruped (Schöner et al., 1990; Collins & Richmond, 1994; Collins & Stewart, 1993a), and hexapod gait models (Collins & Stewart, 1993b; Beer, 1990). Most of them follow these two lines and are based on the coupled, nonlinear oscillator method for modelling.

A Neuromodulatory Approach

As a model for legged locomotion control, Grillner proposed that each limb of an animal is governed by a separate CPG (Grillner, 1975, 1985), and that interlimb coordination is achieved through the actions of interneurons which couple together these CPGs. With this scheme, gait transitions are produced by switching between different sets of coordinating interneurons.

Grillner's strategy has been adopted, in spirit, by some CPGs modelling studies. For instance, Bay and Hemami (1987) used a CPG network of four coupled van der Pol oscillators to control the movements of a segmented biped. Each limb of the biped was composed of two links, and each oscillator controlled the movement of a single link. Bipedal walking and hopping were simulated by using the oscillators' output to determine the angular positions of the respective links. Transitions between out-of-phase and in-phase gaits were generated by changing the nature of the inter-oscillator coupling, e.g., the polarities of the network interconnections were reversed to produce the walk-to-hop transition.

This approach is, in principle, physiologically reasonable. The notion that supraspinal centres may call on functionally distinct sets of coordinating interneurons to generate different gaits is plausible but not yet experimentally established. In addition, from a different but relevant perspective, it is shown that rhythmic neuronal networks can be modulated, e.g., reconfigured, through the actions of neuroamines and peptides, and thereby enabled to produce several different motor patterns (Pearson, 1993).

A Synergetic Approach

Synergetics deals with cooperative phenomena. In synergetics, the macroscopic behaviour of a complex system is characterised by a small number of collective variables which in turn govern the qualitative behaviour of the system elements (Collins, 1995).

Schöner and colleagues used a synergetic approach in a study of quadruped locomotion (Schöner et al., 1990). They analysed a network model that was made up of four coupled oscillators, each representing a limb of a model quadruped. The phase difference among limbs were used as collective variables to characterise the interlimb coordinative patterns of this discrete dynamical system. Gait transitions were simply modelled as phase transition, which could also be interpreted as bifurcations in a dynamical system.

This approach is significant in that it relates system parameter changes and stability issues to gait transitions. Its primary weakness, however, is that the physiological relevance of the relative phase coupling terms is unclear.

A Group-Theoretic Approach

According to the arguments of Collins (1995), the traditional approach to modelling a locomotor CPG has been to set up and analyse, either analytically or numerically, the parameter-dependent dynamics of a hypothesized neural circuit. Motivated by Schöner et al.'s works, Collins et al. dealed wth the CPGs dynamics from the perspective of group theory (Collins & Stewart, 1993a,b). They considered various networks of symmetrically coupled nonlinear oscillators and examined how the symmetry of the respective systems leads to a general class of phase-locked oscillation patterns. In this approach the transitions between different patterns can be modelled as symmetry-breaking Hopf bifurcation. It is well established that, in a Hopf bifurcation, the dynamics of a nonlinear system may change when its parameters are varied. An old limit cycle may disappear and several new limit cycles may emerge depending on how the model parameters change. As the symmetries reach the least level, theoretically the chaotic phenomena may arise.

The theory of symmetric Hopf bifurcation predicts that symmetric oscillator networks with invariant structures can sustain multiple patterns of rhythmic activities. It emphasizes that one intact CPG architecture is sufficient for hosting all possible pattern changes (Golubitsky et al., 1998). This approach is significant in that it provides a novel mechanism for generating gait transitions in locomotor GPGs. The primary disadvantage is that its model-independent feature makes it difficult to provide information about the internal dynamics of individual oscillators.

A Primer to SER and SMER Algorithms

Scheduling by Edge Reversal (SER)

Consider a neighbourhood-constrained system comprised of a set of processes and a set of atomic shared resources represented by a connected graph $G = (N,E)$, where N is the set of processes and E the set of edges defining the interconnection topology. An edge exists between any two nodes if and only if the two corresponding processes share at least one atomic resource.

SER works in the following way: starting from any acyclic orientation ω on G there is at least one sink node, i.e., a node that has all its edges directed to itself. Only sink nodes are allowed to operate while other nodes remain idle. This obviously ensures mutual exclusion at any access made to shared resources by sink nodes. After operation a sink node will reverse the orientation of all of its edges, becoming a source and thus releasing the access to resources to its neighbours. A new acyclic orientation is defined and the whole process is then repeated for the new set of sinks (Barbosa & Gafni, 1989; Barbosa, 1996). Let $\omega' = g(\omega)$ denote this greedy operation. SER can be regarded as the endless repetition of the application of $g(\omega)$ upon G. Assuming that G is finite, it is easy to see that eventually a set of acyclic orientations will be repeated defining a period of length p. This simple dynamics ensures that no deadlock or starvation will ever occur since at every acyclic orientation there is at least one sink, i.e., one node allowed to operate. Also, it has been proven that inside any period every node operates exactly m times (Barbosa & Gafni, 1989), i.e., the value of m is the same for all nodes within any period.

SER is a fully distributed algorithm of graph dynamics. An interesting property of this algorithm lies in its generality in the sense that any topology will have its own set of possible SER dynamics. Figure 1 illustrates the SER dynamics.

Figure 1. A simple graph G under SER, with m=1, operation cycle p=2

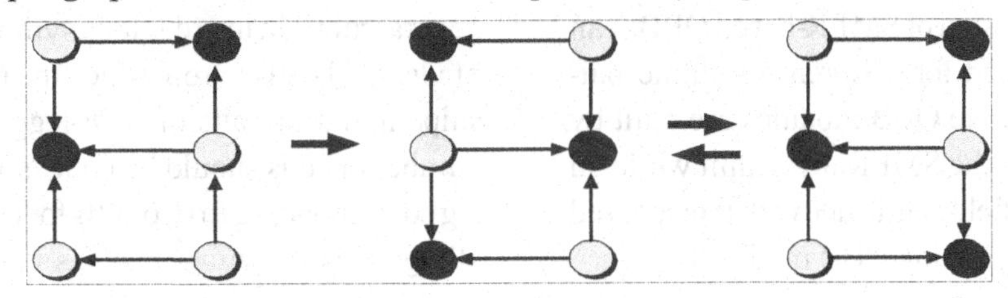

Scheduling by Multiple Edge Reversal (SMER)

SMER is a generalisation of SER where prespecified access rates to atomic resources are imposed on processes in a distributed resource-sharing system that is represented by a multigraph *M(N,E)*. Unlike SER, with SMER a number of oriented edges can exist between any two nodes. Between any two nodes *i* and *j*, $i, j \in N$, there can exist e_{ij} unidirected edges, $e_{ij} \geq 0$. The reversibility of node *i* is r_i, i.e., the number of edges that shall be reversed by *i* toward each of its neighbouring nodes, indiscriminately, at the end of the operation. Node *i* is an r_i-sink if it has at least r_i edges directed to itself from each of its neighbours. Each r_i-sink node *i* operates and reverses r_i edges towards each of its neighbours, the new set of r_i-sinks will operate, and so on. Like sinks under SER, only r_i-sink nodes are allowed to operate under SMER. It is easy to see that with SMER, nodes are allowed to operate more than once consecutively.

The following lemma states a basic topologic and resource configuration constraint toward the definition of *M*, where **gcd** is the *greatest common divisor.*

Lemma 1. *(Barbosa, 1996; França, 1994) Let nodes i and j be two neighbours in* ***M****. If no deadlock arises for any initial orientation of the shared resources between* ***i*** *and* ***j****, then* $e_{ij} = r_i + r_j - \gcd(r_i, r_j)$ *and* $\max\{r_i, r_j\} \leq e_{ij} \leq r_i + r_j - 1$.

It is important to know that there is always at least one SMER solution for any target system's topology having arbitrary prespecified reversibilities at any of its nodes (Barbosa et al., 2001). In the following sections, SER and SMER will be used to construct the artificial CPGs by implementing oscillatory building blocks (OBBs) as asymmetric Hopfield-like networks, where operating sinks can be regarded as firing neurons in purely inhibitory neuronal networks.

NATURALLY INSPIRED DISCRETE AND ANALOG OBB MODULES

Of long-standing interest are questions about rhythm generation in networks of nonoscillatory neurons, where the driving force is not provided by endogenous pacemaking cells. A simple mechanism for this is based on reciprocal inhibition between neurons, if they exhibit the property of *postinhibitory rebound* (PIR) (Wang & Rinzel, 1992). The PIR mechanism (Kuffler & Eyzaguirre, 1955) is an intrinsic property of many central nervous system neurons which is referred to a period of increased neuronal excitability following the cessation of inhibition. It is often included as an element in computational models of neural networks involving mutual inhibition (Perkel, 1976; Roberts & Tunstall, 1990). The ensured mutual exclusion activity between any two neighbouring nodes coupled under SMER suggests a scheduling scheme that resembles anti-correlated firing activity between inhibitory neurons exhibiting PIR. The discrete and the analog version of SMER-based OBBs can thus be customised for different rhythmic patterns, where discrete OBB modules are built by directly adopting the SMER algorithm while an asymmetric Hopfield neural network is employed for implementing analog OBB modules.

Architecture of Gait Models

Out-of-phase (walking and running) and in-phase (hopping) are the major characteristics of observed gaits in bipeds, while in quadrupeds, more gait types were observed and enumerated (Alexander, 1984), as walk, trot, pace, canter, transverse gallop, rotary gallop, bound and pronk. Unlike bipeds and quadrupeds, hexapod locomotion can have more complicated combinations of leg movements. Despite the variety, however, some general symmetry rules should still be obeyed and remained as the basic criteria for gait prediction and construction. For instance, it is a generally accepted view that multiply legged locomotion (usually more than six legs) often display a travelling wave sweeping along the chain of oscillators (Collins & Stewart, 1993a; Golubitsky et al., 1998).

Golubitsky and colleagues (1998) proposed a general rule that in order for a symmetric network of identical cells to reproduce the phase relationships found in gaits of a *2n*-legged animal, the number of cells should be (at least) *4n*, with each leg corresponding to two cells for behaviour modelling. This assertion matches with our coupled inhibitory neurons' representation for a set of legs, organised under the SER/SMER scheme. Following Golubitsky et al.'s schematic diagram of a general gait model, we can continue to get the architecture-isomorphic substitution of their CPG network and use this general model architecture throughout our research.

Discrete OBB Modules

Instead of modelling electrophysiological activities of interconnected neurons based on membrane potential functions, we build an artificial CPG network with SMER-based OBBs for the collective behaviours of a neuron set. A simple discrete OBB module is defined to have a r_i-sink node and a r_j-sink node sharing the number of e_{ij} resources. Two nodes pertain to two neighbour-

Figure 2. The isomorphic transition of a general multiply legged locomotion architecture. (a) A 4n-node model reproduced from the work of Golubitsky et al. (1998). (b) Our general model in which all nodes are coupled with each other through, and receive pattern signals from the CPG.

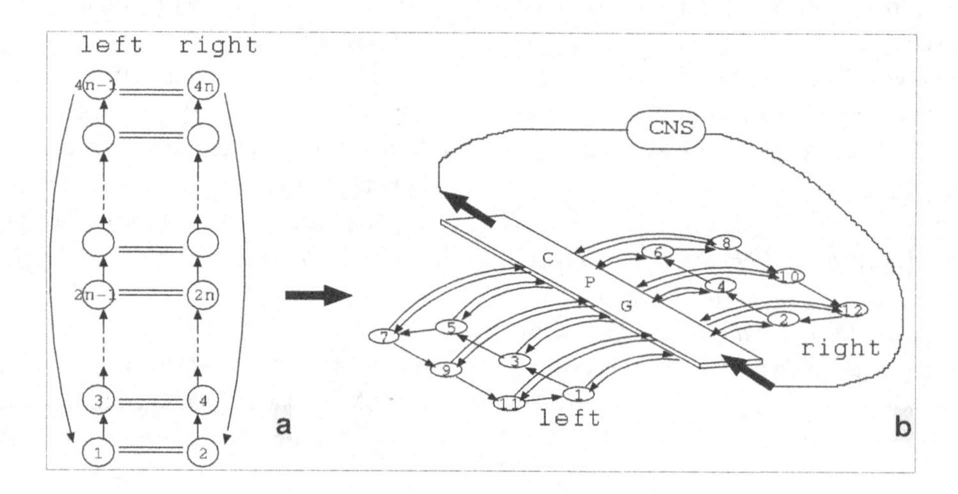

ing motor neurons, respectively, with the shared resources signifying that two motor neurons are interconnected. The exchange of r_i (or r_j) resources between two nodes results in pre-defined firing frequencies of two motor neurons.

The representative cases of three gait patterns of a cockroach, i.e., metachronal, medium-speed, and fast-speed locomotions, according to the prototypes of Collins and Stewart (1993b), are analysed by using the discrete OBBs. A transition between different gait patterns is usually mediated by command signals from CNS under a conscious reflex, or a population of neurons in CPG in direct response to external stimuli under an unconscious reflex. Either case involves modulation of the connection topology and internal parameters of CPG building blocks which can greatly alter network operations (Getting, 1989).

Figure 3 shows a conceptual hexapod's CPG representing gait phase relationship between six legs. Each leg is simplified as having one degree of freedom with a pair of flexor and extensor motor neurons and muscles, though any degrees of freedom of a leg are feasible by assuming more complexity under the proposed strategy. From an anatomical point of view, cockroach leg movement is driven by flexor and extensor motor neurons. Flexor muscles lift a leg from the ground, while extensor muscles do the opposite. This procedure can be well imitated by an OBB module by taking neurons *i* and *j* in the module as flexor and extensor motor neurons, respectively (see Figure 4).

Figure 3. Movement circulation sequences and phase relations among six legs of a cockroach, each leg represented by one electrically compact node. White colour means legs on the ground, extensor neurons and muscles activity; black colour means legs swing, flexor neurons and muscles activity. (a) Metachronal gait. (b) Medium-speed gait. (c) Fast-speed gait.

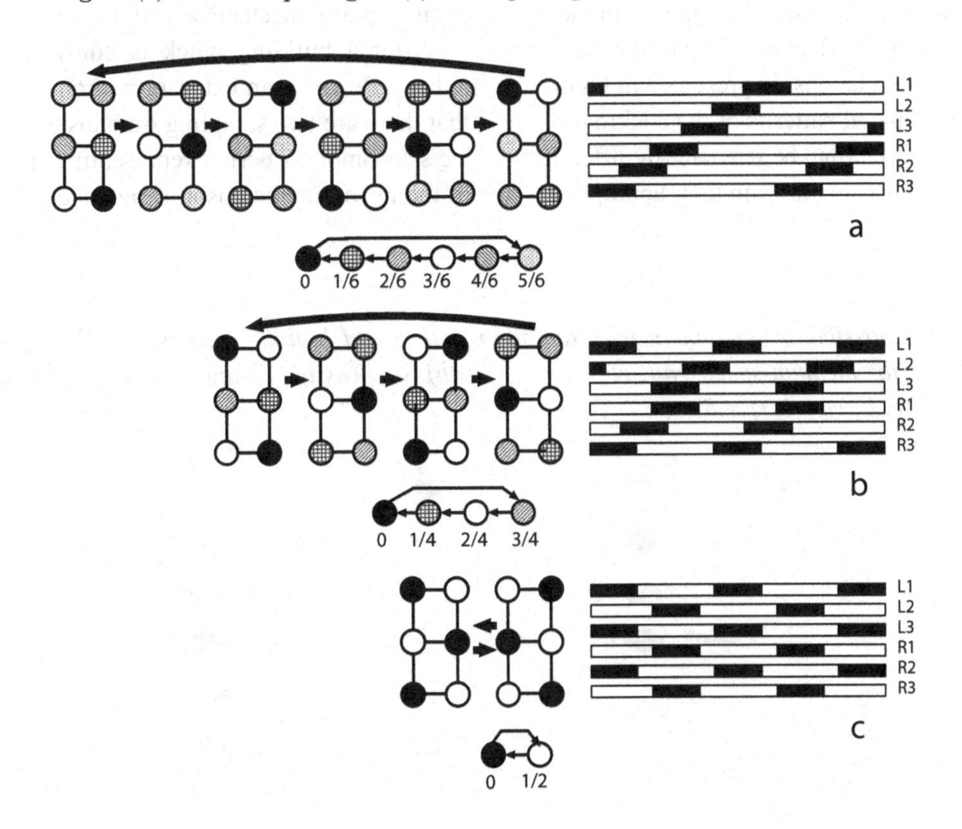

There is a timing relation between the pair of flexor and extensor for the different locomotion speeds. As a cockroach escapes more swiftly, the duty factor for the firing duration of the cockroach's extensor (corresponding to stance) will decrease drastically, while the firing duration of its flexor (corresponding to swing) remains basically unchanged, which is consistent with biological observations (Pearson, 1976). This insight indicates that hexapod speed is determined largely by the extensor firing, i.e., the time duration of a leg contacting the ground.

The state transition of each leg and the corresponding phase relations among different legs are important to simulate the gait model. The phase circulation can be represented by the circulation of OBB modules. In the case of a cockroach's fast-speed gait, simple SER-based OBBs can be applied to initiate oscillation and coordinate the movement of the six legs (see Figure 3c). In order to formulate more complex metachronal movement, an appropriate OBB configuration in the pattern circulation (Figure 4) is chosen for each of six legs, respectively, according to the corresponding phase relationship presented in Figure 3a. The coordinated patterns of the cockroach's metachronal gait can thus be generated by different SMER-based OBB modules in their appropriate configurations. This process is generalisable to modelling the other gaits. An example of how a possible scheme of firing circulation patterns of building blocks can simulate the activity envelope of a pair of flexor and extensor motor neurons is shown in Figure 5.

It is widely recognised that animal's locomotion behaviour is a continuous-time procedure. So far our gait analysis only samples some typical time instant from these continuous, high-dimensional waveforms where the number of dimensions is equal to the number of legs. The samples are the snapshots with at least one leg supporting ground substantially and, ignoring all other snapshots which may contribute little for the model retrieval. Given that a kind of high pass filter characteristic is a common property observed in every real neuron (Bässler, 1986; Matsuoka, 1987), we believe that the sampled time instants can lead to smooth, continuous waveforms for driving motor behaviours since the filtering property can be naturally implemented by the hysteretic phenomena of mechanical oscillation.

After a building block is constructed for either a flexor or an extensor neuron, it is clear that there are two sampling time instants in one leg's locomotion period representing the firing of the flexor and extensor, respectively. It is the

Figure 4. One possible scheme of firing circulation patterns of building blocks. (a) Four possible configurations for medium-speed gait; $ri = 1$, $rj = 3$. (b) Six possible configurations for slow-speed (metachronal) gait; $ri = 1$, $rj = 5$.

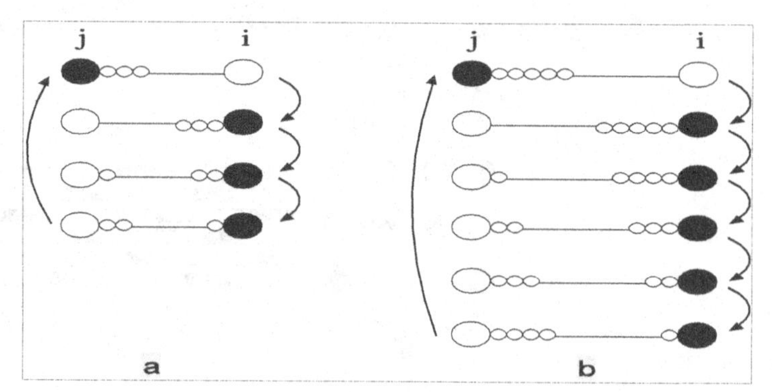

Figure 5. Mimicking CPGs with SER/SMER-driven networks. Activity of flexor (node j) and extensor (node i) motor neurons during walking of a cockroach and its SER/SMER analogous simulation.

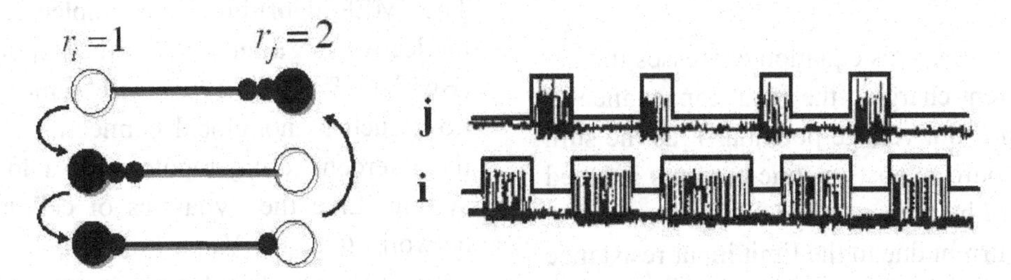

individual OBB modules rather than the integrated OBB network that is governed by an SMER algorithm. Therefore, the methodology is especially useful for digital circuit implementation as the gait model operates asynchronously without the necessity of a global clock.

Dynamic Properties of Analog OBB Modules

SER and SMER have the potential to provide the greatest concurrency among scheduling schemes on resource-sharing systems (Barbosa & Gafni, 1989; Barbosa et al., 2001). The mutual exclusion characteristic between any two neighbouring nodes coupled under SMER makes this scheduling scheme suitably tailored for simulating PIR, a mechanism widely employed for locomotion and other rhythmic activities. The dynamics of SMER depends on the initial allocation of shared resources, different configurations may lead to different cycling behaviours and even deadlock or starvation of the system. This feature leaves us a space on how to mimic CPGs with SMER to simulate numerous rhythmic patterns while avoiding possible wrong design.

Given the ability of SMER to mimic many gait patterns, this kind of strategy, however, is essentially intuitive due largely to the discrete nature of SMER. It is thus more efficient in digital simulation than analogous applications. The analogous behaviours, on the other hand, are ubiquitous in the real world. They are continuous rather than discrete in time domain, better described by analog circuitry rather than digital one. Following the digital version we present a series of novel continuous time OBB structures which are similar to Hopfield networks but governed by the SMER algorithm. These structures can be classified into two major categories, namely simple OBB and composite OBB, depending on the network complexity.

Hopfield Neural Network

In 1982 J. J. Hopfield published his seminal work on the prominent emergent properties of one kind of neural model which rekindled great interest of scientists in neural network analysis. Different from McCulloch-Pitts neural model (1943) in which neurons are intrinsically Boolean comparators with limited inputs, Hopfield studied the biological structure of the large-scale interconnected neural systems and proposed his model with the emphasis on the whole behaviours and properties of a neural system. A single neuron is simplified to a simple electronic device containing only a resistor, a capacitor and a nonlinear component stipulating the input-output relationship. The dynamics of an interacting system of N coupled neurons can be described by a set of coupled nonlinear differential equations governed by Kirchhoff's current law,

$$C_i \frac{dV_i}{dt} = \sum_{j=1}^{N} w_{ij} f_j(V_j) - \frac{V_i}{R_i} + I_i$$

where $i=1,...,N$. This equation expresses the net input current charging the input capacitance C_i of neuron i to a voltage potential V_i as the sum of three sources: postsynaptic currents induced in neuron i by presynaptic activity in neuron j; leakage current due to the limit input resistance R_i of neuron i; and input currents I_i from other neurons external to the circuit. The model retains two important aspects for computation: dynamics and nonlinearity (Hopfield, 1982; Hopfield & Tank, 1986).

Hopfield classified his models into two categories, namely symmetric and asymmetric models. Between any two coupled neurons i and j, if the synaptic weight from neuron i to j is equal to the weight from neuron j to i, then this is a symmetric Hopfield network, otherwise it is named asymmetric Hopfield network. The most prominent property of symmetric model is its auto-association, i.e., the system energy keeps diminishing and finally it reaches the local (or global) minimum energy level as the system state evolves. Unlike the symmetric model, the asymmetric network may exhibit oscillation and chaos when it evolves with time. In some motor systems like CPGs, coordinated oscillation is the desired computation of the circuit. A proper combination of asymmetric synapses can enforce chosen phase relationships between different oscillators (Hopfield & Tank, 1986). In recognition of the properties of the network dynamics, we present a novel methodology to embed the SMER algorithm into the asymmetric Hopfield neural networks. Specifically, we will show that this methodology is useful in simulating all CPG patterns introduced in Golubitsky's symmetric Hopf bifurcation theory (Golubitsky et al., 1998, 1988).

Dynamics of a Simple OBB Module

The SMER algorithm can be implemented schematically using a network like the Hopfield model (Hopfield & Tank, 1985) with some modifications such as nonglobal connections based on the interconnecting topology of a locomotion system. Like the dynamics of cellular neural networks (Chua & Yang, 1988a,b), the input and output voltages of each node in an OBB network are normalised to digital low or high level while the internal potential is continuous within the normalised interval *[0,1]*. The SMER-based OBB modules can be classified into two complexity levels, namely, simple and composite. The simple OBB modules consist of only two interconnected nodes with prespecified reversibilities. Figure 6 shows its circuit representation. The composite OBBs may contain an arbitrary number of cells interconnected with any topology. Both types of OBB modules follow *Lemma 1* for initial shared resources allocation and configuration to avoid possible abnormal operations, e.g., deadlock or starvation, during oscillation.

Now consider a submultigraph of $M(N,E)$, namely M^{ij}, having a pair of coupling nodes n_i and n_j with r_i and r_j as their reversibility, respectively. In the SMER-based, simple OBB module, the postsynaptic membrane potential of neuron i at k instant, $M^i_{PSP}(k)$, depends on three factors, i.e., the potential at the last instant $M^i_{PSP}(k-1)$, the impact of its coupled neuron output $v^j_{out}(k-1)$, and the negative feedback of neuron i itself $v^i_{out}(k-1)$, without considering the external impulse. The selection of system parameters, such as the neuron thresholds and synapse weights, are crucial for modelling. In our model, let $r = \max(r_i, r_j)$ and $r' = h(r)$, where h is a function of getting highest integer level and multiplying it by *10*, e.g., if $r_i = 56$ and $r_j = 381$ then $h(r) = h(\max(56,381)) = h(381) = 10^3$. We can design the neuron i and j's thresholds θ_i and θ_j and their synaptic weights as following,

Figure 6. A circuit representation of a simple OBB module

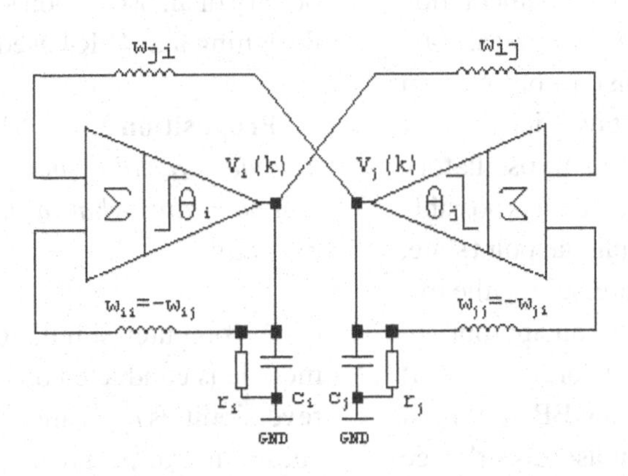

$$\theta_i = \frac{\max(r_i, r_j)}{r_i + r_j - \gcd(r_i, r_j)} \quad (1)$$

$$w_{ij} = \frac{\max(r_i, r_j)}{r'} \quad (2)$$

$$\theta_j = \frac{\min(r_i, r_j) - \gcd(r_i, r_j)}{r_i + r_j - \gcd(r_i, r_j)} \quad (3)$$

$$w_{ji} = \frac{\min(r_i, r_j)}{r'} \quad (4)$$

We arrange the system parameters by comparing two nodes' reversibilities. If $r_i > r_j$, then we have $\theta_i > \theta_j$ and $w_{ij} > w_j$, i.e., a node with smaller reversibility, corresponding to a neuron with lower threshold in an OBB module, will oscillate at a higher frequency than its companion does. This arrangement scheme ensures that the behaviour of SMER-based OBB modules is consistent with its original SMER algorithm. The difference equation in the discrete time domain of this system can be formulated as follows: each neuron's self-feedback strength is $w_{ii} = -w_{ij}$, $w_{ij} = -w_{ji}$, respectively, and the activation function is a sigmoid Heaviside type. It is worth noticing that k is the local clock pulse of each neuron, a global clock is not necessary. Thus we have,

$$\begin{cases} M_i(k+1) = M_i(k) + w_{ji}v_j(k) + w_{ii}v_i(k) \\ M_j(k+1) = M_j(k) + w_{ij}v_i(k) + w_{jj}v_j(k) \end{cases} \quad (5)$$

where,

$$\begin{cases} v_i(k) = \max(0, \mathrm{sgn}(M_i(k) - \theta_i)) \\ v_j(k) = \max(0, \mathrm{sgn}(M_j(k) - \theta_j)) \end{cases} \quad (6)$$

We consider the designed circuit as a conservative dynamical system in an ideal case. The total energy is constant, no loss or complement is allowed. The sum of two cells' postsynaptic potential at any given time is normalised to one. It is clear that this system has the capability of self-organised oscillation with the firing rate of each neuron arbitrarily adjustable. However, like most dynamic systems, our model has a limit in its dynamic range. There exists a singular point as each cell's postsynaptic potential equals to its threshold; in this case, the system may transit to another different oscillation behaviour or even halt. Within its normal dynamic range, the following properties of the OBB module guarantee that this design is appropriate for implementation of the SMER algorithm (see Yang & França, 2003 for proofs of these propositions). In our current design, an OBB module will stay at its singular point state forever. When design an oscillatory neural network using OBB modules, it is possible to avoid the occurrence of any singular point state, i.e., $M_i(k+1) = M_i(k)\ \forall k, \forall i$, by pre-setting the initial membrane potential $M_i(0)$ to an appropriate value so that $w_j v_j(k) + w_i v_i(k) \neq 0$ for $\forall k, \forall i, \forall j$. It is possible to extend our current OBB definition to allow external inputs and noise disturbance to break the operation halt by adding additional connections, however, this is not the theme of this chapter.

Proposition 1. *The circuit representing the fundamental two-node SMER system is a starvation- and deadlock-free oscillation system.*

Another prominent property is about the periodicity of an oscillation system. It is essential for designing a SMER-based CPG system.

Proposition 2. *The SMER-based simple OBB module is a stable and periodic oscillation system no matter what initial potential its neurons may have.*

A computer simulation of this simple OBB module is conducted on a pair of predefined cell reversibilities $r_i = 4$ and $r_j = 3$ with the arbitrarily chosen initial postsynaptic potentials as $M_i(0)=$ 0.8, $M_j(0)=$ 0.2 (see Figure 7). The numerical simulation of OBB dynamics was made using Matlab's Simulink. The system will experience a convergence process upon the chosen initial postsynaptic potentials before it reaches a cycle of periodic oscillation.

Figure 7. A simulation of a simple OBB module based on a SMER case. (a) The state circulation of the module, where steps k=0 to 6 form a period, the output states and internal membrane potentials are shown by $i = v_i(k) / M_i(k)$ and $j = v_j(k) / M_j(k)$ corresponding to two nodes. (b) Waveform of node i, and (c) waveform of node j.

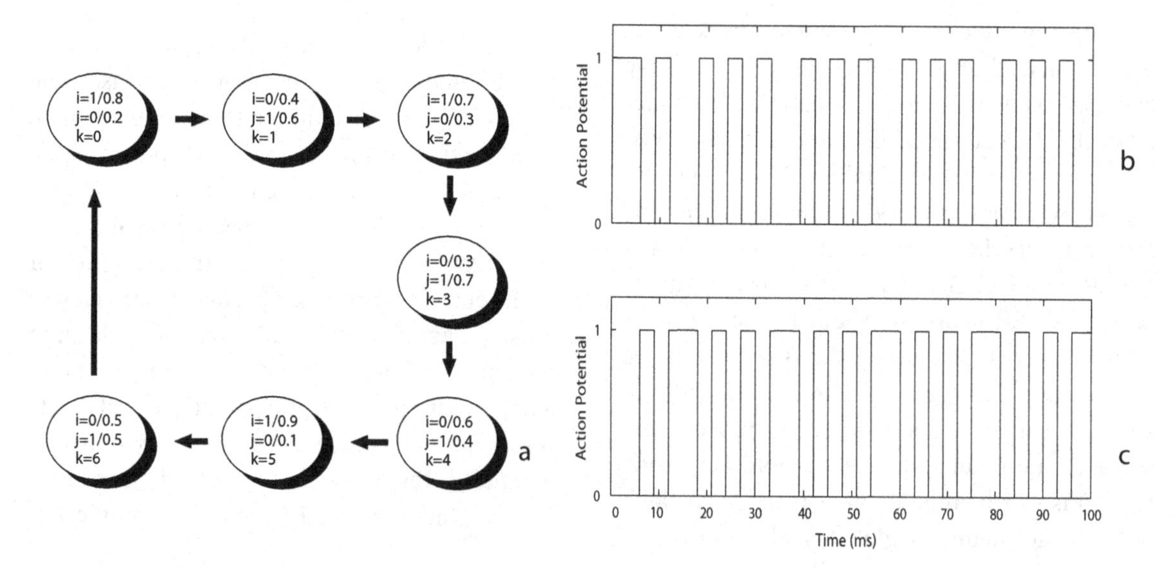

Dynamics of a Composite OBB Module

The composite OBB module is a generalisation of the simple OBB module which can consist of a number of simple OBB modules. A composite OBB can have more nodes and therefore, more complicatedly organised topologies. Suppose a multigraph *M(N,E)* containing a set of *m* nodes $N = \{n_1, n_2, ..., n_m\}$ and a set of edges $E = \langle 0|1 \rangle_{ij}$ where $\forall i, j \in N$, which define the connection topology of this multigraph by using $\langle 1 \rangle_{ij}$ and $\langle 0 \rangle_{ij}$ for with and without an edge between nodes *i* and *j*, respectively. Each node *i* has its own reversibility r_i. There are e_{ij} shared resources on the corresponding edge $\langle 1 \rangle_{ij}$, with their number and configuration stipulated by *Lemma 1.*

Different from a simple OBB module in which there are exactly two nodes coupled with each other, a composite OBB module has more than two nodes, and at least one node of them has connection with at least two nodes. The composite OBB module can be dissected into various simple ones, each node of the composite OBB is split into the corresponding copies which share the same local clock of their maternal node. A copy is a component of a simple module, and the total number of copies is twice as many as the number of edges in the composite OBB module. We then terminologically regard a node in a composite OBB module as a *macroneuron* and a node's copy as its *clone*. Their definitions are followed,

Definition 1. *A macroneuron is defined to be a node i that satisfies* $\forall i \in N$, *where N is the set of nodes in multigraph M(N,E).*

Definition 2. *A clone, which has independent reversibility and represents the coupling characteristics of its maternal macroneuron with one of the neighbouring macroneurons, is a unique component of its maternal macroneuron.*

According to the principles of SMER, a macroneuron of the composite OBB module operates in a "whole-or-none" mechanism in terms of the firing of its clones. It will fire if and only if all its clones fire. Fundamentally, a composite OBB module operates on the basis of its constituent simple OBB modules. A schematic diagram of a composite OBB module is shown in Figure 8.

As a generalised version, a composite OBB module can represent a more complicated oscillating neuronal network and reproduce more rhythmic patterns than a simple one. It does not take any fixed form in terms of the number of constituent clones, the connection structure, etc. Different applications specify different macroneurons and their complexity. Since it is impossible to assign a unified threshold to a macroneuron that may have more than one independent clone, the usual way to analyse this kind of modules is to dissect them into the subsystems of simple modules where equations (1) – (6) are applicable. The output of a macroneuron, n_i, is then determined by all its clones. We have,

$$V_i(k) = \prod_{j=1}^{n} v_i^j(k) \qquad (7)$$

where $\forall i \in N$, $i \neq j$ and v_i^j is the output of clone *j* of the macroneuron *i*, which couples with a corresponding clone of another macroneuron. The superscript sequence *j* = 1, 2, ..., *n* is the clone number of a macroneuron n_i.

It is important to choose the initial postsynaptic potential values properly for the clones which are the simple OBB modules. Different choices will lead to different rhythmic patterns of a composite OBB module. To avoid system halt, no clone should be idle if its macroneuron is designed to be firing by equation (7). Within an appropriate parameter range, a random selection of initial postsynaptic potential values is allowed. After an initial duration whose length is determined by the choice of initial postsynaptic potentials, the system will oscillate periodically. A pseudo-code operation of a macroneuron is given in Table 1.

Figure 8. Two equivalent architectures to illustrate the SMER algorithm, where $r_i = r_m = r_n = 1$; $r_j = 2$ $r_k = 3$. Left, the original SMER graph description. Right, an alternative description for the composite OBB module with the macroneurons and their clones.

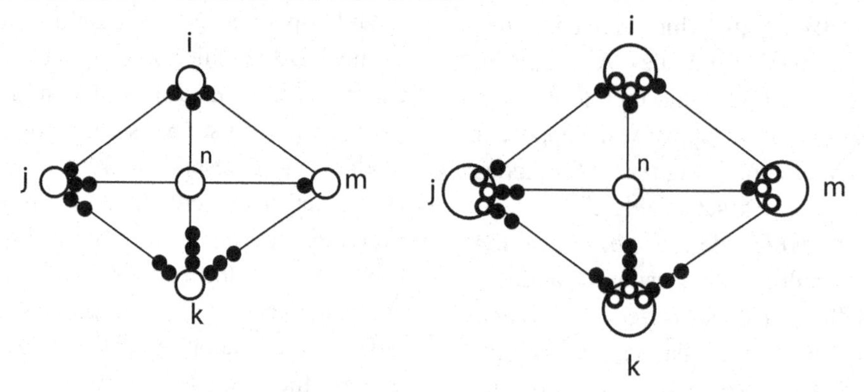

Table 1. Pseudo-code operation of a macroneuron

```
/*Program: Macroneuron_firing_function*/
CONST:
r[i : 1...m] : integer;
threshold[i, j : 1...m], weight[i, j : 1...m] : real;

VAR:
membrane[i, j : 1...m] : real;
firing[i] : boolean;

Initial State:
membrane[i, j] := random;
firing[i] := false;

Action at a local pulse:
if (∃j ∈ (1, m)‖(j!=i) ∧ (weight[i, j]!= 0) ∧ (membrane[i, j] < threshold[i, j]))
      begin
            firing[i] := false;
            await NOT(membrane[i,j] < threshold[i,j]);
            continue;
      end
else
      begin
            firing[i] := true;
            if (∀j ∈ (1, m)‖(j!=i) ∧ (weight[i, j]!= 0))
                  begin
                        membrane[i,j] := membrane[i,j] – weight[i,j];
                        membrane[j,i] := membrane[j,i] + weight[i,j];
                        continue;
                  end
      end
end
```

A GENERAL LOCOMOTION CPG ARCHITECTURE

The Architecture

A general architecture of a locomotion system may have three components: (1) the oscillation system as driving motor, (2) the CPGs as pattern source, and (3) the animal's CNS as command source. The first two parts formulate the proposed locomotion architecture in this chapter and can be constructed by a model with reciprocally inhibitory relationship between the so-called flexor and extensor motor neurons, at each leg as driving motor and the general locomotor CPGs as pattern source. The CPG is a core of the locomotion system which converts different conscious signals from the CNS to the bioelectronics-based rhythmic signals for driving the motor neurons. These motor neurons are essentially relaxation oscillators. They form a bistable system in which each neuron switches from one state to another when one of two internal thresholds is met by one neuron and back when another threshold is met by the other neuron.

For a $2n$-legged animal, the general locomotion model has the topology of a complete graph $M(N,E)$ with $||N||=4n$ and $||E||=C_{4n}^{2}$. Each of the upper layer macroneurons drive a flexor muscle while the lower layer ones drive a extensor muscle, see Figure 9. In this architecture there is a connection between any two macroneurons with the strength in the range $[0,1]$. Different gait pattern has a corresponding, different connection weight set. It sounds reasonable for the gait transition to be controlled by the biological signals from the CNS for avoiding risk or following animal's willingness. The size of signal flow is equal to the number of total couplings depending on the type of gait patterns and the number of animal legs. For those specific models derived from the uniform framework, the biped, quadruped, hexapod animals have 4, 16 and 30 bits of parallel signals from their CNS to CPGs , respectively, in order to control the locomotion and transition of all their gait types. We are not able to say, in our study, that an animal with less signalling bits is a lower-level animal since the meaning of the number of descending CNS signals is twofold. First, even a centipede, which is a relatively low-level insect, has much more CNS signals than in this study, however, these still only occupy a very small percentage of its neural processing part. Second, higher-level animals such as the biped have much more neural activities at other neural processing mechanisms, besides the CNS-CPGs controlling signals, while lower-level animals may have less neural processing activities besides CNS-CPGs signals. Therefore, the number of parallel signals of any animal model in this study is not directly related with its neural processing complexity.

It is known that locomotion speed is defined by both coordinated phase relation and duty factor, which is the relative proportion that a flexor neuron is firing in one period. As an animal's speed increases, the firing time of the extensor (corresponding to stance) will decrease dramatically while the firing time of the flexor (corresponding to swing) keeps basically constant (Pearson, 1976). The duty factor can be modified by changing the reversibilities of two coupled nodes in a simple OBB module as per the coordinated phase relationship.

Simulation of Hexapod Gaits

We have shown the remarkable potential of asymmetric SMER-based neural network on rhythmic pattern formation. Now we can build an artificial CPG example from the general gait pattern model. Some basic criteria on macroneuron interconnection will be followed.

1. Any two macroneurons in a CPG network should be coupled directly if their activities are exactly out of phase.
2. The ipsilateral macroneurons should be connected to form a cyclic undirected ring.

Figure 9. A general locomotor CPG architecture of 2n-legged animal. Each macroneuron couples with all of the rest 4n-1 macroneurons, which is not shown explicitly. The upper is the flexor layer and the lower the extensor layer.

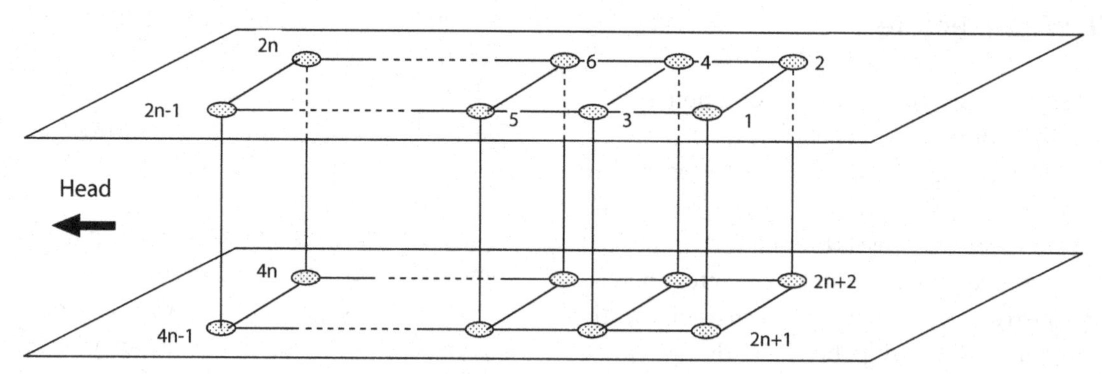

3. If the phase difference between one macroneuron and anyone of its contralateral macroneurons (in the same layer) is less than that between this macroneuron and its ipsilateral neighbours, then two contralateral macroneurons have a connection.
4. Based on the above connection, different arrangement of initial membrane potential of clones may lead to different gait patterns.

From a standing posture of a stick hexapod insect (Figure 10a) which keeps all structural symmetry with all macroneurons in the 12-cell network permutable, eight symmetry-breaking branches are bifurcated to eight primary gait patterns (Buono & Golubitsky, 2001; Golubitsky et al., 1999) as shown in Table 2. Among the total twelve macroneurons *{1,2,...12}* in this system, we take its upper layer subgroup *{1,2,...6}* as the set of flexors driving rear left (RL), rear right (RR), middle left (ML), middle right (MR), front left (FL) and front right (FR) legs, respectively. The complexity of numerical simulation of multiply legged animals increases arithmetically with the number of legs. In order to avoid enumerative listings we only deal with one hexapod gait, the rolling tripod. All other hexapod gaits can be simulated using the general model, in a same way as the rolling tripod gait.

The rolling tripod is a gait that ipsilateral legs are one-third period out of phase while contralateral legs are half period out of phase. A coupled structure with 30 connections, derived from the original all-connected, general gait model, is achieved according to the criteria for macroneuron interconnection (see Figure 10b). Each macroneuron has five clones, with one having their reversibilities. The arrangement of initial membrane potentials of all clones of 12 macroneurons is to make this rolling tripod gait model generate rhythmic patterns in the following kinetic diagram, in which a pair of numbers is a pair of macroneurons firing simultaneously. A fraction in the curly bracket is the phase within a period when the pair of macroneurons fire.

$$\begin{array}{ccccc}
(1,11)\{0\} & \Rightarrow & (1,11)\{0\} & \Leftarrow & (4,10)\{\frac{5}{6}\} \\
 & & \Downarrow & & \Uparrow \\
 & & (6,8)\{\frac{1}{6}\} & & (5,7)\{\frac{2}{3}\} \\
 & & \Downarrow & & \Uparrow \\
 & & (3,9)\{\frac{1}{3}\} & \Rightarrow & (2,12)\{\frac{1}{2}\}
\end{array}$$

Table 2. Hexapodal primary gaits and description

Gait pattern	Pattern description	
	Ipsilateral legs	Contralateral legs
Pronk	in phase	in phase
Pace	in phase	half cycle out
Lurch	half cycle out	in phase
Tripod	half cycle out	half cycle out
Inchworm	one-sixth cycle out	in phase
Metachronal	one-sixth cycle out	half cycle out
Caterpillar	one-third cycle out	in phase
Rolling tripod	one-third cycle out	half cycle out

Current arrangement of system parameter matrices for the thresholds, synaptic weights, initial membrane potentials are θ, *W* and *M*(0), respectively, as shown below, displays a short stage converging to the rolling tripod period. An algorithm is required to acquire the symmetric threshold matrix (see Table 3).

The flexors of different legs of the rolling tripod gait fire in the order of ***RL, FR, ML, RR, FL, MR***; the corresponding extensors fire in the order of ***FL, RR, MR, FR, RL, ML***. A snapshot phase of each macroneuron in the flexor layer is: $RL=0$, $ML=\frac{1}{3}$, $FL=\frac{2}{3}$, $RR=\frac{1}{2}$, $MR=\frac{5}{6}$, $FR=\frac{1}{6}$.

Based on the given hexapod locomotion model, the matrices of activity thresholds, weights and initial membrane potentials of all interacting clones in the rolling tripod gait can be created for numerical simulation (Figure 10c~f) with the computational model developed in the section of OBB modules.

Any parameter with subscript *ij* means that it belongs to macroneuron *i* which is connecting with macroneuron j. In *W* matrix, the diagonal values are negative feedbacks of the clones themselves. In θ matrix, the value of *zero* has different significance. If the diagonal symmetry position of *zero* is *one*, then there is a connection between these two macroneurons while *zero* means this clone's threshold is *zero*. Otherwise *zero* means there is no connection between two macroneurons.

There is an optimal range of initial membrane potential configuration for a gait pattern. If a clone of a macroneuron is designed to fire at start, it should have initial value as,

$$M_{ij}(0) \in \begin{cases} (0,0.1) & \textit{if} \quad w_{ij} = 0.1 \quad \textit{and} \quad \theta_{ij} = 0 \\ (1.0,1.1) & \textit{if} \quad w_{ij} = 0.1 \quad \textit{and} \quad \theta_{ij} = 1 \end{cases}$$

If it is idle at start, then it is assigned with the following initial value,

$$M_{ij}(0) \in \begin{cases} (-0.1,0) & \textit{if} \quad w_{ij} = 0.1 \quad \textit{and} \quad \theta_{ij} = 0 \\ (0.9,1.0) & \textit{if} \quad w_{ij} = 0.1 \quad \textit{and} \quad \theta_{ij} = 1 \end{cases}$$

The initial membrane potential of a clone should have a scope of $2w_{ij}$ around the centre point of its threshold. This configuration guarantees that all clones of all macroneurons do not have convergent stage such that the starting state of the oscillation system is immediately within an oscillation cycle without possible deadlock occurring due to unsuitable choice of initial potentials. A random choice of the membrane potentials is allowed subject to an initial convergent period before entering an oscillation cycle.

Table 3. Pseudo-code of threshold determination

```
/*Program: Threshold determination*/
CONST:
G(N,E);

VAR:
Threshold[i,j : 1...4n] : real;

Action:
for (i := 1; i ≤ 4n; i++)
    for (j := 1; j ≤ 4n; j++)
        if (((i, j) ∈ E) ∧ (i < j))
            begin
                threshold[i,j] := 1.0;
                threshold[j,i] := 0.0;
            end
        else
            begin
                threshold[i,j] := 0.0;
                threshold[j,i] := 0.0;
            end
        end
    end
end
```

DISCUSSIONS AND PROSPECT

We present a general approach from the macroscopic point of view, based on large-scale, parallel computation exercised on legged biological specimens. The significance of this macroscopic method and related model is twofold. First, it is possible to apply the method and related model directly on retrieving single gait of an animal independently. Second, the method emphasizes on providing a general, whole spectrum simulation of any gait types without limitations of continuous mathematical models. In addition, the method is amenable to circuit implementation due to its digital and scalar nature. The sole presumption is that the spatio-temporal structure of an object is known.

Model Assessment

The prominent characteristic of the asymmetric Hopfield network under SMER is that every boolean input of a macroneuron can influence its output because of the multiplication operation between input and output of the neuron. The model has been shown to be able to retrieve all gait patterns proposed by Golubitsky et al.'s general pattern generation theory through calculation of large-scale matrices in discrete equations environment. A mathematical description to construct the oscillatory building blocks is provided, together with a detailed method on how to design a rhythmic system and its critical parameter matrices. The main contributions of this work can be summarised as follows,

1. A general neural network model embedding the SMER algorithm and capable of

Figure 10. The hexapod rolling tripod gait. (a) A statically standing cockroach. (b) The coupled architecture. (c) ~ (h) Numerical results of the upper layer flexor neurons correspond to macroneurons 1 ~ 6, respectively. The extensor macroneurons behave in pair with the flexor macroneurons following the kinetic diagram.

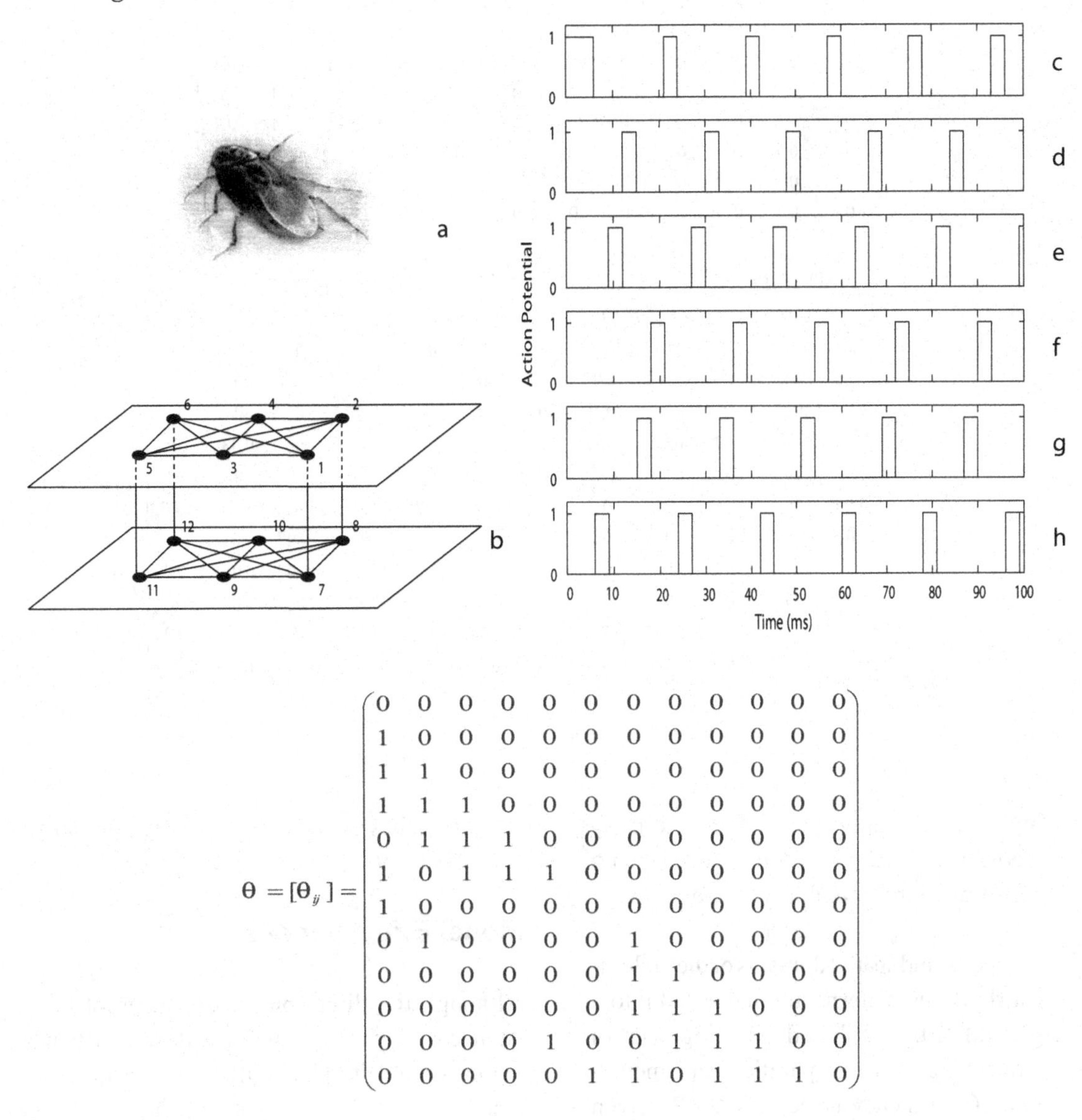

$$\theta = [\theta_{ij}] = \begin{pmatrix} 0&0&0&0&0&0&0&0&0&0&0&0 \\ 1&0&0&0&0&0&0&0&0&0&0&0 \\ 1&1&0&0&0&0&0&0&0&0&0&0 \\ 1&1&1&0&0&0&0&0&0&0&0&0 \\ 0&1&1&1&0&0&0&0&0&0&0&0 \\ 1&0&1&1&1&0&0&0&0&0&0&0 \\ 1&0&0&0&0&0&0&0&0&0&0&0 \\ 0&1&0&0&0&0&1&0&0&0&0&0 \\ 0&0&0&0&0&0&1&1&0&0&0&0 \\ 0&0&0&0&0&0&1&1&1&0&0&0 \\ 0&0&0&0&1&0&0&1&1&1&0&0 \\ 0&0&0&0&0&1&1&0&1&1&1&0 \end{pmatrix}$$

simulating CPGs in a uniform framework is proposed. Some preliminary definitions, including the basic topologic requirements for the target modelling and the characteristics of the novel network are demonstrated.

2. The equivalence between the novel neural network architecture and the plausible functionalities of biological CPGs is produced. Simple and composite OBBs generating different rhythmic patterns were presented. Such scheme was proven to be efficient on

$$W = [w_{ij}] = \begin{pmatrix} -0.1 & 0.1 & 0.1 & 0.1 & 0 & 0.1 & 0.1 & 0 & 0 & 0 & 0 & 0 \\ 0.1 & -0.1 & 0.1 & 0.1 & 0.1 & 0 & 0 & 0.1 & 0 & 0 & 0 & 0 \\ 0.1 & 0.1 & -0.1 & 0.1 & 0.1 & 0.1 & 0 & 0 & 0 & 0 & 0 & 0 \\ 0.1 & 0.1 & 0.1 & -0.1 & 0.1 & 0.1 & 0 & 0 & 0 & 0 & 0 & 0 \\ 0 & 0.1 & 0.1 & 0.1 & -0.1 & 0.1 & 0 & 0 & 0 & 0 & 0.1 & 0 \\ 0.1 & 0 & 0.1 & 0.1 & 0.1 & -0.1 & 0 & 0 & 0 & 0 & 0 & 0.1 \\ 0.1 & 0 & 0 & 0 & 0 & 0 & -0.1 & 0.1 & 0.1 & 0.1 & 0 & 0.1 \\ 0 & 0.1 & 0 & 0 & 0 & 0 & 0.1 & -0.1 & 0.1 & 0.1 & 0.1 & 0 \\ 0 & 0 & 0 & 0 & 0 & 0 & 0.1 & 0.1 & -0.1 & 0.1 & 0.1 & 0.1 \\ 0 & 0 & 0 & 0 & 0 & 0 & 0.1 & 0.1 & 0.1 & -0.1 & 0.1 & 0.1 \\ 0 & 0 & 0 & 0 & 0.1 & 0 & 0 & 0.1 & 0.1 & 0.1 & -0.1 & 0.1 \\ 0 & 0 & 0 & 0 & 0 & 0.1 & 0.1 & 0 & 0.1 & 0.1 & 0.1 & -0.1 \end{pmatrix}$$

$$M(0) = [M_{ij}(0)] =$$

$$\begin{pmatrix} 0 & -0.02 & -0.05 & -0.03 & 0 & -0.08 & -0.05 & 0 & 0 & 0 & 0 & 0 \\ 1.02 & 0 & 0.08 & -0.07 & -0.01 & 0 & 0 & 0.03 & 0 & 0 & 0 & 0 \\ 1.05 & 0.92 & 0 & -0.03 & -0.07 & 0.02 & 0 & 0 & 0 & 0 & 0 & 0 \\ 1.03 & 1.07 & 1.03 & 0 & 0.09 & 0.05 & 0 & 0 & 0 & 0 & 0 & 0 \\ 0 & 1.01 & 1.07 & 0.91 & 0 & 0.03 & 0 & 0 & 0 & 0 & 0.08 & 0 \\ 1.08 & 0 & 0.98 & 0.95 & 0.97 & 0 & 0 & 0 & 0 & 0 & 0 & -0.03 \\ 1.05 & 0 & 0 & 0 & 0 & 0 & 0 & 0.05 & 0.07 & -0.02 & 0 & 0.03 \\ 0 & 0.97 & 0 & 0 & 0 & 0 & 0.95 & 0 & -0.02 & -0.07 & 0.03 & 0 \\ 0 & 0 & 0 & 0 & 0 & 0 & 0.93 & 1.02 & 0 & -0.02 & 0.06 & -0.07 \\ 0 & 0 & 0 & 0 & 0 & 0 & 1.02 & 1.07 & 1.02 & 0 & 0.06 & 0.02 \\ 0 & 0 & 0 & 0 & 0.92 & 0 & 0 & 0.97 & 0.94 & 0.94 & 0 & -0.07 \\ 0 & 0 & 0 & 0 & 0 & 1.03 & 0.97 & 0 & 1.07 & 0.98 & 1.07 & 0 \end{pmatrix}$$

building internally analog and externally Boolean oscillatory neural networks with self-timing and potentially suitable to very large scale implementations.

3. Temporal and spatial domains of the biological rhythmic patterns are integrated into a generalised, distributed updating control strategy for scalable parallel implementations. Consistency aspects of SMER-driven systems with dynamic topologies are accommodated.
4. Numerical simulations have demonstrated the efficiency of the model on the retrieval of biological rhythmic patterns. A complete procedure, from both theoretical and practical aspects, is illustrated. Results are consistent with related mathematical approaches such as group theory (Golubitsky et al., 1998; 1988).

Model Effectiveness

Although the direct physiological proof of existence of CPGs is scarce, many scientists have demonstrated its plausibility. The clinical gait analysis is, for instance, one of the most investigated topics in the medical studies all over the world. As a research result, it is well established that the centre of pressure (CoP) weaves about the centre of mass (CoM) when a person is standing statically. Due to the conduction delays from CNS to the leg musculature, a phase delay between CoM and CoP should be detectable if a central control process is active (Winter et al.,

1998). An experiment was conducted in which, a volunteer stood on a pair of force platforms and voluntarily rocked back and forward slowly in time with a metronome. CoM was measured using a 6-camera Vicon system and full-body BodyBuilder model (Eames et al., 1999). By the above logic, one would expect to see a phase (time) delay, probably around 150-250ms, between CoM and CoP since the motion is consciously generated by the volunteer. However, the experimental results show no time delay between the signals of CoM and CoP.

One reasonable explanation may be that, if the volunteer is conscious on his motion from quiet standing, it seems there should exist some phase lag because of the conduction delay from CNS; otherwise his behaviour may be controlled by some mechanism like CPG, a biologically plausible, preprogrammed pattern organ or equivalent structure most probably located in the spinal cord, in which case the signal conduction phase lag may be trivial. Based on this assumption, the CPG model introduced in this chapter will be especially valuable for retrieving biological patterns as we expected.

As we described, this chapter provides a general architecture and coherent methodology on gait modelling. Some additional modifications may, however, be expected if this method is used for a particular development of gait related issues. For example, on the design of a robot with each leg having multiple degrees of freedom, one may consider the arrangement of an OBB net at each joint and make connections among all OBB nets. A more detailed CPG architecture could be conceived in this way, nevertheless it would be still based on the general asymmetric Hopfield-like neural network under SMER.

FUTURE RESEARCH DIRECTIONS

As a newly proposed approach to CPG design and biological computation, one could see some prospects on the modelling process and its applications. One of the possible novel works is to determine the dynamic features of OBB models in a more detailed mathematical framework. An immediate example would be the investigation of system stability by analysing the symmetry in the system parameter matrices.

Another interesting direction is that of the general methodology of asymmetric Hopfield neural network combining SMER can be applied and improved. One of the goals of this study is to provide a reconfigurable network structure using simple and composite OBBs. A specific coupled CPG architecture should be created according to different specimen under research. Generally speaking, the more detailed information is available on a locomotor CPG architecture, such as the number of neurons or coupling situation, the more insights can be fed back for a better model implementation, such as the better error tolerance. For instance, the bipedal gaits can be retrieved with a very simple model consisting of only four macroneurons for the relaxation oscillation of two legs. However, it is possible and better to work out a much more detailed model consisting of more macroneurons coupled with the different weights. In this case a bipedal CPG architecture may have macroneurons for its toes, heels, ankles, knees, back and even two arms respectively, with the stronggest coupling strength between a pair of toe and heel while the weakest one between a pair of arm and knee. This more detailed model is expected to be more physiologically reasonable in the sense that, intuitively there exist some weak coupling relations between a pair of arm and knee during human walking, and no great impact on locomotion as two arms are amputated except that some kind of instability is introduced.

Another direction for future research of CPGs is in the field of sensorimotor integration. The proposed model requires adaptive legged locomotion; thus the pattern generation parameters must undergo continuous modulation by sensory inputs. The study of insect sensorimotor systems, particularly recent interest in the functional architecture of the insect brain, point towards a hierarchical organisation of behaviour (Webb, 2001; 2002; 2004). The basis of behaviour is the direct reflex. In the context of walking, direct reflexes utilise tactile and proprioceptive information to make immediate adjustments to leg movements. While it has been argued (Cruse, 1991) that local reflexes alone may be sufficient to generate co-ordinated insect gaits, there is also evidence for CPG involvement in walking, and the interaction of these two mechanisms is an interesting problem to explore. Most "insect-inspired" robotics to date has been limited to exploration of reflex control, but in fact there is clear evidence that insect behaviour is modulated by secondary pathways that carry out more sophisticated sensory processing such as adaptive classification and associative learning. An intriguing hypothesis is that one of the main secondary pathways, which passes through the highly organised brain structures known as mushroom bodies, effectively implements a functional equivalent of some learning algorithms. The learning system can underpin the acquisition of anticipatory behaviour. For instance, the learning can capture the visual depth stimuli that predict an impending tactile or proprioceptive stimulus. A learning system can thus make an appropriate motor response, to generate a reflex and (or) CPG modulation of the walking pattern.

CONCLUSION AND FUTURE TRENDS

After a brief review of the history of studies on the legged locomotion and state of the art on computational models of CPGs, we present a simple CPG architecture to generate a full range of rhythmic patterns for a specific legged animal. This general CPG model features some attractive properties. The model is a purely distributed dynamical system without the need of a central clock due to the adoption of the OBB modules. The model is reconfigurable in the sense that it is easy to modify the sharing resources within the OBB modules to change the connection strength between motor neurons, thus to make different gait patterns for a smooth gait transition. The model benefits from the modular OBB modules which are amenable to VLSI implementation.

In principle, two research trends will co-exist in the future on analysing and modeling the biological rhythmic pattern generation mechanisms. The low level, traditional approach uses the stringent mathematical methods such as dynamical system theory to describe and solve the phenomenological problems (Wilson, 1999; Izhikevich, 2007). This approach usually focuses on the microscopic, cellular level aspects of pattern dynamics, with emphasis on the onset, development, propogation, transition and stability of the activity of a neuron population.

Another is a high level, macroscopic approach dedicated to mimicking the observable motion of the animals. This approach usually involves a mathematical description of the target prototype geometry and a simplified neural network model (Golubitsky et al., 1998; Yang & França, 2003) to produce the same dynamical results with the advantage of modularity and quick development without any loss of accuracy. It is a viable bio-inspired methodology to implement a biological dynamical system. For instance, we have used the SMER-based OBB modules to integrate, in a single model, the axial and appendicular movements of a centipede in its fastest gait pattern of locomotion (Braga et al., in press). The similar motor mechanism can also be generalised and applied to the swimming animals such as the lamprey.

ACKNOWLEDGMENT

This research has been supported in part by a Brazilian CNPq grant No.143032/96-8, a British BBSRC grant BBS/B/07217, an EPSRC grant EP/E063322/1 and a Chinese NSFC grant 60673102.

REFERENCES

Alexander, R.McN. (1984). The gait of bipedal and quadrupedal animals. *International Journal of Robotics Research,* 3, 49-59.

Aristotle. De Partibus Animalium, de Incessu Animalium, de Motu Animalium. In A.S. Peek & E.S. Forster (Translators), *In Parts of Animals, Movement of Animals, Progression of Animals.* Cambridge, MA: Harvard University Press, 1936.

Barbosa, V.C., & Gafni, E. (1989). Concurrency in heavily loaded neighbourhood-constrained systems. *ACM Transactions on Programming Languages and Systems,* 11(4), 562-584.

Barbosa, V.C. (1996). *An introduction to distributed algorithms.* Cambridge, MA: The MIT Press.

Barbosa, V.C., Benevides, M.R.F., & França, F.M.G. (2001). Sharing resources at nonuniform access rates. *Theory of Computing Systems,* 34(1), 13-26.

Bässler, U. (1986). On the definition of central pattern generator and its sensory control. *Biological Cybernetics,* 54, 65-69.

Bay, J.S., Hemami, H. (1987). Modeling of a neural pattern generator with coupled nonlinear oscillators. *IEEE Transactions on Biomedical Engineering,* 34, 297-306.

Beer, B.D. (1990). *Intelligence as adaptive behavior: An experiment in computational neuroethology.* San Diego, CA: Academic Press.

Braga, R.R., Yang, Z., & França, F.M.G. (in press). Implementing an artificial centipede CPG: Integrating appendicular and axial movements of the scolopendromorph centipede. *International Conference on Bio-inspired Systems and Signal Processing,* Madeira, Portugal, January, 2008.

Buono, P.L., & Golubitsky, M. (2001) Models of central pattern generators for quadrupedal locomotion: I. Primary gaits. *Journal of Mathematical Biology,* 42, 291-326.

Chua, L.O., & Yang, L. (1988). Cellular neural networks: Theory. *IEEE Transactions on Circuits and Systems, part I,* 35, 1257-1272.

Chua, L.O., & Yang, L. (1988). Cellular neural networks: Application", *IEEE Transactions on Circuits and Systems, part I,* 35, 1273-1290.

Collins, J.J. (1995). Gait transitions. In M.A. Arbib (Ed.), *The handbook of brain theory and neural networks* (pp.420-423). New York: The MIT Press.

Collins, J.J., & Richmond, S.A. (1994). Hard-wired central pattern generators for quadrupedal locomotion. *Biological Cybernetics,* 71, 375-385.

Collins, J.J., & Stewart, I.N. (1993a). Coupled nonlinear oscillators and the symmetries of animal gaits. *Journal of Nonlinear Science,* 3, 349-392.

Collins, J.J., & Stewart, I.N. (1993b). Hexapodal gaits and coupled nonlinear oscillator models. *Biological Cybernetics,* 68, 287-298.

Cruse, H. (1991). Coordination of leg movement in walking animals. *Proceedings of 1st International Conference on Simulation of Adaptive Behaviour* (pp. 105-119), Paris, France.

Dijkstra, E.W. (1971). Hierarchical ordering of sequential processes. *Acta Informatica,* 1(2), 115-138.

Dimitrijevic, M.R., Gerasimenko, Y., & Pinter, M.M. (1998). Evidence for a spinal central pattern generator in humans. *Annals of the New York Academy of Sciences,* 860, 360-376.

Eames, M.H.A., Cosgrove, A., & Baker, R. (1999). Comparing methods of estimating the total body centre of mass in three-dimensions in normal and pathological gaits. *Human Movement Science, 18*, 637-646.

França, F.M.G. (1994). *Neural networks as neighbourhood-constrained systems.* Unpublished doctoral dissertation, Imperial College, London, England.

França, F.M.G., & Yang, Z. (2000). Building artificial CPGs with asymmetric hopfield networks. *Proceedings of International Joint Conference on Neural Networks* (pp. IV:290-295), Como, Italy.

Getting, P.A. (1989). Emerging principles governing the operation of neural networks", *Annual Review of Neuroscience,* 12, 185-204.

Golubitsky, M., Stewart, I., & Schaeffer, D.G. (1988). *Singularities and groups in bifurcation theory, Volume II.* Springer-Verlag.

Golubitsky, M., Stewart, I., Buono, P.L., & Collins, J.J. (1998). A modular network for legged locomotion. *Physica D,* 115, 56-72.

Golubitsky, M., Stewart, I., Buono, P.L., & Collins, J.J. (1999) Symmetry in locomotor central pattern generators and animal gaits. *Nature,* 401, 693-695.

Grillner, S. (1975). Locomotion in vertebrates: Central mechanisms and reflex interaction. *Physiological Reviews,* 55, 247-304.

Grillner, S. (1985). Neurobiological bases of rhythmic motor acts in vertebrates. *Science,* 228, 143-149.

Hodgkin, A.L., & Huxley, A.F. (1952). A quantitative description of membrane current and its application to conduction and excitation in nerve. *Journal of Physiology,* 117(4), 500-544.

Hopfield, J.J. (1982). Neural networks and physical systems with emergent collective properties. *Proceedings of the National Academy of Sciences of the United States of America,* 79, 2554-2558.

Hopfield, J.J., & Tank, D.W. (1985). Neural computation of decisions in optimization problems. *Biological Cybernetics,* 52, 141-152.

Hopfield, J.J., & Tank, D.W. (1986). Computing with neural circuits: A model. *Science,* 233, 625-632.

Izhikevich, E.M. (2007). *Dynamical systems in neuroscience: The geometry of excitability and bursting.* Cambridge, MA: The MIT Press.

Kuffler, S.W., & Eyzaguirre, C. (1955). Synaptic inhibition in an isolated nerve cell. *Journal of General Physiology,* 39, 155-184.

Linkens, D.A., Taylor, Y., & Duthie, H.L. (1976). Mathematical modeling of the colorectal myoelectrical activity in humans. *IEEE Transactions on Biomedical Engineering,* 23, 101-110.

Matsuoka, K. (1987). Mechanisms of frequency and pattern control in the neural rhythm generators. *Biological Cybernetics,* 56, 345-353.

McCulloch, W.S., & Pitts, W. (1943). A logical calculus and the ideas immanent in the nervous activity. *Bulletin of Mathematical Biophysics,* 5, 115-133.

Pearson, K.G. (1976) The control of walking. *Scientific American,* 235, 72-86.

Pearson, K.G. (1993). Common principles of motor control in vertebrates and invertebrates. *Annual Review of Neuroscience,* 16, 265-297.

Perkel, D.H. (1976). A computer program for simulating a network of interacting neurons: I. Organization and physiological assumptions. *Computers and Biomedical Research,* 9, 31-43.

Roberts, A., & Tunstall, M.J. (1990). Mutual re-excitation with post-inhibitory rebound: A simulation study on the mechanisms for locomotor rhythm generation in the spinal cord of Xenopus embryos. *European Journal of Neuroscience,* 2, 11-23.

Schöner, G., Jiang, W.Y., & Kelso, J.A.S. (1990). A synergetic theory of quadrupedal gaits and gait transitions. *Journal of Theoretical Biology,* 142, 359-391.

Stein, P.S.G. (1978). Motor systems with specific reference to the control of locomotion. *Annual Review of Neuroscience,* 1, 61-81.

Taga, G., Yamaguchi, Y., & Shimizu, H. (1991). Self-organized control of bipedal locomotion by neural oscillators in unpredictable environment. *Biological Cybernetics,* 65, 147-159.

Taga, G. (1995). A model of the neuro-musculo-skeletal system for human locomotion – I. Emergence of basic gait. *Biological Cybernetics,* 73, 97-111.

Taft, R. (1955). An introduction: Eadweard Muybridge and his work. In E. Muybridge (Ed.), *The human figure in motion* (pp.7-14). New York: Dover Publications.

Tsutsumi, K., & Matsumoto, H. (1984). A synaptic modification algorithm in consideration of the generation of rhythmic oscillation in a ring neural network. *Biological Cybernetics,* 50, 419-430.

Turing, A.M. (1952). The chemical basis of morphogenesis. *Philosophical Transactions of the Royal Society B,* 237, 37-72.

Wang, X.J., & Rinzel, J. (1992). Alternating and synchronous rhythms in reciprocally inhibitory model neurons. *Neural Computation,* 4, 84-97.

Webb, B. (2002). Robots in Invertebrate Neuroscience. *Nature,* 417, 359-363.

Webb, B. (2001). Can robots make good models of biological behaviour? *Behavioural and Brain Sciences,* 24, 1033-1050.

Webb, B. (2004). Neural mechanisms for prediction: Do insects have forward models? *Trends in Neurosciences,* 27, 278-282.

Wilson, H.R. (1999). *Spikes, decisions, and actions: dynamical foundations of neuroscience.* New York: Oxford University Press.

Winter, D.A., Prince, F., & Patla, A.E. (1998). Stiffness control of balance during quiet standing. *Journal of Neurophysiology,* 80, 1211-1221.

Yang, Z., & França, F.M.G. (2003). A generalized locomotion CPG architecture based on oscillator building blocks. *Biological Cybernetics,* 89, 34-42.

ADDITIONAL READING

Alexander, R.M. (1989) Optimization and gaits in the locomotion of vertebrates. *Physiological Reviews,* 69, 1199-1227.

Arena, P. (2000). The central pattern generator: A paradigm for artificial locomotion. *Soft Computing – A Fusion of Foundations, Methodologies and Applications,* 4(4), 251-266.

Bucher, D., Prinz, A.A., & Marder, E. (2005). Animal-to-animal variability in motor pattern production in adults and during growth. *Journal of Neuroscience,* 25, 1611–1619.

Butt, S.J., & Kiehn, O. (2003). Functional identification of interneurons responsible for left-right coordination of hindlimbs in mammals. *Neuron,* 38, 953–963.

Buono, P.L. (2001). Models of central pattern generators for quadruped locomotion. *Journal of Mathematical Biology,* 42(4), 327-346.

Cangiano, L., & Grillner, S. (2005). Mechanisms of rhythm generation in a spinal locomotor network deprived of crossed connections: the lamprey hemicord. *Journal of Neuroscience,* 25, 923–935.

Cohen, A.H. (1992). The role of heterarchical control in the evolution of central pattern generators. *Brain, Behavior and Evolution,* 40, 112-124.

Cruse, H. (2002). The functional sense of central oscillations in walking. *Biological Cybernetics,* 86(4), 271-280.

Delcomyn, F. (1999). Walking robots and the central and peripheral control of locomotion in insects. *Autonomous Robots,* 7(3), 259-270.

Ermentrout, G.B., & Chow, C.C. (2002). Modeling neural oscillations. *Physiology & Behavior,* 77, 629-633.

Ghigliazza, R.M., & Holmes, P. (2004). A Minimal model of a central pattern generator and motoneurons for insect locomotion. *SIAM Journal of Applied Dynamical Systems,* 3(4), 671-700.

Grillner, S. (2003). The motor infrastructure: from ion channels to neuronal networks. *Nature Review Neuroscience,* 4, 573–586.

Grillner, S., Markram, H., De Schutter, E., Silberberg, G., & LeBeau, F.E. (2005). Microcircuits in action—from CPGs to neocortex. *Trends in Neurosciences,* 28, 525–533.

Hatsopoulos, N. (1996). Coupling the neural and physical dynamics in rhythmic movements. *Neural Computation,* 8, 567-581.

Ijspeert, A.J., & Kodjabachian, J. (1999). Evolution and development of a central pattern generator for the swimming of a Lamprey. *Artificial Life,* 5(3), 247-269.

Ijspeert, A.J. (2001). A connectionist central pattern generator for the aquatic and terrestrial gaits of a simulated salamander. *Biological Cybernetics,* 84(5), 331-348.

Jing, J., Cropper, E.C., Hurwitz, I., & Weiss, K.R. (2004). The construction of movement with behavior-specific and behavior-independent modules. *Journal of Neuroscience,* 24, 6315–6325.

Jones, S.R., & Kopell, N. (2006). Local network parameters can affect inter-network phase lags in central pattern generators. *Journal of Mathematical Biology,* 52(1), 115-140.

Katz, P.S., Sakurai, A., Clemens, S., & Davis, D. (2004). Cycle period of a network oscillator is independent of membrane potential and spiking activity in individual central pattern generator neurons. *Journal of Neurophysiology,* 92, 1904–1917.

Kiehn, O., & Butt, S.J. (2003). Physiological, anatomical and genetic identification of CPG neurons in the developing mammalian spinal cord. *Progress in Neurobiology,* 70, 347–361.

Latash, M.L. (1993). *Control of human movement.* Champaign, IL:Human Kinetics Publishers.

Lewis, M.A., Etienne-Cummings, R., Hartmann, M.J., Xu, Z.R., & Cohen, A.H. (2003). An in silico central pattern generator: Silicon oscillator, coupling, entrainment, and physical computation. *Biological Cybernetics,* 88(2), 137-151.

Magill, R.A. (2001). *Motor learning: Concepts and applications.* New York: McGraw-Hill Companies, Inc.

Marder, E., & Calabrese, R.L. (1996). Principles of rhythmic motor pattern generation. *Physiological Reviews,* 76, 687–717.

Marder, E., Bucher, D., Schulz, D.J., & Taylor, A.L. (2005). Invertebrate central pattern generation moves along. *Current Biology,* 15: R685–R699.

McCrea, D.A. (2001). Spinal circuitry of sensorimotor control of locomotion. *Journal of Physiology,* 533(1), 41-50.

McGeer, T. (1993). Dynamics and control of bipedal locomotion. *Journal of Theoretical Biology,* 163, 277-314.

Nakada, K., Asai, T., & Amemiya, Y. (2003). An analog CMOS central pattern generator for interlimb coordination in quadruped. *IEEE Transactions on Neural Networks,* 14(5), 1356-1365.

Norris, B.J., Weaver, A.L., Morris, L.G., Wenning, A., García, P.A., & Calabrese, R.L. (2006).

A central pattern generator producing alternative outputs: Temporal pattern of premotor activity. *Journal of Neurophysiology*, 96, 309-326.

Nusbaum, M.P., & Beenhakker, M.P. (2002). A small-systems approach to motor pattern generation. *Nature*, 417, 343-350.

Olree, K.S., & Vaughan, C.L. (1995). Fundamental patterns of bilateral muscle activity in human locomotion. *Biological Cybernetics*, 73(5), 409-414.

Pearson, K., & Gordon, J. (2000). Locomotion. In Kandel, E., Schwartz, J., & Jessel, T. (Eds.), *Principles of neural science* (pp.737-755). New York: McGraw-Hill Companies, Inc.

Pinto, C.M.A., & Golubitsky, M. (2006). Central pattern generators for bipedal locomotion. *Journal of Mathematical Biology*, 53(3), 474-489.

Prentice, S.D., Patla, A.E., & Stacey, D.A. (1998). Simple artificial neural network models can generate basic muscle activity patterns for human locomotion at different speeds. *Experimental Brain Research*, 123(4), 474-480.

Raibert, M.H. (1986). *Legged robots that balance*. Cambridge, MA: MIT Press.

Rinzel, J., Terman, D., Wang, X., & Ermentrout, B. (1998). Propagating activity patterns in large-scale inhibitory neuronal networks. *Science*, 279, 1351-1355.

Ryckebusch, S., Wehr, M., & Laurent, G. (1994). Distinct rhythmic locomotor patterns can be generated by a simple adaptive neural circuit: Biology, simulation, and VLSI implementation. *Journal of Computational Neuroscience*, 1(4), 339-358.

Schmidt, R.A., & Lee, T.D. (1999). *Motor control and learning: A behavioral emphasis*. Champaign, IL: Human Kinetics Publishers.

Stein, P.S.G., Grillner, S., Selverston, A.I., & Stuart, D.G. (Eds.). (1997). *Neurons, Networks, and Motor Behavior*. Cambridge, MA: MIT Press.

Stein P.S.G. (2005). Neuronal control of turtle hindlimb motor rhythms. *Journal of Comparative Physiology A: Neuroethology, sensory, neural and behavioral physiology*, 191(3), 213-229.

Schöner, G., & Kelso, J.A. (1988). Dynamic pattern generation in behavioural and neural systems. *Science*, 239, 1513-1520.

Vogelstein, R.J., Tenore, F., Etienne-Cummings, R., Lewis, M.A., & Cohen, A.H. (2006). Dynamic control of the central pattern generator for locomotion. *Biological Cybernetics*, 95(6), 555-566.

Zhang, X., Zheng, H., & Chen, L. (2006). Gait transition for a quadrupedal robot by replacing the gait matrix of a central pattern generator model. *Advanced Robotics*, 20(7), 849-866.

ENDNOTES

[a] Current contact: School of Engineering and Electronics, Edinburgh University, Edinburgh EH9 3JL. Email: Zhijun.Yang@ed.ac.uk

Section III
Real-Life Applications with Biologically Inspired Models

Chapter XIII
Genetic Algorithms and Multimodal Search

Marcos Gestal
University of A Coruña, Spain

José Manuel Vázquez Naya
University of A Coruña, Spain

Norberto Ezquerra
Georgia Institute of Technology, USA

ABSTRACT

Traditionally, the Evolutionary Computation (EC) techniques, and more specifically the Genetic Algorithms (GAs), have proved to be efficient when solving various problems; however, as a possible lack, the GAs tend to provide a unique solution for the problem on which they are applied. Some non global solutions discarded during the search of the best one could be acceptable under certain circumstances. Most of the problems at the real world involve a search space with one or more global solutions and multiple local solutions; this means that they are multimodal problems and therefore, if it is desired to obtain multiple solutions by using GAs, it would be necessary to modify their classic functioning outline for adapting them correctly to the multimodality of such problems. The present chapter tries to establish, firstly, the characterisation of the multimodal problems will be attempted. A global view of some of the several approaches proposed for adapting the classic functioning of the GAs to the search of multiple solutions will be also offered. Lastly, the contributions of the authors and a brief description of several practical cases of their performance at the real world will be also showed.

INTRODUCTION

Following a general prospect, the GAs (Holland, 1975) (Goldberg, 1989) try to find a solution using a population: a randomly generated initial set of individuals. Every one of these individuals –who represent a potential solution to the problem-, will evolve according to the theories proposed by Darwin (Darwin, 1859) about natural selection and they will be more adapted to the required solution as generations pass.

Nevertheless, the traditional GAs find certain restrictions when the search space where they work has, either more than a global solution, or an unique global solution and multiple local optima. When faced with such scenarios, a classical GA tends to focalise the search on the environment of the global solution; however, it might be interesting to know the higher possible number of solutions due to different reasons: exact search space knowledge, implementation ease of local solutions compared with the global one, interpretation ease of some solutions compared with other ones, etc. To get this, an iterative process will be performed until the desired goals might be achieved. The process will start with the individuals grouping into species that will search independently a solution into their related environments. Following the later, the crossover operation will involve individuals of different species in order not to leave search space areas unexplored. The process will be repeated according to the achievement of the desired goals.

MULTIMODAL PROBLEMS

The multimodal problems can be defined as those problems that have either multiple global optima or multiple local optima (Harik, 1995).

For this type of problems, it is interesting to obtain the greatest number of solutions due to several reasons; on one hand, when there is not a total knowledge of the problem, the solution obtained might not be the best one, as it can not be stated that no better solution could be found at the search space not explored yet. On the other hand, although being certain that the best solution has been achieved, there might be other equally fitted or slightly worse solutions that might be preferred due to different factors (easier application, simpler interpretation, etc.) and therefore considered globally better.

One of the most characteristic multimodal functions used in lab problems is the Rastrigin function (see Figure 1) which offers an excellent graphical point of view about what multimodality means.

Providing multiple optimal (and valid) solutions, and not only a unique global solution, is crucial in multiple environments. Usually, it is very

Figure 1. Rastrigin function: 3D and 2D representations

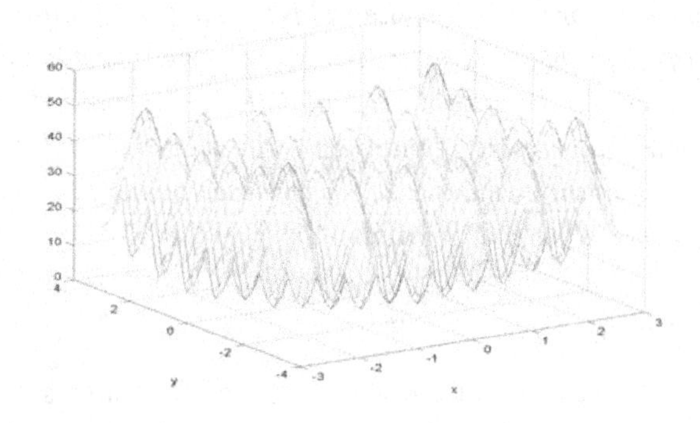

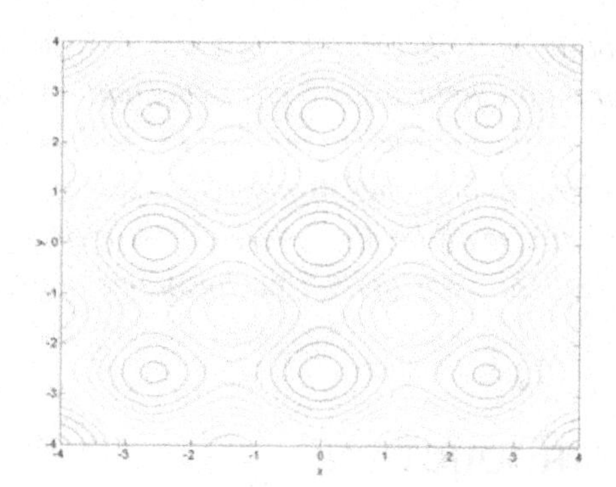

complex to implement in practice the best solution, so it can offer multiple problems: computational cost too high, complex interpretation, difficulty in the acquisition of information, etc.

In these situations it is useful to have a choice of several valid solutions and, although they are not the best solution to the problem, they might offer a level of acceptable adjustment, be simpler to implement, to understand or to distribute than the ideal global one.

EVOLUTIONARY COMPUTATION AND MULTIMODAL PROBLEMS

As it has been mentioned, the application of EC techniques to the resolution of multimodal problems sets out the difficulty that this type of techniques shows since they tend to solely provide the best of the found solutions and to discard possible local optima that might have been found throughout the search. Quite many modifications have been included in the traditional performance of the GA in order to achieve good results with multimodal problems.

A crucial aspect when obtaining multiple solutions consists on keeping the diversity of the genetic population, distributing as much as possible the genetic individuals throughout the search space.

Classic Approaches

Niching methods allow GAs to maintain a genetic population of diverse individuals, so it is possible to locate multiple optimal solutions within a single population.

In order to minimise the impact of homogenisation, or at least to restrict it to later states of searching phase, several alternatives have been designed, based most of them on heuristics. One of the first alternatives for promoting the diversity was the applications of scaling methods to the population in order to emphasize the differences among the different individuals. Other direct route for avoiding the diversity loss involves focusing on the elimination of duplicate partial high fitness solutions (Landgon, 1996).

Some other of the approaches try to solve this problem by means of the dynamic variation of crossover and mutation rates (Ursem, 2002). In that way, when diversity decreases, a higher amount of mutations are done in order to increase the exploration through the search space; when diversity increases, the mutations decrease and crossovers increase with the aim of improving exploitation in optimal solution search. There are also proposals of new genetic operators or variations of the actual ones. For example, some of the crossover algorithms that improve diversity and that should be highlighted are **BLX** (Blend Crossover) (Eshelman & Schaffer, 1993), **SBX** (Simulated Binary Crossover) (Deb & Agrawal, 1993), **PCX** (Parent Centric Crossover) (Deb, 2003), **CIXL2** (Confidence Interval Based Crossover using L2 Norm) (Ortiz, Hervás & García, 2005) or **UNDX** (Unimodal Normally Distributed Crossover) (Ono & Kobayashi, 1999).

Regarding replacement algorithms, schemes that may keep population diversity have been also looked for. An example of this type of schemes is ***crowding*** (DeJong, 1975) (Mengshoel & Goldberg, 1999). Here, a newly created individual is compared to a randomly chosen subset of the population and the most closely individual is selected for replacement. Crowding techniques are inspired by Nature where similar members in natural populations compete for limited resources. Likewise, dissimilar individuals tend to occupy different niches and are unlikely to compete for the same resource, so different solutions are provided.

Fitness sharing was firstly implemented by Goldberg and Richardson for being used on multimodal functions (Goldberg & Richardson, 1987). The basic idea involves determining, from the fitness of each solution, the maximum number of individuals that can remain around it, awarding the individuals that exploit unique areas of the domain. The ***dynamic fitness sharing*** (Miller & Shaw, 1995) was proposed in order to correct the dispersion of the final distribution of the individuals into niches with two components: the distance function, which measures the overlapping of individuals, and the comparison function, which results "1" if the individuals are identical, but gets closer to "0" as much different they are.

The ***clearing*** method (Petrowski, 1996) is quite different from the previous ones, as the resources are not shared, but assigned to the best individuals, who will be then kept at every niche.

The main disadvantage of the techniques previously described lies in the fact that they add new parameters that should be configured according the process of execution of GA. This process may be disturbed by the interactions among those parameters (Ballester & Carter, 2003).

Own Proposals

Once detected the existing problems they should be solved, or at least, minimised. With this goal, the Artificial Neural Network and Adaptive System (RNASA) group have developed two proposals that use EC techniques for this type of problems. Both proposals try to find the final solution but keeping partial solutions within the final population.

The main ideas of the two proposals, together with the problems used for the tests are explained at the following sections.

Hybrid Two-Population Genetic Algorithm

To force a homogeneous search throughout the search space, the approach proposed here is based on the addition of a new population (genetic pool) to a traditional GA (secondary population). The genetic pool will divide the search space into sub-regions. Each one of the individuals of the genetic pool has its own fenced range for gene variation, so every one of these individuals would represent a specific sub-region within the global search space. On the other hand, the group of individual ranges in which any gene may have its value, is extended over the possible values that a gene may have. Therefore, this genetic pool would sample the whole of the search space.

Figure 2. Structure of populations of hybrid two-population genetic algorithm

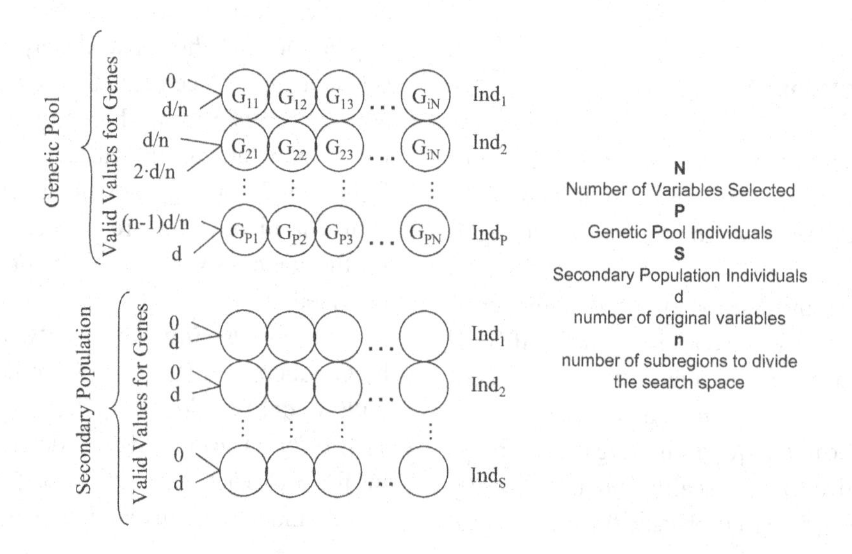

It should be borne in mind that a traditional GA performs its search considering only one sub-region (the whole of the search space). Here the search space will be divided into different sub regions or intervals according to the number of genetic individuals in the genetic pool.

Since the individuals in the genetic pool have restrictions in their viable gene values, one of these individuals would not be provided with a valid solution. In addition to the genetic pool, another population is then used (the secondary population), where a classical GA would develop its individuals in an interactive fashion with those individuals of the genetic pool.

Differently from the genetic pool, the genes of individuals of secondary population provide solutions as they may adopt values throughout the whole of the search space; whereas, the genetic pool would act as a support, keeping search space homogeneously explored.

Next, both populations, which are graphically represented in Figure 2, will be described in detail.

The Populations: Genetic Pool

As it has been previously mentioned, every one of the individuals at the genetic pool represents a sub-region of the global search space. Therefore, they should have the same structure or gene sequence than when using a traditional GA. The difference lies in the range of values that these genes might have.

When offering a solution, traditional GA may have any valid value, whereas in the proposed GA, the range of possible values is restricted. Total value range is divided into the same number of parts than individuals in genetic pool, so that a sub-range of values is allotted to each individual. Those values that a given gene may have will remain within its range for the whole of the performance of the proposed GA.

In addition to all that has been said, every individual at the genetic pool will be responsible of the genes that correspond to the best found solution up to then (meaning whether they belong to the best individual at secondary population). This Boolean value would be used to avoid the modification of those genes that, in some given phase of performance, are the best solution to the problem.

Furthermore, every one of the genes in an individual has an I value associated which indicates the relative increment that would be applied to the gene during a mutation operation based only on increments and solely applied to individuals of the genetic pool. It is obvious that this incremental value should be lower than the maximum range in which gene values may vary. The structure of the individuals at genetic pool is shown at Figure 3.

As these individuals do not represent global solutions to the problem that has to be solved, their fitness value will not be compulsory but it will reduce the complexity of the algorithm and, of course, it will increase the computational efficiency of the final implementation.

Figure 3. Genetic pool individual i

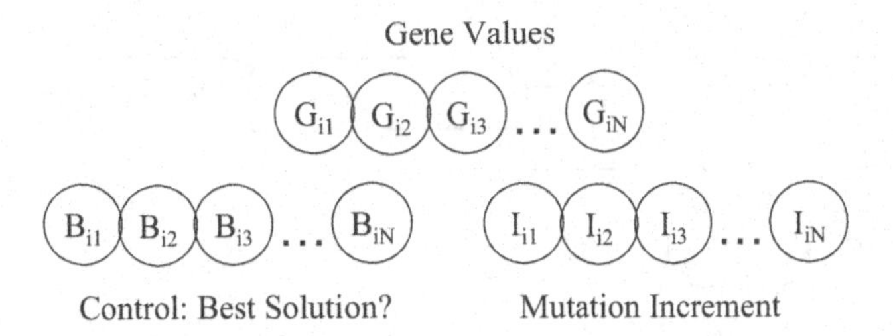

The Populations: Secondary Population

The individuals of the secondary population are quite different from the previous. In this case, the genes of the individuals of the secondary population can take any value throughout the whole space of possible solutions, a fact that makes them able to offer global solutions to the problem. This is not possible in genetic pool because their genes are restricted to different sub-ranges.

The evolution of the individuals at the genetic pool will be carried out by traditional GA rules. The main difference lies in the operator crossover. In this case a modified crossover will be used. As the information is stored in isolated populations, the parents who produce the new offspring will not belong to the same population. Hence, the genetic pool and secondary population are combined instead. In this way information of both populations will be merged to produce the most fitted offspring.

The Genetic Operators: Crossover

As it was pointed before, the crossover operator recombines the genetic material of individuals of both populations. This recombination involves a random individual from secondary population and a *representative* of the genetic pool.

This *representative* will represent a potential solution offered by the genetic pool. As a unique individual can not verify this requirement, the *representative* will be formed by a subset of genes of different individuals on the genetic pool. Gathering information from different partial solutions will result in producing a valid global solution.

Therefore, the value for every gene of the representative will be randomly chosen among all the individuals in the genetic pool. After a value is assigned to all the genes, this new individual represents not a partial, unlike every one of the individuals separately, but a global solution.

Now, the crossover operator will be applied. This crossover function will keep the secondary population diversity, so the offspring will contain

Figure 4. Crossover operator for hybrid two-population genetic algorithm

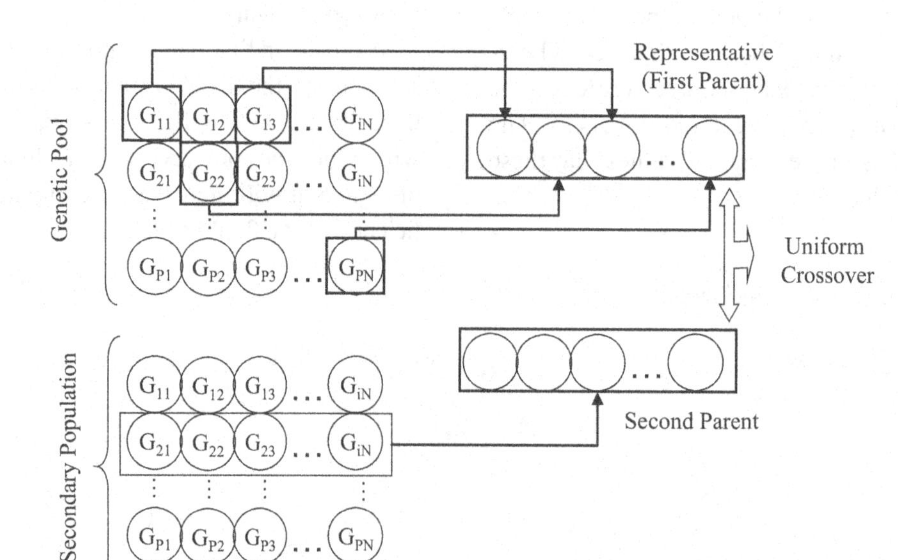

Figure 5. Pseudo code for nutation: Delta initialization

```
IF (not Bi)
   Gi = Gi + Ii
   IF (Gi > LIM_SUP_GEN)
      Gi = LIM_INF_GEN
      Ii = Ii – Delta
   ENDIF
ENDIF
```

$$\text{Delta} = \frac{(LIM\_SUP\_IND)-(LIM\_INF\_IND)}{IND\_POOL}$$

values from the genetic pool. Therefore the GA would be able to maintain multiple solutions in the same population. The crossover operator does not change the genetic pool because the last one only acts as an engine to keep the diversity. This process is summarised in Figure 4.

The Genetic Operators: Mutation

The mutation operator increases the value of individual genes in the genetic pool. It introduces new information into the genetic pool for the representative to use it and finally, by means of the crossover operator, introduces it in secondary population.

It should be noted that the new value will have upper limit, so when it is reached the new gene value will be reset to the lower value.

When generations advance the increment amount is reduced, so the increment applied to the individuals in the genetic pool will take lower values. The different increments between iterations are calculated bearing in mind the lower value for a gene (LIM_INF_IND), the upper value for that gene (LIM_SUP_IND) and the total number of individuals in the genetic pool (IND_POOL), as Figure 5 summarises. In such way, first generations will explore the search space briefly (coarse-grain search) and, as the search process advance, it is intended to do a more exhaustive route through all the values that a given gene may have (fine-grain search).

Genetic Algorithm with Division into Species

Another proposed solution is an adaptation of the niching technique. This adaptation consists on the division of the genetic population into different and independent subspecies. In this case the criterion that determines the specie for a specific individual is done according to genotype similarities (similar genotypes will form isolated species). This classical concept has been provided with some improvements in order to, not only decrease the number of iterations needed for obtaining solutions, but also increase the number of solutions kept within the genetic population.

Throughout evolution the individuals tend to gather into different species; as every one these species will be adapted to an environment and will evolve differently, there will be the same number of optimal individuals than of species. Sometimes the individuals also split up from the initial group due to different reasons and they undergo crossover in other place with individuals of other groups.

Broadly speaking, the grouping technique into species divides the initial population into groups—the species—where the individuals have similar characteristics. In this way, each group will be specialised in a given area of search space; each species will be responsible for searching the solutions that exist in its area, which are also different from other solutions provided by other species. The evolution of every species will be performed by means of a classical GA; also the crossover and mutation operations will be applied as usual but the descendants will be discarded if they do not fulfil the criteria required for belonging to the species of their progenitors, meaning if their genotype is too different from the genotypes of the species.

The natural distribution of individuals into species and their separate evolution is tried to be mimicked. In nature, there are individuals adapted to cold, dry or hot environments; each group keeps its life into a given environment by means of adapting specific characteristics that distinguish it form other groups.

Nevertheless, this technique is not free of disadvantages; certain requirements are needed for a good functioning and they are not directly achieved during the initial arrangement of the problem. For instance, it should be desirable that the population could be perfectly distributed throughout the whole search space, as well as that the groups were fairly distributed along that space and in a quantity in accordance with the number of total solutions of the problem. Unexplored areas might exist if these characteristics were not present; in contrast, there might be another areas highly explored where, depending on the grouping procedure, several species might coexist. The most of these problems can be avoided by means of an automated mechanism for the number of existing species. It seems clear that if the number of solutions is different for different problems, the GA, the responsible for finding solutions, should be the one who manages the number of evolved species in each generation. In order to do this, the GA will be allowed to expand the starting number of species as generations advance.

The increase of the species number will be achieved by performing crossover on individuals from different species throughout several generations. As the resulting descendants mix the knowledge from the species of their ancestors, a new species can be created in a different location from the ones of their progenitors. In this way, the species stagnation can be avoided and the exploration of new areas, together with the appearance of new knowledge, may be obtained; in short, the environment diversity is achieved. As individuals might migrate or be expelled and afterwards create new species with other compatible individuals, the performance of individuals is again modelled in their natural environment.

The crossover operations between the species are applied similarly to what happens in biology. It origins, on one hand, the crossovers between similar individuals are preferred (as it was done at the previous step using GAs) and on the other, the crossovers between different species are enabled, although in a lesser rate.

From an initial population generated for these techniques to be implemented, some steps - following described and graphically represented in Figure 6- take place:

1. Random creation of the genetic population
2. Organisation into species (a species comprises individuals that have high similarities among their genotypes)
3. Application of the Genetic Algorithm on every one of the defined species
4. Introduction of new individual coming from the crossover on different species. These individuals are located in another area of the search space. A variant of the functioning implies the elimination of the individuals that, after several generations, do not create a species large enough.
5. Verifying whether the number of evolutions reaches the maximum allowed or if the population reaches a top level of individuals (defined at the beginning of the execution). If some of these conditions are not fulfilled, the algorithm execution finishes; if they are so, the individuals of the existing population will be again arranged into species (step 2).

The individuals generated after these crossovers could, either be incorporated to an already existing species or, if they analyse a new area of the search space, create themselves a new species.

Finally, the GA provides as much solutions as species remain actives over the search space.

PRACTICE

The different proposals provided have been tested on classic lab problems, as Ackley or Rastrigin functions, and on real problems as the following described, related to juice samples. Due to the complexity of this last problem, the following subsections will be focus on its description and on the results achieved.

Variable Selection for the Classification of Apple Beverages

The importance of juice beverages in daily food habits makes juice authentication an important issue, for example, to avoid fraudulent practices.

A successful classification model can be used for checking two of the most important cornerstones of the quality control of juice-based beverages: monitor the amount of juice and monitor the amount (and nature) of other substances added to the beverages. Particularly, sugar addition is a common and simple adulteration, though difficult to characterise. Other adulteration methods, either alone or combined, involve addition of water, pulp wash, cheaper juices, colorants, and other undeclared additives (intended to mimic the compositional profiles of pure juices) (Saavedra, García & Barbas, 2000).

Infrared spectrometry (IR) is a fast and convenient technique to perform screening studies in order to assess the quantity of pure juice in commercial beverages. The interest lies in developing, from the spectroscopy data, classification methods that might enable the determination of the amount of natural juice contained in a given sample.

However, the information gathered from the IR analyses has some fuzzy characteristics (random noise, unclear chemical assignment, etc.), so analytical chemists tend to use techniques like Artificial Neural Networks (ANN) to develop ad-hoc classification models. Previous studies (Gestal et al, 2005) showed that ANN classify apple juice beverages according to the concentration of natural juice they contained and that ANN had advantages over classical statistical methods, such as faster model development and easy application of the methodology on R&D laboratories. Disappointingly, the large number of variables derived from IR spectrometry requires too much ANN training time and, the most important, it makes very difficult to establish relationships between these variables and analytical knowledge.

Several approaches were used to reduce this number of variables to a smaller subset, which should retain the classification capabilities. In such way, the ANN training process and the interpretation of the results will be highly improved.

Furthermore, a previous variable selection will produce others advantages: cost reduction (if the classification model requires the use of a reduced amount of data, the time needed to

Figure 6. Overview of the proposed system

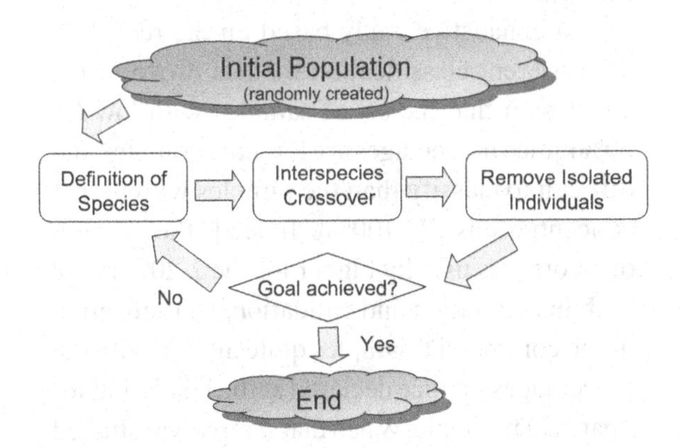

obtain them will be shorter than the required by methods using the original data); increased efficiency (if the system needs to process less information also less time for its processing will be required); improved understanding (when two models solve the same task, but one of them use less information, this one will be more thoroughly understood. Furthermore a simpler model, implies easier knowledge extraction, easier understanding and easier validation).

Nevertheless, after numerous tests, it was proved that it was a highly multimodal problem, since there were several combinations of variables (obtained using different methods) that offered similar results when classifying the samples.

Data Description

In the present practical application, the small spectral range that was measured by means IR technique (wavelengths from 1250 cm-1 to 900 cm-1) provides 176 absorbance values (variables that measure light absorption). Figure 7 shows a superposition of several typical profiles obtained with IR spectroscopy.

The main goal is the establishment of the amount of sugar of a sample, using the absorbance values returned from the IR measure as data. But the amount of data extracted from a sample by IR is huge, so the direct application of mathematical and/or computational methods (although possible) requires a lot of time. It is important to establish whether all those data provide the same amount of information for sample differentiation, so the problem becomes an appropriate case for the use of variable selection techniques.

Previously to variable selection, data sets should be created for models to be developed and validated.. Samples with different amounts of pure apple juice were prepared at the laboratory. Besides, 23 apple juice-based beverages sold in Spain were analysed (the declared amount of juice printed out on their labels was used as input data). The samples were distributed in 2 ranges: samples containing less than 20% pure juice (Table 1) and samples with more than 20% pure apple juice (Table 2). IR spectra were obtained all over the samples.

This data was split into independent and different data sets for extracting the rules (ANN training) and for validation. The commercial samples were used to further check the performance of the model. It is worth noting that if a predicted value does not match with that given on the labels of the commercial products it might be owing to either a wrong performance of the model (classification error) or an inaccurate labelling of the commercial beverage.

Classification Test Considering all the Original Variables

The first test involved all the variables given by the IR spectroscopy. ANN used all the absorbance values related with the training data to obtain a reference classification model. Later, the results obtained with this model were be used to compare the performance of the proposals over the same data.

Different classification techniques (PLS, SIMCA, Potential Curves, etc.) were used too (Gestal et al, 2004), with very similar results. But the best results were achieved using ANN (Freeman & Skapura, 1991), which are very useful for addressing the processes of variable selection employing GA with ANN-based fitness functions.

An exhaustive study based on the results of the different classification models drove to the conclusion that the set of samples with low (2-20%) juice percentages are far more complex and difficult to classify than the samples with higher concentrations (25-100%). Indeed, the number of errors was usually higher for the 2-20% range both in calibration and validation. Classification of the commercial samples quite agreed with the percentages of juice declared at the labels, but for a particular sample. When that sample was studied

Figure 7. Typical IR spectrums

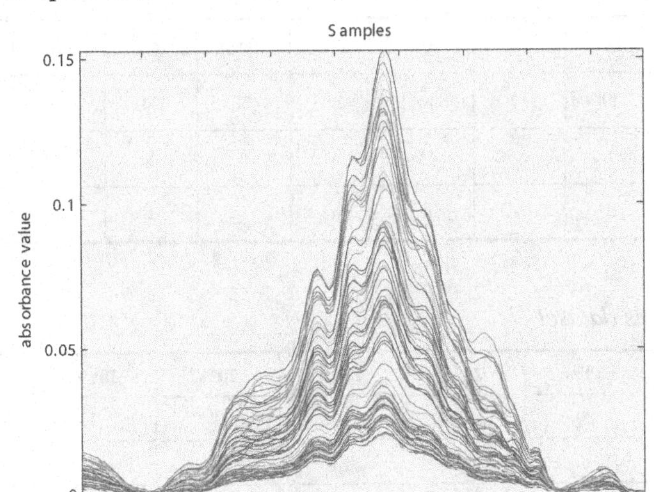

in detail it was observed that its spectrum was slightly different from the usual ones. This would suggest that the juice contained an unusually high amount of added sugar(s).

Variable Selection Approaches

Once the data are available, a variable selection process can be performed in order to optimise the ANN based model.

First, two simple approaches will be briefly described: pruned and fixed search. Both approaches are based on a traditional GA and use ANN for fitness calculation. As the results show, both techniques offer good solutions but present a common problem: an execution of each method provides only a valid solution (discarding the optimal ones). Later, this will be addressed with the two approaches described in the previous sections.

All the approaches use a GA to guide the search by evaluating the prediction capabilities of different ANN models developed employing different sets of IR variables. The problem to be addressed by the different approaches is to find out a small set of IR variables that, when combined with an ANN model, might be capable of classifying apple juice-based beverages properly (according to the amount of pure apple juice they contain).

Once this subset of IR variables is proposed, their associate absorbance values are used as input patterns to the ANN. So, ANN will consider as much input processing elements (PEs) as variables. The output layer has one PE per category (6 for the lower range and 5 for the higher concentrations). After several previous trials considering several hidden layers (from 1 to 4), each with different PEs (from 1 to 50), a compromise was established between the final fitness level reached by the ANN and the time required for its training. This compromise was essential because although better results were obtained with more hidden layers the time required for training was much higher as well. It was decided not to extensively train the network but to get a good approximation to its real performance and to elucidate whether the input variables are really suitable for the accurate classification of the samples.

Table 1. Low concentrations dataset

Juice Concentration	2%	4%	6%	8%	10%	16%	20%	Total
Training	19	17	16	22	21	20	19	134
Validation	1	1	13	6	6	6	6	39
Commercial	0	0	0	0	1	1	0	2

Table 2. High concentrations dataset

Juice Concentration	20%	25%	50%	70%	100%	Total
Training	20	19	16	14	7	86
Validation	6	18	13	1	6	44
Commercial	0	2	0	0	19	21

Table 3. Classification with ANNs using all the variables: percent accuracy

	Training	Validation	Commercial
Low Concentrations	Training (134) **134 (100%)**	Validation (39) **35 (89.74%)**	Comercial (2) **2 (100%)**
ANN Configuration for low concentrations *Topology: 176 / 50 / 80 / 5*	*learning rate: 0.0005*	*stop criterion: mse=5 or epochs=500.000*	
High Concentrations	Training (86) **86 (100%)**	Validation (44) **43 (97.72%)**	Commercial (21) **21 (100%)**
ANN Configuration *Topology: 176 / 8 / 5 / 5*	*learning rate: 0.0005*	*stop criterion: mse=1 or epochs=500.000*	

The goal is to determine which solutions, among those provided by the GA, represent good starting points. Therefore, it would be enough to extend the ANN learning up to the point where it starts to converge. For this particular problem, convergence started after 800 cycles; in order to warrant that this step is reached, 1000 iterations were fixed.

The selected variables were later used as inputs of an external ANN responsible of providing the final classification results.

Pruned Search

This approach starts by considering all variables from groups where are gradually discarded. The GA will steadily reduce the amount of variables that characterise the objects from the original data set, until getting an optimal subset that might enable a satisfactory general classification. The remaining set of variables is used to classify the samples, and the results are used to determine how relevant the discarded wave numbers were for the classification. This process can be continued as long as the classification results are equal, or at least similar, to those obtained using the overall set of variables. Therefore, the GA determines how many and which wave numbers will be considered for the classification.

In this approach, each individual in the genetic population is described by n genes, each representing one variable. With the binary encoding each gene might be either 0 or 1, indicating whether the gene is active or not and, therefore, if the variable should be considered for classification.

The evaluation function guides the pruning process in order to get individuals with a low number of variables. To achieve this, the function should help those individuals that, besides classifying accurately, make use of fewer variables. In this particular case the function increments the percentage of active genes to the MSE (medium square error) obtained by the ANN, so the individuals with less active genes – and with a similar classification performance – will have a lower fitness and consequently better chances to survive.

Table 4 shows the results of several runs of pruned search. It should be borne in mind that each solution was provided in a different execution and that, within the same execution, only

Table 4. Classification with pruned search: percent accuracy

Low Concentrations	Training (134)	Validation (39)	Comercial (2)
Run1: Selected Variables [42 77]	129 (96.27%)	23 (58.97%)	1 (50%)
Run2: Selected Variables [52 141]	115 (85.82%)	22 (56.41%)	0 (0%)
Run3: Selected Variables [102 129]	124 (92.54%)	25 (64.10%)	0 (0%)
ANN Configuration for low concentrations *Topology: 2/ 10 / 60 / 7*	*learning rate: 0.0001*	*stop criterion: mse=5 or epochs=500.000*	
High Concentrations	**86**	**43**	**21**
Run1: Selected Variables [42 77]	83 (96.51%)	34 (79.07%)	21 (100%)
Run2: Selected Variables [52 141]	82 (95,35%)	35 (81.40%)	20 (95.24%)
Run3: Selected Variables [102 129]	83 (96.51%)	33 (76.74%)	21 (100%)
ANN Configuration *Topology: 2 / 10 / 60 / 5*	*learning rate: 0.0001*	*stop criterion: mse=1 or epochs=500.000*	

Table 5. Classification with fixed search: percent accuracy

Low Concentrations	Training (134)	Validation (39)	Comercial (2)
Run1: Selected Variables [12 159]	125 (93.28%)	23 (58.97%)	0 (0%)
Run2: Selected Variables [23 67]	129 (96.27%)	26 (66.66%)	1 (50%)
Run3: Selected Variables [102 129]	129 (96.27%)	25 (64.10%)	0 (0%)
ANN Configuration for low concentrations *Topology: 2/ 10 / 60 / 7*	*learning rate: 0.0001*	*stop criterion: mse=5 or epochs=500.000*	
High Concentrations	**86**	**43**	**21**
Run1: Selected Variables [12 159]	82 (95.35%)	31 (72.09%)	19 (90.47%)
Run2: Selected Variables [23 67]	81 (94.18%)	31 (72.09%)	19 (90.47%)
Run3: Selected Variables [102 129]	81 (94.18%)	33 (76.74%)	21 (100%)
ANN Configuration *Topology: 2 / 10 / 60 / 5*	*learning rate: 0.0001*	*stop criterion: mse=1 or epochs=500.000*	

one solution provided valid classification rates. As it can be noted, classification results are very similar, although the variables used for performing the classification are different.

ANNs models obtained are slightly worse than those obtained using 176 wave numbers, although the generalisation capabilities of the best ANNs model are quite satisfactory, as there is only one error when commercial beverages are classified.

Fixed Search

This approach uses a real codification in the chromosome of the genetic individuals. The genetic population consists of individuals with n genes, where n is the amount of wave numbers that are a priori considered enough for the classification according to some external criterion. Each gene represents one of the 176 wave numbers considered in the IR spectra. The GA tries to find the subsets of variables that perform the best classification model using only the number of variables predefined in the genotype.

In this case, the final number of variables is given in advance, so some external criterion will be needed. In order to simplify the comparison of the results, in this work the final number of variables is defined by the minimum number of principal components that can describe the data set: two variables.

As the amount of wave numbers remains constant in the course of the selection process, this approach defines the fitness of each genetic individual as the mean square error reached by the ANN at the end of the training process.

Table 5 shows the results of several runs of pruned search. It should be highlighted that different runs provide only one valid solution each time.

Since the genetic individuals contain only two genes and the crossover operator has to use the unique available crossing point (between the genes), only half of the information from each parent would be transmitted to its offspring , This converts the fixed search approach into a random search when only two genes constitute the chromosome.

Hybrid Two-Population Genetic Algorithm

As discussed previously, the main disadvantage of non-multimodal approaches is that they discard optimal solutions while a final or global solution is preferred. But there are situations where the final model is extracted after analysing different solutions of the same problem.

Table 6. Classification with hybrid two-population genetic algorithm

Low Concentrations	Training (134)	Validation (39)	Comercial (2)
Run1: Selected Variables [89 102]	127 (95.77%)	29 (74.36%)	0 (0%)
Run1: Selected Variables [87 102]	130 (97.01%)	28 (71.79%)	0 (0%)
Run1: Selected Variables [88 89]	120 (89.55%)	29 (74.36%)	0 (0%)
ANN Configuration for low concentrations *Topology: 2/ 10 / 60 / 7*	*learning rate: 0.001*	*stop criterion: mse=2 or epochs=500.000*	
High Concentrations	**86**	**43**	**21**
Run1: Selected Variables [89 102]	83 (96.51%)	35 (81.39%)	21 (100%)
Run1: Selected Variables [87 102]	83 (96.51%)	36 (83.72%)	21 (100%)
Run1: Selected Variables [88 89]	82 (95.35%)	39 (90.69%)	21 (100%)
ANN Configuration *Topology: 2 / 10 / 60 / 5*	*learning rate: 0.001*	*stop criterion: mse=2 or epochs=500.000*	

For example, after analysing the different solutions provided by one exaction of the classification task with Hybrid Two-Population Genetic Algorithm three valid (and similar) models were obtained (see Table 6). Furthermore the results were clearly superior to those obtained with the previous alternatives. Besides, it was observed that the solutions concentrate along specific spectral areas (around the 88 wave number); It would not be possible with the previous approaches.

Genetic Algorithm with Division into Species

As in the previous section, several classification models (all of them valid) can be analysed after an execution of the classification algorithm, therefore allowing the extraction of more information about the areas where solutions are focused.

Table 7 shows the different solutions obtained in one run of the method. Each solution is provided by different species into the GA. In this search there were different executions: for selecting

Table 7. Genetic algorithm with division into species

Low Concentrations	Training (134)	Validation (39)	Comercial (2)
Run1: Selected Variables [88 97]	131 (957.76%)	32 (82.05%)	1 (50%)
Run1: Selected Variables [96 124]	132 (98.51%)	30(76.92%)	2 (100%)
Run1: Selected Variables [3 166]	130 (97.01%)	28 (71.79%)	0 (0%)
ANN Configuration for low concentrations *Topology: 2/ 10 / 60 / 7*	*learning rate: 0.001*	*stop criterion: mse=2 or epochs=500.000*	
High Concentrations	**86**	**43**	**21**
Run1: Selected Variables [9 171]	86 (100%)	40 (93.02%)	18 (85.71%)
Run1: Selected Variables [12 15]	83 (96.51%)	34 (79.07%)	18 (85.71%)
Run1: Selected Variables [161 165]	83 (96.51%)	33 (76.74%)	17 (80.95%)
ANN Configuration *Topology: 2 / 10 / 60 / 5*	*learning rate: 0.001*	*stop criterion: mse=2 or epochs=500.000*	

variables related with low concentrations and with high concentrations.

FUTURE TRENDS

Due to the fact that ANN are sometimes excessively used, the quality of the variable selection is quite emphasised, although the time required for performing such selection is not a critical factor. As the higher computational load lies in the training of the inner ANN, several trainings could be performed simultaneously by means of the proposed distributed execution.

The proposed system also uses a back propagation learning based-ANN for evaluating the samples. Other variable selection models, arisen from other type of networks (LVQ, SOM, etc.), are being developed for performing a more complete study.

FUTURE RESEARCH DIRECTIONS

Most of the effort would be focused on the development and test of new fitness functions. These functions would be able to assess the quality of a sample and would try to require less computational requirements than the ANN actually used.

Another option to preserve the ANN use would consist on the distribution of the genetic individual's evaluation. It would be interesting to compare both approximations.

CONCLUSION

Several conclusions can be drawn from the different proposals for variable selection:

First of all, the classification results were, at least, acceptable, so different techniques based on the combination of GA and ANN demonstrated that they are valid approaches and can be used to perform variable selection.

Best results were obtained using a multimodal GA, as it was expected, due to its ability to maintain the genetic individuals homogeneously distributed over the search space. Such diversity not only induces the appearance of optimal solutions, but also avoids the search to stop on a local minimum. This option does not provide only a solution but a group of them that have similar fitness. This allows Chemistry scientists to select a solution with a sound chemical background and extract additional information.

REFERENCES

Ballester, P.J., & Carter, J.N. (2003). Real-Parameter Genetic Algorithm for Finding Multiple Optimal Solutions in Multimodel Optimization. *Proceedings of Genetic and Evolutionary Computation* (pp. 706-717)

Darwin, C. (1859). *On the Origin of Species*. John Murray, London.

Deb, K. (2003). *A Population-Based Algorithm-Generator for Real-Parameter Optimization. (Tech. Rep. No.* KanGAL 2003003)

Deb, K., & Agrawal, S. (1995). Simulated binary crossover for continuous search space. *Complex Systems 9(2)* (pp. 115-148)

DeJong, K.A., (1975). An Analysis of the Behaviour of a Class of Genetic Adaptive Systems. Phd. Thesis, University of Michigan, Ann Arbor.

Eshelman, L.J., & Schaffer, J.D. (1993). Real coded genetic algorithms and interval schemata. *Foundations of Genetic Algorithms (2)* (pp. 187-202)

Freeman, J. A., & Skapura, D.M. (1991). *Neural Networks: Algorithms, Applications, and Programming Techniques*, Addison-Wesley, Reading, MA.

Gestal, M., Gómez-Carracedo, M.P., Andrade, J.M., Dorado, J., Fernández, E., Prada , D. &

Pazos, A. (2005). Selection of variables by Genetic Algorithms to Classify Apple Beverages by Artificial Neural Networks. *Applied Artificial Intelligence* (pp. 181-198)

Gestal, M., Gómez-Carracedo, M.P., Andrade, J.M., Dorado, J., Fernández, E., Prada , D. & Pazos, A. (2004). Classification of Apple Beverages using Artificial Neural Networks with Previous Variable Selection. *Analytica Chimica Acta* (pp. 225-234)

Goldberg, D. E., & Richardson J. (1987). Genetic Algorithms with Sharing for Multimodal Function Optimization. *Proceedings of 2nd International Conference on Genetic Algorithms (ICGA)*, (pp. 41-49), Springer-Verlag.

Goldberg, D.E (1989) *Genetic Algorithms in Search, Optimization & Machine Learning.* Addison-Wesley.

Harik, G. (1995). Finding multimodal solutions using restricted tournament selection. *Proceedings of the Sixth International Conference on Genetic Algorithms*, (ICGA) (pp. 24-31).

Holland, J. H. (1975). *Adaptation in Natural and Artificial Systems.* University of Michigan Press, Ann Arbor.

Landgon, W. (1996). *Evolution & Genetic Programming Populations.* (Tech. Rep. No. RN/96/125). London: University College.

Mengshoel, O.J., & Goldberg, D.E. (1999). Probabilistic Crowding: Deterministic Crowding with Probabilistic Replacement. *Proceedings of Genetic and Evolutionary Computation* (pp. 409-416)

Miller, B., & Shaw, M., (1995). *Genetic Algorithms with Dynamic Niche Sharing for Multimodal Function Optimization.* (Tech. Rep. No. IlliGAL 95010). University of Illinois.

Ono, I., & Kobayashi, S. (1999). A real-coded genetic algorithm for function optimization using unimodal normal distribution. *Proceedings of International Conference on Genetic Algorithms* (pp. 246-253)

Ortiz, D., Hervás, C., & García, N. (2005). CIXL2: A crossover operator for evolutionary algorithms based on population features. *Journal of Artificial Intelligence Research.*

Petrowski, A. (1996). A Clearing Procedure as a Niching Method for Genetic Algorithms. *International Conference on Evolutionary Computation.* IEEE Press. Nagoya, Japan.

Saavedra, L., García, A. & Barbas, C. (2000). Development and validation of a capillary electrophoresis method for direct measurement of isocitric, citric, tartaric and malic acids as adulteration markers in orange juice. *Journal of Chromatography 881(1-2)* (pp. 395-401)

Ursem, R. K. (2002). Diversity-Guided Evolutionary Algorithms. *Proceedings of VII Parallel Problem Solving from Nature* (pp. 462-471). Springer-Verlag.

ADDITIONAL READING

Ashurst, P.R. (1998). *Chemistry and Technology of Soft Drinks and Fruit Juices,* Sheffield Academic Press Ltd.

Beasley, D., Bull, D.R., & Martin, R.R. (1993). An overview of genetic algorithms: Part 1, Fundamentals. University Computing, vol. 15, nº 2, (pp. 58-69).

Beasley, D., Bull, D.R., & Martin, R.R. (1993). An overview of genetic algorithms: Part 2, Research Topics. University Computing, vol. 15, nº 4, (pp. 170-181).

Bishop, C.M. (1995). *Neural Networks for Pattern Recognition.* Oxford University Press.

Cantú-Paz, E., Newsam, S., & Kamath, C. (2004). Feature selection in scientific applications. In *Pro-*

ceedings of the Tenth ACM SIGKDD international Conference on Knowledge Discovery and Data Mining. New York, ACM Press, (pp. 788-793).

DeJong, A.K. & Spears, W.M. (1993). On the State of Evolutionary Computation. *Proceedings of the International Conference on Genetic Algorithms*, (pp. 618-623).

Dreyfus, G. (2005). *Neural Networks: Methodology and Applications*. Springer.

Fogel, D.B. (2006). *Evolutionary Computation: Toward a New Philosophy of Machine Intelligence 3rd Edition*. Piscataway, NJ, IEEE Press.

Fogel, L.J., Owens, A.J., & Walsh, M.J. (1966). *Artificial Intelligence Through Simulated Evolution*. Wiley Publishing, New York.

Fuleki, T., Pelayo, E., & Palabay, R.B. (1995). Carboxylic acid composition of varietal juices produced from fresh and stored apples. *Journal of Agriculture and Food Chemistry*. vol. 43, nº 3, (pp. 598-607).

Gestal, M., Cancela, A., Gómez-Carracedo, M.P., Andrade, J.M. (2006). Several Approaches to Variable Selection by means of Genetic Algorithms. *Artificial Neural Networks in Real-Life Applications*. (pp. 141-164). Hershey, Idea Group Publishing.

Goldberg, D.E. (2002). *The Design of Innovation: Lessons from and for Competent Genetic Algorithms*. Reading, MA, Addison-Wesley.

Guyon, I., & Elisseeff, A. (2003). An Introduction to Variable and Feature. *JMLR Special Issue on Variable and Feature Selection (Kernel Machines Section)*.

Haykin, S. (1999). *Neural Networks: A Comprehensive Foundation*, Prentice Hall.

Koza, J.R. (1992). *Genetic Programming: On the Programming of Computers by Means of Natural Selection*. The MIT Press. Cambridge, MA.

Krzanowski. W.J. (1990). *Principles of multivariate analysis: a user's perspective*. Oxford, Clarendon Pres.

Matlab: Genetic Algorithm and Direct Search Toolbox 2. Available at: www.mathworks.com/academia/student_version/r2007a_products/gads.pdf

Matlab: Neural Network Toolbox. Available at: www.mathworks.com/products/neuralnet/

Pavan, M., Consomni, V., & Todeschini, R. (2003). Development of Order Ranking Models by Genetic Algorithm Variable Subset Selection (GA-VSS). *Proceedings of Conferentia Chemometrica*. (pp. 27-29).

Rabuñal, J.R., Dorado, J., Gestal, M., & Pedreira, N. (2005). Diversity and Multimodal Search with a Hybrid Two-Population GA: An Application to ANN Development. *Computacional Intelligence and Bioinspired Systems*. (pp. 382-390). Springer-Verlag.

Roboards, K. & Antolovich, M. (1995). Methods for assessing the authenticity of orange juice, a review. *The Analyst*, vol. 120, (pp. 1-28).

Skapura, D.M. (1996). *Building Neural Networks*. New York, ACM Press (Addison-Wesley Publishing Company).

Tomassini, M. (1995). A Survey of Genetic Algorithms. *Annual Reviews of Computational Physics* Vol. III, World Scientific, (pp. 87-117).

Yang J., & Honavar, V. (1998). Feature subset selection using a genetic algorithm. *IEEE Intelligent Systems (Special Issue on Feature Transformation and Subset Selection)*, 13(2), (pp. 44-49).

Chapter XIV
Biomolecular Computing Devices in Synthetic Biology

Jesús M. Miró
Universidad Politécnica de Madrid (UPM), Spain

Alfonso Rodríguez-Patón
Universidad Politécnica de Madrid (UPM), Spain

ABSTRACT

Synthetic biology and biomolecular computation are disciplines that fuse when it comes to designing and building information processing devices. In this chapter, we study several devices that are representative of this fusion. These are three gene circuits implementing logic gates, a DNA nanodevice and a biomolecular automaton. The operation of these devices is based on gene expression regulation, the so-called competitive hybridization and the workings of certain biomolecules like restriction enzymes or regulatory proteins. Synthetic biology, biomolecular computation, systems biology and standard molecular biology concepts are also defined to give a better understanding of the chapter. The aim is to acquaint readers with these biomolecular devices born of the marriage between synthetic biology and biomolecular computation.

INTRODUCTION

Molecular biology, biochemistry and genetics have been and are the most important drivers of cellular biology. However, new disciplines examining biological processes have emerged. They have ventured new viewpoints and offered a wide range of possibilities for both generating and applying knowledge in other areas. Most of these new fields are multidisciplinary. They feed off different areas like physics, chemistry, mathematics or computing. Of these new disciplines, this chapter will examine synthetic biology and biomolecular computation, but not separately, together, linking the two fields. The goal of this union is to study, design, simulate and implement new biomolecular systems capable of making computations *in vivo*, that is, able to process information inside living systems.

In this chapter, we will look at several simple examples that will give readers an overview of this marriage. These examples will be biomolecular devices based on biological principles that process information. In fact, we will look at NOT and AND logic gates built using genetic circuits (Weiss, 2003); the logic AND built using just nucleic acids that works thanks to competitive hybridization and toeholds (Fontana, 2006; Seelig, 2006), and a nanodevice composed of a DNA hairpin that opens and closes through genetic regulation. Finally, we will consider an automaton that diagnoses disease by checking for the presence of different RNA molecules associated with the disease in question and releases the drug if the diagnosis is positive (Benenson, 2004). This is a beautiful example of the marriage of biomolecular computation and synthetic biology with a promising application in biomedicine.

Other devices, noteworthy because they were the first to merge biomolecular computing and synthetic biology, are a genetic toggle switch with two stable states (Gardner, 2000), and the repressilator, which is an oscillatory gene network composed of three repressors (Elowitz, 2000). Both of these devices were designed *in vivo*. Another circuit with two stable states was designed *in vitro* (Kim, 2006). Being extracellular, this circuit achieves better control of the circuit parameters.

SYNTHETIC BIOLOGY, BIOMOLECULAR COMPUTATION AND SYSTEMS BIOLOGY

The definition of synthetic biology is changing and the borders of this discipline are fuzzy. In the following, we will try to give a comprehensive definition that includes the most relevant research that is being developed and could be considered to be part of this field. We will also describe biomolecular computation and systems biology. Systems biology is also a relative new and emerging field that can help to move synthetic biology and biomolecular computation forward.

Synthetic biology: is a discipline half-way between science and engineering (Benner, 2005; De Lorenzo, 2006; ETC group, 2007). It is concerned with:

- The design, construction and modification of biomolecular systems and organisms to perform specific functions.
- To get a better understanding of biological mechanisms.

The operation of these synthetic biomolecular systems is based on the processes of the central dogma of molecular biology, that is, DNA replication, and especially DNA transcription and translation. But there are also designs that are based on more specific biological processes like the competitive hybridization of nucleic acids, the operation of certain enzymes, etc.

There are at present two trends, bottom-up and top-down, in synthetic biology projects (Benner, 2005):

- The bottom-up trend takes the form of a hierarchy inspired by computer engineering (Andrianantoandro, 2006). The building blocks are DNA, RNA, proteins, lipids, amino acids and the other metabolites. These building blocks interact with each other through biochemical reactions to form simple devices. The devices are linked to form modules that can do more complex tasks. These modules are connected to set up biomolecular networks. These networks can be integrated into a cell and change the cell's behaviour. The DNA sequences with special functions, like the operator, the promoter, etc., are called BioBricks in synthetic biology. And there are proposals for recording standard biological parts (http://parts.mit.edu). The bottom-up trend emerged because synthetic biology focuses on creating genetic circuits. The operation of these circuits is based on gene expression regulation through transcription control that primarily involving genes and proteins (Benner, 2005; Hasty, 2002). Therefore, these biomolecules play the key role in device design and construction. However, recent studies on RNA's important cell regulating functions are encouraging its use in the design and construction of synthetic biology devices (Isaacs, 2006; Rinaudo, 2007).

- The top-down trend isolates or reduces parts of the biological systems to a minimum. Its objective is to be able to understand these parts and use them to build more complex synthetic biological systems. One example of this tactic is projects aiming to discover the least number of genes needed for bacteria to survive (Glass, 2004; Hutchison, 1999). The goal is to use these bacteria as a basic mould to build new synthetic organisms by adding the genes required to perform special functions, such as generating alternative fuels, to this skeleton genome. Another example is a project on the synthesis of the artemisinin antimalarial drug in engineered yeast (Ro, 2006).

There are other ways of pigeonholing synthetic biology projects, such as *in vivo* and *in vitro* projects (Forster, 2007).

Another type of less common synthetic biology devices whose operation is based on more specific processes are:

- Logic circuits of nucleic acids based on competitive hybridization (Seelig, 2006; Takahashi, 2005).
- Biomolecular automata that are based on competitive hybridization and on restriction enzyme operation (Benenson, 2001; Benenson, 2004).
- DNA nanodevices based on competitive hybridization (Bath, 2007; Liedl, 2007; Simmel, 2005).
- Devices based on virus structure (Ball, 2005; Cello, 2002).
- New materials based on the properties of nucleic acids and proteins (Ball, 2005).

The goals of synthetic biology are to generate new scientific knowledge to explain biological processes, find new biological circuit design principles (now based on gene expression regulation) and their application in other disciplines, like systems biology, biomolecular computation, medicine, pharmacology and bionanotechnology.

Biomolecular computation: is a scientific discipline that is concerned with processing information encoded in biological macromolecules like DNA, RNA or proteins, although the most important advances have been made using DNA (Rodríguez-Patón, 1999). The first paper on *in vitro* biomolecular computation focused on DNA and solved what is known as the Hamiltonian path problem (Adleman, 1994). The next paper used a filtering model to solve the SAT (Boolean satisfiability problem) (Lipton, 1995). These papers demonstrated that DNA computation was massively parallel, offering a huge information density as just one test tube could contain in the order of 10^{20} DNA strands to encode information. Additionally, the operations are executed in parallel on all the strands. These properties of DNA computation have been exploited to develop bioalgorithms for cryptography (Gehani, 1999), memories (Baum, 1995) and autonomous molecular computing (Sakamoto, 2000; Stojanovic, 2003). But despite the advances, the scientific community realized that DNA computers are no competitor for electronic computers in terms of computational problem solving speed. The strength of biomolecular computing, and specifically DNA computation, is that it is perfect option for processing and handling biological information and can be applied to solve problems within the field of biomedicine (Benenson, 2004; Condon, 2004; Rianudo, 2007; Shapiro, 2006).

Another important branch of biomolecular computing is the so-called Membrane Computing or P Systems (Paun, 2000; Paun, 2002). P Systems are distributed computational models inspired in the living cell. In last years this abstract and theoretical model has been used like a new framework for developing computational models of complex biological processes and systems (Ciobanu, 2006).

Originally, biomolecular computation targeted mathematical and computational problem solving (Rodríguez-Patón, 1999). However, more recent papers on biomolecular automata for intelligent drugs design (Benenson, 2004) or gene networks for boolean logic gates (Weiss, 2003) are broadening horizons towards the application of these systems in other disciplines like medicine, pharmacology and bionanotechnology. These are the newest and most promising 'killer applications' of DNA computing.

Systems biology: is a scientific discipline examining biological processes, systems and its complex interactions using robust and precise mathematical and physical models. The goal of these models is to describe, understand and predict the dynamic behaviour of these biological processes and systems (Kitano, 2002). Key concepts in systems biology are:

- **Complexity:** the properties and dynamic behaviour of biological systems are hard to understand because of the huge number of molecules and interactions between molecules in different networks (genetic, metabolic, etc.) and on different time scales. (Amaral, 2004; Csete, 2004; Goldenfeld, 1999).
- **Modularity:** biological systems have the property of containing separable parts that are, even so, able to fulfil the function that they performed as part of the rest of the biological system. These separable parts are called modules. (Alon, 2007; Hartwell, 1999).
- **Robustness:** this is the persistency of the biological system's properties in face of the perturbations and under different noisy conditions. Feedback control and redundancy play an exceedingly important role in system robustness (Mcdams, 1999; Kitano, 2004; Stelling, 2004).

The goals of systems biology are to generate new scientific knowledge to explain the biological processes.

Relationship between systems biology and synthetic biology: In the coming years we are very likely to see a special relationship grow up between systems biology and synthetic biology. Systems biology is an ideal tool for helping synthetic biology to model the design and construction of biomolecular devices. In return, synthetic biology can lead systems biology to a better understanding of the dynamic behaviour of biological systems by creating biologically-inspired devices to improve the models. This device modelling (*in info* or *in silico*) and construction (*in vitro* or *in vivo*) cycle will boost and further refine systems biology models and the synthetic biology devices (Church, 2005; Di Ventura, 2006).

Some prominent centres:
In USA:

- Adam Arkin's group at California University, Berkeley: *Arkin Laboratory*, (http://genomics.lbl.gov/).
- George M. Church's group at Harvard University: *Center for Computational Genetics*, (http://arep.med.harvard.edu/).
- James J. Collins's group at Boston University: *Applied Biodynamics Laboratory*, (http://www.bu.edu/abl/).

- Jay Keasling's group at California University, Berkeley: *Synthetic Biology Engineering Research Center* (SynBERC), (http://www.synberc.org).
- Tom Knight's group in the MIT, Cambridge: *Computer Science and Artificial Intelligence Laboratory* (CSAIL), (http://knight.openwetware.org/).
- Andrew Murray's group at Harvard University: *Harvard FAS Center for Systems Biology*, (http://www.sysbio.harvard.edu/csb/).
- J. Craig Venter's group in Rockville: *J. Craig Venter Institute*, (http://www.jcvi.org/).
- Ron Weiss's group at Princeton University: *weisslab*, (http://weisswebserver.ee.princeton.edu/).

In Europe:

- Uri Alon's group in the Weizmann Institute of Science, Rehovot (Israel): *Uri Alon Laboratory*, (http://www.weizmann.ac.il/mcb/UriAlon/).
- Victor de Lorenzo's group in the Centro Nacional de Biotecnología (CNB), Madrid (Spain): *Molecular Environmental Microbiology Laboratory* (MEML), (http://www.cnb.csic.es/~meml/).
- Luis Serrano's group in the Centre for Genomic Regulation (CRG), Barcelona (Spain): *Design of Biological Systems*, (http://pasteur.crg.es/portal/page/portal/Internet).
- Friedrich C. Simmel's group at Munich University: *Biomolecular nanoscience*, (http://www.e14.physik.tu-muenchen.de/.

Some prominent conferences:

- BioSysBio 2008: Computational Systems Biology, Bioinformatics, Synthetic Biology. April 20-22, 2008. London (United Kingdom)
- Workshop on Computation in Genetic and Biochemical Networks. September 4-8, 2006. York (United Kingdom).
- The 14th International Meeting on DNA Computing. June 2-6, 2008. Prague, (Czech Republic).
- Cold Spring Harbor Meeting on Engineering Principles in Biological Systems. December 3-6, 2006. Cold Spring Harbor, New York (USA).
- First International Conference on Synthetic Biology 1.0 (SB1.0). June 10-12, 2004. MIT, Cambridge (USA).
- Second International Conference on Synthetic Biology 2.0 (SB2.0). May 20-22, 2006. University of California, Berkeley (USA).
- Third International Conference on Synthetic Biology 3.0 (SB3.0). June 24-26, 2007. Zurich (Switzerland).

MOLECULAR BIOLOGY DEFINITIONS

The following are definitions of biomolecules and standard molecular biology processes that will be useful throughout this chapter.

- **Nucleoside:** Molecule composed of a nitrogenous base linked to a sugar called pentose. The bases in the DNA nucleoside are adenine (A), guanine (G), cytosine (C) and thymine (T). The pentose is a deoxyribose. The bases in the RNA nucleoside are adenine, guanine, cytosine and uracil (U). The pentose is a ribose.
- **Nucleotide:** Molecule composed of a nucleoside and one or more phosphate groups linked by means of ester bonds with the sugar.
- **DNA or RNA strands:** Polymers of nucleotides linked by covalent bonds.
- **DNA or RNA hybridization:** Two single DNA or RNA strands with complementary base sequences that bind (hybridize) under suitable temperature and pH conditions to form a double strand. Hybridization takes place through complementary base pairing

across Hydrogen Bonds. Thymine binds to adenine and guanine to cystosine in DNA, and uracil binds to adenine in the case of RNA. The single strands have different ends called 3' and 5'. This implies that the hybridization of the two single strands is antiparallel, i.e. the 3' end of one strand binds to the 5' end of the other and vice versa.

- **DNA double helix:** DNA in living cells generally has a double helix structure formed by two interwoven strands that are linked by complementary bases. The length per turn is 34 Angstroms and its diameter is 20 Angstroms.
- **Gene:** A gene is the building block of inheritance in living beings. Physically, it is portion of one of the two strands of DNA making up the double helix and is composed of a nucleotide sequence that contains the information for RNA molecule synthesis.

Additionally, a gene has neighbouring portions with special functions (Figure 1):

- **Promoter:** region to which the RNA polymerase (RNAp) enzyme binds to start transcription. It is positioned in front of the gene.
- **Operator:** region to which the regulatory proteins bind to stimulate or inhibit gene transcription. It can be positioned in front of, behind or inside the promoter.
- **Terminator:** region where gene transcription ends. It is positioned at the end of the gene.
- **Regulatory sequence:** region where the proteins regulating gene transcription are coded. There may be several sequences positioned at either the start or end of the gene. These sequences can be far away from the gene and also in different double helix strands. Note that the first regulatory sequence in Figure 1 is not in the same strand as the gene.
- **Regulatory proteins:** are proteins that bind to the gene operator and act differently depending on whether they are repressors or activators. Repressor proteins directly

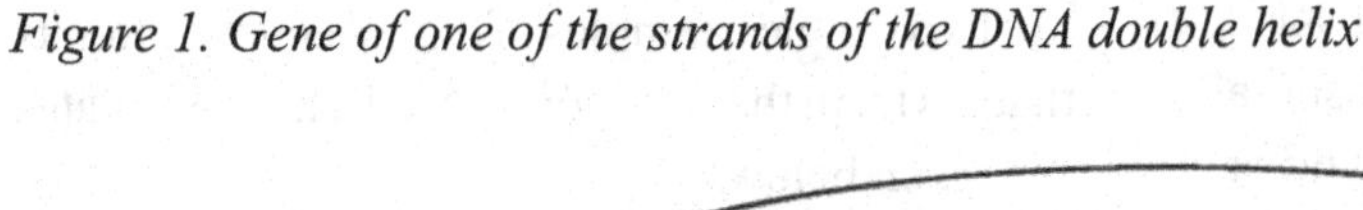

Figure 1. Gene of one of the strands of the DNA double helix

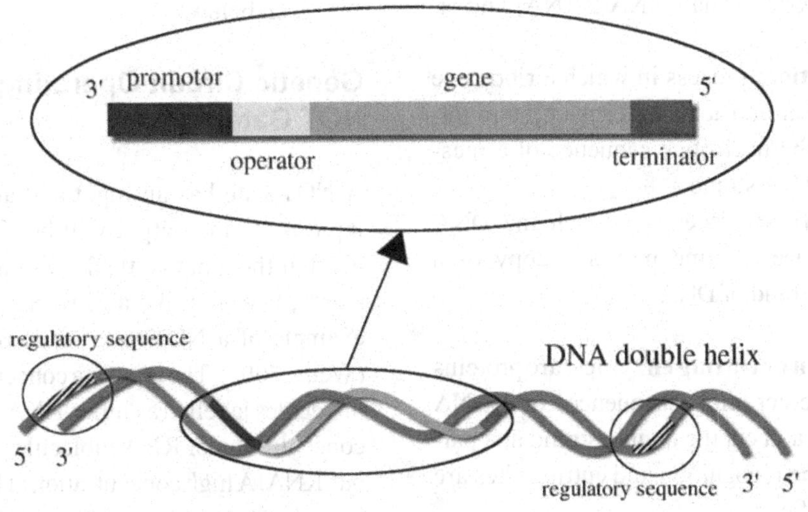

prevent the RNAp from binding to the promoter or block its progress, inhibiting translation in both cases (inhibited gene). On the other hand, the activator proteins help the RNAp to bind to the promoter and promote translation (activated gene). One important property of the regulatory proteins is that they can be activated or deactivated by means of small molecules called inducers. Note that a regulatory sequence is another gene with its respective promoter, operator, terminator and regulatory sequences. These latter regulatory sequences are again other genes with their promoters, operators, etc. And all this makes up a complicated genetic network.

- **Operon:** in bacteria, the promoter, operator, gene and terminator complex is termed operon. A standard example is the *Escherichia coli* bacterium's *lac* operon that codifies the proteins required for lactose transportation and metabolism. In this example, lactose actually works as an inducer molecule.
- **Transcription:** process in which the RNA polymerase enzyme copies a DNA strand to a complementary RNA strand. The different types of RNA are messenger RNA (mRNA), transfer RNA (tRNA), ribosomal RNA (rRNA), small RNA (sRNA) and ribozymes.
- **Translation:** process in which a ribosome picks up amino acids to form a protein following the nucleotide sequence of a messenger RNA strand.
- **Replication:** process in which the DNA polymerase enzyme makes a copy of a double strand of DNA.

Restriction or cutting enzymes: are proteins that recognize certain subsequences of a DNA double strand and cut the double strand at a particular site. The recognition and cutting sites are enzyme specific.

EXAMPLES OF DIGITAL GENETIC CIRCUITS

Genetic circuits are the most common devices in synthetic biology. The operation of these circuits is based on genetic transcription regulation. Most of these devices are digital circuits, where the logical '1s' and '0s' are equivalent to high and low biomolecule concentrations.

One problem with genetic circuits is the assembly of circuits with each other. The reason is that, unlike electronic circuits where the input and output signal is always voltage, the molecule type representing the output signal of one circuit might not be the same as the molecule type representing the input signal of the next circuit, and the two cannot be connected. The MIT has put forward an idea to solve this problem. This idea is to get a 'universal' measurement unit, called PoPS (polymerase per second). PoPS represents the number of mRNA molecules of a gene transcripted per second.

Another problem is isolating the genetic circuit when it is introduced into a cell to prevent parts of the circuit interacting with parts of the cell or prevent the biomolecules belonging to the circuit propagating in the cell.

Two simple examples of digital genetic circuits that perform the function of NOT and AND gates are given below.

Genetic Circuit Operating as a Logic NOT Gate

A NOT gate has an input and an output. If the input is '1', the output will be '0'. On the other hand, if the input is '0', the output is '1'. In other words, it works like an inverter. Figure 2 is an example of a NOT gate made of biomolecules (Weiss, 2002). The input is a concentration of RNA molecules labelled as input-RNA. The output is a concentration of RNA molecules labelled as output-RNA. A high concentration of RNA molecules is equivalent to a logical '1' and a low concentration is equivalent to a logical '0'. The logic gate is composed of the gene (with its promoter, operator and terminator), the concentration of repressor proteins and the other biomolecules required for translation and transcription (like, for example, the ribosomes that synthesize the repressor protein from the input-RNA or the RNA polymerase that synthesizes the output-RNA).

In the following we explain how this NOT gate works. Figure 2-a illustrates the case of the input '1' and output '0'. The input '1' is equivalent to a high concentration of input-RNA that results in a high concentration of gene repressor protein that stops the RNA polymerase from progressing. Consequently, the gene is inhibited for a longer time and produces a low concentration of output-RNA, or, in other words, we get '0' at the output.

Note that concentrations are said to be low not zero because repressor proteins generally never manage to inhibit the gene completely and a minimum concentration of RNA is always formed. This is termed the basal concentration.

Figure 2-b illustrates the case of the input '0' and output '1'. The input '1' is equivalent to a low concentration of input-RNA that results in a low concentration of repressor protein. Consequently, the gene is activated for a longer time and produces a high concentration of output-RNA, or, in other words, we get '1' at the output.

Genetic Circuit Operating as a Logic AND Gate

An AND gate can have more than one input and an output. If all inputs are '1', the output will be '0'. In all other cases the output is '0'. Figure 3 is an example of an AND gate with two inputs made of biomolecules (Hasty, 2002; Weiss, 2003). It represents the case where the two inputs are '1'. One of the inputs is the concentration of an inducer molecule (Input 1). The other is the concentration of an activator protein (Input 2). The output is the concentration of mRNA resulting from the gene transcription. As in the case of the NOT gate, a high concentration of biomolecules matches a logical '1' and a low concentration a logical '0'. The logic gate is composed of a gene (with its promoter, operator and terminator) and the biomolecules required for transcription (unlike in Figure 2, Figure 3 illustrates RNAp).

In the following we explain how this AND gate works. The protein is at first inactive. It needs to bind to the inducer to be activated and be able to bind to the operator. When the concentrations of the inducer and the inactive protein are high, i.e. when there is a logical '1' at both inputs, the active protein concentration is high. Then the gene is activated for a longer time. This promotes the binding of the RNAp to the promoter and yields a high concentration of mRNA, i.e. we have a logical '1' at the output.

On the other hand, when the concentration of the inactive protein, the concentration of the inducer or both concentrations are low, i.e. when

Figure 2. NOT gate

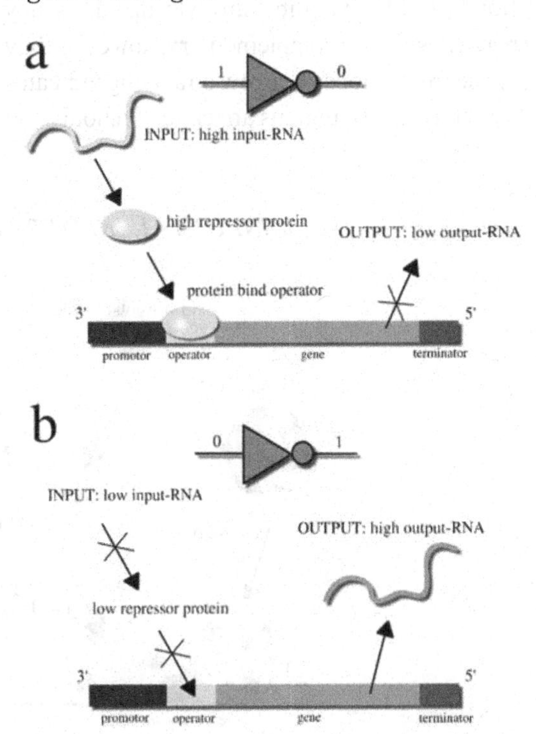

either or both of the inputs is a logical '0', the active protein concentration is low. Then the gene is activated for a shorter time, and the RNAp is less likely to bind to the promoter. Consequently, the concentration of mRNA is low, i.e. we have a logical '0' at the output.

EXAMPLE OF AN AND GATE BASED ON COMPETITIVE HYBRIDIZATION

DNA's and RNA's property of competitive hybridization can be exploited to build a range of nanosystems, like, for example, logic gates of the order of nanometres and made exclusively of nucleic acids. In the following we explain the process of competitive hybridization and the AND gate operation, but, beforehand, we define toeholds.

Toeholds

Strands *A* and *B* in Figure 4 are hybridized. As they are not absolutely complementary, however, they expose free regions (an arrow on a string indicates its 3' end). These regions are called toeholds and are useful as a recognition site and anchor point for other strands with more complementary bases (Fontana, 2006; Seelig, 2006). Toeholds play a key role in the process of competitive hybridization.

Competitive Hybridization

The process of competitive hybridization takes place between DNA and RNA strands and involves the displacement of a strand by another with bases that are more complementary and, therefore, have higher affinity to the disputed strand.

Figure 5 is an example of competitive hybridization between the strands *A'* and *B* to bind to strand *A*. *A'* is absolutely complementary to strand *A*. The process is as follows. We add strand *A'* to the hybridized strands *A* and *B* (Figure 5-1). *A'*

Figure 4. Toeholds with two strands of hybridized DNA or RNA

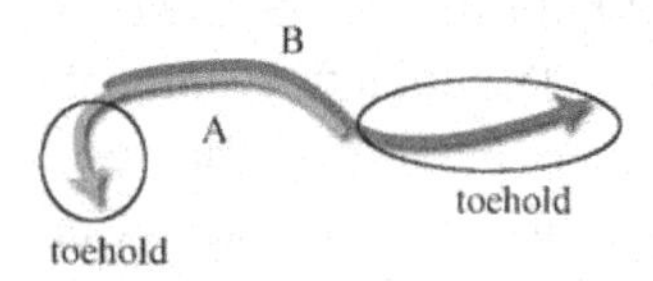

Figure 3. AND gate for INPUT 1=1, INPUT 2=1

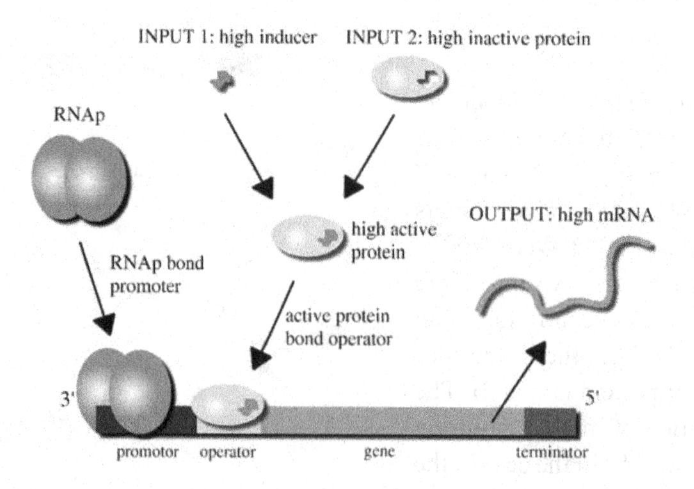

Figure 5. Example of the competitive hybridization process

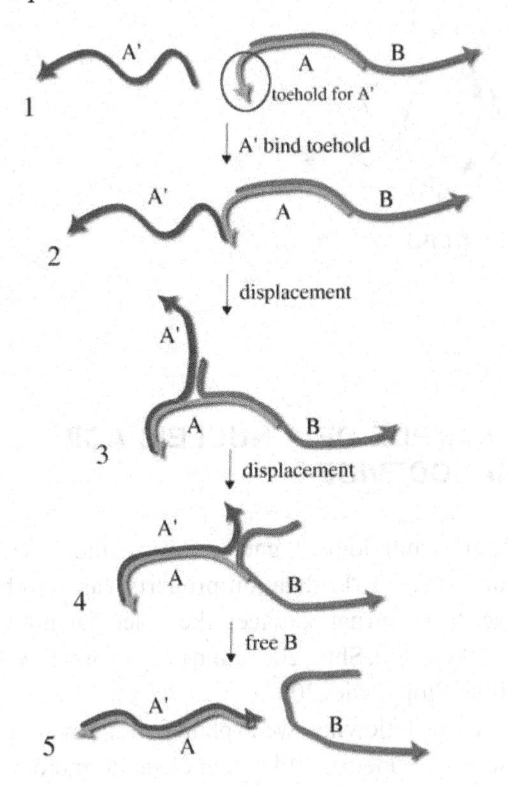

Figure 6. AND gate with two inputs for the case INPUT 1=1, INPUT 2=1

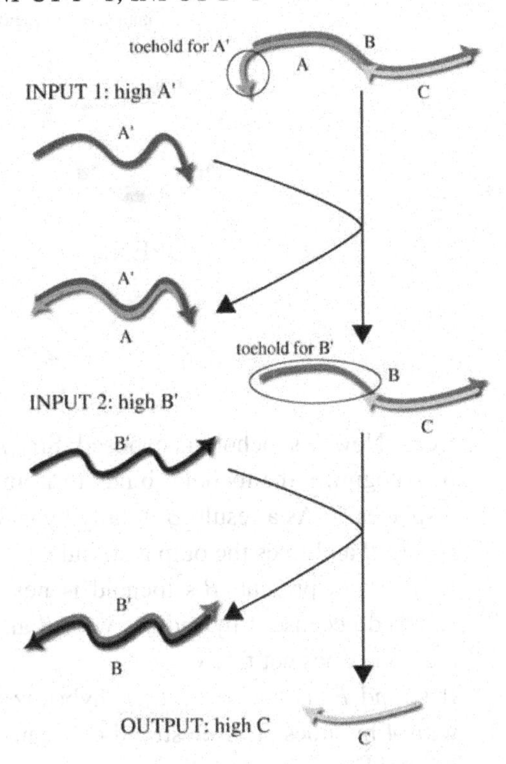

recognizes and binds to the toehold of *A* (Figure 5-2). This sets off the competition between *A'* and *B* to bind to *A* (Figure 5-3). *A'* has greater affinity to *A* and displaces *B* (Figure 5-4). Finally *A'* completely hybridizes with *A* and frees *B* (Figure 5-5).

Other structures, like loops, also play an important role in hybridization and are very useful for designing logic gates (Seelig, 2006; Takahshi, 2005). This is because a loop of two hybridized strands reduces the strength of the bond between the two strands and boosts the competitive hybridization process.

AND Gate Operation

Figure 6 is an example of the logic AND gate with two inputs made of nucleic acids only (Fontana, 2006; Seelig, 2006). It illustrates the case where both inputs are '1'. One input is the concentration of strand *A'* (Input 1). The other is the concentration of strand *B'* (Input 2). The output is the concentration of strand *C*. The gate is the concentration of the complex formed by the hybridization of strands *A*, *B* and *C*. The input strands *A'* and *B'* are absolutely complementary to strands *A* and *B*, respectively.

For individual molecules, the AND gate works as follows:

- Let *A'* and *B'* both be present. At first the toehold of strand *A* of the gate is exposed. When strand *A'* recognizes and binds to the toehold, it displaces strand *B*. The result is that *A'* totally hybridizes to *A*, and the remainder of the gate (*B* linked to *C*) is

Figure 7. Nucleic acid hairpin

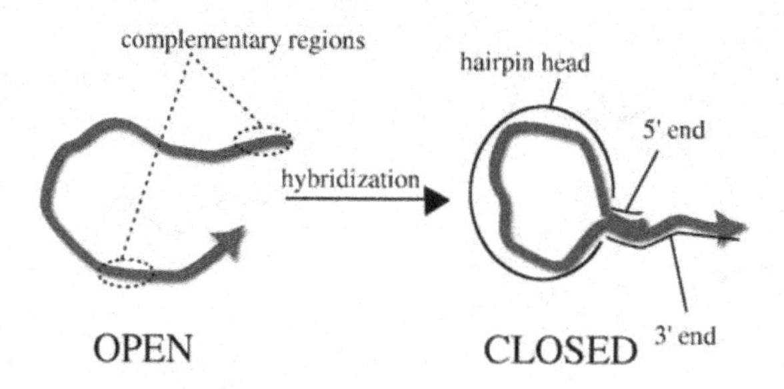

freed. Now *B'*s toehold is exposed. Strand *B'* recognizes the toehold, binds to it and displaces *C*. As a result, *B'* totally hybridizes to *B* and frees the output strand *C*.

- If *A'* is not present, *B'*s toehold is never exposed because *A* hybridizes with *B* and therefore does not free *C*.
- If strand *B'* is not present, *A'* hybridizes with *A* but does not free strand *C* because it hybridizes with *B*.
- Finally, if neither *A'* nor *B'* are present, *C* is not freed because it hybridizes to *B*.

This explanation of how the AND gate works is for individual molecules. Remember though that it is the molecule concentrations that are equivalent to the logical '1' and '0'. If the concentration of *A* is high, INPUT 1= '1', we will have a high concentration of gates with exposed toeholds of *B*. If the concentration of *B* is also high, INPUT 2= '1', we will have a high concentration of free *C*, OUTPUT= '1'. On the other hand, if either or both of the input concentrations are low, the concentration of free *C* will be low, OUTPUT= '0'.

Unlike circuits based on genetic regulation, competitive hybridization-driven circuits are more prone to modularity and scalability because the only biomolecules involved are nucleic acids, i.e. the input and output signals are of the same type (Seelig, 2006).

EXAMPLE OF A NUCLEIC ACID NANODEVICE

Apart from logical gates, DNA's and RNA's competitive hybridization property can also be used to design nanodevices like molecular motors (Yurke, 2000; Shin, 2004) and biosensors (Beyer, 2006; Stojanovic, 2004).

In the following, we explain a simple nanodevice that operates like a molecular motor. It consists of a hairpin fuelled by two strands of nucleic acids with open and closed states. The importance of this nanodevice is that it can be controlled by means of genetic regulation and work like a biosensor.

Figure 7 is a strand of nucleic acids with two complementary regions, situated at the 5' end and 3' end. In energy terms, the hybridization between the two regions is favourable. This produces a hairpin-shaped strand. This structure is very useful for computation with nucleic acids (Benenson, 2004; Beyer, 2006; Sakamoto, 2000).

In the following we explain how this DNA nanodevice works. The hairpin switches from closed to open when the gene *F* is activated and transcribes a fuel-strand *F*, as shown in Figure 8. The strand *F* recognizes and binds to the hairpin toehold *H*. *F* has a greater affinity to the 3' end and displaces and releases the 5' end. This opens the hairpin.

Figure 8. Hairpin operation

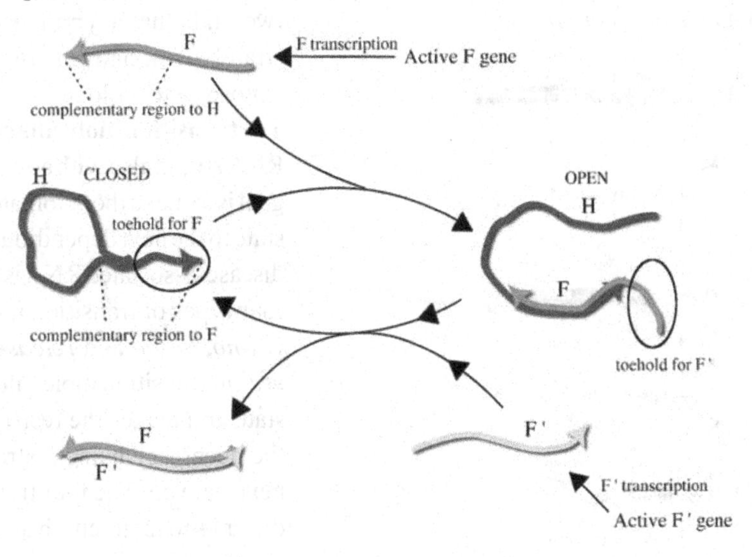

The hairpin switches from open to closed when gene *F'* is activated and transcribes the fuel-strand *F'*, which is completely complementary to *F*. Then, *F'* recognizes and binds to the toehold of *F*. It then displaces the 3' end and totally hybridizes to *F* to free the hairpin. The 3' and 5' ends hybridize and the hairpin closes. Note that each open and close cycle yields a double-stranded waste product *F-F'*. This device can operate as a biosensor that measures the concentration of proteins activating the *F* gene, for example. The concentration of the open and closed hairpins indicates the result of the measurements.

Nanodevices similar to the hairpin, like tweezers with two bonded fluorophores that indicate whether the tweezers are open or closed, have been designed and built (Yurke, 2000). When they are closed the fluorescence intensity emitted by the fluorophores is less than when they are open. Additionally, tweezers open or closed states are also controlled by genetic regulation (Dittmer, 2004; Dittmer, 2005). These new devices can measure the concentration of proteins or other biomolecules and work like biosensors.

EXAMPLE OF A BIOMOLECULAR AUTOMATON

It is now possible to program biological computers, namely, computers made mostly of DNA that work like automata, i.e. like programmed nanomachines (Benenson, 2001; Benenson, 2004; Stojanovic, 2003). A finite automaton is a computer with a finite number of states that uses transition rules to process a finite number of symbols and pass from one state to the next.

Figure 9 is a biological computer that work like an intelligent drug (Benenson, 2004). This automata is programmed to measure the presence of three different types of RNA molecules associated with a disease (RNA 1, RNA 2 and RNA 3), and, if all three are present, free an RNA strand that acts like a drug. The computer works like a finite automaton with two states: *yes* and *no*. The automaton's state is represented by the hairpin's exposed toehold in Figure 9. The possible exposed toeholds are:

Figure 9. Operation of the biomolecular automaton if all three RNA molecules are present

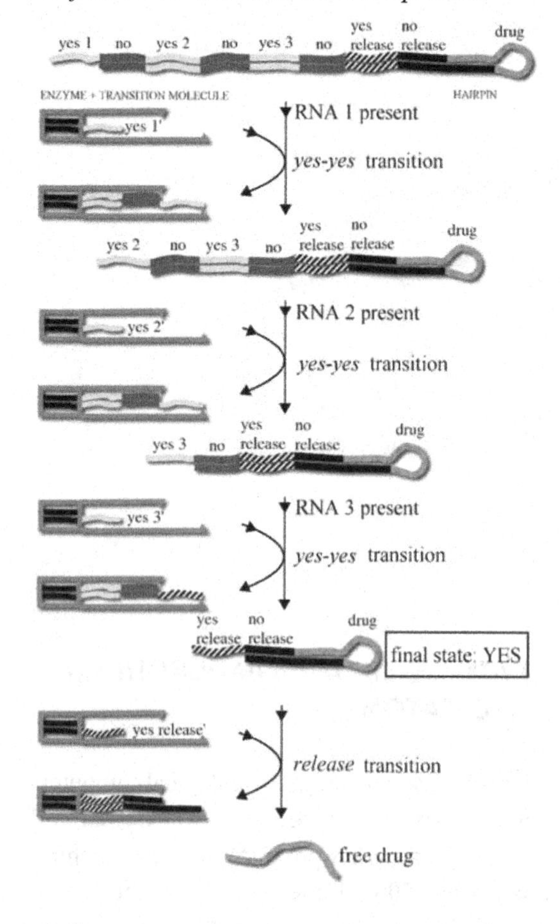

- *yes 1*, *yes 2*, *yes 3* and *yes release* that represent the state *yes*.
- *no* and *no release* that represent the state *no*.

At the end of the computation, the hairpin can only expose one of the following toeholds: *yes release* or *no release*. If it exposes *yes release*, the computation ended with the *yes* state and the drug is released. If it exposes *no release*, the computation ended with the *no* state and the drug is not released.

During the computation process, the automaton moves from one state to another through a **complex** formed by a restriction enzyme linked to a transition molecule.

- The **restriction enzyme** is a protein with two ends that asymmetrically cut the two strands of the hairpin to assure that it always exposes a toehold.
- The **transition molecule** consists of a double RNA strand also with an exposed toehold. Its goal is to make the automaton move from one state to the next depending on whether or not disease-associated RNA is present. There are four types of transition molecules: *yes-yes*, *yes-no*, *no-no* and *release*. For example, a *yes-no* transition molecule recognizes a *yes* state and makes the restriction enzyme cut the necessary hairpin strands to make the next state *no*. The four transition molecules differ as to their length and the toehold they expose.
    - The length has the function of making the enzyme ends the right size to cut the hairpin so that the toehold it exposes represents the next state.
    - The toehold has the function of recognizing the current state of the automaton from the hairpin toehold and binding to this toehold. The toehold of a transition molecule can be: *yes 1'*, *yes 2'*, *yes 3'*, *yes release'* and *no'*. These toeholds are complementary to the possible hairpin toeholds: *yes1*, *yes 2*, *yes 3*, *yes release* and *no* respectively. There is no transition molecule with a *no release'* toehold that is complementary to the hairpin's *no release* toehold. This assures that no enzyme can cut the hairpin in this case and the drug is not released.

In the following we explain how this biomolecular automaton works. In Figure 9 the hairpin's original state is *yes*, represented by the exposed toehold *yes 1*. If RNA 1 is present, a cutting enzyme linked to a *yes-yes* transition molecule with the *yes 1'* toehold appears due to different competitive hybridization processes, which, for simplicity's sake, we will not discuss at length here. This complex binds to the hairpin when it recognizes the *yes 1* toehold. The enzyme then cuts the sequence at the right site to make the next state yes represented by the *yes 2* toehold. If RNA 2 is present, a cutting enzyme linked to a *yes-yes* transition molecule with the *yes 2'* toehold appears. This binds to and cuts the hairpin to expose *yes 3*. If RNA 3 is present, the process is repeated again and the hairpin is left with an exposed *yes release* toehold. The computation ends with the *yes* state. In this case, an enzyme linked to a release transition molecule with the *yes release'* toehold appears. This complex recognizes and binds to the hairpin's *yes release* toehold. It cuts the hairpin and releases the drug. Finally, this drug specifically binds to messenger RNA causing the disease and prevents its transcription.

If any of the three RNA molecules indicating the disease are not present, a cutting enzyme linked to the *yes-no* transition molecule appears. This complex cuts the hairpin and exposes the *no* toehold. There is only one enzyme, linked to a *no-no* transition molecule, that recognizes this *no* toehold. This complex is always present and cuts the hairpin so that the next toehold is still *no*. It goes on to act like another *no-no* transition complex, and so on, until the hairpin exposes the *no release* toehold. The computation ends with the *no* state. The *no release* toehold is not associated with any transition molecule and the drug is not released.

The experiments measure the concentration levels of RNA indicators of disease rather than using individual molecules. If the concentration of all these RNA molecules is high, a high concentration of hairpins switching from one *yes* state to another is formed. This releases a high concentration of the drug. At the other end of the scale, a high concentration of hairpins that end in the *no* state and do not release the drug is formed.

This biomolecular automaton behaves like a biosensor that also measures concentrations, emits a response depending on the measurement taken. Finally, note that, in the experiments run biomolecular automata like this, no outside energy was consumed during computation (Shapiro, 2006).

FUTURE TRENDS

Possible short-term goals of the devices created by synthetic biology and biomolecular computation are:

- Diagnose and treat diseases *in vivo* by means of biomolecular information-processing devices and nanodevices.
- Assemble genetic circuits to produce devices performing more complex functions inspired by computer engineering.
- Design and construct biomolecular systems exclusively to get a better understanding of cellular processes and to verify systems biology models.
- Design and construct new devices that merge conventional electronics with the new biomolecular devices to lay the foundation of a new discipline: bionanoelectronics.

Working together with other disciplines, like systems biology, the key long-term goals are:

- Develop comprehensive models for individual cells, multicellular systems, tissues, organs and, finally, organisms.
- Design and construct devices able to oversee the operation of human tissues and take action as necessary.

2020 Science is an interesting web site about "the evolution, challenges and potential of computer science and computing in scientific research in the next fifteen years" (http://research.microsoft.com/towards2020science). Some of the milestones included are:

- Understanding complex biological systems, from cells and organisms to ecosystems.
- A computational theory of synthetic biology.
- Personalised in situ molecular-computer 'smart drug'.
- Computer science concepts now core discipline in natural sciences (as mathematics are today in physical sciences).
- Molecular computers increasingly used to design alternative 'smart' medical therapies.

CONCLUSION

The purpose of this chapter is to study the new devices that synthetic biology offers to biomolecular computation. Both disciplines come together when it comes to creating information-processing devices based on biological principles. One proof of the close relationship there is between the two disciplines is that research on synthetic biology focuses on the development of genetic circuits, which clearly fall within the field of biomolecular computation.

In this chapter, we detail a number of examples inspired by devices that have been implemented in the laboratory and published in important journals. These are three logic gates, a DNA nanodevice and a biomolecular automaton. The operation of these devices is based on gene expression regulation and other more specific biological processes, like competitive hybridization, which are also explained at length.

From an engineering viewpoint, the key goal is to design biomolecular devices inspired by the computer engineering hierarchy to perform increasingly complex tasks. And these new devices can be very useful in other disciplines like:

- Cell biology and systems biology, with a view to getting a better understanding of the operation of cell biology processes.
- Pharmacology and medicine, with a view to producing new drugs inspired by these devices. The biomolecular automaton in this chapter is an example.

FUTURE RESEARCH DIRECTIONS

Some of the most interesting future research lines in our group within synthetic biology and biomolecular computation are:

- Design new programmable biomolecular automata with new biological information-processing capabilities, like stochastic automata and biomolecular automata networks.
- Description and mathematical formalization of biomolecular processes using automata and formal language theory.
- Distributed biomolecular computation using microfluidic systems. These systems are composed of microchannels and microchambers that contain biomolecular devices capable of working together on a task.
- Modelling of cellular and neurobiological processes developing new P systems.

ACKNOWLEDGMENT

This research has been partially funded by the Spanish Ministry of Science and Education (MEC) under project TIN2006-15595, by BES-2007-16220 and by the Comunidad de Madrid (grant No. CCG06-UPM/TIC-0386 to the LIA research group)."

REFERENCES

Adleman, L. M. (1994). Molecular computation of solutions to combinatorial problems. *Science*, 226, 1021-1024.

Adrianantoandro, E., Basu, S., Karig, D. K., & Weiss, R. (2006). Synthetic biology: new engineering rules for an emerging discipline. *Molecular Systems Biology*, doi: 10.1038/msb4100073.

Amaral, L. A. N., Díaz-Guilera, A., Moreira, A. A., Goldberger, A. L., & Lipsitz, L. A. (2004). Emergence of complex dynamics in a simple model of signaling networks. *Proceedings of the National Academy of Sciences of the United States of America*, 101, 15551-15555.

Alon, U. (2007). *An introduction to systems biology: design principles of biological circuits*. Chapman & Hall/CRC - Taylor & Francis Group.

Ball, P. (2005). Synthetic biology for nanotechnology. *Nanotechnology*, 16, R1-R8.

Bath, J., & Turberfield, A. J. (2007). DNA nanomachines. *Nature Nanotechnology*, 2, 275-284.

Baum, E. B. (1995). Building an associative memory vastly larger than the brain. *Science*, 268, 583-585.

Benenson, Y., Gil, B., Ben-Dor, U., Adar, R., & Shapiro, E. (2004). An autonomous molecular computer for logical control of gene expression. *Nature*, 429, 423-429.

Benenson, Y., Paz-elizur, T., Adar, R., Keinan, E., Liben, Z., & Shapiro, E. (2001). Programmable and autonomous computing machine made of biomolecules. *Nature*, 414, 430-434.

Benner, S. A., & Sismour, A. M. (2005). Synthetic biology. *Nature Reviews Genetics*, 6, 533-543.

Beyer, S., & Simmel, F. C. (2006). A modular DNA signal translator for the controlled release of a protein by an aptamer. *Nucleic Acids Research*, 34, 1581-1587.

Cello, J., Paul, A. V., & Wimmer, E. (2002). Chemical synthesis of poliovirus cDNA: generation of infectious virus in the absence of natural template. *Science*, 297, 1016-1018.

Church, G. M. (2005). From systems biology to synthetic biology. *Molecular Systems Biology*, doi: 10.1038/msb4100007.

Ciobanu, G., Pérez-Jiménez, M. J., & Paun, G. (2006). Applications of Membrane Computing Series: Natural Computing Series. Berlin, Springer.

Condon, A. (2004). Automata make antisense. *Nature*, 429, 351-352.

Csete, M., & Doyle J. (2004). Bow ties, metabolism and disease. *Trends in Biotechnology*, 22, 446-450.

De Lorenzo, V., Serrano, L., & Valencia, A. (2006). Synthetic biology: challenges ahead. *Bioinformatics*, 22, 127-128.

Di Ventura, B., Lemerle, C., Michalodimitrakis, K., & Serrano, L. (2006). From *in vivo* to *in silico* biology and back. *Nature*, 443, 527-533.

Dittmer, W. U., Kempter, S., Rädler, J. O., & Simmel, F. C. (2005). Using gene regulation to program DNA-based molecular devices. *Small*, 1, 709-712.

Dittmer, W. U., & Simmel, F. C. (2004). Transcriptional control of DNA-Based Nanomachines. *Nano Letters*, 4, 689-691.

Elowitz, M. B., & Leibler, S. (2000). A synthetic oscillatory network of transcriptional regulators. *Nature*, 403, 335-338.

ETC group. (2007). Extreme genetic engineering: An introduction to synthetic biology. Published online January , 2007, http://www.etcgroup.org/upload/publication/pdf_file/602

Fontana, W. (2006). Pulling strings. *Science*, 314, 1552-1553.

Forster, A. C., & Church, G. M. (2007). Synthetic biology projects *in vitro*. *Genome Research*, 17, 1-6.

Gardner, T. S., Cantor, C. R, & Collins, J. J. (2000). Construction of a genetic toggle switch in *Eschiria coli*. *Nature*, 403, 339-342.

Gehani, A., LaBean, T. H., & Reif , J. H. (1999). *DNA-based cryptography*. Paper presented at 5th DIMACS Workshop on DNA Based Computers, MIT, Cambridge.

Glass, J. I., Alperovich, N., Assad-Garcia, N., Baden-Tillson, H., Khouri, H., Lewis, M., Nierman, W. C., Nelson, W. C., Pfannkoch, C., Remington, K., Yooseph, S., Smith, H. O., & Venter, J. C. (2004). Estimation of the minimal mycoplasma gene set using global transposon mutagenesis and comparative genomics. *Genomics: GTL II*, 51-52.

Goldenfeld, N., & Kadanoff, L. A. (1999). Simple lessons from complexity. *Science*, 284, 87-89.

Hasty, J., McMillen, D., & Collins, J. J. (2002). Engineered gene circuits. *Nature*, 420, 224-230.

Hartwell, L. H., Hopfield, J. J., Leibler, S., & Murray, A. W. (1999). From molecular to modular cell biology. *Nature*, 402, c47-c52.

Hutchison, C. A., Peterson, S. N., Gill, S. R., Cline, R. T., White, O., Fraser, C. M., Smith, H. O., & Venter J. C. (1999). Global transposon mutagenesis and a minimal mycoplasma genome. *Science*, 286, 2165-2169.

Isaacs, F. J., Dwyer, D. J., & Collins, J. J. (2006). RNA synthetic biology. *Nature Biotechnology*, 24, 545-554.

Kim, J., White, K. S., & Winfree, E. (2006). Construction of an *in vitro* bistable circuit from synthetic transcriptional switches. *Molecular Systems Biology*, doi: 10.1038/msb4100099.

Kitano, H. (2002). Systems biology: a brief overview. *Science*, 295, 1662-1664.

Kitano, H. (2004). Biological robustness. *Nature Reviews Genetics*, 5, 826-836.

Liedl, T., Sobey, T. L., & Simmel, F. C. (2007). DNA-based nanodevices. *Nanotoday*, 2, 36-41.

Lipton, R. J. (1995). DNA solutions of hard computational problems. *Science*, 268, 542-545.

McAdams, H. H., & Arkin, A. (1999). It's a noisy business! Genetic regulation at the nanomolar scale. *Trends in Genetic*, 15, 65-69.

Paun, G. (2000). Computing with membranes. *Journal of Computer and System Sciences*, 61, 108-143.

Paun, G. (2002). *Computing with membranes. An introduction*. Berlin, Springer.

Rinaudo, K., Bleris, L., Maddamsetti, R., Subramanian, S., Weiss, R., & Benenson, Y. (2007). A universal RNAi-based logic evaluator that operates in mammalian cells. *Nature Biotechnology*, doi: 10.1038/nbt1307.

Ro, D., Paradise, E. M., Ouellet, M., Fisher, K. J., Newman, K. L, Ndungu, J. M., Ho, K. A., Eachus, R. A., Ham, T. S., Kirby, J., Chang, M. C. Y., Withers, S. T., Shiba, Y., Sarpong, R., & Keasling, J. D. (2006). Production of the antimalarial drug precursor artemisinic acid in engineered yeast. *Nature*, 440, 940-943.

Rodríguez-Patón, A. (1999). *Variantes de la concatenación en computación con ADN*. PhD thesis, Universidad Politécnica de Madrid.

Sakamoto, K., Gouzu, H., Komiya, K., Kiga, D., Yokoyama, S., Yokomori, T., & Hagiya, M. (2000). Molecular computation by DNA hairpin formation. *Science* , 288, 1223-1226.

Seelig, G., Soloveichik, D., Zhang, D. Y., & Winfree, E. (2006). Enzyme-free nucleic acid logic circuits. *Science*, 314, 1585-1588.

Shapiro, E., & Benenson, Y. (2006). Bringing DNA computers to life. *Scientific American*, 294, 44-51.

Shin, J. S., & Pierce N. A. (2004). A synthetic DNA walker for molecular transport. *Journal of the American Chemical Society*, 126, 10834-10835.

Simmel, F. C., & Dittmer, W. U. (2005). DNA nanodevices. *Small*, 1, 284-299.

Stelling, J., Sauer, U., Szallasi, Z., Doyle III, F., & Doyle, J. (2004). Robustness of cellular functions. *Cell*, 118, 675-685.

Stojanovic, M. N., & Kolspashchikov, D. M. (2004). Modular aptameric sensors. *Journal of the American Chemical Society*, 126, 9266-9270.

Stojanovic, M. N., & Stefanovic, D. (2003). A deoxyribozyme-based molecular automaton. *Nature Biotechnology*, 21, 1069-1074.

Takahashi, K., Yaegashi, S., Kameda, A., & Hagiya., M. (2005). *Chain reaction systems based on loop dissociation of DNA*. Paper presented at the DNA 11, Eleventh International Meeting on DNA Based Computers, Preliminary Proceedings, London, Ontario, Canada.

Weiss, R., & Basu, S. (2002) *The device physics of cellular logic gates. First Workshop on Non-Silicon Computing*. Boston, MA.

Weiss, R., Basu, S., Hooshangi, S., Kalmbach, A., Karig, D., Mehreja, R., & Netravali, I. (2003). Genetic circuit building blocks for cellular computation, communications, and signal processing. *Natural Computing*, 2, 47-84.

Yurke, B., Turberfield, A. J., Mills, Jr A. P., Simmel, F. C., & Neumann J. L. (2000). A DNA-fuelled molecular machine made of DNA. *Nature*, 406, 605-608.

ADDITIONAL READING

Alberts, B., Johnson, A., Lewis, J., Raff, M., Roberts, K., & Walter, P. (2002). *Molecular biology of the cell*. New York: Garland.

Alon, U. (2007). Simplicity in biology. *Nature*, 446, 497.

Brent, R., & Bruck, J. (2006). Can computers help to explain biology?. *Nature*, 440, 416-417.

Dirks, R. M., Lin, M., Winfree, E., & Pierce, N. A. (2004). Paradigms for computational nucleic acid design. *Nucleic Acids Research*, 32, 1392-1403.

Foster, I. (2006). A two-way street to science's future. *Nature*, 440, 419.

Freitas, R. A. Jr. (2005). Current status of nanomedicine and medical nanorobotics. *Journal of Computational and Theoretical Nanoscience*, 2, 1-25.

Gibbs, W. W. (2004). Synthetic life. *Scientific American*, 290, 74-81.

Green, S. J., Lubrich, D., & Tuberfield, A. J. (2006). DNA hairpins: fuel for autonomous DNA devices. *Biophysical Journal*, 91, 2966-2975.

Klipp, E., Herwig, R., Kowald, A., Wierling, C., & Lehrach, H. (2005). *Systems biology in practice. Concepts, implementation, and application*. Weinheim: Wiley-VCH.

Mao, C., Labean, T. H., Reif, J. H., & Seeman, N. C. (2000). Logical computation using algorithmic self-assembly of DNA triple-crossover molecules. *Nature*, 407, 493-496.

Paun, G., Rozenberg, G., & Salomaa, A. (1998). *DNA computing, new computing paradigms*. Berlin, Springer.

Strogatz, S. (2000). *Nonlinear dynamics and chaos: with applications to physics, biology, chemistry and engineering*. Reading, MA: Perseus Publishing.

Szallasi, Z., Stelling, J., & Periwal, V. (Eds.). (2006). *System modeling in cellular biology: from concepts to nuts and bolts*. Cambridge, MA: The MIT Press.

Chapter XV
Guiding Self-Organization in Systems of Cooperative Mobile Agents

Alejandro Rodríguez
University of Maryland, USA

Alexander Grushin
University of Maryland, USA

James A. Reggia
University of Maryland, USA

ABSTRACT

Drawing inspiration from social interactions in nature, swarm intelligence has presented a promising approach to the design of complex systems consisting of numerous, simple parts, to solve a wide variety of problems. Swarm intelligence systems involve highly parallel computations across space, based heavily on the emergence of global behavior through local interactions of components. This has a disadvantage as the desired behavior of a system becomes hard to predict or design. Here we describe how to provide greater control over swarm intelligence systems, and potentially more useful goal-oriented behavior, by introducing hierarchical controllers in the components. This allows each particle-like controller to extend its reactive behavior in a more goal-oriented style, while keeping the locality of the interactions. We present three systems designed using this approach: a competitive foraging system, a system for the collective transport and distribution of goods, and a self-assembly system capable of creating complex 3D structures. Our results show that it is possible to guide the self-organization process at different levels of the designated task, suggesting that self-organizing behavior may be extensible to support problem solving in various contexts.

INTRODUCTION

The term *swarm intelligence*, initially introduced by Beni, 1988 in the context of cellular robotics, refers to a collection of techniques inspired in part by the behavior of social insects, such as ants, bees, termites, etc., and of aggregations of animals, such as flocks, herds, schools, and even human groups and economic models (Bonabeau, 1999; Kennedy, 2001). These swarms possess the ability to present remarkably complex and "intelligent" behavior, despite the apparent lack of relative complexity in the individuals that form them. These behaviors can include cooperative synchronized hunting, coordinated raiding, migration, foraging, path finding, bridge construction, allocation of labor, and nest construction. Past discoveries (Deneubourg, 1989) have led investigators to the belief that such behaviors, although in part produced by the genetic and physiological structure of the individuals, are largely caused by the *self-organization* of the systems they form (Aron, 1990; Bonabeau, 1996). In other words, out of the direct or indirect local interactions between the individuals, the collective behavior emerges in a way that may have the appearance of being globally organized, although no centralized control or global communication actually exists. It is precisely this self-organization that artificial swarm intelligence systems try to achieve, by infusing the components, homogeneous or heterogeneous, of a system with simple rules. Swarm intelligence presents a novel and promising paradigm for the design and engineering of complex systems, increasingly found in many fields of engineering and science, where the number of elements and the nature of the interactions among them make it considerably difficult to model or understand the system's behavior by traditional methods.

Several methodological approaches to swarm intelligence have been explored, but they often share a common feature: collections of simple entities (simulated birds, ants, vehicles, etc.) move autonomously through space, controlled by forces or interactions exerted locally upon each other, either directly or through the environment. These local interactions are often governed in a simple manner, via small sets of rules or short equations, and in some cases the sets of reactive agents used are best characterized as *particle systems* where each agent is viewed as a particle. This provides swarming systems with a sets of properties that includes scalability, fault tolerance, and perhaps more importantly, self-organization. Through this latter property, there is no need for a system to be controlled in a hierarchical fashion by one or more central components that determine the required behavior of the system. Instead, collective behavior emerges from the local interactions of all the components, and the global system behaves as a super-organism of loosely connected parts that react "intelligently" to the environment.

In our view, the self-organizing feature of swarm systems represents its main advantage and also its main disadvantage: the resulting global behavior is often hard to predict based solely on the local rules, and in some cases it can be hard to control the system, that is, to obtain a desired behavior by imposing local rules on its components. This not only can require prolonged, trial-and-error style tweaking and fine tuning, but even limits the kinds of problems that can be tackled by these essentially reactive systems.

In our ongoing research in swarm intelligence (Grushin, 2006; Lapizco, 2005; Rodriguez, 2004; Winder, 2004), we have proposed, and shown to be partially successful, an approach to overcome these limitations: the introduction of layered controllers into the previously purely reactive particles or components of a system. The layered controllers allow each particle to extend its reactive behavior in a more goal-oriented style, switching between alternative behaviors in different contexts, while retaining the locality of the interactions and the general simplicity of the system. In this way, by providing a larger, more complex set of behaviors for the particles and finer control over them, the resulting system remains self-organizing, but a form of self-guided behavior is available to the particles, producing a guided self-organizing system. This is achieved by equipping particles with an internal state, which affects their reactive behaviors. State changes can be triggered by locally detected events, including particular observations of states of nearby particles. Once particle systems have been extended with this top-down control of the particles, while keeping the bottom-up approach for the system in general, particle systems can be investigated as a means for general problem-solving.

In this chapter we present three examples of systems designed using this approach to demonstrate that it can work effectively. The first system involves foraging in a 2D world by competing teams of agents that use collective movements or flocking. This scenario presents teams of homogeneous agents with the opportunity and necessity to self-assign to smaller sub-teams that undertake specialized behavior, while working for a collective goal and competing for resources against rival teams. The second system solves the problem of collection and distribution of objects by mobile agents, also in a 2D world. This problem closely resembles the problem of collective transport, common in the robotics and mobile multi-agent systems literature due in part to the fact that it strictly requires cooperation, and optionally coordination, among the agents. The third and final system presents the self-assembly of predetermined 3D structures by blocks that move in a continuous environment with physical properties and constraints approximating those of the real world. Although no single block/agent has an explicit representation of the target structure nor has an assigned location within this structure, the global structure emerges through the systemic properties of the collection of blocks that communicate among themselves through the use of stigmergy, that is, through modification of the environment. Basic versions of these three problems, foraging, collective transport and self-assembly, are commonly found in nature among social insects, and nature has provided simple, effective strategies that work despite the relatively simple cognitive levels of such insects, which have served as inspiration for human-engineered systems that solve these and similar problems.

In all cases presented here, the particles or agents not only collectively move in a continuous world, but also collectively solve the problem at hand, leveraging the underlying self-organizing processes to achieve specific goals. These systems clearly show that our approach is apt to produce problem-solving capabilities suitable for complex, goal-oriented behavior, and that the approach is general enough to tackle widely different problems.

BACKGROUND

Interacting particle systems were initially introduced by Reynolds, 1987 as a method for the behavioral simulation of aggregations of animals, particularly fish schools and flocking birds. By modeling each individual animal as a reactive particle whose movements depend only on interactions with nearby particles, Reynolds achieved realistic simulation of hundreds of thousands of individuals that move in a life-like manner, without scripting the movement of any of them. This clearly showed that it was possible to achieve sophisticated, seemingly organized behavior in a complex system without exerting global control on its components. Flocking systems using interacting particles have been found to be useful in the study of crowds (Braun, 2003; Helbing, 2000) and in the control of mobile robots (Balch, 1998; Gaudiano, 2003; Mataric, 1995), especially when formation is an advantage, as in the case of cooperative sensing (Parunak, 2002).

After the success of particle systems in the simulation of biological systems, researchers found applications for problem solving. One of the best known examples of the application of swarm intelligence to engineering problems is particle swarm optimization (PSO). PSO (Eberhart, 1995; Kennedy, 1995; Kennedy, 2001) extends particle systems to high-dimensional abstract/cognitive spaces based on a social model of the interactions between agents. PSO performs numerical optimization by spreading a group of agents or particles in the solution space of the problem and letting all particles move simultaneously through the space looking for an optimal solution. The movement of the particles is influenced by their individual and collective memories (particles tend to return to the best observed solutions). This simple strategy produces an organized, parallelized heuristic search of the space that has proven effective in many problems (Eberhart, 2001; Kennedy, 2001). A different but equally successful application of swarm intelligence is represented by ant colony optimization (ACO), which solves combinatorial optimization problems by imitating the mechanism used by ants for path-optimization (Beckers, 1992; Burton, 1985). When presented with multiple paths between the colony and a food source, ants will eventually settle on the shortest path. They do this without knowledge of the entire path or paths by any individual ant, but by collectively marking the environment with chemical signals, pheromones, that other ants can recognize, creating a form of communication by the modification of the environment, known as stigmergy. The idea of several entities collaborating through stigmergy inspired a meta heuristic for solving problems of combinatorial optimization (Dorigo, 1996; Dorigo, 1999). The key points of this heuristic could be briefly summarized as having a set of agents that individually take decisions probabilistically, and create marks based on the appropriateness of these decisions that can later be used by the same and other agents to take later decisions. This heuristic has been successfully applied to problems like the Traveling Salesperson Problem (Dorigo, 1996), which is a well-known study case for combinatorial optimization, but has been more commonly applied, along with similar algorithms, to routing, load balancing and related problems in networks (Dorigo, 1999; Caro, 1998; Schoonderwoerd, 1997).

COMPETITIVE FORAGING

Inspired by the task of foraging, traditionally employed in robotic systems as a test to their cooperation abilities, competitive foraging calls for adversarial teams of agents to collect small units of resources, or minerals, concentrated in *deposits* throughout a continuous 2D surface and store them in the team's *home* base, while preventing rival teams from raiding their home or monopolizing the minerals (Figure 1).

In this problem each agent is independently governed by a two-layer controller architecture. The bottom layer (local information plus movement dynamics) controls the reactive behavior of the agent, making instantaneous decisions about the actions to be performed. It takes input solely from the local environment at a given instant, and outputs a corresponding movement based on the immediate goals of the agent and current (local) state of the environment, resembling past particle systems (Reynolds, 1987, Tu, 1994). The top layer, not found in past particle systems, consists of a very limited memory and a finite state machine (FSM) that directs agent behavior in a top-down fashion, modifying the movement dynamics used over time. For example, if the FSM decides that it is time for the agent to go home, it will switch to the state *carrying* and provide the bottom layer with the target location of its home. The bottom layer will then determine at each step the steering actions needed to properly navigate from the current location to the home. Since the bottom layer is mostly reactive it can temporarily override the long term goal of going home for a more pressing need, such as avoiding a competing agent or obstacle.

The state of each agent represents its high-level goals and intentions and varies according to a FSM (Figure 2), switching in response to events

Figure 1. A small section of a 300x300 continuous world with a single "mineral deposit" on the right and two different teams (dark and light arrows) exploiting it. A unit of distance is the maximum distance an agent can move during a time-step. Teams' homes are denoted by small solid squares and mineral units as spots. The dark team agents are returning home carrying minerals (spots adjacent to arrows). Most agents of the light team are returning to the deposit after unloading minerals at home, but some of them are simply exploring.

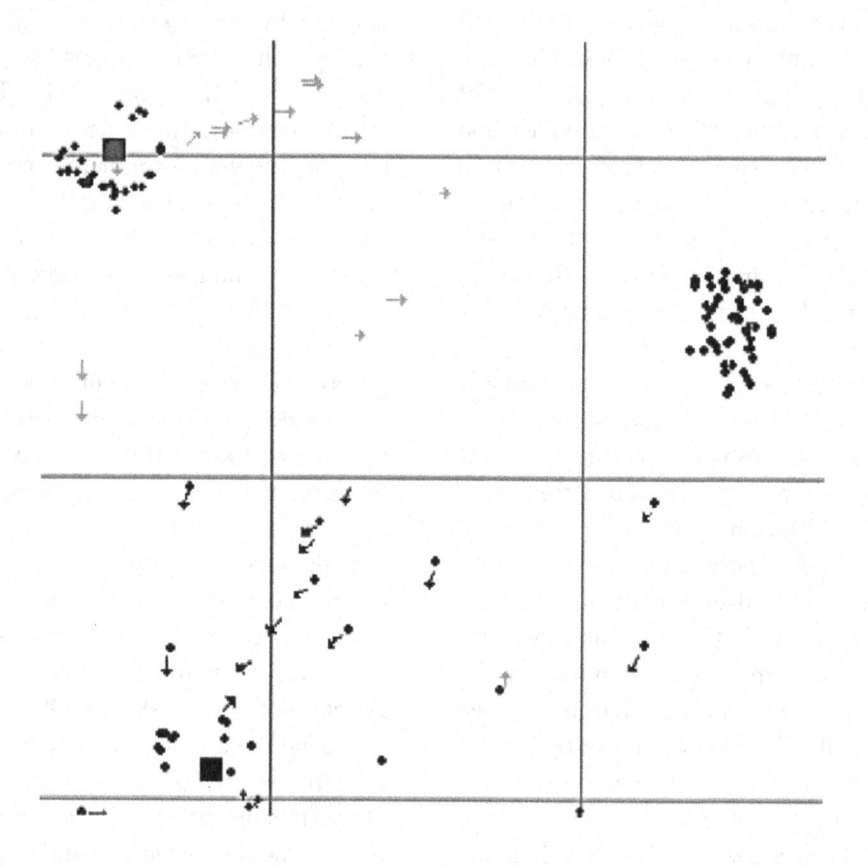

in the environment or the achievement of goals. This state governs the low-level reactive behaviors and dynamics, assigning different controllers to different situations. Movements are governed by a set of individual influences inspired by the flocking behavior of migratory birds (Reynolds, 1987, Reynolds, 1999) (avoiding an obstacle or competing agent, staying with the flock, keeping the same distance from other agents in the flock, etc.) that produce an instantaneous velocity in a desired direction and are combined in order of priority. By changing which individual influences are combined and their relative weights, a large variety of movement behaviors can be implemented, each one associated with a different state or goal, guiding the underlying self-organizing process (Rodriguez, 2004). Notice that the memory requirements of an agent are limited to the space required to store a predetermined number of locations, e.g. the agent's home and the location of the most recently observed deposit of minerals.

This two-layer controller allows the designer of the system to directly incorporate the logic of

Figure 2. FSM of an agent showing its states and the movement behaviors associated with each state. States are represented by circles labeled by <State/associated controller>, while arrows represent transitions labeled by the triggering event. The initial state is marked by a filled circle.

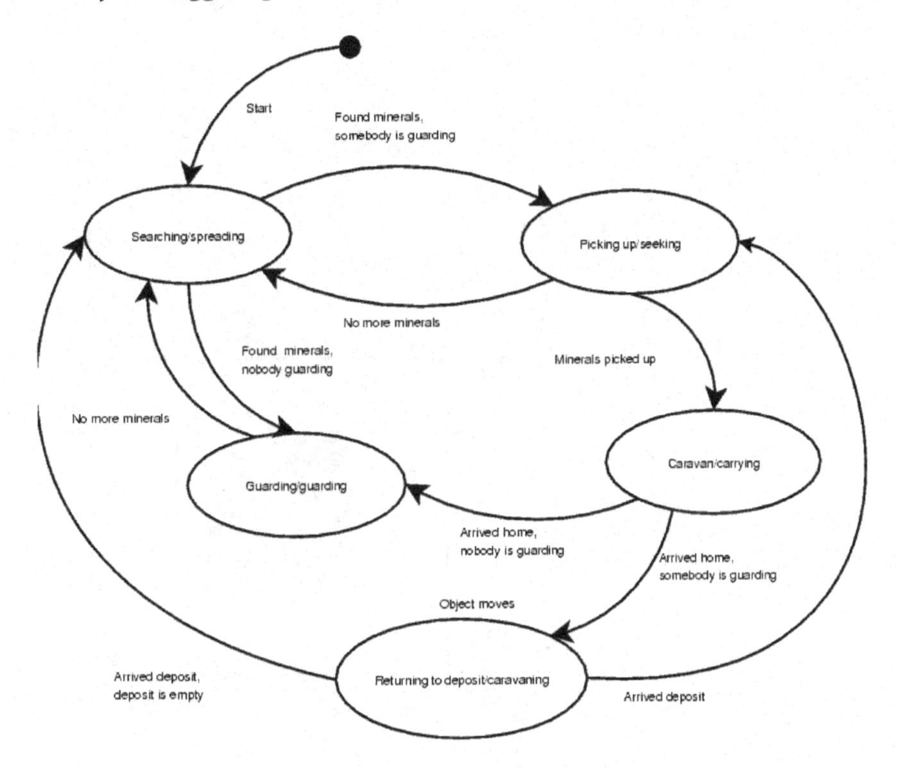

the desired global behavior into each individual agent, guiding the self-organizing process without using global interactions or global controls. That is, by indicating how the short-term goals of an agent should change according to its general (but local) situation, it is possible to guide groups of agents into following a general plan or step-by-step strategy from a higher level view of the problem. Although the process of manually designing an FSM controller requires not only good knowledge of the problem, but also to account for hard-to-predict interactions of the agents that can affect the global behavior, being able to directly represent the logic of the solution from a top level greatly eases the difficulty of producing a desired system-wide behavior.

In this particular case, to each state in the FSM is associated one of four different *movement behaviors* (Rodriguez, 2004). Briefly, *spreading* tends to form a flock slightly similar to the broad, V-shaped formations that migratory birds adopt (Figure 3a). This formation is ideal for searching for deposits, since upon seeing a deposit, an agent will tend to "pull" the rest of the flock to it via local interactions. *Seeking agents* go after a target, for example a unit of mineral, while avoiding obstacles and other agents (Figure 3b). *Caravaning agents* move in a pattern similar to spreading agents, but they are homing on a particular point in space (e.g., a previously visited deposit) and thus have a narrower front (Figure 3c). Finally, *guarding agents* patrol a deposit (Figure 3d), distributing themselves about evenly through the deposit. The interaction between nearby guarding agents and even non-guarding agents make them keep moving, effectively orbiting the deposit.

Figure 3. Collective movements. (a) Spreading agents (arrows), are located roughly side by side. (b) Two seeking agents head to the closest mineral unit (gray spots). (c) Caravanning agents tending to align. (d) Five guarding agents patrolling a deposit.

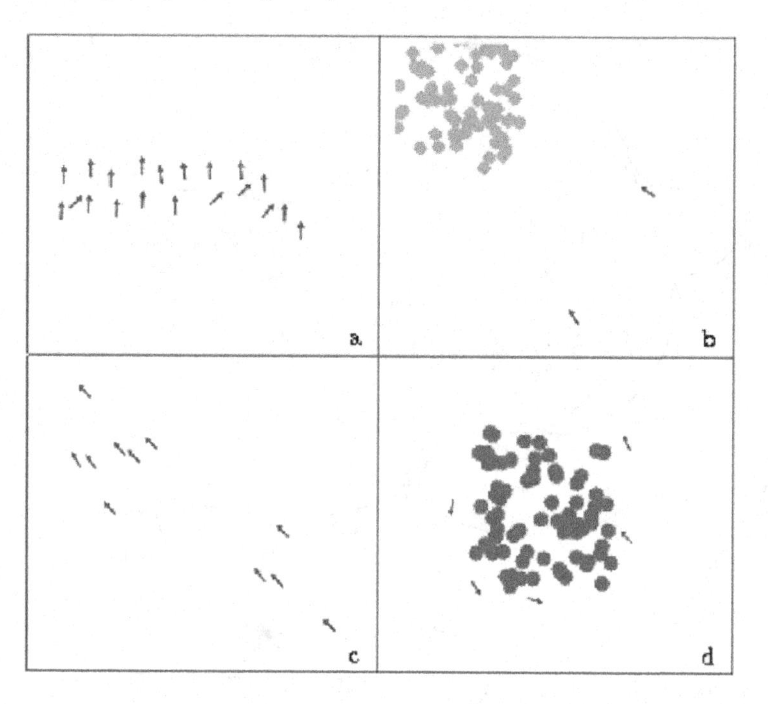

To assess the effectiveness of collective movement behaviors, we allowed the flocking team to compete against independent movement teams in simulations. The independent moving agents follow the same model explained above, with the important exception that they do not take into account other agents in their dynamics (although they still avoid collisions). Additionally, we implemented three different strategies employed by the agents to protect the minerals and compete against their rivals: (1) Full guarding, in which agents guard their homes and any deposits they find. (2) Home-guarding, where agents will not guard a deposit, but will still guard their own home. (3) Non-guarding, the strategy in which agents do not have a guarding state.

When all teams compete against each other simultaneously, the clear winner is the home-guarding flocking (hf) team. Experimental details are omitted here due to space constraints, but they are presented in full in Rodriguez, 2004. As shown in Figure 4, the amount of resources collected by this team increases monotonically over time. Early in simulations (during the first 5,000 steps), both this and the non-guarding flocking team (nf) collect minerals faster than any other team. After the first few thousand steps, the explored area of each team is wide enough for teams to find each others' homes.

Accordingly, the amount of mineral decreases in subsequent iterations for most teams, especially the non-flocking teams. The differences in the mean amount of collected minerals by each team are statistically significant at the level of 95% according to a two-way ANOVA, both in sociality (flocking vs. independent) and guarding strategy (full-guarding, home-only and none). These data suggest two main hypotheses. First, teams of col-

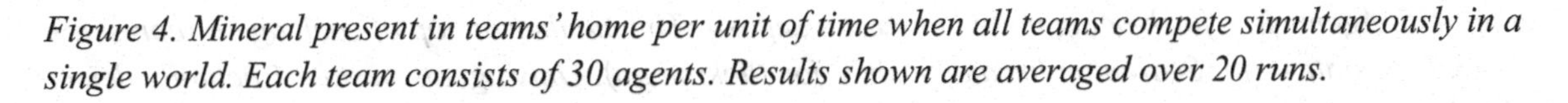

Figure 4. Mineral present in teams' home per unit of time when all teams compete simultaneously in a single world. Each team consists of 30 agents. Results shown are averaged over 20 runs.

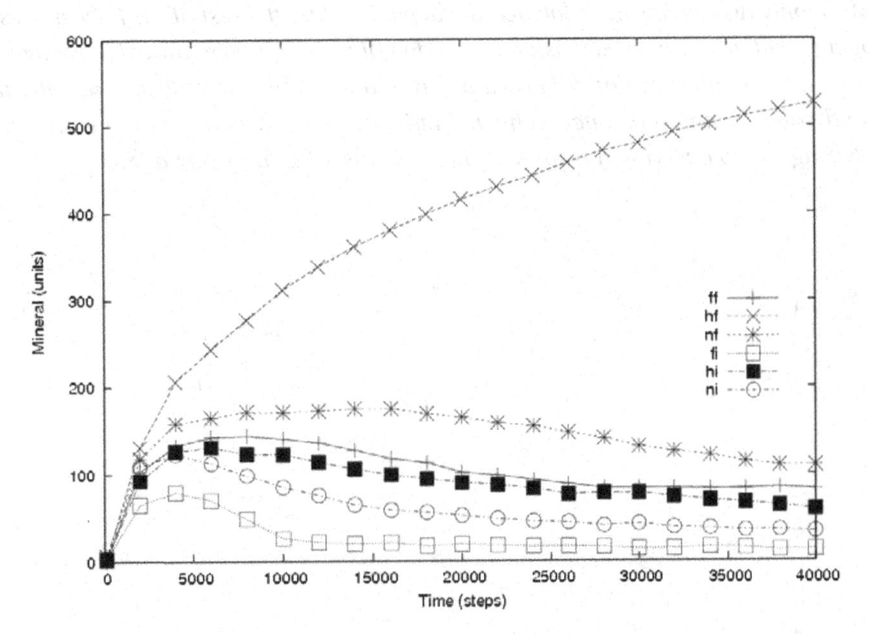

lectively moving agents are more effective at this task than corresponding teams of independently moving agents. With collectively moving agents, whenever a deposit was discovered by an agent, numerous other agents that were unaware of the deposit were immediately nearby and thus pulled in by local inter-agent influences to help collect the discovered minerals (e.g., see Figure 5). Second, for both collectively and independently moving agent teams, agents that guarded only their home did better than non-guarding agents, who in turn did better than full-guarding agents. Presumably allocating agents to guard resources, especially multiple deposits, has a large cost: it removes these agents from collecting minerals, and reduces the size and effectiveness of sub-teams of raiders. This loss is not adequately compensated for by any protective influences they exert through their blocking actions.

The impact of collective versus independent movements on agent teams can be clarified by varying just that factor between two competing teams present by themselves. Simulations of this sort (not illustrated here) make it clear that the flocking teams are always faster in accumulating minerals. Even more striking, the independently moving teams are not sufficiently effective in protecting their homes from being looted by the flocking team, and their collected minerals considerably decrease over time.

COLLECTIVE TRANSPORT

Our second example also uses flocking-like movements controlled by an FSM, but now examines the effectiveness of this architecture when applied to the problem of collective transport. In this problem, as in the competitive foraging task, agents need to collect a particular material and deposit it in predetermined locations, but this time the units of the product to be collected must be pushed or pulled simultaneously by more than one agent in order to move, which makes this a problem

Figure 5. Agents in a flock being pulled toward a deposit. The number on top of each agent represents its current state (0 for Searching for a deposit, 1 for Picking up). Only agents in state 1 actually detect the deposit. At a, only two agents have located the deposit, while the rest of the flock moves northward. At b and c, agents that are near the deposit but that do not yet see it turn toward those agents that have seen the deposit and are already going toward it. From d to f, the whole flock gradually turns toward the deposit and collects minerals. Such behavior indicates an advantage of collective movements in recruiting other agents to carrying a resource when it is discovered by just a few.

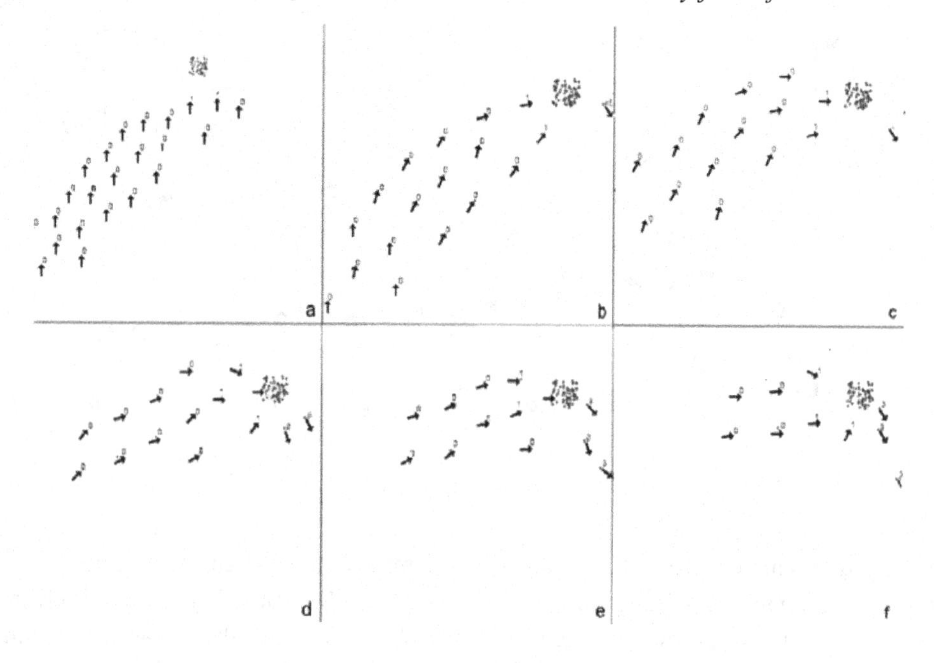

that strictly requires cooperation. Additionally, there are multiple destinations, surrounded by obstacles, where the product might be taken, forcing the agents to choose a destination for every unit of product in a way that will distribute the product equally among the destinations (Figure 6).The product (shaped as two-dimensional rods) is generated by sources at regular intervals and deposited in a random location inside the source's neighborhood. Thus, an infinite supply of product is available for the agents.

As before, an FSM and a simple memory control the high-level decisions of an agent while a basic set of reactive behaviors is associated with every state. This memory works basically as a look-up table, with the positions of the destinations preloaded in the rows of the table at the beginning of the simulation. Along with the position of a destination, agents store their individual beliefs about the amount of product present in a given destination and a time-stamp of the moment when the information was collected. This information is acquired by directly counting the amount of product in the neighborhood when an agent visits the destination, or by exchanging memories with neighboring agents, preserving the most recent memories on a row by row basis. With this information, agents select as destination for the product the destination with the lowest amount of it, which is believed to be most in need.

The FSM controlling the agents is shown in Figure 7. It essentially indicates that agents move in a semi-broad searching pattern looking for product. Once they find it, they try to transport it to a chosen a destination, which will be the destination with the lowest level of product according to the beliefs of the first agent to pick up the product. Of particular interest is the stagnation recovery process implemented by states *moving to destination*, *retrying to move* and *moving away.* If the product does not move, the agent will retry for a short time to move the object. This time is usually long enough to allow other agents to join the effort or to gradually change the direction they are pulling. However, if the problem persists, for example if the product has hit an obstacle or agents are pulling in different directions, the agent will drop the product and move away from it. When an agent detaches from an object, depending on the relative position of the product to the agent, the cohesion behavior often pulls the agent towards other agents carrying the product and makes the former go back to the product. This results in agents reallocating along the product and altering the forces applied to it. This strategy has been observed in ants when recovering from deadlocks and stagnation during the cooperative transport of large prey (Kube, 2000; Sudd, 1960; Sudd, 1965).

The collective transport problem presents several opportunities to model collective behaviors, including movements (Rodriguez, 2005). While transporting the product, agents not only need to agree on the destination, they also have to agree on the path to it. Since they are located at different positions around the object, they might

Figure 6. Layout of the obstacles, sources and destinations in a collective transport world. Sources are spots marked with an 'S' while destinations are marked with a 'D'. Lines are obstacles.

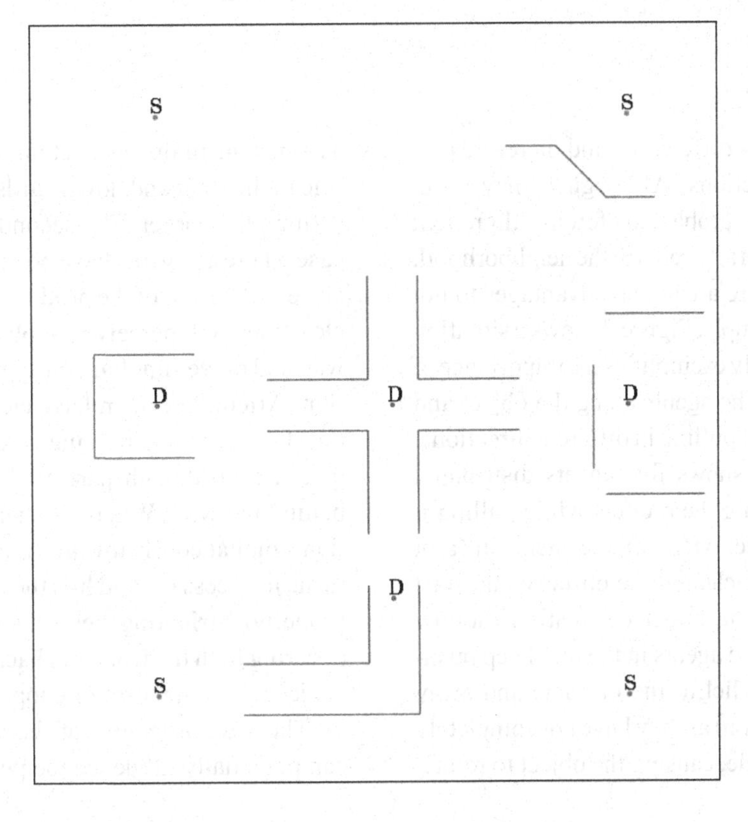

Figure 7. FSM of agents. States are shown as labeled circles while transitions are depicted as arrows. Each transition is labeled as Event/action where event triggers the transition while action is an optional action to be performed by the agent only once during the transition. The initial state is marked as a filled circle.

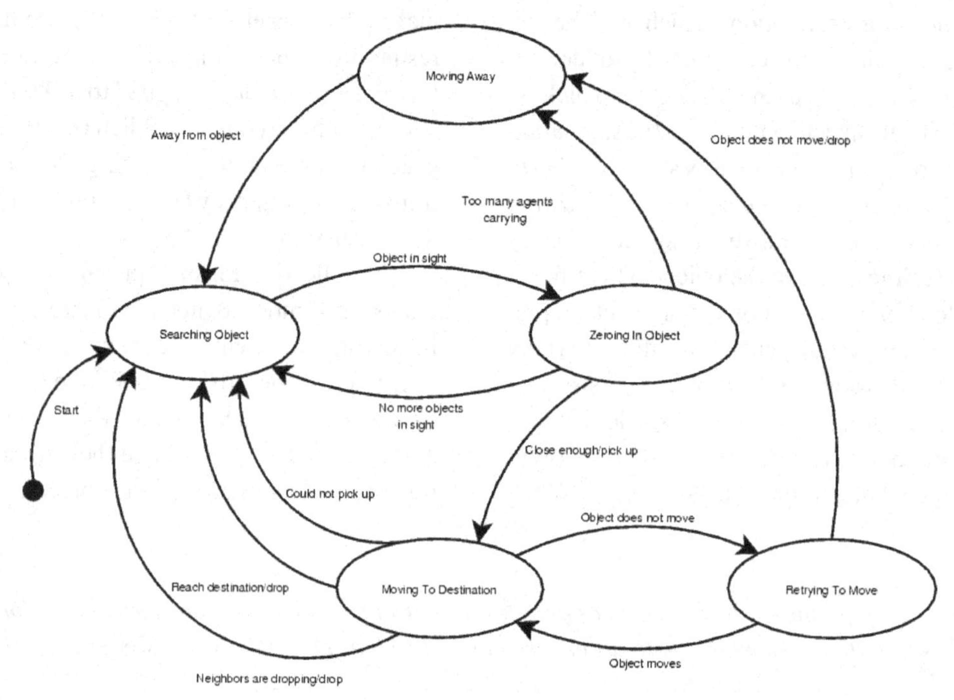

perceive obstacles differently, and therefore pull in different directions. Although it may seem trivial to solve this problem by forcing alignment (same direction) in all agents in the neighborhood, we found that there are actual advantages to not doing so. For example, Figure 8 shows a situation that simultaneously exemplifies the importance of the allocation of the agents along the object and the contribution of pulling in different directions. The left column shows five agents distributed about equally along the product while pulling it around an obstacle. After (b), the agents in front of the object have cleared the obstacle and start pulling towards the target destination (hollow circle); however the agents in the rear keep pushing forward (parallel to the obstacle and away from the destination) as they have not completely cleared the obstacle, causing the object to rotate. This results in the product moving in a diagonal line to the right and downwards (c), while slowly turning the corner. The second column shows a case where all agents have positioned themselves close to the front of the product. Since they are so close they will perceive the obstacle in the same way and move simultaneously in the same direction. After all of them have cleared the obstacle (f), they will start moving directly towards the target, even though part of the product is still behind the wall. When all agents change direction simultaneously towards the target, the object turns in excess (g) and hits the wall (h). Thus, by properly distributing themselves along the object, covering both the front and back end of it, agents achieve a simple form of cooperation.

The size, or length, of the pieces of product can potentially influence the performance of the

Figure 8. Agents moving a product around an obstacle. The product is represented as a hollow bar, while the target destination is a small circle. The agents pushing the bar are represented as small arrows superimposed on the bar. The first column (a-d) and the second column (e-h) show two different time sequences (see text).

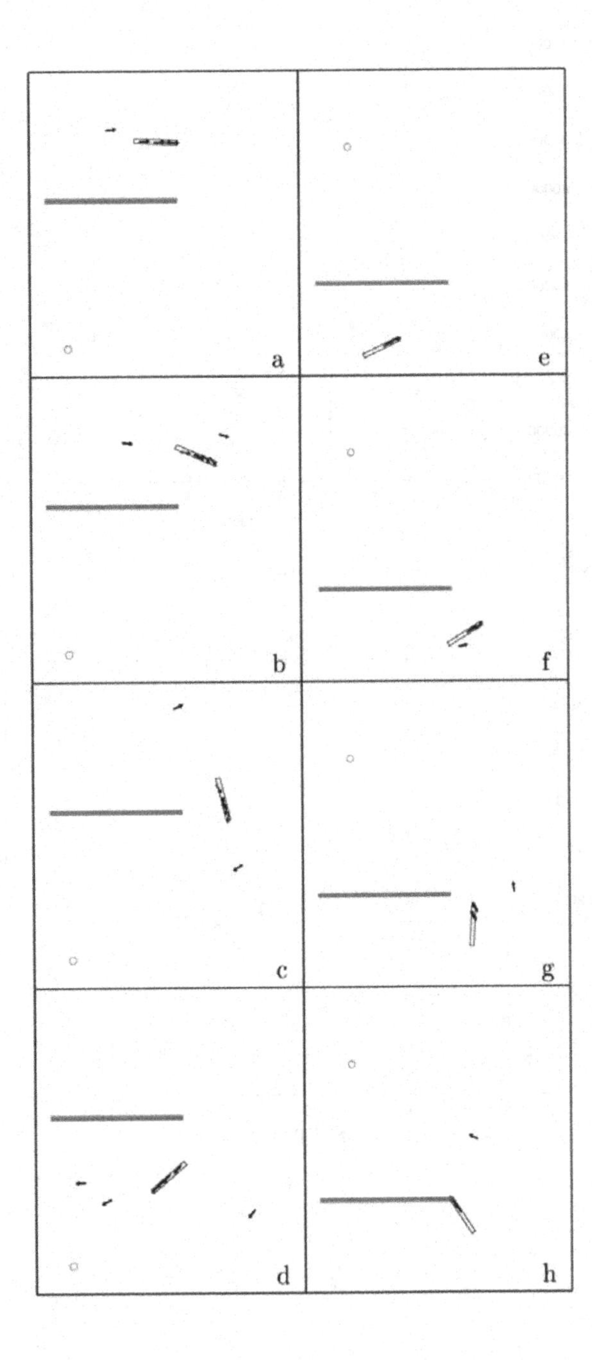

Figure 9. Results of simulation for teams of 60 agents with different sizes of product, averaged over 20 runs. (a) Time required to complete the task (shown with standard deviation) vs. product size. (b) Number of undelivered pieces by the end of the simulation (shown with standard deviation) vs. product size.

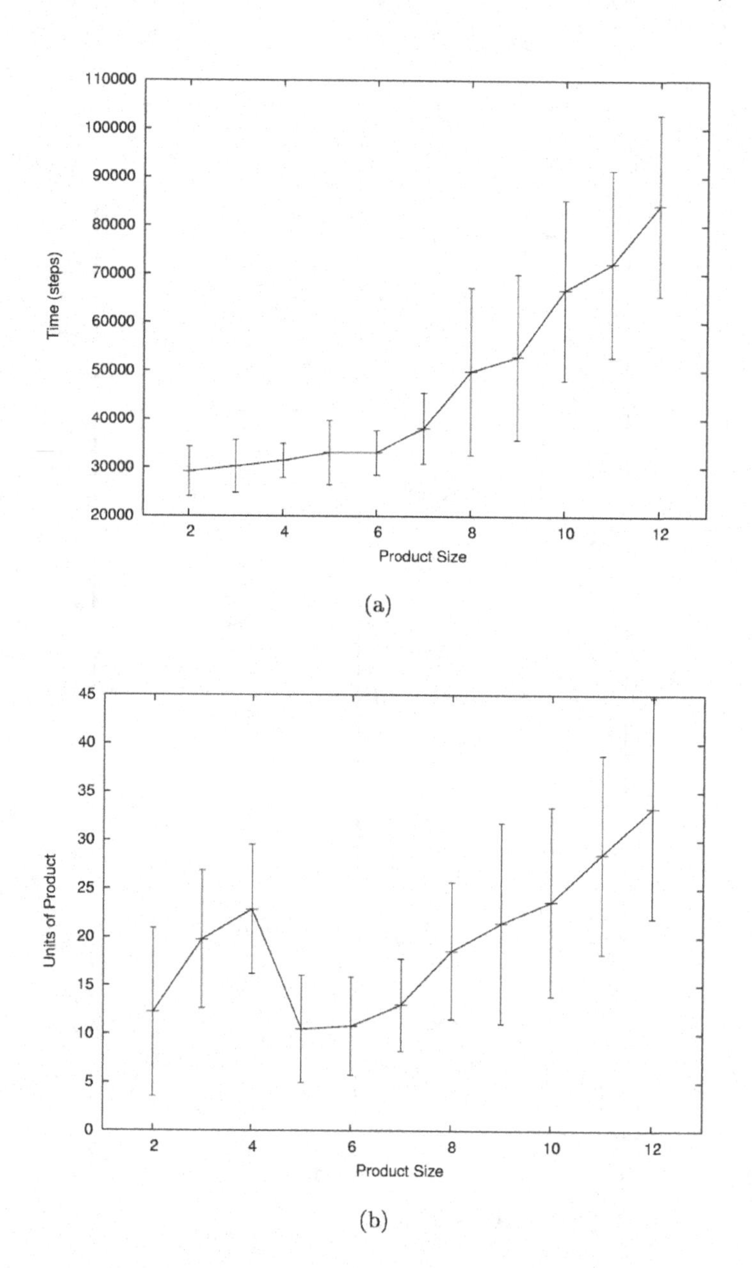

agents in two different ways: (1) by affecting the weight of a piece (proportional to its size), and (2) by affecting the number of agents that can evenly attach to it.

In simulations, when agents are tasked with delivering at least 10 pieces of product to each destination, the time required to complete the task increases depending on the size of the product (Figure 9a), as expected due to the increased weight of the pieces. However, the number of pieces dropped at unintended locations (failed deliveries) is higher for short pieces than for medium-length pieces (Figure 9b). This is probably caused by the fact that larger pieces can be carried by a larger number of agents whose distribution along the product gives them essentially different visual fields, allowing the kind of coordination explained before, making medium-sized pieces easier to transport than short pieces. Experimental details are given at Rodriguez, 2005.

SELF-ASSEMBLY

Our final example, the self-assembly problem, entails designing a set of control mechanisms that individual agents/components can follow in order to form some desired 3D *target structure*. This problem has received considerable attention in recent years (Adam, 2005, Fujibayashi, 2002, Jones, 2003, Klavins, 2002, White, 2005), as its study can provide a better understanding of the fundamental relationship between individual actions and emergent, systemic properties (Camazine, 2001).

An important class of existing approaches to self-assembly is inspired by the nest construction behaviors of paper wasps (Bonabeau, 1999, Theraulaz, 1995, Theraulaz, 1999), where the agents and components are physically distinct. While an individual wasp is not believed to "know" the global properties of the nest that it builds, this nest emerges via *stigmergy*, an implicit form of communication through the environment, where existing local arrangements of nest elements can trigger the deposition of additional material, which, in turn, creates new arrangements, resulting in a self-coordinating construction process. While a modified form of stigmergy has been successfully used for assembling a wide range of structures in an idealized, discrete space (Adam, 2005, Jones, 2003), extending these methods to more complex environments having continuous motion and physical constraints has proved to be much more difficult. In order to overcome this challenge we followed the same underlying approach as in previous sections, and integrated low-level stigmergic behaviors with reactive movement dynamics and with higher level, FSM-like coordination mechanisms (Grushin, 2006).

The basic agents/components of the self-assembly processes that we study are small (1×1×1), medium (2×1×1) and large (4×1×1) blocks, which operate in a continuous world. Initially scatted at random on the ground, blocks arrive and come to rest adjacent to a *seed* or other already stationary blocks. Individual blocks have no prespecified locations within a structure, making blocks of the same size interchangeable. Furthermore, a block's decisions must be based strictly on local information, since blocks can only detect other blocks if their centers fall within a local neighborhood.

The task of assembling some desired structure is made more challenging by the presence of physical constraints within the simulated environment. Importantly, blocks are impenetrable. Furthermore, they are subject to a gravity-like constraint in the form of a restriction on vertical movement, necessitating the assembly, and subsequent disassembly of temporary *staircases*. A block can increase or decrease its vertical position by a single unit by climbing onto or descending from a stationary block.

Subject to these constraints, and similar to the agents in the preceding sections, a block influences its movement by defining an internal "force", which directly affects its acceleration, velocity and position. The net force is computed as a non-linear, weighted combination of a number of component forces, such as cohesion, alignment, various repulsive influences (for avoiding obstacles), random forces for resolving situations of persistent interference, attractive/repulsive forces for guiding blocks up/down staircases, and attractive forces for approaching local goal locations.

These goal locations are determined via the use of *stigmergic rules* (Adam, 2005, Bonabeau, 1999, Theraulaz, 1995) whose antecedents are matched against local arrangements of blocks already part of the emerging structure. While no block b has an explicit representation of the target structure or a preassigned unique location within it, this structure is implicitly encoded by these rules. If a part of the emerging structure falls within b's local neighborhood, b will match it against the antecedents of stigmergic rules, to determine whether there is some adjacent location where b itself can belong. For example, consider the rule illustrated on the left side of Figure 10, which is used by medium blocks for the construction of a temporary staircase (shown on the right side) employed during the assembly of a building-like structure. The two locations in the center of the rule are unmarked, denoting that they must be empty, in order to leave room for a depositing medium block (solid rectangle). Solid line squares indicate that stationary blocks must exist just "behind" this goal location, both at the same level and above. Dashed lines denote "wildcard" locations, where a block can be present or absent. This rule creates a natural procedure (somewhat similar to Hosokawa, 1998) for stair assembly, where the steps are deposited as needed: since a

Figure 10. The use of stigmergy for staircase construction. On the left, a stigmergic rule is illustrated, with the antecedent (to the left of the arrow) showing the arrangement around a potential goal location before the matching block deposits itself, and the consequent (to the right of the arrow) showing the arrangement immediately after. The rule is drawn in three horizontal cross sections above (top), at the same level as (center), and below (bottom) the goal location. Further explanations are given in the text. On the right, the block marked with an asterisk (*) is about to deposit itself into a goal location determined by this rule, as a step of the currently assembling layer of the staircase.

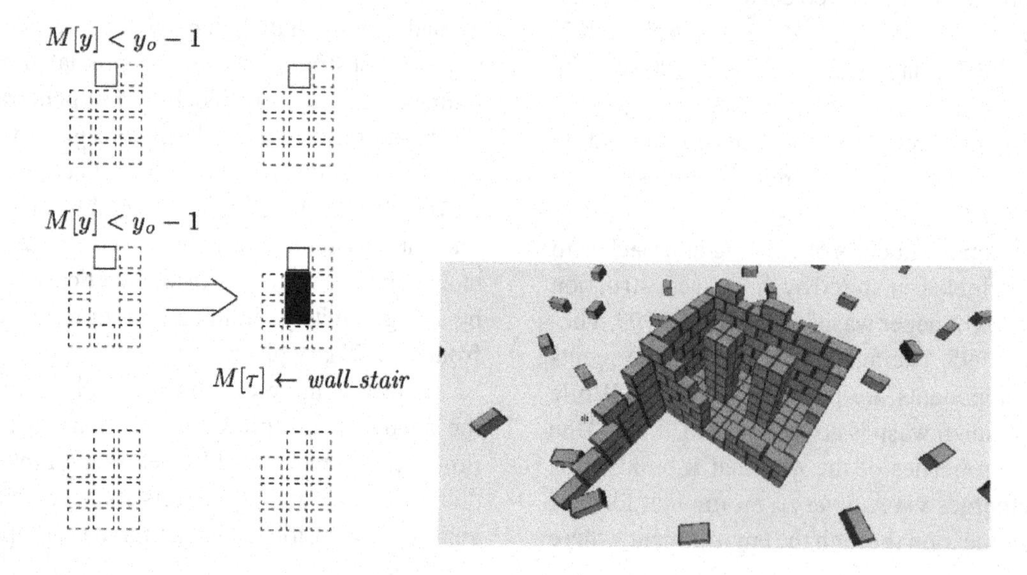

barrier that is two or more blocks in height cannot be scaled, a block deposits itself just in front of that barrier, creating a step that other blocks can climb. Other steps will be deposited using this same rule, and in this way a new "layer" of the staircase is laid down, beginning with the top step and propagating down the emerging staircase, as shown on the right side of Figure 10. Irrespective of its height, the assembly of a staircase can thus be accomplished in a very parsimonious fashion, with a single rule.

It should be noted that the staircase will not be built at a wall location other than what is shown in Figure 10, due to the requirement that the three vertical grid locations to the "upper left" of the goal location must be empty. However, it is still possible that blocks will begin to assemble staircases adjacent to the internal columns, rather than the wall. This illustrates a limitation of a stigmergic approach that relies strictly on geometric pattern matching. We developed an extension of earlier stigmergic models (Bonabeau, 1999, Theraulaz, 1995) to arbitrary structures, by allowing a block to have a limited memory *M*, and augmenting it with three integer variables $M[x]$, $M[y]$ and $M[z]$, for construction in three dimensions. A seed block is assigned arbitrarily chosen constants x_0, y_0 and z_0 as values. As another block *b* deposits itself adjacent to the emerging structure, it reads the memory of a nearby stationary block *b*' and sets its own variables to the correct relative position within the overall structure; for example, if *b* places itself directly to the left of the seed block, it can set $M[x]=x_0-1$, $M[y]=y_0$ and $M[z]=z_0$. A coordinate system thus emerges in this fashion, through simple, purely local interactions (Grushin, 2006, Werfel, 2006). The columns' positions relative to the seed are such that their blocks have values of $M[y]$ that are either y_0 or y_0+1. We thus prevent the placement of staircase blocks next to the columns by imposing the *memory condition* $M[y]<y_0-1$ on blocks adjacent to the goal location, within the rule's antecedent (Figure 10). The rule's consequent states that when a block reaches a goal location that was matched by this rule, and becomes stationary, it sets the value of the *substructure type* variable $M[\tau]$ to *wall_stair*. This allows blocks to determine the part of the structure to which another block belongs. As blocks come to rest at different locations in the environment, their memory variables become markers for other blocks, somewhat similar to the pheromone markers employed in other systems. In this way a form of stigmergy is achieved.

The presence of physical constraints such as gravity and block impenetrability imposes higher-level ordering constraints on the self-assembly process; for example, the columns of the building must be assembled prior to the walls. Furthermore, the assembly and disassembly of staircases has to be appropriately sequenced. To address this, we once again integrate low-level behaviors (namely, stigmergic pattern matching and movement) with a higher level coordination scheme. Specifically, these behaviors are influenced by a block's *mode*. Mode changes generally occur in response to the completion of certain parts of the structure. These locally detected completion events are communicated via *message* variables: when a block sees that a message variable of a nearby block is set to some particular value, it may choose to set its own message variable to that value. Such modifications to *M* are governed by *variable change rules*, which are triggered by observations of particular variable values within the memory of *b* itself, as well as the memories of nearby blocks. Implicitly, these rules define a finite state machine, somewhat akin to those discussed in previous sections. Each state is a mode value that corresponds to the specific subset of stigmergic rules that is applied when attempting to match a pattern, as well as a particular form of movement dynamics.

The methodology above has been successfully applied in simulations towards the self-assembly of a number of specific target structures. Full details are presented at Grushin, 2006. While different structures required the design of distinct sets

Figure 11. The self-assembly of a building at various points in time: (a) t = 151 time steps: floor assembling; (b) t = 919: columns assembling on completed floor; (c) t = 2253: columns complete and left column staircase disassembling; (d) t = 2727: lower parts of door frame almost complete; (e) t = 4273: walls assembling (note staircase on lower left); (f) t = 7914: wall assembly continues with a large block just laid over the door and another already in place for the roof; (g) t = 12205: roof assembling; (h) t = 14383: wall staircase disassembling.

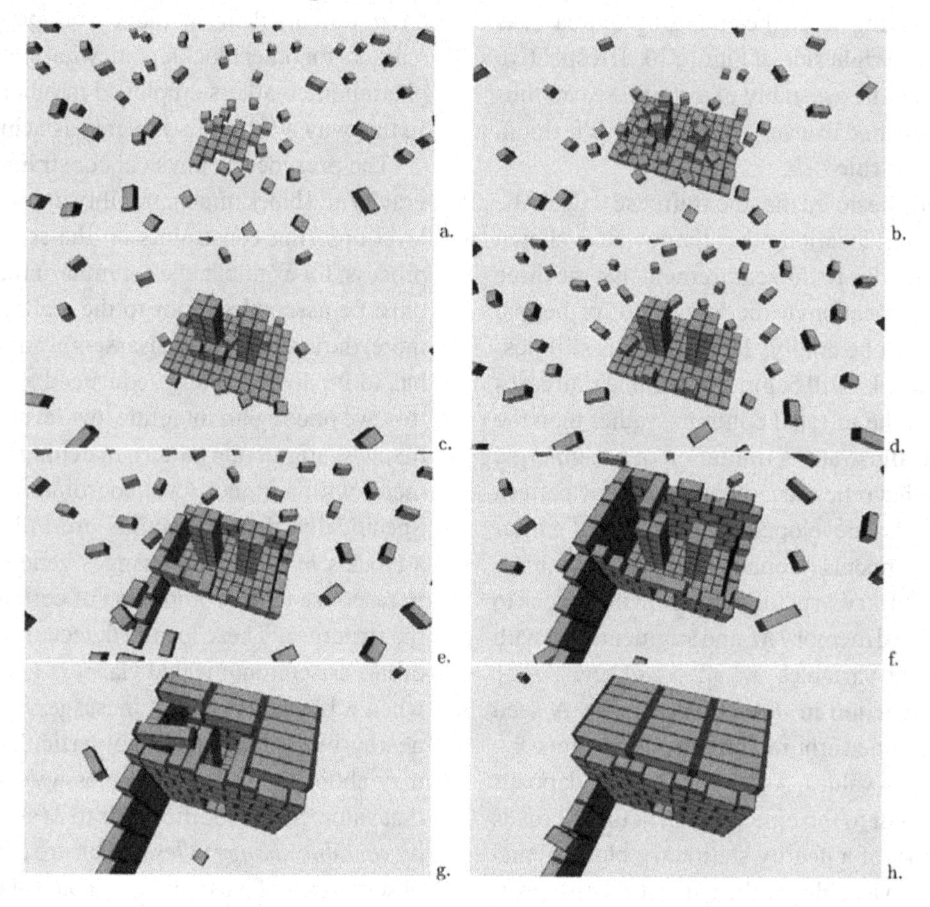

of rules, movement control mechanisms can be effectively reused between structures with only minor modifications. As an example, we discuss here the self-assembly of the building structure shown partially complete in Figure 10, which consists of a total of 289 blocks (52 extra blocks were used to allow temporary structures to assemble). The entire building is assembled using 66 stigmergic rules (6 of which are used for disassembly) and 77 variable change rules. A short video at http://www.cs.umd.edu/~reggia/grushin.html provides an animation of some of the stages of the building's self-assembly via these rules. Figure 11a shows the floor assembling. Two columns then begin emerging from specific locations in the floor, as shown in Figure 11b. As these columns grow in height, some small blocks assemble into staircases that partially wind around the columns in order to allow other small blocks to reach higher levels. Once a column is complete, its staircase disassembles (Figure 11c). After the floor and the columns are complete, the door is marked by depositing the lower sides of its frame, consisting of a small and a medium block each as shown in Figure 11d. The walls, which consist of medium blocks, are then begun along with the wall staircase discussed earlier (Figure 11e). As these walls reach an appropriate height, large blocks begin to climb the staircase, and one such block eventually places itself over the door, to form the top part of the door frame (Figure 11f). When the walls are finished, large blocks climb the staircase once again to build the two-layer roof. The second roof layer (Figure 11g) forms perpendicularly to the roof blocks below; it begins near the edges farthest from the staircase and extends towards the staircase. When the roof is complete and there are no blocks left moving above it (variable change rules are in place to ensure that this is the case), the wall staircase disassembles (Figure 11h), resulting in the final structure.

One factor that affects completion time is the availability of different sized blocks in regions where they are needed. We hypothesized that, as in competitive foraging and collective transport, block availability can be enhanced via the use of collective, flock-like movement behaviors, which allow a block that found the general region of construction to "pull'" other members of the flock there as well. We compared the use of cohesion and alignment to create this "flocking" versus simulations in which such forces were absent. These forces indeed reduce the mean completion time from 14,604.1 to 13,461.0, which is statistically significant. The improvement is considerably more pronounced in the absence of any global information, as demonstrated by running further trials where blocks have no knowledge of the relative direction of the center of the world. In this case, the reduction is from 67,666.5 to 43,338. While blocks must spend a much greater amount of time searching for the region of the world where they are needed, a flock-like dynamic can partially, yet significantly, compensate for the lack of global guidance.

Interestingly, there is sometimes a tradeoff between the goals of increasing block availability and reducing interference. We found that the latter can be particularly severe between blocks of different sizes, and it is sometimes necessary to coordinate the process (via variable change rules) such that only blocks of one particular size are actively assembling at a given point in time; however, this means that blocks of other sizes are temporarily unavailable.

We have thus shown an integrated methodology for controlling the self-assembly of complex structures in a non-trivial environment. While the methodology has been shown to be successful, the design of stigmergic and variable change rules for assembling some desired structure has proven to be a time-consuming and error-prone task. The present focus of our research is on the development of approaches for generating such rules automatically, given a specification of the target structure.

FUTURE TRENDS

Recent years have witnessed an increased degree of interest in the engineering of large complex systems consisting of numerous components interacting in a non-linear fashion. The intrinsically complex nature of such systems renders conventional methods for their analysis inappropriate and raises interests in alternative approaches. One promising approach is presented by guided self-organization. Inspired by examples of self-organizing systems in nature, such as swarming insects or social animals, these systems are characterized by the property that the global, system-wide behavior emerges from local interactions among their components, with no single component in control of the system or aware of its global state. Guiding this process into desired, goal-oriented behavior is a promising artificial intelligence method for problem solving. We expect to see in the near future an expansion of research in this field, providing better understanding of the underlying principles of self-organization. This, along with the availability of high-performance computing machinery, will eventually contribute to the increase in meaningful, real world applications in various areas, especially robotics.

DISCUSSION

The results discussed in this chapter provide substantial support for the hypothesis that self-organizing particle systems can be extended to exhibit behavior more general than just collective movements. Specifically, by giving a few behavioral states to normally purely reflexive agents often found in particle systems, a simple finite state transition graph or set of rules that governs state changes, and a simple memory for storing spatial information, the resulting agent team has the ability to collectively solve a variety of problems, through a process of "guided self-organization".

Three specific problems were used to study the behavior of the resulting agents. In the first problem, search-and-retrieve, an agent team could routinely search for, collect, and return discovered resources to a predetermined home location, all the while retaining movement as a "flock" of individuals. Further, it was found in simulations that a team of agents that moved collectively was more effective in solving search-and-retrieve problems than otherwise similar agents that moved independently. This was because when one or a few agents on a collectively-moving team discovered a site with plentiful resources, they would automatically pull other team members toward that site, greatly expediting the acquisition of the discovered resource. Thus, a benefit of underlying collective movements of particle systems which, to our knowledge, has not been appreciated in past work, is that they have the potential to automatically recruit additional agents to complete a task when one agent detects the necessity of that task.

A second set of experiments, dealing with the collective transport problem, extends the difficulty of the first problem not only by adding obstacles, but also by requiring every object to be transported by more than one agent. The collectively moving agents proved able to complete this new task, confirming that such agents can solve more challenging problems. Further, these simulations showed that simple collective movements can produce cooperation between self-organized sub-teams, as seen for example when a group of agents coordinated to make a bar-shaped object turn as it went around a corner.

Our final problem, that of self-assembly, does not require the agents to transport or manipulate external objects. However, it presents new challenges, requiring these agents (which are embodied as blocks) to position themselves in precise locations in order to form a desired 3D structure, while subject to constraints which include simulated gravity, in addition to impenetrability. In spite of this complexity, it was once again possible to control the spatial self-organization of the agents via an integration of higher-level, state-based coordination mechanisms together with low-level, reflexive behaviors. The latter include not only collective movements, which proved yet again to benefit the efficiency of the process, but also, stigmergic pattern matching, which allows for an implicit form of communication between the agents, and complements the explicit message passing that triggers state changes.

In all of the problems studied, the simulations exhibited group-level decisions not just about which type of movements to make, but also about how to more effectively distribute individual members, and which destinations or goals to pursue. Our results, as well as related recent work (Winder, 2004), show that the reflexive agents of contemporary particle systems can readily be extended to successfully support goal-directed problem solving while still retaining their purely-reflective collective movement behaviors. We believe that these results have important implications for robotics and robotic teams, a belief that is supported by recent work involving manipulation and transportation of objects by multiple physical robots (Chaimowicz, 2001, Sutton, 1998, Yamashita, 2003) and supporting theoretical work (Kazadi, 2002).

ACKNOWLEDGMENT

This work was supported by NSF Award IIS-0325098.

REFERENCES

Adam, J. (2005). *Designing Emergence.* Doctoral dissertation, University of Essex.

Aron, S., Deneubourg, J.L., Goss, S., & Pasteels, J.M. (1990), *Functional Self-Organization Illustrated by Inter-Nest Traffic in the Argentine Ant Iridomyrmex Humilis.* Lecture Notes in BioMeathematics, 89, 533-547.

Balch, T. & Arkin, R. (1998). Behavior-based formation control for multi-robot teams. *IEEE Transaction on Robotics and Automation.* 14:6, 926-939.

Beckers, R., Deneubourg, J. L. & Goss, S. (1992). Trails ant u-turns in the selection of a path by the and lasius niger. *Journal of Theoretical Biology,* 159, 397-415.

Beni, G. (1988). The concept of cellular robotic system. In *Proceedings of the IEEE International Symposium on Intelligent Control* (pp. 57-62).

Bonabeau, E., & Cogne, F. (1997). Self-organization in social insects. *Trends in Ecology and Evolution*, 12, 188-193.

Bonabeau, E., Dorigo, M., & Theraulz, G. (1999). *Swarm Intelligence: From natural to artificial systems.* Oxford University Press.

Braun, A., Musse, S. R., de Oliveira, L. P. L. & Bodmann, B. E. J. (2003). Modeling individual behaviors in crowd simulation. *International Conference on Computer Animation and Social Agents.* (pp. 143-148).

Burton, J. L. & Franks, N. R. (1985). The foraging ecology of the army ant eciton rapax : An ergonomic enigma? *Ecological Entomology*, 10, 131–141.

Camazine, S., Deneubourg, J.L., Franks, N., Sneyd, J., Theraulaz, G., & Bonabeau, E. (2001). *Self-Organization in Biological Systems.* Princeton University Press.

Caro G. D. & Dorigo, M. (1998). Antnet: Distributed stigmergetic control for communications networks. *Journal of Artificial Intelligence Research*, 9, 317–365.

Chaimowicz, L., Kumar, V., & Campos, M.F. (2001). Framework for coordinating multiple robots in cooperative manipulation tasks. In *Proceedings SPIE Sensor Fusion and Decentralized Control in Robotic Systems IV* (pp. 120-127).

Deneubourg, J.L., Goss, S., Franks, N. R., & Pasteels, J. M. (1989). The blind leading the blind: Modeling chemically mediated army ant raid patterns. *Insect Behavior*, 2(), 719-725.

Dorigo, M., Caro, G. D. & Gambardella, L. (1999). Ant algorithms for discrete optimization. *Artificial Life*, 5(2), 137–172.

Dorigo, M., Maniezzo, V. & Colorni, A. (1996). The Ant System: Optimization by a colony of cooperating agents. *IEEE Transactions on Systems, Man, and Cybernetics Part B: Cybernetics*, 26(1), 29–41.

Eberhart, R. & Kennedy, J. (1995). A new optimizer using particle swarm theory. In *Proceedings of the Sixth International Symposium on Micro Machine and Human Science* (pp. 39–43).

Eberhart,R. & Shi, Y. (2001). Particle swarm optimization: Developments, applications and resources. In *Proceedings of the 2001 Congress on Evolutionary Computation*, (pp. 81-86).

Fujibayashi, K., Murata, S., Sugawara, K., & Yamamura, M. (2002). Self-organizing formation algorithm for active elements. In *21st IEEE Symposium on Reliable Distributed Systems* (pp. 416-421).

Gaudiano, P., Shargel, B. & Bonabeau, E. (2003). Control of UAV swarms: What the bugs can teach us. In *2nd American Institute of Aeronautics and Astronautics "Unmanned Unlimited" Conf. and Workshop and Exhibit.*

Grushin, A., & Reggia, J.A. (2006). Stigmergic self-assembly of prespecified artificial structures in a constrained and continuous environment. *Integrated Computer-Aided Engineering*, 13(4), 289-312.

Helbing, D., Farkas, I. & Vicsek, T. (2000). Simulating dynamical features of escape panic. *Nature*. 407, 487-490.

Hosokawa, K., Tsujimori, T., Fujii, T., Kaetsu, H., Asama, H., Kuroda, Y., & Endo, I. (1998). Self-organizing collective robots with morphogenesis in a vertical plane. In *IEEE International Conference on Robotics and Automation* (pp. 2858-2863).

Jones, C., & Mataric, M. (2003). From local to global behavior in intelligent self-assembly. In *IEEE International Conference on Robotics and Automation* (pp. 721-726).

Kazadi, S., Abdul-Khaliq, A., & Goodman, R. (2002). On the convergence of puck clustering systems. *Robotics and Autonomous Systems*, 38(2), 93-117.

Kennedy, J. & Eberhart, R. (1995). Particle swarm optimization. In *IEEE International Conference on Neural Networks Vol.* 4 (pp. 1942–1948). Piscataway, NJ.

Kennedy, J., & Eberhart, R., & Shi, Y. (2001). *Swarm Intelligence*. Morgan Kaufmann.

Klavins, E. (2002). Automatic synthesis of controllers for distributed assembly and formation forming. In *IEEE International Conference on Robotics and Automation.*

Kube, R.C., & Bonabeau, E. (2000). Cooperative transport by ants and robots. *Robotics and Autonomous Systems*, Volume 30(1,2), 85-101.

Lapizco-Encinas, G., & Reggia, J.A. (2005). Diagnostic problem solving using swarm intelligence. In *IEEE Swarm Intelligence Symposium* (pp. 365-372).

Mataric, M. (1995). Designing and understanding adaptive group behavior. *Adaptive Behavior. 4(1),* 51-80.

Parunak, H., Purcell, M. & O'Connell, R. (2002). Digital pheromones for autonomous coordination of swarming UAV's. In *Proceedings 1st UAV Conference.*

Reynolds, C. (1987). Flocks, herds, and schools: A distributed behavioral model. *Computer Graphics*, 21(4), 25-34.

Reynolds, C. (1999). Steering behaviors for autonomous characters. In *Proc. Game Developers Conference* (pp. 763-782).

Rodriguez, A., & Reggia, J.A. (2004). Extending self-organizing particle systems to problem solving. *Artificial Life*, 10(4), 379-395.

Rodriguez, A., & Reggia, J.A. (2005). Collective-movement teams for cooperative problem solving. *Integrated computer-aided engineering*, 12(3), 217--235.

Schoonderwoerd, R., Holland, O. & Bruten, J.. Ant-like agents for load balancing in telecommunications networks. In *AGENTS '97: Proceedings of the first international conference on Autonomous agents*, (pp. 209–216). New York, NY.

Sudd, J.H. (1960). The transport of prey by an ant pheidole crassinoda. *Behavior*, 15, 295-308.

Sudd, J. H. (1965). The transport of prey by ant. *Behavior*, 25, 234-271.

Sutton, R.S., & Barto, A.G. (1998). *Reinforcement Learning: An Introduction.* The MIT Press.

Theraulaz, G., & Bonabeau, E. (1999). A brief history of stigmergy. *Artificial Life*, 5(2), 97-116.

Theraulaz, G., & Bonabeau, E. (1995). Coordination in distributed building. *Science*, 269(), 686-688.

Tu, X., & Terzopoulos, D. (1994). Artificial fishes: Physics, locomotion, perception, behavior. In *Computer Graphics 28 Annual Conference Series* (pp. 43-50).

Werfel, J., & Nagpal, R. (2006). Extended stigmergy in collective construction. *IEEE Intelligent Systems*, 21(), 42-48.

White, P., Zykov, V., Bongard, J., & Lipson, H. (2005). Three dimensional stochastic reconfiguration of modular robots. In *Proc. Robotics: Science and Systems*, (pp. 161-168).

Winder, R., & Reggia, J.A. (2004). Using distributed partial memories to improve self-organizing collective movements. *SMC-B*, 34(4), 1697-1707.

Yamashita, A., Arai, T., Ota, J., & Asama, H. (2003). Motion planning of multiple mobile robots for cooperative manipulation and transportation. *IEEE Transactions on Robotics and Automation*, 19(2), 223-237.

ADDITIONAL READING

Arbuckle, D. & Requicha, A. (2004). Active self-assembly. In *IEEE International Conference on Robotics and Automation,* 896-901.

Barabási, A. (2002). *Linked: The New Science of Networks.* Perseus Publishing, Cambridge, MA.

Bishop, J., Burden, S., Klavins, E., Kreisberg, R., Malone, W., Napp, N. & Nguyen, T. (2005). Self-organizing programmable parts. In *International Conference on Intelligent Robots and Systems.*

Bonabeau, E., Dorigo, M., & Theraulz, G. (1999). *Swarm Intelligence: From natural to artificial systems.* Oxford University Press.

Brooks, R. A. (1986). A robust layered control system for a mobile robot. *IEEE Journal of Robotics and Automation. 1(2)*, 14-23.

Eberhart, R. & Shi, Y. (1999). Empirical study of particle swarm optimization. In *Proceedings of the 1999 Congress on Evolutionary Computation..* 3, 1950.

Flake, G. (1998). *The Computational Beauty of Nature: Computer Explorations of Fractals, Chaos, Complex Systems, and Adaptation.* MIT Press.

Franks, N. (1986). Teams in social insects: group retrieval of prey by army ants (Eciton burchelli, Hymenoptera: Formicidae). *Behavioral Ecology and Sociobiology.* 18(6), 425-429.

Glotzer, S. (2004). Some assembly required. *Science*, 419-420.

Gross, R., Bonani, M., Mondada, F. & Dorigo, D. (2006). Autonomous self-assembly in a swarm-bot. In *Proc. of the 3rd Int. Symp. on Autonomous Minirobots for Research and Edutainment*, (pp. 314-322).

Gross, R. & Dorigo, M. (2004). Cooperative transport of objects of different shapes and sizes. In *Ant Colony Optimization and Swarm Intelligence, 4th International Workshop (pp.* 107-118).

Kennedy, J., & Eberhart, R. (Ed.). (2001). *Swarm Intelligence.* Morgan Kaufmann.

Mataric, M. (1995). Designing and understanding adaptive group behavior. *Adaptive Behavior. 4(1),* 51-80.

Nembrini, J., Reeves, N., Poncet, E., Martinoli, A. & Winfield, A. (2005), Mascarillons: Flying swarm intelligence for architectural research. In *IEEE Swarm Intelligence Symposium (*pp. 225-232).

Payton, D., Estkowski, R. & Howard, M (2003). Compound behaviors in pheromone robotics. *Robotics and Autonomous Systems*, 44(3-4), 229-240.

Reeves, R. (1983). Particle systems: A technique for modeling a class of fuzzy objects. In *Proceedings of the 10th annual conference on Computer graphics and interactive techniques. (pp. 359-375).*

Resnick, M. (1994). *Turtles, Termites and Traffic Jams.* MIT Press.

Sauter, J., Matthews, R., Parunak, H. D., Brueckner, S. (2005). Performance of digital pheromones for swarming vehicle control. *In Proceedings of the fourth international joint conference on Autonomous agents and multiagent systems.* 903-910.

Shen, W. M., Will, P. & Khoshnevis, B. (2003). Self-assembly in space via self-reconfigurable robots, in *IEEE International Conference on Robotics and Automation. (*pp. 2516-2521).

Chapter XVI
Evaluating a Bio-Inspired Approach for the Design of a Grid Information System: The SO-Grid Portal

Agostino Forestiero
Institute of High Performance Computing and Networking CNR-ICAR, Italy

Carlo Mastroianni
Institute of High Performance Computing and Networking CNR-ICAR, Italy

Fausto Pupo
Institute of High Performance Computing and Networking CNR-ICAR, Italy

Giandomenico Spezzano
Institute of High Performance Computing and Networking CNR-ICAR, Italy

ABSTRACT

This chapter proposes a bio-inspired approach for the construction of a self-organizing Grid information system. A dissemination protocol exploits the activity of ant-inspired mobile agents to replicate and reorganize metadata information on the basis of the characteristics of the related Grid resources. Resource reorganization emerges from simple operations of a large number of agents, in a "swarm intelligence" fashion. Moreover, a discovery protocol allows Grid clients to locate useful resources on the Grid through a semi-informed approach. This chapter also describes the SO-Grid Portal, a simulation portal through which registered users can simulate and analyze the ant-based protocols. This portal can be used by researchers to perform "parameter sweep" studies, as it allows for the graphical comparison of results obtained in previous sessions. We believe that the deployment of the SO-Grid portal, along with the definition and discussion of the protocols presented in this chapter, can foster the understanding and use of swarm intelligence, multi-agent and bio-inspired paradigms in the field of distributed computing.

INTRODUCTION

To support the design and execution of complex applications, modern distributed systems must provide enhanced services such as the retrieval and access to content, the creation and management of content, and the placement of content at appropriate locations. In a Grid, these services are offered by a pillar component of Grid frameworks, the *information system*. This chapter discusses a novel approach for the construction of a Grid information system which allows for an efficient management and discovery of information. The approach, proposed in (Forestiero et al., 2005) in its basic version, exploits the features of (1) epidemic mechanisms tailored to the dissemination of information in distributed systems (Peterson et al., 1997, Eugster & al., 2004) and (2) self organizing systems in which "swarm intelligence" emerges from the behavior of a large number of agents which interact with the environment (Bonabeau & al., 1999, Dasgupta, 2004).

The proposed ARMAP protocol (*Ant-based Replication and MApping Protocol*) disseminates Grid resource *descriptors* (i.e., metadata documents) in a controlled way, by spatially mapping these descriptors according to their semantic classification, so to achieve a logical reorganization of resources. A resource descriptor can be composed of a syntactical description of a Grid service (e.g. a Web Services Description Language - WSDL - document) and/or a semantic description of the capabilities of the service.

Descriptor reorganization results from pick and drop operations performed by a large number of agents. Each ARMAP agent travels the Grid through P2P interconnections among Grid hosts, and uses simple probability functions to decide whether or not to *pick* descriptors from or *drop* descriptors into the current Grid host. This approach is inspired by the activity of some species of ants and termites that cluster and map items within their environment (Bonabeau & al., 1999).

Furthermore, a self-organization approach based on ants' pheromone (Van Dyke & al., 2005) enables each agent to regulate its activity, i.e. its operation *mode*, only on the basis of local information. Indeed, each agent initially works in the *copy* mode: it can generate new descriptors and disseminate them on the Grid. However, when it realizes from its own past activity that a sufficient number of replicas have been generated, it switches to the *move* mode: it only moves descriptors from one host to another without generating new replicas. This switch is performed when the level of a pheromone variable, which depends on agent's activity, exceeds a given threshold.

The ARMAP protocol can effectively be used to build a Grid information system in which (1) resource descriptors are properly replicated and (2) the overall entropy is reduced. A balance between these two features is achieved by regulating the pheromone threshold, i.e., by shortening or extending the time interval in which agents operate under the *copy* mode.

A semi-informed discovery protocol exploits the logical resource organization achieved by ARMAP. Indeed, whenever a large number of descriptors of a specific class are accumulated in a restricted region of the Grid, it becomes convenient to drive query messages (issued by users to locate descriptors of this class) towards this region, in order to maximize the number of discovered descriptors and minimize the response time. While this chapter focuses on enhancements and performance of the ARMAP protocol, the discovery protocol, namely ARDIP (*Ant-Based Resource Discovery Protocol*) is here shortly discussed, whereas an extensive analysis can be found in (Forestiero & al., 2007).

This chapter also introduces the SO-Grid (*Self Organizing Grid*) portal, a Web portal which gives remote users access to an event-based simulator written in Java. This portal, available at the URL *http://so-grid.icar.cnr.it*, allows for the experimental reproduction of simulation results and can be used by researchers to perform "parameter sweep" studies. Through the portal front-end, a user can (1) register to the portal and create a private space in the file system that will contain the results of all the performed simulations; (2) run a new simulation, after setting network and protocol parameters, and graphically monitor the performance indices and (3) choose a subset of simulation data to graphically compare the related performance results.

We hope that the SO-Grid portal will help foster the understanding and use of swarm intelligence, multi-agent and bio-inspired paradigms in the field of distributed computing.

BACKGROUND

The main purpose of the protocol presented in this chapter is the dissemination of metadata documents and their intelligent reorganization. These two objectives are correlated, since an intelligent dissemination of information can increase efficiency management and facilitate discovery, as discussed in (Forestiero & al., 2007).

Information dissemination is a fundamental and frequently occurring problem in large, dynamic and distributed systems whose main purpose is the management and delivery of content. In (Iamnitchi & Foster, 2005) it is proposed to disseminate information selectively to groups of users with common interests, so that data is sent only to where it is wanted. In our chapter, instead of classifying users, the proposal is to exploit the classification of resources: resource descriptors are replicated and disseminated with the purpose of creating regions of the network that are specialized in specific classes of resources. In (Aktas & al., 2007) information dissemination is combined with the issue of effective replica placement, since the main interest is to place replicas in the proximity of requesting clients by taking into account changing demand patterns. Specifically, a metadata document is replicated if its demand is higher than a defined threshold and each replica is placed according to a multicast mechanism that aims to discover the data server which is the closest to demanding clients.

The ARMAP protocol, proposed in this chapter, is inspired by biological mechanisms, and in particular exploits the self-organizing and swarm intelligence characteristics of several biological systems, ranging from ant colonies to wasp swarms and bird flocks. In these systems, a number of small and autonomous entities perform very simple operations driven by local information, but from the combination of such operations a complex and intelligent behavior emerges (Dasgupta, 2005): for example, ants are able to establish the shortest path towards a food source; birds travel in large flocks and rapidly adapt their movements to the ever changing characteristics of the environment.

These biological systems can be quite naturally emulated in a Grid through the multi-agent paradigm (Sycara, 1998): the behavior of insects and birds is imitated by mobile agents which travel through the hosts of a Grid and perform their operations. Multi-agent computer systems can inherit interesting and highly beneficial properties from their biological counterparts, namely: (1) self-organization, since decisions are based on local information, i.e., without any central coordinator; (2) adaptivity, since agents can flexibly react to the ever-changing environment; (3) stigmergy awareness (Grassè, 1959), since agents are able to interact and cooperate through the modifications of the environment that are induced by their operations.

The ARMAP protocol is specifically inspired to ant algorithms, a class of agent systems which aim to solve very complex problems by imitating the behavior of some species of ants (Bonabeau & al., 1999). A technique based on ant pheromone is exploited to tune the behavior of ARMAP agents, making them able to autonomously switch from the *copy* to the *move* mode. This kind of approach is discussed in (Van Dyke & al., 2005), where a decentralized scheme, inspired by insect pheromone, is used to control the activity of a single agent and better accomplish the system goal.

The ARMAP pick and drop probability functions are inspired by probability functions that were introduced in (Deneubourg & al., 1990), and later elaborated and discussed in (Bonabeau & al., 1999), to emulate the behavior of some species of ants that build cemeteries by aggregating corpses in clusters. In ARMAP, such functions have been adapted through the following main modifications: (i) descriptors are not only moved and aggregated, as are corpses in the mentioned papers, but also *replicated*; (ii) descriptors are reorganized according to the class to which they belong, so as to reduce the overall entropy of the system.

DISSEMINATION OF RESOURCE DESCRIPTORS

The aim of the ARMAP protocol (Forestiero & al., 2005) is to achieve a logical organization information on the Grid by spatially mapping resource descriptors according to the semantic classification of related resources. It is assumed that the resources have been previously classified into a number of classes *Nc*, according to their semantics and functionalities (Crespo & Garcia Molina, 2002).

The ARMAP protocol has been analyzed in a Grid in which hosts are connected in a P2P fashion and each host is connected to at most 8 neighbor peers. The Grid has a dynamic nature, and hosts can disconnect and rejoin the network. When connecting to the Grid, a host generates a number of agents given by a discrete Gamma stochastic function, with average *Ngen*, and sets the life time of these agents to *PlifeTime*, which is the average connection time of the host, calculated on the basis of the host's past activity. This mechanism allows for controlling the number of agents that operate on the Grid: indeed, the number of agents is maintained to a value which is about *Ngen* times the number of hosts.

Periodically each ARMAP agent sets off from the current host and performs a number of hops through the P2P links that interconnect the Grid hosts. Then the agent uses random *pick* and *drop* functions in order to replicate and move descriptors from one peer to another. More specifically, at each host an agent must decide whether or not to *pick* the descriptors of a given class, and then carry them in its successive movements, or to *drop* the descriptors that it has previously picked from another host.

As a consequence of pick and drop operations, each host can maintain descriptors of local resources, which are never removed, as well as descriptors of resources published by other hosts, which can be picked up and discarded by agents. In the following, when distinction is relevant, such descriptors will respectively be referred to as *local* and *remote* descriptors.

Pick and drop probability functions are discussed in the following.

Pick Operation

Whenever an ARMAP agent hops to a Grid host, it must decide, for each resource class, whether or not to *pick* the descriptors of this class which are managed by the current host, unless it already carries some descriptors of the same class. In order to achieve replication and mapping functionalities, a *pick* random function is defined with the intention that the probability of picking the descriptors of a class decreases as the local region of the Grid accumulates these descriptors and vice versa. This assures that as soon as the equilibrium condition is broken (i.e., descriptors belonging to different classes begin to be accumulated in different regions), a further reorganization of descriptors is favored.

The basic *Ppick* probability function is shown in formula (1). The agent evaluates this function, generates a real number between 0 and 1, from a uniform random distribution, and then it performs the pick operation if the generated number is lower than the value of the *Ppick* function. This function is the product of two factors, which take into account, respectively, the *absolute* accumulation of descriptors of a given class and their *relative* accumulation (i.e., with respect to other classes). While the absolute factor is inspired by the pick probability defined in (Bonabeau & al., 1999), the relative factor was introduced here to achieve the spatial separation of descriptors.

$$P_{pick} = \left(\frac{fa}{k1 + fa}\right)^2 \cdot \left(\frac{k2}{k2 + fr}\right)^2 \quad (1)$$

The *fa* fraction is computed as the number of *local* descriptors of the class of interest, maintained by the hosts located in the *visibility region*, out of the overall number of descriptors of the class of interest (including both *local* and *remote*) that are maintained by the same hosts. The value of *fa* is comprised between 0 and 1, as well as the value of *fr*, defined below. The *visibility region* includes all the hosts that are reachable from the current host with a given number of hops, i.e. within the *visibility radius*, which is an algorithm parameter. As more *remote* descriptors of a class are accumulated in the visibility region, *fa* decreases, the first factor of the pick function decreases as well, and the probability that the agent will picks the descriptors becomes lower. This facilitates the formation of accumulation regions. Conversely, the pick probability is large if a small number of *remote* descriptors are located in the local region of the Grid.

The *fr* fraction is computed as the number of descriptors of the class of interest, accumulated in the hosts located in the visibility region, divided by the overall number of descriptors, of all classes, that are accumulated in the same region. As the local region accumulates more descriptors of a class, with respect to other classes, *fr* increases and accordingly the value of the pick probability for this class decreases, and vice versa. This facilitates the spatial separation of descriptors of different classes. The constants *k1*, *k2* are set to the value 0.1, as in analogous formulas defined in (Bonabeau et al., 1999).

The pick operation can be performed with two different *modes*. If the *copy* mode is used, the agent, when executing a pick operation, leaves the descriptors on the current host, generates a replica of them, and carries such replicas until it will drop them in another host. Conversely, with the *move* mode, as an agent picks the descriptors, it removes them from the current host (except for the *local* descriptors, which cannot be removed), thus preventing an excessive proliferation of replicas.

Drop Operation

As well as the pick function, the drop function is first used to break the initial equilibrium and then to strengthen the mapping of descriptors of different classes in different Grid regions. Whenever an agent gets to a new Grid host, it must decide, if it is carrying some descriptors of a given class, whether or not to *drop* such descriptors in the current host. As opposed to the pick operation, the drop probability function *Pdrop*, shown in formula (2), is directly proportional to the relative accumulation of descriptors of the class of interest in the visibility region. Therefore, as accumulation regions begin to emerge, the drop operations are and more and more favored in these regions. In (2) the threshold constant *k3* is set to 0.3, as in (Bonabeau & al., 1999).

$$P_{drop} = \left(\frac{fr}{k3 + fr}\right)^2 \quad (2)$$

A Pheromone Mechanism for Reduction of System Entropy

A spatial entropy function, based on the well known Shannon's formula for the calculation of information content, is defined to evaluate the effectiveness of the ARMAP protocol. For each peer *p*, the local entropy *Ep* gives an estimation of the extent to which the descriptors have already been mapped within the visibility region centered in *p*. *Ep* has been normalized, so that its value is comprised between 0 and 1. As shown in formula (3), the overall entropy *E* is defined as the average of the entropy values *Ep* computed at all the Grid hosts. In (3), *fr(i)* is the fraction of descriptors of class *Ci* that are located in the visibility region with respect to the overall number of descriptors located in the same region.

$$Ep = \frac{\sum_{i=1..Nc} fr(i) \cdot \log_2 \frac{1}{fr(i)}}{\log_2 Nc}, \quad E = \frac{\sum_{p \varepsilon Grid} Ep}{Np} \qquad (3)$$

In (Forestiero et al., 2005) it was shown that the overall spatial entropy can be minimized if each agent exploits both the ARMAP modes, i.e. *copy* and *move*. In the first period of its life, each agent *copies* the descriptors that it picks from a Grid host, but when it realizes from its own activeness that the mapping process is at an advanced stage, it begins simply to *move* descriptors from one host to another, without creating new replicas.

In fact, agents cannot always operate under the *copy* mode, since eventually every host would be assigned a very large number of descriptors of all classes, thus weakening the efficacy of descriptor mapping. The protocol is effective only if each agent, after replicating a number of descriptors, switches from *copy* to *move*. A self-organization approach based on ants' pheromone mechanism enables an agent to perform this mode switch only on the basis of local information. This approach is inspired by the observation that agents perform more operations when the system entropy is high, whereas operation frequency gradually decreases as descriptors are properly reorganized. In particular, at given time intervals, i.e. every 2,000 seconds, each agent counts up the number of times that it has evaluated the pick and drop probability functions, and the number of times that it has actually performed *pick* and *drop* operations, as the generated random number is lower than the value of the probability function. At the end of each time interval, the agent makes a deposit into its pheromone base, by adding a pheromone amount equal to the ratio between the number of "unsuccessful" (not actually performed) operations and the overall number of operation attempts. At the end of the i-th time interval, the pheromone level Φi is computed with formula (4).

$$\Phi i = Ev \cdot \Phi i - 1 + \varphi i \qquad (4)$$

In this formula, φi is the fraction of unsuccessful operations performed in the last time interval. An evaporation mechanism is used to give a higher weigh to recent behavior of the agent, and the evaporation rate *Ev* is set to 0.9. With these settings, the value of Φi is always comprised between 0 and 10. As soon as the pheromone level exceeds a pheromone threshold *Tf* (whose value is also in the range [0,10]), the agent realizes that the frequency of *pick* and *drop* operations has remarkably reduced, so it switches its protocol mode from *copy* to *move*. The value of *Tf* can be used to tune the number of agents that work in *copy* mode and are therefore able to create new descriptor replicas, as discussed in the next section.

PERFORMANCE EVALUATION

The performance of the ARMAP protocol has been evaluated with an event-based simulator written in Java, which will be described in the chapter section related to the SO-Grid Portal.

Simulation runs have been performed with the following setting of network and protocol parameters. The number of peers *Np*, or Grid size, is set to 2500. The average connection time of a specific peer, *Plifetime*, is generated according to a Gamma distribution function, with an average value set to 100,000 seconds. The use of the Gamma function assures that the Grid contains very dynamic hosts, that frequently disconnect and rejoin the network, as well as much more stable hosts. Every time a peer disconnects from the Grid, it loses all the *remote* descriptors previously deposited by agents, thus contributing to the removal of obsolete information. The average number of Grid resources owned and published by a single peer is set to 15. Grid resources are classified in a number of classes *Nc*, which is equal to 5. The mean number of agents that travel the Grid is set to *Np*/2: this is accomplished, as explained in the previous section, by setting the mean number of agents generated by a peer, *Ngen*, to 0.5. The average time *Tmov* between two successive agent movements (i.e., between two successive evaluations of *pick* and *drop* functions) is 60 s. The maximum number of P2P hops that are performed within a single agent movement, *Hmax*, is set to 3. The visibility radius *Rv*, defined in the previous section and used for the evaluation of pick and drop functions, is set to 1. Finally, the pheromone threshold *Tf* ranges from 3 to 10.

It is worth noting that the So-Grid replication algorithm is very robust with respect to variations of the above mentioned parameters. The values of algorithm parameters, e.g., the number of resources per peer, the number of resource classes, the average connection time of a peer, etc., can affect the rapidity of the process, and in some cases can slightly influence the steady values of performance indices, but the qualitative behavior is always preserved. This can be verified by performing "parameter sweep" simulation through the SO-Grid Portal, as discussed in the next section.

The following performance indices are used. The overall entropy *E*, discussed in the precious section, is used to estimate the effectiveness of the ARMAP protocol in the reorganization of descriptors. The *Nrpr* index is defined as the mean number of descriptors that are generated for each resource. Since new descriptors are only generated by ARMAP agents that work in the *copy* mode, the number of such agents, *Ncopy*, is another interesting performance index.

Prior to the numerical analysis, a graphical description of the behavior of the replication algorithm is given in Figure 1. For the sake of

Figure 1. Accumulation and reorganization of resource descriptors, belonging to 3 resources classes, in a region of the Grid

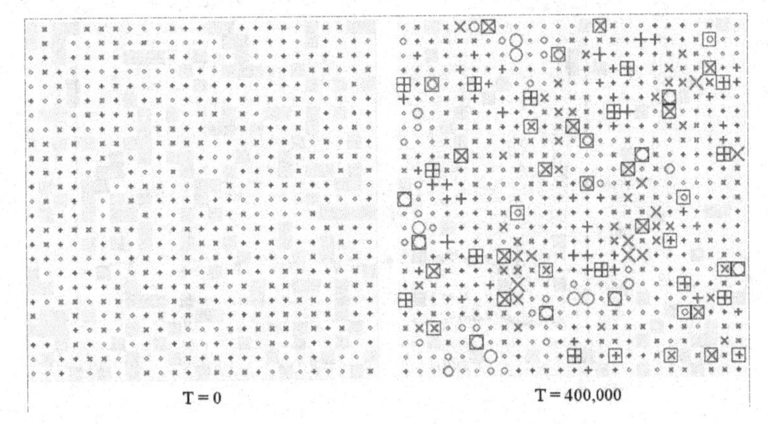

clearness, here the number of classes is set to 3, and the peers are arranged in a 2-dimensional mesh. The figure shows only a portion of the Grid, though the simulation was performed on a network with 2500 hosts.

In this figure, different symbols are associated with the three classes. Each peer is visualized by the symbol that corresponds to the class to which the largest number of descriptors, maintained by this peer, belong. Furthermore, the symbol size is proportional to the number of descriptors of the dominant class. Two snapshots of the network are depicted: the first is taken when the ARMAP process is initiated, at time 0, the second is taken 400,000 seconds later, in a quite steady situation. This figure shows that descriptors are initially distributed in a completely random fashion, but subsequently they are accumulated and reorganized by agents in separate regions of the Grid, according to their class. The peers with a marked border are *representative* peers that work as attractors for query messages, as briefly described in the next section.

A set of simulation runs have been performed to evaluate the performance of the basic pick and drop probability functions and the effectiveness of the pheromone mechanism that drives the mode switch of agents. These simulations were performed with a Grid size *Np* equal to 2500 and a number of classes *Nc* equal to 5. Figure 2 reports the number of agents that work in *copy* mode, *Ncopy* (also called *copy agents* in the following), versus time, for different values of the pheromone threshold *Tf*. When ARMAP is initiated, all the agents (about 1250, half the number of peers) are generated in the *copy* mode, but subsequently several agents switch to *move*, as soon as their pheromone value exceeds the threshold *Tf*. This corresponds to the sudden drop of curves that can be observed in Figure 2. This drop does not occur if *Tf* is equal to 10 because this value can never be reached by the pheromone (see formula (4)); therefore, with *Tf*=10 all agents remain in *copy* along all their lives.

After the first phase of the ARMAP process, an equilibrium is reached because the number of new agents which are generated by hosts (such agents always set off in *copy* mode) and the number of agents that switch from *copy* to *move* get balanced. Moreover, if the pheromone threshold *Tf* is increased, the average interval of time in which an agent works in *copy* becomes longer, because the pheromone level takes more time to reach this threshold; therefore the average value of *Ncopy*

Figure 2. Number of agents in copy mode for different values of the pheromone threshold Tf

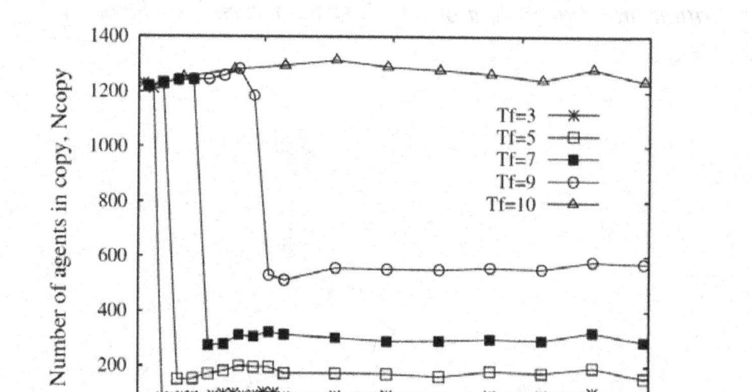

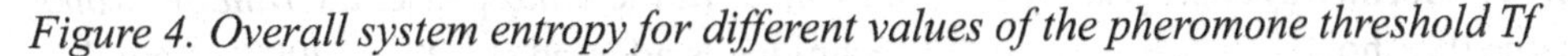

Figure 3. Mean number of replicas per resource for different values of the pheromone threshold Tf

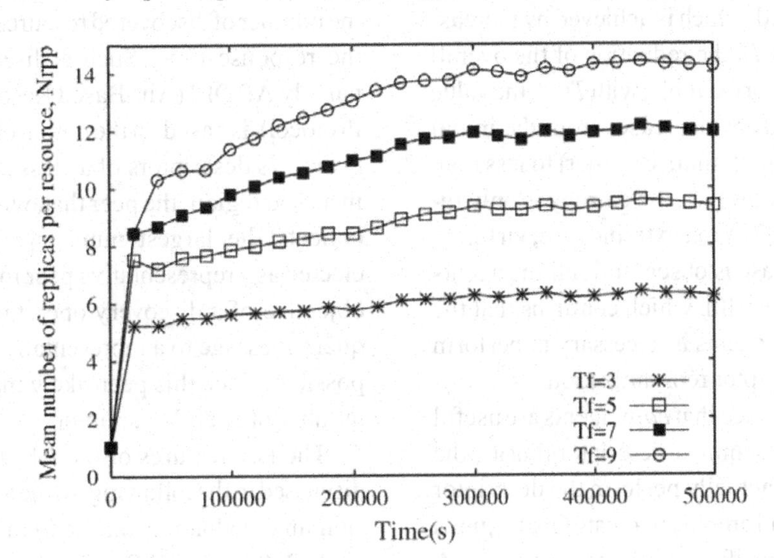

Figure 4. Overall system entropy for different values of the pheromone threshold Tf

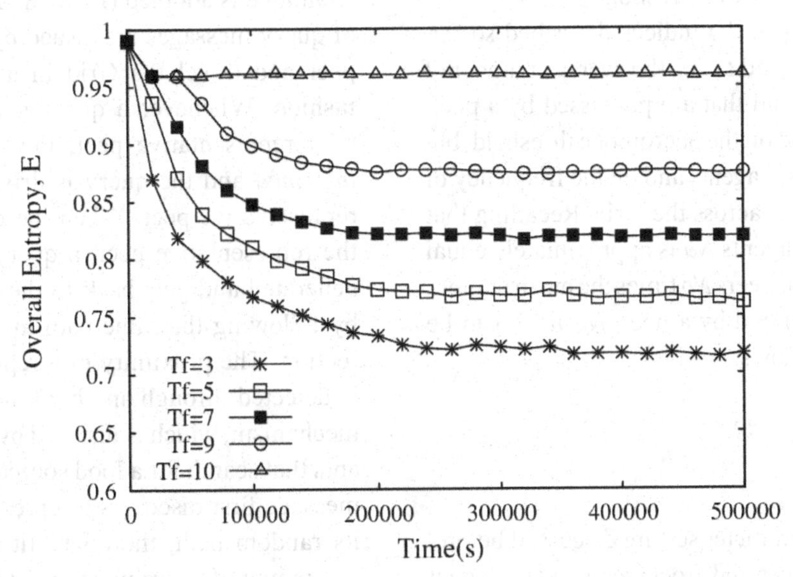

at the equilibrium becomes larger. In conclusion, the setting of *Tf* is a simple and efficient method to tune the number of *copy* agents, and therefore the rate of generation of new replicas.

Figure 3 shows the mean number of descriptor replicas per resource and confirms that descriptor dissemination is more intense if the pheromone threshold is increased, because a larger number of *copy* agents operate on the network. However, a more intense dissemination is not always associated to a better reorganization, i.e. to a more effective spatial separation of descriptors belonging to different classes.

Figure 4 shows that when the number of *copy* agents is increased, which is achieved by increasing the threshold *Tf*, the reduction of the overall entropy is lower. For example, with *Tf*=3, the value of the overall entropy decreases from the initial value of about 1 (maximum disorder) to less than 0.72. With *Tf*=9, however, the entropy is only reduced to about 0.87. As an extreme case, virtually no entropy decrease is observed if all the agents operate in *copy* (*Tf*=10), which confirms that the presence of *move* agents is necessary to perform an effective descriptor reorganization.

It can be concluded that *copy* agents are useful to replicate and disseminate descriptors but it is the *move* agents that actually perform the descriptor reorganization and are able to create Grid regions specialized in specific classes of resources. A balance between the two features (replication and reorganization) can be performed by appropriately tuning the pheromone threshold.

As opposed to the indices described so far, the processing load *L*, i.e., the average number of agents per second that are processed by a peer, does not depend on the pheromone threshold, but on the number of agents and on the frequency of their movements across the Grid. Recalling that the number of agents *Na* is approximately equal to the number of peers *Np* times the mean number of agents generated by a peer, *Ngen*, *L* can be obtained as follows:

$$L = \frac{\mathrm{Na}}{\mathrm{Tmov} \cdot \mathrm{Np}} \approx \frac{\mathrm{Ngen}}{\mathrm{Tmov}}$$

With the parameter setting discussed before, each peer receives and processes about one agent every 120 seconds, which can be considered an acceptable load.

RESOURCE DISCOVERY

The logical resource organization achieved by ARMAP can be exploited through a semi-informed discovery protocol, in order to maximize the number of discovered resources and minimize the response time. Such a discovery protocol, namely ARDIP (Ant-Based Resource Discovery Protocol), is based on the notion of *representative peers.* As descriptors of a class are accumulated in a Grid region, the peer that, within this region, collects the largest number of descriptors is elected as a representative peer for this class. The objective of a discovery operation is to direct a query message to a representative peer as fast as possible, since this peer likely maintains a large number of useful descriptors.

The key features of the ARDIP protocol are discussed in the following, while a detailed discussion and evaluation can be found in (Forestiero & al., 2007). With ARDIP, a discovery operation is performed in two phases, one *blind* and one *informed.* In the *blind* phase, the *random walks* technique is adopted (Lv & al., 2002): a number of query messages are issued by the requesting peer and travel the Grid in a *blind* (random) fashion. Whenever a query gets close enough to a representative peer, the search becomes *informed* and the query is driven towards this representative peer. When the query arrives at the representative peer, a queryHit message is generated and gets back to the requesting peer by following the same path in the opposite direction. The proximity of a representative peer is detected through another kind of pheromone mechanism, which is inspired by the behavior of ants that search for a food source. When a query message first discovers a representative peer in its random path, the queryHit message leaves an amount of pheromone in the first peers of its return journey, and this pheromone can be later recognized by other peers that casually approach this representative peer.

The semi-informed protocol aims to combine the benefits of both blind and informed resource discovery approaches which are currently used in P2P networks (Tsoumakos & Roussopoulos, 2003). In fact, a pure blind approach (e.g. using

Figure 5. SO-Grid portal: Overall architecture

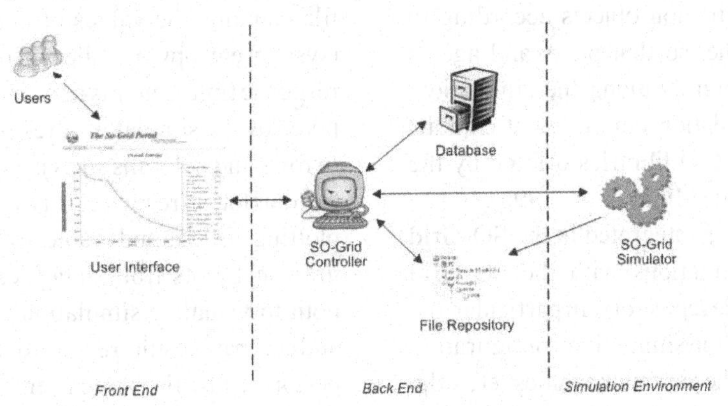

flooding or random walks techniques) is simple to implement but has limited performance and can cause an excessive network load, whereas a pure informed approach generally requires a very structured resource organization which is impractical in a large, heterogeneous and dynamic Grid.

THE SO-GRID PORTAL

The ICAR research institute of the Italian CNR recently developed the So-Grid (Self Organizing Grid) Portal, available at the URL *http://so-grid.icar.cnr.it*, in order to allow remote users to experience and evaluate the bio-inspired protocols described in this chapter. The portal, designed according to the Java J2EE technology and built upon the JBoss application server, provide users with a Web interface through which it is possible to run a remote simulation, after setting its parameters, graphically monitor the simulation execution at run time and compare the results of several simulations in order to perform "parameter sweep" analysis.

The overall architecture of the portal is depicted in Figure 5. The portal is composed of three main tiers: (1) the portal front-end, which gives access to the Web interface; (2) the portal back-end, which includes the So-Grid controller, a database and a file repository and (3) the So-Grid event-based simulator.

In the following, the portal components will be briefly described, in a right to left order, starting from the simulator.

The SO-Grid Simulator

The performances of the ARMAP and ARDIP protocols have been evaluated through an event-based simulator written in Java. Simulation *objects* are used to emulate Grid hosts and bio-inspired agents. Each object reacts to external *events* according to a finite state automaton and responds by performing specific operations and/or by generating new messages/events to be delivered to other objects.

For example, a peer visited by an agent gives it information about the descriptors that this peer maintains; afterwards the agent uses the probability functions define in formulas (1) and (2) to decide whether or not to *pick* descriptors from or *drop* descriptors into the peer. Finally, the agent sets the simulation time in which it will perform its next movement on the Grid and creates a related event that will be delivered at the specified time to the peer to which the agent will move. Events are ordered in a general queue by a simulation *manager* component, and they are delivered to corresponding destination objects according to their expiration time, so that peers and agents can operate concurrently along the simulation. Moreover, the simulation environment exploits the visual facilities and libraries offered by the Swarm environment (Minar & al., 1996).

The simulator was integrated in the SO-Grid portal through interactions with the SO-Grid Controller and the file repository. In particular, the Controller prepares the simulation configuration file which contains the parameter values set by the user, sends this file to the simulator and starts the simulation. As the simulation proceeds, the partial results are stored in the file repository, so as to let the Controller access them. When the simulation terminates, the simulator informs the Controller, which in turn updates database information pertaining to the running simulations.

File Repository and Database

Both the file repository and the database contain information used by the Controller to manage registered users, running simulations and simulation results.

Specifically, the file repository is organized in folders. Each folder is assigned to a registered user, and essentially contains configuration files and results files. A configuration file contains a set of parameter-value associations and is given as input to the simulator. Parameters can be related either to the simulated network (number of nodes, average connection time of nodes, number of resources, etc.) or to the bio-inspired protocols (number of agents, values of probability function constants, etc.). Each parameter has a default value that can be modified by the user: the resulting configuration file is saved in the user folder, in order to maintain all information about simulation settings, for future use.

Moreover, a user folder stores the files which contain the results of the simulations performed by the user. The name of a results file is chosen by the user before starting the simulation. A results file contains the values of performance indices (system entropy, number of replicas, discovery response times, etc.) as they are periodically computed by the simulator. A results file is organized in rows and columns so as it can be given as input to Gnuplot, a freeware software that offers batch plotting utilities and is able to automatically build graphical plots from data files. Gnuplot is used both to visualize simulation results at run time and compare results related to several simulations performed by the same user.

The database stores two kinds of information. The first is the account data (name, password, e-mail address etc.) of registered users, which is checked when a user requests to log on the portal. Furthermore, the database maintains information about the running simulations; in particular the database associates the process id of each running simulation to the user which has requested it and to the name of the file that will contain the results. This helps the management of simulations: for example, after the log in procedure of a user, the portal checks if there is a simulation started by this user which is still running, and in this case the user is redirected to the portal form which shows the results in real time. Moreover, the database is used to control the number of ongoing simulations: indeed, to maintain the performance of the simulator at an acceptable level, each user can execute only one simulation at a time and no more than 5 simulations can be executed concurrently by different users.

SO-GRID CONTROLLER

The SO-Grid Controller is the component that receives, through the graphical interface, user requests and interacts with other components to satisfy them. It is built upon the J2EE technology and its main functionalities are performed by Java Servlets. Some of the tasks that the Controller performs are the following:

- It manages the user log on procedure, by checking account data of users stored on the database;
- It manages the user choice of simulation parameters, and creates the correspondent configuration file, which is passed as input to the simulator;
- It interacts with the simulator to start and stop simulations on behalf of users, and updates related information on the database;
- It accesses the file repository to retrieve simulation results and show them to users, according to their preferences, through the graphical interface;
- When requested by a user, it builds a Gnuplot file for the comparison of results related to different simulation runs, and passes it to the graphical interface.

SO-Grid GUI

The portal front-end provides the user GUI through which users can operate on the portal, execute simulations and visualize results. In this section, we show some of the GUI Web forms, in order to illustrate how the portal can be actually used.

Figure 6 depicts the login form that is presented to users when they want to enter the simulation area. Notice that the other areas, i.e. the "Home" area and the areas that give access to information concerning documents and involved people, is available to all users and does not need a log on procedure.

Once in the simulation area, the user can choose whether to run a new simulation or visualize the results of one or more simulations executed previously. If a simulation started in a previous session is still running, the user can also monitor its execution.

Before running a new simulation, the user accesses a Web form in which he/she can set the duration of the simulation and the values of network and simulation parameters, and can choose the name of the results file, as shown in Figure 7. For each parameter, the user can accept the default value or decide to change and set it within a given range. For example, the default number of classes is 5, but values between 3 and 12 are admitted.

While the simulation proceeds, the user can visualize, at run time, the values related to the different performance indices, by means of plots built by the Gnuplot software. As an example, Figure 8 shows a snapshot of the SO-Grid interface, in which the user can observe the trend of the overall entropy, plotted versus time. Whenever he/she wants, the user can decide to stop a running

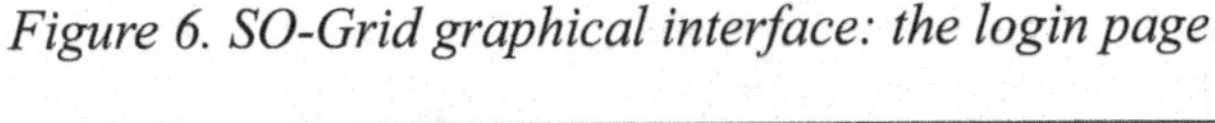

Figure 6. SO-Grid graphical interface: the login page

Figure 7. SO-Grid graphical interface: Web form for parameter setting

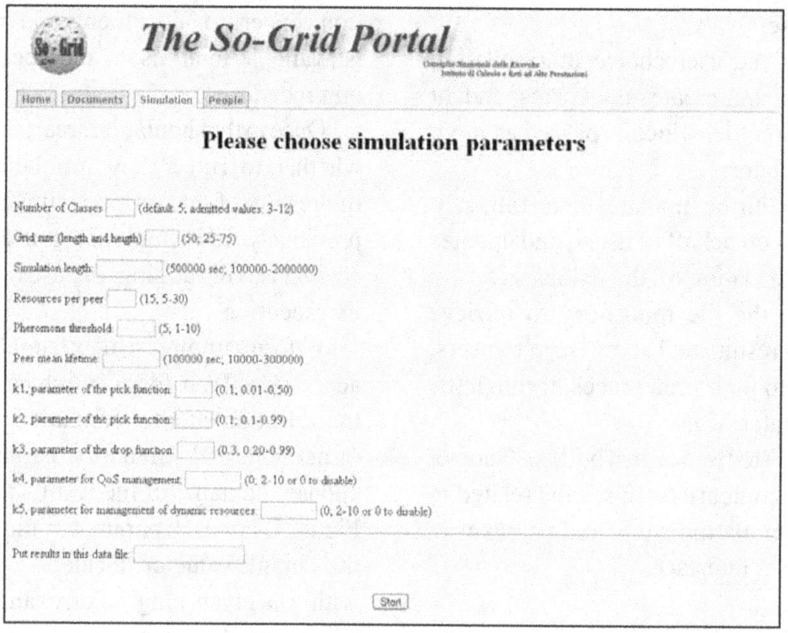

The So-Grid Portal
Home
Documents
Simulation
People
Please choose simulation parameters
Number of Classes (default 5, admitted values 3-12)
Grid size (length and heigth) (50, 25-75)
Simulation length (500000 sec, 100000-2000000)
Resources per peer (15, 5-30)
Pheromone threshold (5, 1-10)
Peer mean lifetime (100000 sec, 10000-300000)
k1, parameter of the pick function (0.1, 0.01-0.50)
k2, parameter of the pick function (0.1; 0.1-0.99)
k3, parameter of the drop function (0.3, 0.20-0.99)
k4, parameter for QoS management (0, 2-10 or 0 to disable)
k5, parameter for management of dynamic resources (0, 2-10 or 0 to disable)
Put results in this data file
Start

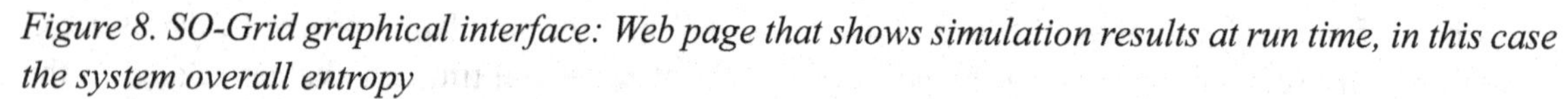

Figure 8. SO-Grid graphical interface: Web page that shows simulation results at run time, in this case the system overall entropy

The So-Grid Portal
Home
Documents
Simulation
People
Overall Entropy
Entropy
Time(s)
Performance indices
Time 62000
Overall entropy 0.8362
No. of replicas per resource 7.462
No. of agents in copy 184
No. of representative peers 196
Avg QoS of resources 5.0360
Avg age of resources (s) 70746
Perc. of updated resources 38.9
Network traffic (msgs/s) 0.0200
Perc. of striking queries 0.9416
Avg no. of results per query 566
Avg response time (s) 0.91
Avg QoS of results 5.02
Avg age of results (s) 73671
Stop Simulation
Main menu

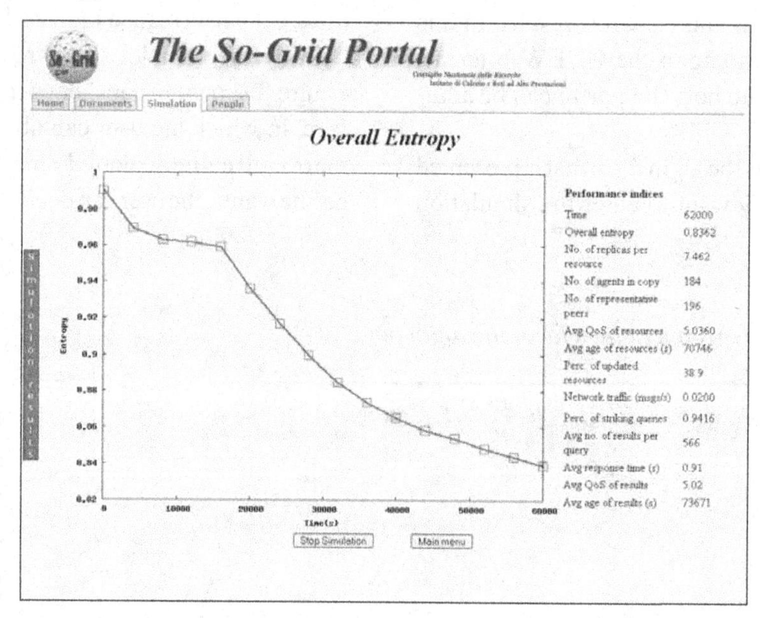

Figure 9. SO-Grid graphical interface: Web form that allows the user to compare results related to different simulations

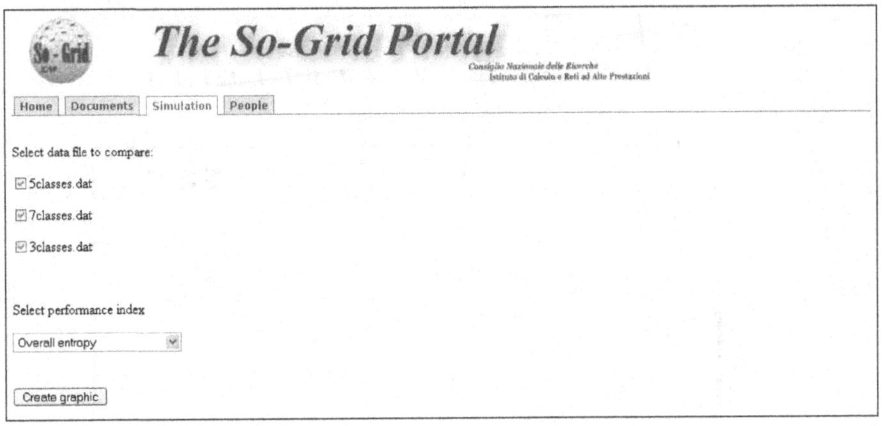

simulation, otherwise this will terminate when the simulation time expires; in both cases the results are automatically saved in the file whose name was specified in the parameter setting phase.

The user can also configure and visualize a plot that compares results related to different simulations. Through the form shown in Figure 9, he/she can check the results files corresponding to past simulations (in this case results obtained with different numbers of classes), and the performance index of interest. After clicking on the "Create graphic" button, the user will be shown a plot that compares the values of the chosen performance index obtained with the selected simulations. This is a very user friendly and powerful tool that allows users to carry out "parameter sweep" analysis of the bio-inspired protocols presented in this chapter. For example, Figure 10 depicts the values of the overall entropy obtained in three simulations in which the number of classes is set to 3, 5 and 7.

CONCLUSION AND FUTURE WORK

This chapter proposes a bio-inspired approach for the construction of a Grid information system. The ant-inspired ARMAP protocol allows for the dissemination and reorganization of resource descriptors according to the characteristics of related resources. This in turn enables the use of another ant-inspired protocol, ARDIP, which copes with the discovery of resources on behalf of users. ARMAP is executed by a number of ant-like mobile agents that travel the Grid through P2P interconnections among hosts. Agents disseminate metadata documents on the Grid and aggregate information related to similar resources in neighbor Grid nodes, so contributing to decrease the overall system entropy. Resource replication and reorganization can be tuned by appropriately setting a pheromone threshold in order to foster or reduce the activeness of agents. Simulation results show that the ARMAP protocol is able to achieve the mentioned objectives, and is inherently scalable, as agent operations are driven by self-organization and fully decentralized mechanisms, and no information is required about the global state of the system.

To foster the understanding and use of bio-inspired algorithms in distributed systems, this chapter also illustrates the SO-Grid Portal, a simulation portal through which registered users can simulate and experience the behaviour of

Figure 10. SO-Grid graphical interface: comparison of overall entropy values obtained when the number of classes is set to 3, 5 and 7

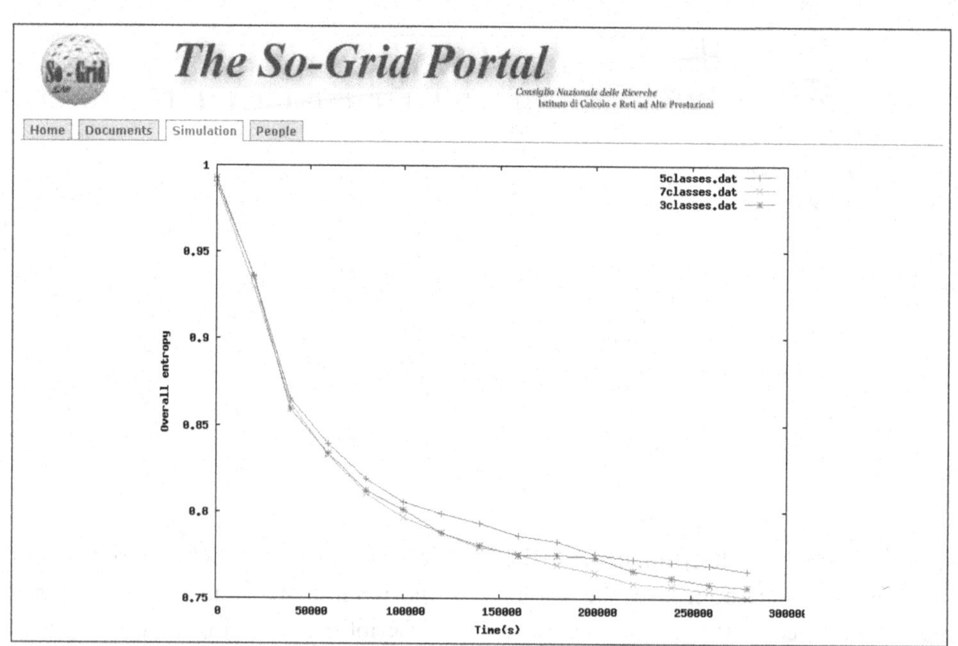

the discussed protocols. This portal is specifically designed to carry out "parameter sweep" analysis, since a user can graphically compare the results of different simulations executed in previous sessions.

The exploitation of bio-inspired algorithms in distributed systems surely needs more analysis and evaluation in future years. An important issue is that results are always obtained empirically and tuned by experience: a very important step forward would be the construction of an analytical model that relates the operations of simple agents to the overall results obtained by their collective work.

A set of research avenues could also be explored to enhance the efficiency of the discussed protocols. For example, in a future version proposed in this chapter, agents could replicate and disseminate descriptors on the basis of the quality of service (QoS) of the corresponding resources. If the descriptors pertaining to high quality and recently updated resources were disseminated more rapidly than other descriptors, the users would be likely given access to more useful and appropriate resources.

Other improvements could be obtained by better driving the movements of agents: for example, an agent that has reached an accumulation region on the Grid could give other agents this information so as to make their work more effective. This kind of interaction could be inspired by another class of bio-inspired mechanisms, which try to emulate the movements of birds in a flock. Another possible enhancement could be obtained by reorganizing the P2P overlay of Grid hosts according to the "small-world" paradigm (Kleinberg, 2000). A small number of short distance P2P connections should be substituted by long-distance connections, that would directly link hosts located in distant regions of the Grids. It is conceivable that this approach, without significantly altering the load of the system, would allow agents to move on the network more rapidly and efficiently, which could fasten the dissemination and reorganization of descriptors.

REFERENCES

Aktas, M. S., Fox, G. C., & Pierce, M. (2007). Fault tolerant high performance Information Services for dynamic collections of Grid and Web services. *Future Generation Computer Systems*, 23(3), 317-337.

Bonabeau, E., Dorigo, M., & Theraulaz, G. (Eds.). (1999). *Swarm Intelligence: From Natural to Artificial Systems*, New York, USA: Oxford University Press.

Crespo, A., & Garcia-Molina, H. (2002). *Routing indices for peer-to-peer systems.* Paper presented at the 22nd International Conference on Distributed Computing Systems (ICDCS 2002), Wien, Austria.

Dasgupta, P. (2004). *Intelligent Agent Enabled P2P Search Using Ant Algorithms*, Paper presented at the 8th International Conference on Artificial Intelligence, Las Vegas, USA.

Deneubourg, J. L., Goss, S., Franks, N., Sendova-Franks, A., Detrain, C., & Chretien, L. (1990) *The dynamics of collective sorting robot-like ants and ant-like robots.* Paper presented at the 1st International Conference on Simulation of Adaptive Behavior on From Animals to Animats, Cambridge, USA.

Eugster, P., Guerraoui, R., Kermarres, A.M., & Massoulieacute, L. (2004). *Epidemic Information Dissemination in Distributed System*, IEEE Computer, 37(5), 60-67

Forestiero, A., Mastroianni, C., & Spezzano, G. (2005). *A Multi Agent Approach for the Construction of a Peer-to-Peer Information System in Grids*, Paper presented at the International Conference on Self-Organization and Adaptation of Multi-agent and Grid Systems (SOAS 2005), Glasgow, United Kingdom.

Forestiero, A., Mastroianni, C., & Spezzano, G. (2007). *A Decentralized Ant-Inspired Approach for Resource Management and Discovery in Grids*, Multiagent and Grid Systems - An International Journal, 3(1), 43-63.

Grassè, P. (1959). *La reconstruction du nid et les coordinations inter-individuelles chez belicositermes natalensis et cubitermes sp. la theorie de la stigmergie: Essai d'interprtation du comportement des termites constructeurs*, Insectes Sociaux 6, 41–84.

Iamnitchi, A., & Foster, I. (2005). *Interest-aware information dissemination in small-world communities*, Paper presented at the 14th IEEE International Symposium on High Performance Distributed Computing (HPDC 2005), Research Triangle Park, USA.

Kleinberg, J. (2000). Navigation in a Small World , Nature 406(2000), 845.

Lv, C., Cao, P. Cohen, E. Li, L., & Shenker, S. (2002). Search and *replication in unstructured peer-to-peer networks*, ACM, Sigmetrics.

Minar, N., Burkhart, R., Langton, C., & Askenazi, M. (1996). *The swarm simulation system, a toolkit for building multi-agent simulations*, from http://www.santafe.edu.

Petersen, K., Spreitzer, M., Terry, D., Theimer, M., & Demers, A. (1997). *Flexible Update Propagation for Weakly Consistent Replication*, Paper presented at the 16th ACM Symposium on Operating System Principles, Saint-Malo, France.

Sycara, K. (1998). *Multiagent systems*, Artificial Intelligence Magazine, Association for the Advancement of Artificial Intelligence, 10(2), 79–93.

Tsoumakos, D., & Roussopoulos, N. (2003). *A Comparison of Peer-to-Peer Search Methods*, Paper presented at the 6th International Workshop on the Web and Databases (WebDB), San Diego, USA.

Van Dyke Parunak, H., Brueckner, S. A., Matthews, R., & Sauter, J. (2005). *Pheromone Learning for Self-Organizing Agents*, IEEE Transactions on Systems, Man, and Cybernetics, Part A: Systems and Humans, 35(3).

ADDITIONAL READING

To better understand the concepts concerning bio-inspired techniques for the management of resources in distributed systems, the reader can refer to "Swarm Intelligence: From Natural to Artificial Systems" authored by Bonabeau, Dorigo and Theraulaz, pioneers of so called *ant optimization* and the simulation of social insects. With their book , they provide an overview of the state of the art in swarm intelligence. They go further by outlining future directions and areas of research where the application of biologically inspired algorithms seems to be promising. The book describes several phenomenona in social insects that can be observed in nature and explained by models. The book also provides examples of the latter being transferred successfully to algorithms, multi-agent systems and robots, or at least describes promising approaches for doing this.

Another interesting book is "Ant Colony Optimization" authored by Dorigo and Stützle. This book presents an overview of this rapidly growing field, from its theoretical inception to practical applications, including descriptions of many available ACO algorithms and their uses. The book first describes the translation of observed ant behavior into working optimization algorithms. The ant colony metaheuristic is then introduced and viewed in the general context of combinatorial optimization. This is followed by a detailed description and guide to all major ACO algorithms and a report on current theoretical findings. The book surveys ACO applications now in use, including routing, assignment, scheduling, subset, machine learning, and bioinformatics problems.

Other suggested works in the field of swarm intelligence are the following:

Camazine, S., Franks, N. R., Sneyd, J., Bonabeau, E., Deneubourg, J.-L., and Theraula, G. (2001). Self-Organization in Biological Systems. Princeton University Press, Princeton, NJ, USA.

Deneubourg, J. L., Goss, S., Franks, N., Sendova-Franks, A., Detrain, C., & Chretien, L. (1990) *The dynamics of collective sorting robot-like ants and ant-like robots*. Paper presented at the 1st International Conference on Simulation of Adaptive Behavior on From Animals to Animats, Cambridge, USA.

Grassè, P. (1959). *La reconstruction du nid et les coordinations inter-individuelles chez belicositermes natalensis et cubitermes sp. la theorie de la stigmergie: Essai d'interprtation du comportement des termites constructeurs*, Insectes Sociaux 6, 41–84.

Martin, M., Chopard, B., and Albuquerque, P. (2002). Formation of an ant cemetery: swarm intelligence or statistical accident? Future Generation Computer Systems 18, 7, 951–959.

Parunak, H. V. D., Brueckner, S., Matthews, R. S., and Sauter, J. A. (2005). Pheromone learning for self-organizing agents. IEEE Transactions on Systems, Man, and Cybernetics, Part A 35, 3, 316–326.

Van Dyke Parunak, H., Brueckner, S. A., Matthews, R., & Sauter, J. (2005). *Pheromone Learning for Self-Organizing Agents*, IEEE Transactions on Systems, Man, and Cybernetics, Part A: Systems and Humans, 35(3).

Additional readings concerning the management of resources both in Grid environment and in peer to peer networks, are:

Anand, S., Padmanabhan, G., Shaowen, R. and Wang, B. (2005). A self-organized grouping (sog) method for efficient grid resource discovery. In Proc. of the 6th IEEE/ACM International Workshop on Grid Computing. Seattle, Washington, USA. ACM Journal Name, Vol. V, No. N, April 2007.

Chakravarti, A. J., Baumgartner, G., and Lauria, M. (2005). The organic grid: self-organizing computation on a peer-to-peer network. IEEE Transactions on Systems, Man, and Cybernetics, Part A 35, 3, 373–384.

Cheema, A. S., Muhammad, M., and Gupta, I. (2005). Peer-to-peer discovery of computational resources for grid applications. In Proc. of the 6th IEEE/ACM International Workshop on Grid Computing. Seattle, WA, USA, 179–185.

Cohen, E. and Shenker, S. 2002. Replication strategies in unstructured peer-to-peer networks. In Proc. of the Special Interest Group on Data Communication ACM SIGCOMM'02. Pittsburgh, Pennsylvania, USA.

Foster, I. and Kesselman, C. (2003). The Grid 2: Blueprint for a New Computing Infrastructure. Morgan Kaufmann Publishers Inc., San Francisco, CA, USA.

Iamnitchi, A., Foster, I., Weglarz, J., Nabrzyski, J., Schopf, J., and Stroinski, M. (2003). A peer-to-peer approach to resource location in grid environments. In Grid Resource Management. Kluwer Publishing.

Petersen, K., Spreitzer, M. J., Terry, D. B., Theimer, M. M., and Demers, A. J. (1997). Flexible update propagation for weakly consistent replication. In Proc. of the Sixteenth ACM symposium on Operating systems principles SOSP '97. ACM Press, New York, NY, USA, 288–301.

Ran, S. (2003). A model for web services discovery with qos. ACM SIGecom Exchanges 4, 1, 1–10.

Sharma, P., Estrin, D., Floyd, S., and Jacobson, V. (1997). Scalable timers for soft state protocols. In Proc. of the 16th Annual Joint Conference of the IEEE Computer and Communications Societies, INFOCOM'97. Vol. 1. IEEE Computer Society, Washington, DC, USA,222–229.

Stoica, I., Morris, R., Karger, D., Kaashoek, M. F., and Balakrishnan, H. (2001). Chord: A scalable peer-to-peer lookup service for internet applications. In Proc. of the Conference on Applications, technologies, architectures, and protocols for computer communications SIGCOMM'01. ACM Press, New York, NY, USA, 149–160.

TheGlobusAlliance. (2006). The web services resource framework. http://www.globus.org/wsrf/.

Trunfio, P., Talia, D., Fragopoulou, P., Papadakis, C., Mordacchini, M., Pennanen, M., Popov, K., Vlassov, V., and Haridi, S. (2006). Peer-to-peer models for resource discovery on grids. Tech. Rep. TR-0028, Institute on System Architecture, CoreGRID Network of Excellence. March. For an extended version, see "Peer-to-Peer resource discovery in Grids: Models and systems", Future Generation Computer Systems, 2007.

Tsoumakos, D. and Roussopoulos, N. (2003a). Adaptive probabilistic search for peer-to-peer networks. In Proc. of the Third IEEE International Conference on P2P Computing P2P'03. 102–109.

Vu, L. H., Hauswirth, M., and Aberer, K. (2005). QoS-based Service Selection and Ranking with Trust and Reputation Management. In Proc. of the International Conference on Cooperative Information Systems CoopIS'05, 31 Oct - 4 Nov 2005, Agia Napa, Cyprus. Vol. 3760. 446–483.

Finally, we propose the following readings that discuss bio-inspired approaches for the dissemination and discovery of resources in Grid environment:

Erdil, D. C., Lewis, M. J., and Abu-Ghazaleh, N. (2005). An adaptive approach to information dissemination in self-organizing grids. In Proc. of the International Conference on Autonomic and Autonomous Systems ICAS'06. Silicon Valley, CA, USA.

Eugster, P., Guerraoui, R., Kermarres, A.M., & Massoulieacute, L. (2004). *Epidemic Information Dissemination in Distributed System*, IEEE Computer, 37(5), 60-67

Forestiero, A., Mastroianni, C., & Spezzano, G. (2005). *A Multi Agent Approach for the Construction of a Peer-to-Peer Information System in Grids*, Paper presented at the International Conference on Self-Organization and Adaptation of Multi-agent and Grid Systems (SOAS 2005), Glasgow, United Kingdom.

Forestiero, A., Mastroianni, C., & Spezzano, G. (2007). *A Decentralized Ant-Inspired Approach for Resource Management and Discovery in Grids*, Multiagent and Grid Systems - An International Journal, 3(1), 43-63.

Forestiero, A., Mastroianni, C., and Spezzano, G. (2006). An agent based semi-informed protocol for resource discovery in grids. In Proc. of the International Conference on Computational Science(4) ICCS'06. 1047–1054.

Chapter XVII
Graph Based Evolutionary Algorithms

Steven M. Corns
Iowa State University, USA

Daniel A. Ashlock
University of Guelph, Canada

Kenneth Mark Bryden
Iowa State University, USA

ABSTRACT

This chapter presents Graph Based Evolutionary Algorithms. Graph Based Evolutionary Algorithms are a generic enhancement and diversity management technique for evolutionary algorithms. These geographically inspired algorithms are different from other methods of diversity control in that they not only control the rate of diversity loss at low runtime cost but also allow for a means to classify evolutionary computation problems. This classification system enables users to select an algorithm a priori that finds a satisfactory solution to their optimization problem in a relatively small number of fitness evaluations. In addition, using the information gathered by evaluating several problems on a collection of graphs, it becomes possible to design test suites of problems which effectively compare a new algorithm or technique to existing methods.

INTRODUCTION

Graph based evolutionary algorithms (GBEAs) are a diversity management technique for evolutionary algorithms which places the members of the evolving population on the vertices of a combinatorial graph. They also provide a generic improvement by reducing the number of fitness evaluations required to find an acceptable solution at exceedingly low runtime cost through the selection of the correct graph as a population structure. The combinatorial graph provides a geography for the evolving population of solutions. Different graphs yield different levels of diversity preservation. The level of preservation needed for effective solution is problem specific.

The amount of diversity present in a population initially is related to population size, but, even in a large population, this initial diversity can disappear rapidly. Graph based evolutionary algorithms impose restrictions on mating and placement within the evolving population (new structures are placed in their parents' graph neighborhood). This simulates natural obstacles (like geography) or social obstacles (like the mating dances of some bird species) to mating (Kimura & Crow, 1963; Wright, 1986).

The GBEA technique differs from other AI enhancements to search and optimization in that they are a minor algorithmic modification that can be added to any evolutionary algorithm without requiring complex pre-calculation, costly additional data structures, or extensive modifications of the algorithm being enhanced. Only the adjacency matrix of a combinatorial graph need be added, and it is used only to limit mate choice – typically replacing a single call to a randomization algorithm. GBEAs are an exceptionally simple enhancement, provided the correct graph for a given problem is known.

When Evolutionary Algorithms (EAs) are used as an optimization process, populations of potential solutions are evolved to search a solution space for an acceptable answer. This often avoids problems with convergence to local optima that are found in gradient search methods (Goldberg, 1989) and enables the search of discrete spaces. These methods have provided novel and superior solutions for a wide range of problems in physics, engineering, biology, economics and many other areas (Blaize, Knight, & Rasheed, 1998; Fabbri, 1997; Keane, 1994; Vavak, Jukes, & Fogarty, 1997). A key issue when applying EAs is how to maintain sufficient diversity. When diversity loss is too rapid, an ineffective search of the solution space may occur, leading to premature convergence. When diversity is preserved too aggressively, the algorithm may converge slowly or not at all. The right amount of diversity is problem specific. A rule of thumb is that the need to preserve diversity increases with problem complexity and deceptiveness (Bryden, Ashlock, Corns, & Willson, 2006).

There are several available methods for preserving diversity, including EcoGAs (Davidor, 1993), island GAs (Whitley & Starkweather, 1990), niching (Goldberg, 1989), and taboo search (Goldberg, 1989). EcoGAs and Island GAs are essentially the same as a type of GBEA, albeit using a less diverse class of graphs. EcoGAs use a grid shaped graph, and Island GAs use clusters of complete graphs connected through controlled migration events. Niching and taboo search require comparison of solutions in the evolving population either to all other solutions in the population or a taboo list, and so incur additional computational costs. GBEAs have exceedingly low computational overhead because of the fashion in which they mimic passive methods of diversity preservation found in nature.

Carefully choosing the connectivity of the combinatorial graphs used permits the level of diversity preservation to be increased or decreased (Bryden et al., 2006). Greater diversity causes more candidate solutions to evolve within the population. They may compete or cooperate (by juxtaposing compatible sub-structures) as they explore the search space. This can also result in finding multiple solutions for multi-modal problems.

An unexpected and welcome outcome of research on GBEAs has been that, by examining the effects that different graph structures have on the time to solution for different problems, it was observed that similar problems grouped together. This makes it possible to use the performance on different graphs as a taxonomic character, giving a similarity measure with which to compare problems. These graph-based characters of problems make it possible to generate and evaluate test suites that ensure a thorough evaluation of new techniques or algorithms. In addition, having a database of problems that documents the best graph for each problem allows the user to make informed decision when selecting a graph for use on a new problem.

BACKGROUND

A *combinatorial graph* or *graph* G is a collection of vertices V(G) and edges E(G), where E(G) is a set of unordered pairs from V(G). Two vertices of the graph are *neighbors* if they are members of the same edge. We say an edge is *incident* on the two vertices it contains. The *degree* of the vertex is the number of edges containing (incident upon) that vertex. If all vertices in a graph have the same degree, the graph is said to be *regular*, and if the common degree of a regular graph is *k*, then the graph is said to be *k-regular*. If you can go from any vertex to any other vertex traveling along vertices and edges of the graph, the graph is *connected*. A *cycle* is a path going from any vertex back to itself. The *diameter* of a graph is the longest that the most direct path between any two of the vertices can be.

List of Graphs

The mathematical definitions of the graphs used here are given, as well as those of other graphs required to properly describe them.

Definition 1: *The* complete graph *on n vertices, denoted K_n, has n vertices and all possible edges. An example of a complete graph is shown in Figure 1(a).*

Definition 2: *The* complete bipartite graph *with n and m vertices, denoted $K_{n,m}$, has vertices divided into disjoint sets of n and m vertices and all possible edges that have one end in each of the two disjoint sets. The 3-pre-Z graph shown in Figure 1(e) is the complete bipartite graph $K_{4,4}$.*

Definition 3: *The* n-cycle, *denoted C_n, has vertex set Z_n. Edges are pairs of vertices that differ by 1 (mod n) so that the vertices form a ring with each vertex having two neighbors.*

Definition 4: *The* n-hypercube, *denoted H_n, has the set of all n character binary strings as its set of vertices. Edges consist of pairs of strings that differ in exactly one position. A 4-hypercube is shown in Figure 1(d).*

Definition 5: *The* n x m-torus, *denoted $T_{n,m}$, has vertex set $Z_n \times Z_m$. Edges are pairs of vertices that differ either by 1 (mod n) in their first coordinate or by 1 (mod m) in their second coordinate, but not both. These graphs are n x m grids that wrap (as tori) at the edges. A T_{12_6} graph is shown in Figure 1(c).*

Definition 6: *The generalized* Petersen graph *with parameters n,k, denoted $P_{n,k}$ has vertex set [0, 1, . . . , 2n – 1]. The two sets of vertices are both considered to be copies of Z_n. The first n vertices are connected in a standard* n-cycle. *The second n vertices are connected in a cycle-like fashion, but the connections jump in steps of size k (mod n). The graph also has edges joining corresponding members of the two copies of Z_n. The graph $P_{32,5}$ is shown in Figure 1(b).*

Definition 7: *A* tree *is a connected graph with no cycles. Degree zero or one vertices are termed leaves of the tree. A balanced regular tree of degree k is a tree constructed in the following manner. Begin with a single vertex. Attach k neighbors to that vertex and place these neighbors in a queue. Processing the queue in order, add k-1 neighbors to the vertex most recently removed from the queue and add these neighbors to the end of the queue. Continue in this fashion until the tree has the desired number of vertices. We denote these graphs RBT(n,k) where n is the number of vertices. (Notice that not all n are possible for a given k.)*

Definition 8: *The* graph Z *is created by starting with a bipartite graph and then simplexifying the entire graph to reach the desired number of vertices. Simplexification is a graph operation defined in (Bryden et al., 2006). Two of the steps leading to the* graph Z *are shown in Figure 1(e) and 1(f).*

In addition, four classes of random graphs are used. A random graph is specified by a randomized algorithm which corresponds to a type of random graph (a probability distribution on some set of graphs). We use three instances of each type of random graph.

Definition 9: *An edge move is performed as follows. Two edges {a, b} and {c, d} are found that have the property that none of {a, c}, {a, d}, {b, c}, or {c, d} are themselves edges. The edges {a, b} and {c, d} are deleted from the graph, and the edges {a, c} and {b, d} are added. Notice that edge moves preserve the regularity of a graph if it is regular.*

Definition 10: *We generate random regular graphs by the following algorithm. Start with a regular graph and then repeatedly perform 3,000 edge moves on vertices selected uniformly at random from those that are valid for edge moves. For 3-regular random graphs of population size 512, use $P_{256,1}$ as the starting point. For 4-regular random graphs of population size 512, use $T_{16,32}$ as the starting point. For 9-regular random graphs of population size 512, use H_9 as the starting point. These graphs are denoted R(n, k, i), where n is the number of vertices, k is the regular degree, and i = 1, 2, 3 is the instance of the graph in this study.*

Definition 11: *We generate random toroidal graphs as follows. A desired number of points are placed onto the unit torus (unit square wrapped at*

Figure 1. Examples of graphs used in this study

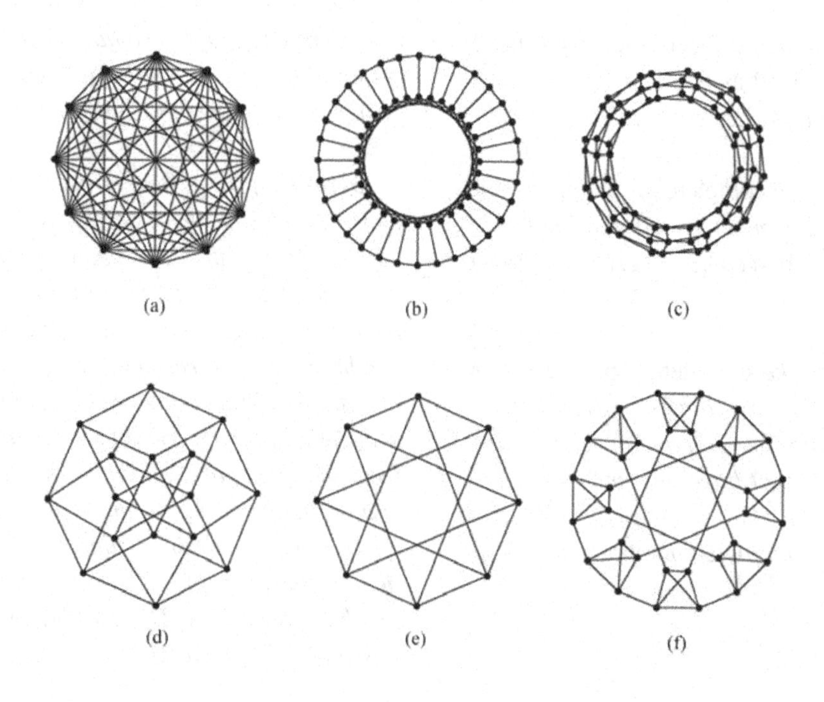

the edges), and edges are created between those within a specified distance from one another. This distance is chosen to give an average degree of about 6. After generation, the graph is checked to see if it is connected. Graphs that are not connected are rejected. Hence, the graphs used were chosen at random from connected random toroidal graphs. These graphs are denoted R(r, i), where r is the radius for edge creation, and i = 1, 2, 3 is the instance of the graph in this study.

A graph used to constrain mating in a population is called the *population structure* of a GBEA. It specifies the geography on which a population lives, permitting mating only between neighbors. The goal is to find graphs that preserve diversity without hindering progress. To date, twenty-six graphs have been tested to investigate how the different underlying structures affect information spread (Corns, Bryden, & Ashlock, 2003; Corns, Bryden, & Ashlock, 2005a; Corns, Bryden, & Ashlock, 2005b).

GRAPH BASED EVOLUTIONARY ALGORITHMS

This section carefully defines graph based evolutionary algorithms. GBEAs are a simple extension of evolutionary algorithms that distribute the population members on a combinatorial graph, placing one solution on each vertex of the graph. It is only possible for population members to mate with or, after mating has generated new solutions, replace graph neighbors. For the local mating rules used here, a steady state evolutionary algorithm is used, in which evolution occurs one mating event at a time (Reynolds, 1992; Syswerda, 1991; Whitley, 1989). A mating event is performed by randomly selecting a population member (the *parent*), and then using fitness proportional selection to determine a mate (the *co-parent*) from the parent's neighbors. A single child is generated and compared to the parent. The child replaces the parent if it is at least as fit. Figure 2 gives a flowchart representation of the algorithm.

Graph connectivity is roughly inversely related to the graph's ability to preserve diversity. Highly connected graphs provide the highest rate of information transfer and thus lower diversity preservation. The complete graph is the most highly connected graph, which permits a GBEA to simulate a standard evolutionary algorithm. Highly connected graphs have superior time to convergence when used for simple uni-modal problems, such as the one-max problem (Bryden et al., 2006). As the number of optima and deceptiveness of the problem increases, graphs which are more sparsely connected have superior mean acceptable solution times. Sparse graphs have been found to perform well in deceptive genetic programming problems, e.g. the PORS problem (Bryden et al., 2006), and the problem of locating DNA error correcting codes (Ashlock, Guo, & Qiu, 2002). To date, use of the correct graph has resulted in a decrease in time to convergence to an optimal solution by as much as 63-fold for the most difficult problem tried. The impact varies substantially with the problem, and about half of all problems tested show a benefit from using a GBEA.

Closely related to the graph connectivity is population size, since both can impact population diversity. For a given type of graph, diameter increases with population size; graphs of low degree exhibit the sharpest increase. When the results for all problems were plotted, it was found that, for all but the most trivial problems, there was an optimal population size for each type of graph, such as in the PORS 16 problem (Figure 3). Using a sparse graph typically preformed well for small population sizes as the initial amount of diversity doesn't have to be as large when a diversity preserving graph is used. By using a diversity preserving graph on problems with expensive fitness evaluations, the time to convergence can be greatly reduced. In addition, when the proper graph is combined with

Figure 2. Algorithm used for graph based evolutionary algorithms

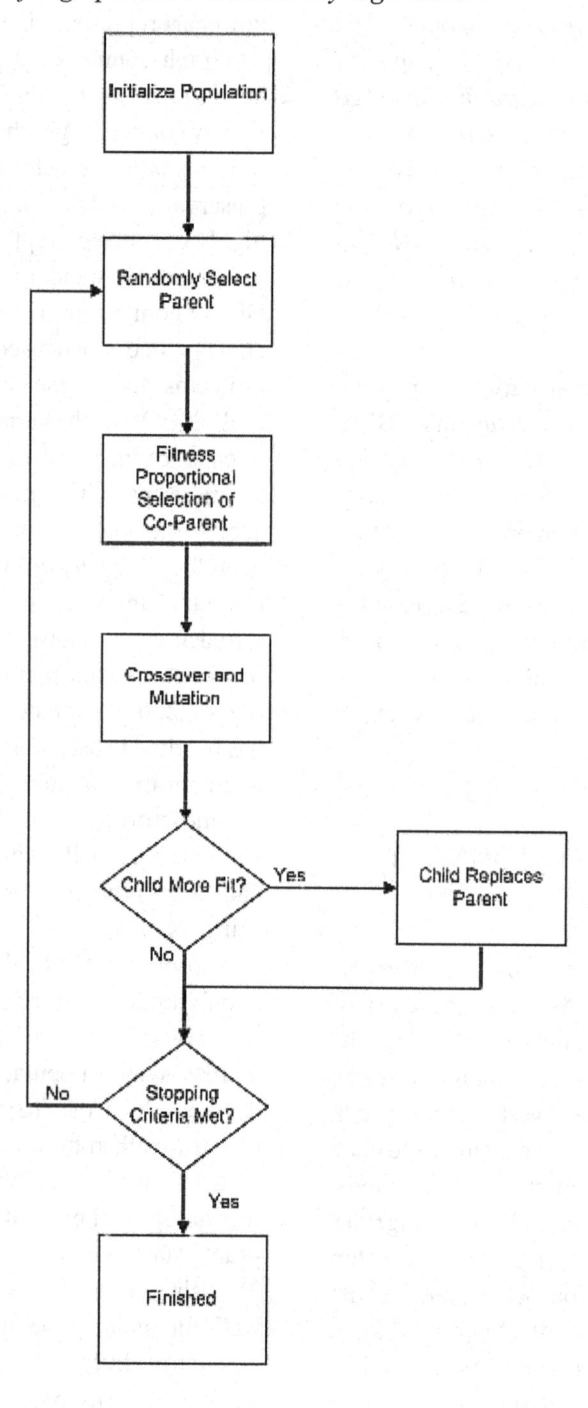

Figure 3. The average number of mating events to solution as a function of population size and graph for the PORS 16 problem

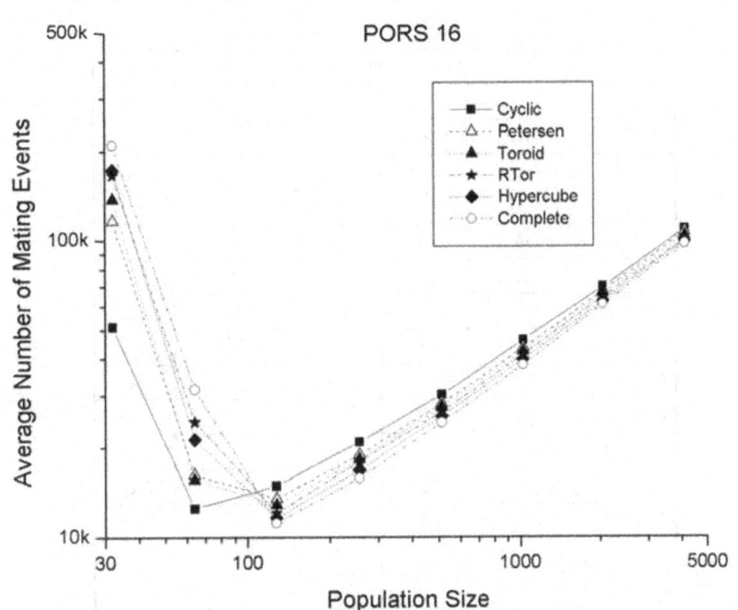

the best known population size, average solution time decreases as much as 94-fold for deceptive problems, an additional performance increase of nearly 50% of that obtained by selecting the right type of graph.

Proper graph selection can also allow users to obtain multiple distinct solutions. It has been found in both test problems (Corns, Ashlock, McCorkle, & Bryden, 2006a) and real world applications (Xiao, Bryden, Engelbrecht, Huang, & McCorkle, 2004) that, while there is a graph that will give a satisfactory solution in the shortest amount of time, using a graph that has a higher level of diversity preservation allows distinct designs to remain in the population. This has been established by making direct comparisons between the population members at the conclusion of experiments when different graph structures were used to solve highly multi-modal problems. Multiple distinct solutions are advantageous for engineering design, where it is typically desirable to have several design alternatives available.

TEST PROBLEMS

Initially 40 problems were used to evaluate the impact that graph choice had on time to solution. For 39 of these test problems, 5,000 simulations were performed for each of the graphs listed in Table 1. To obtain better resolution on the relative ranking of the graph Graphs used in this study with their index name for use in figures. The graphs listed here are all population size 512. performances, 10,000 simulations were performed for the differential equation problem. For each simulation, the number of mating events required to find an acceptable solution were recorded. These results were then used to find the mean and 95% confidence interval to allow comparison of different graph performances. The problems fell into four categories: string problems, real variable problems, genetic programming problems (Koza, 1992; Banzhaf, Nordin, Keller, & Francone, 1998), and a miscellaneous category.

Table 1. Graphs used in this study with their index name for use in figures. The graphs listed here are all population size 512.

Graph	Index Name	Regularity	Diameter	Mean Degree	Random
C_{512}	C512	2	256	2	No
H_9	H9	9	9	9	No
K_{512}	Complete	511	1	511	No
$P_{256,1}$	P256_1	3	129	3	No
$P_{256,17}$	P256_17	3	18	3	No
$P_{256,3}$	P256_3	3	46	3	No
$P_{256,7}$	P256_7	3	22	3	No
$R(512,3,1)$	RAND3_1	3	11	3	Yes
$R(512,3,2)$	RAND3_2	3	10	3	Yes
$R(512,3,3)$	RAND3_3	3	10	3	Yes
$R(512,4,1)$	RAND4_1	4	8	4	Yes
$R(512,4,2)$	RAND4_2	4	7	4	Yes
$R(512,4,3)$	RAND4_3	4	7	4	Yes
$R(512,9,1)$	RAND9_1	9	4	9	Yes
$R(512,9,2)$	RAND9_2	9	4	9	Yes
$R(512,9,3)$	RAND9_3	9	4	9	Yes
$R(0.07,1)$	RTor07_1	No	19	7.445	Yes
$R(0.07,2)$	RTor07_2	No	20	7.805	Yes
$R(0.07,3)$	RTor07_3	No	16	7.52	Yes
$T_{16,32}$	T16_32	4	24	4	No
$T_{4,128}$	T4_128	4	66	4	No
$T_{8,64}$	T8_64	4	36	4	No
Z	Z	4	19	4	No
$RBT(512,3)$	RT1n512d3	3,1	16	1.996	No
$RBT(512,4)$	RT1n512d4	4,1	11	1.996	No
$RBT(510,5)$	RT1n510d5	5,1	9	1.996	No

String Problems

The first and simplest problem examined is the one-max problem. This problem uses a string of 20 bits for a chromosome, with the fitness of the string being the number of 1's in the string. Two point crossover and single bit-flip mutation are used in the problem. This results in a very simple uni-modal problem with an easy to visualize fitness landscape.

The second string problem is the self-avoiding walk (SAW) problem. The self-avoiding walk (SAW) problem uses a string over the alphabet {up; down; left; right} as its chromosome which dictate moves made on a grid. Problem difficulty is related to the grid size, with grid sizes of 3x3, 3x4, 4x4, 4x5, 5x5, 5x6, and 6x6 used so far. The length of the chromosome is equal to the number of cells in the grid minus one. A fitness evaluation is performed by starting in the lower left corner of the grid and using the chromosome to dictate moves. The chromosome specifies a *walk*, with moves leaving the grid ignored. The fitness of a chromosome is the number of cells visited at least once. Since there are a number of moves equal to the number of cells minus the first, and since revisiting a cell earns no additional fitness, the optimal solution does not revisit any squares and is thus *self-avoiding*. Every problem has a known best fitness equal to the number of cells, making detection of optimal solutions simpler than previous string problems of variable difficulty such as n-k landscapes (Kauffman & Levin, 1987). For the instances of SAW problems in our studies, two-point crossover and single character mutation (which replaces a character selected at random with a random character) are used.

Two ordered gene problems are used in this study; sorting a list into order and maximizing the period of a permutation (Ashlock, 2006). Both crossover and mutation operators are used in the problem, with the crossover operator retaining the entries after the crossover point, but changing their *order of occurrence* to match the co-parent's. The mutation operator changes two entries in the permutation chosen uniformly at random. For both problems, the crossover and mutation rates are set at 100%. The list ordering problem was run for lists of length 8, 9 and 10, and the period maximization problem used lengths of 30, 32, 34 and 36.

Real Variable Problems

The real-variable problems examined are the five functions proposed by DeJong (denoted F1 – F5) as well as the Griewangk function (Whitley, Mathias, Rana, & Dzubera, 1996) in 3-7 dimensions. The function F1 is a three dimensional bowl. DeJong's function F2 is a fourth degree bivariate polynomial surface featuring a broad sub-optimal peak. Function F3 is a sum of integer parts of five independent variables, creating a function that is flat where it is not discontinuous, a kind of six dimensional ziggurat. Function F4 is a fourth-order paraboloid in thirty dimensions, with distinct diameters in different numbers of dimensions, made more complex by adding Gaussian noise. Function F5 is the "Shekel's Foxholes" function with many narrow local optima placed on a grid. A detailed description can be found in (Bryden et al., 2006). The Griewangk function is a sum of quadratic bowls with cosine terms to add noise. The functions are translated to be a positive function, and one is described in each dimension. The cosine term inserts many local optima, but as the dimension of the function increases it becomes smoother and hence less challenging (Whitely et al., 1996). For this reason we include this function in five cases of relatively low dimension, N = 3, 4 . . . 7.

Genetic Programming Problems

Three genetic programming problems were studied: a differential equation solution problem, the plus-one-recall-store (PORS) problem (Ashlock, 2006), and the parity problem (Koza, 1992). In addition to solving a differential equation within a GBEA, we also modify the usual GP technique by extracting the derivatives needed to compute fitness symbolically. This method is described more completely in (Kirstukas, Bryden, & Ashlock, 2004).

We solve the differential equation:

$$y'' - 5y' + 6y = 0 \quad (1)$$

which is a simple homogeneous equation with the two dimensional solution space,

$$y = Ae^{2t} + Be^{3t} \quad (2)$$

for any constants A, B. The parse tree language used has operations and terminals given in Table 2. Trees were initialized to have six total operations and terminals. Fitness for a parse tree coding a function f(x) was computed as the sum over 100 equally spaced sample points in the range [-2, 2] of the error function $E(x) = (f''(x)-5f'(x)+6f(x))^2$. This is essentially the squared deviation from agreement with the differential equation. This function is to be minimized, and the algorithm continues until 1,000,000 mating events have taken place (this did not happen in practice) or until the fitness function summed over all 100 sample points drops below 0.001. The initial population is composed of randomly generated trees. Crossover was performed by the usual subtree exchange (Koza, 1992). If this produced a tree with more than 22 nodes, then a subtree of the root node iteratively replaced the root node until the tree had n nodes or less. This operation is called *chopping.* Mutation was performed by replacing a subtree picked uniformly at random with a new random subtree of the same size for each new tree produced. A constant mutation was also applied to each new parse tree. For a tree that contains one or more constants in either a terminal or as part of a unary scaling operation, one of the constants is selected with a uniform distribution and a number uniformly distributed in the region [-0.1, 0.1] is added to it. These constants are initialized in the range of [-1, 1], but may go outside this range by the constant mutation operator. Local elite fitness proportional mating was used for the differential equation solution problem.

The PORS problem is a maximization problem in genetic programming with a small operation set and a calculator style memory (Ashlock & Lathrop, 1998). The goal is to find a parse tree with n nodes that generates the largest possible integer value when it is evaluated. There are two operations: integer addition and a store function that places the current value in an external memory location. There are also two terminals: the integer one and a terminal that recalls the value from the external memory. There are four distinct building blocks using these nodes: one that evaluates to an integer value of 2, one that evaluates to an integer value of 3, one that doubles the input value, and one that triples the input value. By properly combining these building blocks, the maximum value for a given number of nodes can be found. The difficulty of the PORS problem varies by which building blocks are necessary and is tied to the congruence class (mod 3) of the number of nodes permitted. The cases for n = 15, n = 16 and n=17 nodes, respectively representatives of the hardest, easiest and intermediate difficulty classes, were used in this study. The n=15 case has a single solution comprised of smaller building blocks, giving it a deceptive nature, while the PORS 16 problem has several solutions that are relatively easy to find. The initial population was composed of randomly generated trees with exactly n nodes. A successful individual was defined to be a tree that produces the largest possible number (these numbers are computed in (Ashlock & Lathrop, 1998)). Crossover was performed by the usual subtree exchange (Koza, 1992). If this produced a tree with more than n nodes, then a *chopping* operation was performed. Mutation was performed by replacing a subtree picked uniformly at random with a new random subtree of the same size.

The third genetic programming problem examined was the odd-parity problem (Banzhaf et al., 1998). The objective of the problem is to find the truth value (boolean) of the proposition "an odd number of the input variables are true." Two representations were used to investigate this problem: simple parse trees and function stacks. The simple parse tree representation used no ADFs, logical operations, and the same variation operators as the PORS problem. The function stack representation is derived from Cartesian Genetic Programming (Miller & Thomson, 2000), where the parse tree is replaced with a directed acyclic graph that has its vertices stored in a linear chromosome. For both representations, the fitness is the number of cases of the odd-parity problem computed correctly.

Other Problems

Finding Steiner systems is another task examined (Ashlock, Bryden, & Corns, 2005). Steiner systems are used in statistical design of experiments. A *k-tuple* for a given set *V* of n objects has every pair included in one but only one of the *k*-subsets. For the set {A, B, C, D, E, F, G}, an example of a Steiner 3-tuple (triple system) would be: {{A, B, D}, {B, C, E}, {C, D, F}, {D, E, G}, {A, E, F}, {B, F, G}, {A, C, G}}. Using difference sets to establish the length of the chromosome, *k*-tuples are adjacent blocks of *k* integers in the chromosome. Fitness is evaluated by summing the number of distinct differences between members of the same tuple, so that a repeated pair in a tuple does not contribute to the fitness score. Six instances of the Steiner problem ($S(k,n)$) were used: S(3,49), S(3, 55), S(4,37), S(4,49), S(4,61) and S(5,41). For labeling purposes, triple systems are denoted STS*n*, quadruples SQS*n* and quintuples SQu*n*.

The final problem was the DNA barcode problem (Qiu, Guo, Wen, Ashlock, & Schnable, 2003). DNA barcodes are used as embedded markers for genetic constructs so they can be identified when sequencing pooled genetic libraries. They operate as an error correcting code over the DNA alphabet {C, G, A, T} able to correct relative to the edit metric (Gusfield, 1997). Fitness is the size of the code located by Conway's lexicode algorithm (Qiu et al., 2003). The algorithm studied finds six-letter DNA words that are at a mutual distance of at least three.

TAXONOMY

Early work in comparing different graph structures in GBEAs showed that different problem types preferred different families of graphs. Given the large amount of data compiled in these studies, it became apparent that in addition to solving similar problems efficiently, these preferences could be used to taxonomically classify problems.

A taxonomy is a hierarchical classification of a set. Linnaeus established the first definite hierarchy used to classify living organisms. Each organism was assigned a kingdom, phylum, class, order, family, genus, and species. This hierarchy gave a tree-like structure to taxonomy, the enterprise of classifying living creatures. Modern taxonomy of living organisms has nineteen levels of classification, extending Linnaeus' original seven. This can be displayed using a cladogram, a tree diagram showing the evolutionary relationship among various taxonomic groups (Mayr & Ashlock, 1991).

Taking our cue from this taxonomy of living organisms, we construct trees to find and visualize relationships among members of other groups by using a similar method on the distinctive characteristics of those group members. Hierarchical clustering can produce a tree-structured classification of any set of data given only a similarity measure on members of the set and some sort of averaging procedure for members of the set. Here we extract a set of measurements or taxonomic characters from our test problems and use hierarchical clustering to produce a cladogram that classifies the problems as more and less similar. Hierarchical clustering starts with the members of a set, thought of as singleton subsets. It then joins the closest pair and replaces them with their average in character space.

The choice of taxonomic characters used is critical. Imagine, for example, that you are attempting to derive a family tree for a group of twenty insect species. All have six legs and so "number of legs" is a useless character, at least within the group, for classifying the insects. The size of the insects varies a great deal with the weather, which determines the amount of food they can find. Thus, size is a bad character for classification. The ratio of thorax length to abdomen length turns out both to vary within the group and remain the same in different years. The color of the larva also varies within the group and does not depend on the weather. These latter two characters may be useful. One is a real number, the other is a class variable taking on the values red, white, orange, or yellow. If we are performing a numerical taxonomic analysis, we will need to assign numbers in a somewhat arbitrary fashion to the colors. The preceding brief discussion gives only a taste of the difficulty of choosing good taxonomic characters. Readers familiar with automatic classification, decision trees, and related branches of machine learning will recognize that those fields also face similar issues choosing decision variables. Any taxonomic character or decision variable must be relevant to the decision being made, vary across the set of objects being classified, and be cleanly computable for all members of the set of objects being classified.

GBEAs yield a source of taxonomic characters that are computable for any evolutionary computation problem that has a detectable solution or end point. These characters are numerical. These characters are objective in the sense that they do not favor any particular choice of representation or parameter setting. In outline, these characters are computed in the following fashion. The time-to-solution for a problem varies in a complex manner with the choice of graphical connection topology. This complexity is the genesis of our taxonomic characters. The taxonomic characters used to describe a problem are the normalized mean solution times for the problem on each of a variety of graphs. While this presents a set of objective characters that enable automatic classification, these are not necessarily the "right" or only characters. Using results from the 40 problems described, the mean number of mating events to solution were normalized to yield the taxonomic characters for the problems. Normalization consisted of subtracting the minimum average time from each average time and then dividing through by the maximum among the resulting reduced times. The taxonomic characters for each problem are thus numbers in the set [0, 1] for each graph. In this way a method for comparing the similarity of the problems was introduced with the problem's relative hardness removed. This information can then be used to construct a cladogram using neighbor-joining or displayed in two-dimensional space using nonlinear projection. It should be noted that the randomized graphs were indistinguishable from the graphs they were derived from, and so were not used in the taxonomy process. Also, the random toroid graphs had similar behaviors, so only one instance was used for classification, making the total number of graphs used 15.

Neighbor-Joining Taxonomy

For each the 40 problems examined (P), a real vector m(P) was constructed using the normalized mean solution times for the 15 graphs examined. For each pair of problems P and Q, the Euclidean distance d(P,Q) between the vectors m(P) and m(Q) was computed using the formula:

$$d(P,Q) = \sqrt{\sum_{i=1}^{23} \left[m(P,G_i) - m(Q,G_i) \right]^2}$$

d(P,Q) was interpreted as the distance between the problems P and Q.

To transform this distance data into a cladogram (or tree), clustering was done using UPGMA (Unweighted Pair Group Method with Arithmetic mean) (Sneath & Sokal, 1973), a method which is especially reliable when the distances examined have a uniform meaning. Given a collection of taxa and distance dij between taxa i and j, the method starts with the two taxa x and y that are the least distant and merges them into a new unit z representing the average of x and y. For all taxa other than x and y, a new distance from z (diz) is then computed using the averages of their distance from x and y, completely replacing x and y with z. This procedure is then repeated for the next set of least distant pairs until the last two taxa are merged, weighting the computed average distances by the number of original taxa they are formed from.

The distances found using UPGMA are then used to form a cladogram representing how distant the corresponding problem vectors are in Euclidean space (Figure 4). The horizontal distances on this figure are proportional to the distance between problems or the averages of problems that form the interior nodes. Problems that are separated by a small horizontal distance are similar to one another, while problems that have a wider separation are dissimilar.

Nonlinear Projection

Nonlinear projection (Ashlock & Schonfeld, 2005) is used to create a two-dimensional representation of an n-dimensional data set. The Euclidean two-space distances in the projection are as similar as possible to the Euclidean n-space distances, making the method similar in application to the principle components method. Nonlinear projection is more general because all projections on the data points in two dimensional real space are searched rather than just the linear ones. Figure 5 shows a nonlinear projection of the 40 problems in this study, resulting in a topology showing similarity/dissimilarity.

EVALUATING TEST SUITES

Another application of GBEAs is in the development of test suites. When a new method or algorithm is developed, it is necessary to compare it to existing methods. To do this comparison in a meaningful way, a suitable test suite of problems needs to be solved with both the proposed algorithm and the pre-existing methods. A good test suite is composed of problems that among other things are widely different from each other (Whitely et al., 1996). Using the taxonomy of problems constructed by comparing graph performance, it is possible to select problems that have dissimilar characteristics to obtain an effective test suite. While the taxonomy is by no means complete, it already provides useful information on problem relatedness. Referring to Figures 4 and 5, it can be seen that many of the problems that were used in the early development of evolutionary algorithms, such as the DeJong and Greiwangk functions, show a great deal of similarity. This matches (Whitely et al., 1996). While this does not completely discount their value in a test suite, it does show that they do not adequately represent the diverse set of problems that can be examined by evolutionary computation techniques. Of the problems listed earlier, the PORS and SAW problems represent the most diverse groups of test problems. The instances of these problem studied exhibit a broad diversity of best and worst graphs, indicating that each form a broad set of test problems and may be acceptable for use in a test suite.

OTHER APPLICATIONS OF GBEAs

Graph based evolutionary algorithms have been applied successfully to several engineering and biological problems. They have been applied to stove designs (Bryden, Ashlock, McCorkle, & Urban, 2003), DNA error correcting codes (Ashlock et al., 2002), a hydraulic mixing nozzle problem (Xiao et al., 2003) and to a multi-objective problem for finding antibiotic treatment regimens (Corns, Hurd, Hoffman, & Bryden, 2006b) that benefit swine growth and minimize antibiotic resistance in potentially hazardous GI tract bacteria. Each of these problems received the most benefit from using a different graph.

The stove design problem is a uni-modal design problem, and was best solved using a complete or hypercube graph. Like most uni-modal problems, the solution is at the top of a hill in design space, and so, if the problem is non-deceptive, the faster the algorithm converges the better. The DNA error correcting code problem is quite difficult and

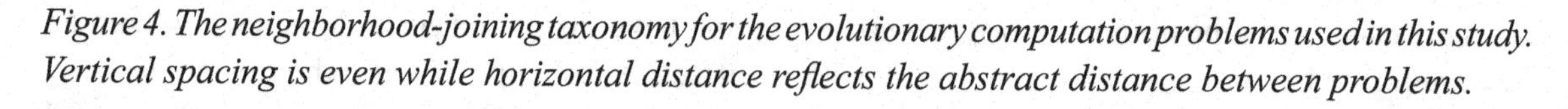

Figure 4. The neighborhood-joining taxonomy for the evolutionary computation problems used in this study. Vertical spacing is even while horizontal distance reflects the abstract distance between problems.

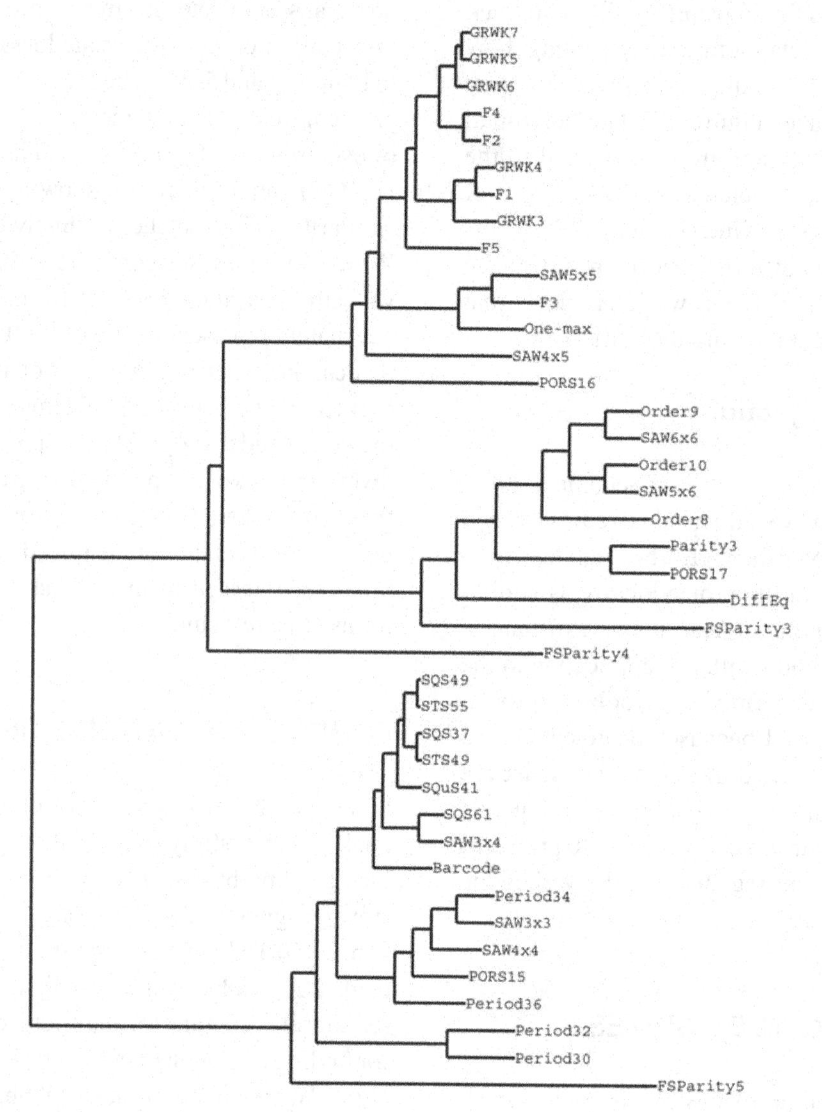

highly multi-modal. It does best with a diversity preserving graph, the cycle, giving a performance increase of approximately 50% compared to the complete graph, corresponding to about 6000 fewer mating events.

The hydraulic mixing nozzle and the antibiotic resistance problems are both still under investigation, but preliminary results show that they both benefit from some diversity preservation, although they also benefits from some sharing of information. The hydraulic mixing nozzle problem provides a single solution using a highly connected graph, and up to three solutions appeared when a sparse graph was used. The antibiotic resistance problem is a multi-objective function that has two competing fitness functions. Because of the

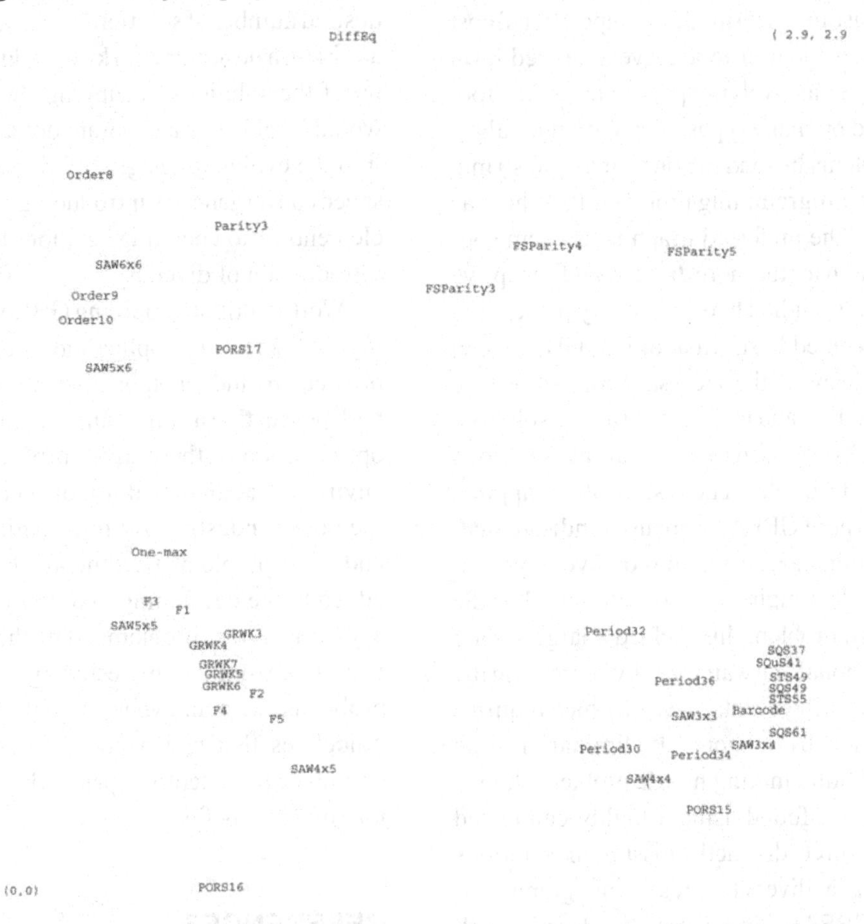

Figure 5. A nonlinear projection in two dimensions of the problem distance in 15 dimensions. The actual position of a problem on the diagram is the beginning of the string representing its name. The points at the corner give the scale of the projection.

more complicated fitness landscape, a graph that preserves a moderate amount of diversity has so far given the best compromise between the two fitness function.

FUTURE TRENDS

Given that the number of available graphs increases exponentially as population size increases, there are a nearly limitless number of graphs available for use in GBEAs. Many of the graphs used in the current graph set are commonly found graph structures or derivations from those simple graphs. As data is compared from different problems sets, it is possible to winnow the graph set to remove graphs that have performance characteristics similar to others in the set. This is offset by the addition of other graphs that are found to perform better than the existing graphs for problems that have been examined. In this way, a taxonomy of graphs can also be developed in a fashion similar to the problems that have been studied so far. Using this co-evolving strategy aids in the understanding of underlying similarities between evolutionary algorithms. In addition, it allows for the evaluation of the graph set to improve how well this method works on problems yet to be analyzed.

CONCLUSION

Graph Based Evolutionary Algorithms provide a simple-to-use tool with modest computational cost which allows the user to achieve a desired level of diversity in an evolving population. This tool can be used on many types of evolutionary algorithm problems beyond the bit string, real string, and genetic programming problems used here as examples. The preferred graph is problem specific, although for the more difficult and deceptive test problems studied here a diversity preserving graph performed best. In addition, GBEAs can be used to decrease the necessary population size to find a solution and to decrease time to solution. They can also be used to find multiple solutions if such are desirable. The results of the applied problems where GBEAs were used indicate that, at a minimum, modest amounts of diversity preservation yield a higher quality solution. For the stove design problem this yielded a large saving of computational and wall times by decreasing the number of mating events, each of which requires lengthy fitness evaluations. Preliminary results for the hydraulic mixing nozzle problem show a single solution found using a highly connected graph and three distinctly dissimilar solutions when using a diversity preserving graph. In addition, GBEAs can be used to classify both evolutionary computation problems and evolutionary computation methods. By classifying evolutionary computation problems, it is possible to select an algorithm or method a priori that will give the desired result more rapidly. Taxonomic classification of evolutionary computation problems, enabled by GBEAs, permits objective evaluation of new techniques.

FUTURE RESEARCH DIRECTIONS

One useful tool would be the inclusion of a method to discern the level of population diversity while the graph based evolutionary algorithm is solving the problem. This would give a designer the ability to verify the number of distinctly different solutions in an evolving population to ensure that the desired number of solutions is being delivered. In addition, a designer with knowledge of the make-up of the solutions occupying the search space would be able to incorporate domain knowledge into the evolving designs. This may be done to speed convergence by introducing known solution elements or to encourage exploration by manual introduction of diversity.

Work continues applying GBEAs to engineering and scientific applications. Some of these projects include responsible antibiotic use in both production animals and humans, interactive optimization of thermal fin profiles in a virtual environment, and ideal use of fuel resources for the power industry. As information from these and other problems become available, it will be added to the developing taxonomy of evolutionary computation problems. The theories derived from this work can be tested against newer test problems, with an eventual goal of establishing guidelines that can be used to select a graph and other parameters a priori that achieve near maximum benefits.

REFERENCES

Ashlock, D. & Lathrop, J. I. (1998). A Fully Characterized Test Suite for Genetic Programming. In *Evolutionary Programming VII*, (pp. 537-546). New York, NY: Springer-Verlag.

Ashlock, D., Guo, L., & Qiu, F. (2002). Greedy Closure Genetic Algorithms. In *Proceedings of the 2002 Congress on Evolutionary Computation* (pp. 1296-1301). Piscataway, NJ: IEEE Press.

Ashlock, D. & Schonfeld, J. (2005). Nonlinear Projection for the Display of High Dimensional Distance Data. In *Proceedings of the 2005 Congress on Evolutionary Computation*, vol. 3 (pp. 2776-2783). Piscataway, NJ: IEEE Press.

Ashlock, D.A., Bryden, K.M., & Corns, S.M. (2005). Graph Based Evolutionary Algorithms Enhance the Location of Steiner Systems. In *Proceedings of the 2005 Congress on Evolutionary Computation* (pp. 1861-1866). Piscataway, NJ: IEEE Press.

Ashlock, D. A. (2006). *Evolutionary Computation for Modeling and Optimization.* New York, NY: Springer Science+Business Meida, Inc.

Banzhaf, W. G., Nordin, P., Keller, R. E., & Francone, F. D. (1998). *Genetic Programming: An Introduction.* San Francisco, CA: Morgan Kaufmann.

Blaize, M., Knight, D., & Rasheed, K. (1998). Automated Optimal Design of Two-Dimensional Supersonic Missile Inlets. *Journal of Propulsion and Power* 14, 890-898.

Bryden, K. M., Ashlock, D. A., McCorkle, D. S., & Urban, G. L. (2003). Optimization of Heat Transfer Utilizing Graph Based Evolutionary Algorithms. *Computer Methods in Applied Mechanics and Engineering,* 192(44-46), 5021-5036.

Bryden, K. M., Ashlock, D. A., Corns, S. M., & Willson, S. J. (2006). Graph Based Evolutionary Algorithms. *IEEE Transactions on Evolutionary Computations*, Vol. 10:5, 550-567.

Corns, S. M., Bryden, K. M., & Ashlock, D. A. (2003). Evolutionary Optimization Using Graph Based Evolutionary Algorithms. In *Proceedings of the 2003 International Mechanical Engineering Congress and Exposition* (pp. 315-320). ASME Press.

Corns, S. M., Bryden, K. M., & Ashlock, D. A. (2005a). Solution Transfer Rates in Graph Based Evolutionary Algorithms. In *Proceedings of the 2005 Congress on Evolutionary Computation* (pp. 1699-1705). Piscataway, NJ: IEEE Press.

Corns, S. M., Bryden, K. M., & Ashlock, D. A. (2005b). The Impact of Novel Connection Topologies on Graph Based Evolutionary Algorithms. In C. H. Dagli (Ed.), *Smart Engineering System Design: Neural Networks, Evolutionary Programming, and Artificial Life*, vol. 15 (pp. 201-209). ASME Press.

Corns, S. M., Ashlock, D. A., McCorkle, D. S. & Bryden, K. M. (2006a). Improving Design Diversity Using Graph Based Evolutionary Algorithms. In *Proceedings of the 2006 IEEE World Congress on Computational Intelligence* (pp. 1037-1043). Piscataway, NJ: IEEE Press.

Corns, S. M., Hurd, H. S., Hoffman, L. J. & Bryden, K. M. (2006b). Evolving Antibiotic Regimens to Minimize Bacterial Resistance in Swine. In *11th Annual AIAA/ISSMO Multidisciplinary Analysis and Optimization Conference*, Portsmouth, VA, On-line proceedings.

Davidor, Y., Yamada, T. and Nakano, R. (1993). "The ECOlogical framework II: Improving GA Performance at Virtually Zero Cost. In *Proceedings of the Fifth International Conference on Genetic Algorithms* (pp. 171-175). Morgan Kaufman.

Fabbri, G. (1997). A Genetic Algorithm for Fin Profile Optimization. *International Journal of Heat and Mass Transfer*, Vol. 40, No. 9, pp. 2165-2172.

Goldberg, D. E. (1989). *Genetic Algorithms in Search, Optimization, and Machine Learning.* Reading, MA: Addison-Wesley Publishing Company, Inc.

Gusfield, D. (1997). *Algorithms on Strings, Trees and Sequences: Computer Science and Computational Biology.* Cambridge, MA: Cambridge University Press.

Kauffman, S. A. & Levin, S. (1987). Towards a General Theory of Adaptive Walks on Rugged Landscapes. *Journal of Theoretical Biology*, vol. 128, 11-45.

Keane, A.J. (1994). Experiences with Optimizers in Structural Design. In *Proceedings of the 1st Conference on Adaptive Computing in Engineering Design and Control* (pp. 14-27). Plymouth, UK: Published by the University of Plymouth.

Kimura, M. & Crow, J. (1963). On the Maximum Avoidance of Inbreeding. *Genetic Research*, vol. 4, 399-415.

Kirstukas, S. J., Bryden, K. M., & Ashlock, D. A. (2004). A Genetic Programming Technique for Solving Systems of Differential Equations. In C. H. Dagli (Ed.), *Intelligent Engineering Systems Through Artificial Neural Networks*, vol. 14 (pp. 241-246). ASME Press.

Koza, J. R. (1992). *Genetic Programming.* Cambridge, MA: The MIT Press.

Mayr, E. & Ashlock, P. D. (1991). *Principles of Systematic Zoology.* New York, NY: McGraw-Hill.

Miller, J. F. & Thomson, P. (2000). Cartesian Genetic Programming. *Proceedings of the European Conference on Genetic Programming* (pp. 121-132). London, UK: Springer-Verlag.

Qiu, F., Guo, L., Wen, T.J., Ashlock, D.A. & Schnable, P.S. (2003). DNA Sequence-Based Barcodes for Tracking the Origins of Ests from a Maize CDNA Library Constructed Using Multiple MRNA Sources. *Plant Physiology*, vol. 133, 475-481.

Reynolds, C. (1992). An Evolved, Vision-Based Behavioral Model of Coordinated Group Motion. In J. Meyer, H. L. Roiblat, & S. Wilson (Eds.), *From Animals to Animats 2* (pp. 384-392). Cambridge, MA: The MIT Press.

Sneath, P. H. A. & Sokal, R. R. (1973). *Numerical Taxonomy; the Principles and Practice of Numerical Classification.* San Francisco, CA: W.H. Freeman.

Syswerda, G. (1991). A Study of Reproduction in Generational and Steady State Genetic Algorithms. *Foundations of Genetic Algorithms* (pp. 94-101). San Francisco, CA: Morgan Kaufmann.

Vavak, F., Jukes, K., & Fogarty, T. C. (1997). Adaptive Combustion Balancing in Multiple Burner Boiler Using a Genetic Algorithm with Variable Range of local Search. In *The proceedings of the 7th International Conference on Genetic Algorithms* (pp. 719-726). San Francisco, CA: Morgan Kaufmann.

Whitley, D. (1989). The Genitor Algorithm and Selection Pressure: Why Rank Based Allocation of Reproductive Trials is Best. In *Proceedings of the 3rd ICGA* (pp. 116-121). San Francisco, CA: Morgan Kaufmann.

Whitley, D., Mathias, K., Rana S., & Dzubera, J. (1996). Evaluating Evolutionary Algorithms. *Artificial Intelligence*, vol. 85(1-2): 245-276.

Whitley, D. & Starkweather, T. (1990). GENITOR II: A Distributed Genetic Algorithm. *Journal of Experimental and Theoretical Artificial Intelligence*, vol. 2, 189-214.

Wright, S. (1986). *Evolution.* Chicago, IL: University of Chicago Press. Edited and with Introductory Materials by William B. Provine.

Xiao, A., Bryden, K. M., Engelbrecht, J., Huang, G. & McCorkle, D. S. (2004). Acceleration Methods in the Interactive Design of a Hydraulic Mixing Nozzle Using Virtual Engineering Tools. In *Proceedings of the 2004 International Mechanical Engineering Congress and Exposition.* ASME Press.

ADDITIONAL READING

Ashlock, D. A. (2006). *Evolutionary Computation for Modeling and Optimization.* New York, NY: Springer Science+Business Meida, Inc.

Ashlock, D., Guo, L. and Qiu, F. (2002). Greedy Closure Genetic Algorithms. In *Proceedings of the 2002 Congress on Evolutionary Computation* (pp. 1296-1301). Piscataway, NJ. IEEE Press.

Ashlock, D. A., Walker, J., & Smucker, M. (1999). Graph Based Genetic Algorithms, In *Proceedings of the 1999 Congress on Evolutionary Computation* (pp. 1362-1368). Morgan Kaufmann, San Francisco.

Balakrishnan, K. and Honavar, V. (1996). On Sensor Evolution in Robotics. *Genetic Programming 98*, 455-460.

Banzhaf, W. G., Nordin, P., Keller, R. E., & Francone, F. D. (1998). *Genetic Programming: An Introduction.* Morgan Kaufmann, San Francisco.

Blaize, M., Knight, D. & Rasheed, K. (1998). Automated Optimal Design of Two-Dimensional Supersonic Missile Inlets. *Journal of Propulsion Power* 14, 890-898.

Bryden, K. M., Ashlock, D. A., Corns, S. M., & Willson, S. J. (2006). Graph Based Evolutionary Algorithms. *IEEE Transactions on Evolutionary Computations*, Vol. 10:5, 550-567.

Davidor, Y., Yamada, T. and Nakano, R. (1993). The ECOlogical framework II: Improving GA Performance at Virtually Zero Cost. In *Proceedings of the Fifth International Conference on Genetic Algorithms* (pp. 171-175). Morgan Kaufman.

DeJong, K. A. (1975). *An analysis of the behavior of a class of genetic adaptive systems.* (Doctoral dissertation, University of Michigan, 1975). Dissertation Abstracts International, 36(10), 5140B.

Fabbri, G. (1997). A Genetic Algorithm for Fin Profile Optimization, *International Journal of Heat and Mass Transfer*, Vol. 40, No. 9, 2165-2172.

Fogel, L. J., Owens, A. J., & Walsh, M. J. (1966). *Artificial Intelligence Through Simulated Evolution.* Wiley, New York.

Goldberg, D. E. & Deb, K. (1991). A Comparative Analysis of Selection Schemes Using in Genetic Algorithms. In G. J. E. Rawlins (Ed.), *Foundations of Genetic Algorithms* (pp. 69-93). Morgan Kaufmann, San Mateo, CA.

Hart, E., Ross, P, & Nelson, J. A. D. (1999). Scheduling Chicken Catching – An Investigation into the Success of a Genetic Algorithm in a Real-World Scheduling Problem. *Annals of Operation Research*, vol. 92, 363-380.

Jang, M. & Lee, J. (2000). Genetic Algorithm Based Design of Transonic Airfoils Using Euler Equations. *Collection of Technical Papers-AIAA/ASME/ASCE/ASC Structures, Structural Dynamics and Materials Conference*, 1(2), 1396-1404.

Keane, A.J. (1994). Experiences with Optimizers in Structural Design. In *Proceedings of the 1st Conference on Adaptive Computing in Engineering Design and Control* (pp. 14-27). Published by the University of Plymouth, UK.

Langdon, W. B. & Treleaven, P. C. (1997). Scheduling maintenance of electrical power transmission networks using genetic programming, In K. Warwick, A. Ekwue, & R. Aggarwal (eds.), *Artificial Intelligence Techniques in Power Systems* (pp. 220-237).

Parmee, I. C. (2001). *Evolutionary and Adaptive Computing in Engineering Design.* Springer-Verlag London Limited.

Parmee, I. C. & Watson, A. H. (1999). Preliminary Airframe Design Using Co-Evolutionary Multi-objective Genetic Algorithms. In W. Banzhaf, J. Daida, A. E. Eiben, M. H. Garzon, V. Honavar, M. Jakiela, & R. E. Smith (eds), *Proceedings of the Genetic and Evolutionary Computation Conference (GECCO'99)*, vol. 2 (pp. 1657–1665).

Rudolph, G. (2000). Takeover Times and Probabilities of Non-Generational Selection Rules. In *Proceedings of the Genetic and Evolutionary Computation Conference* (pp. 903-910).

Stephens, C. R. & Poli, R. (2004). EC Theory - 'in Theory' Towards a Unification of Evolutionary Computation Theory, In Menon, A (Ed.), *Frontiers of Evolutionary Computation, volume II* (pp. 129-155).

Teller, A. (1994). The Evolution of Mental Models. In K. Kinnear (ed.), *Advances in Genetic Programming*. The MIT Press.

Urban, G. L., Bryden, K. M., & Ashlock, D.A. (2002). Engineering Optimization of an Improved Plancha Stove. *Energy for Sustainable Development*, 6(2), 5-15.

Wolpert, D. H. and Macready, W. G. (1997). No Free Lunch Theorems for Optimization. *IEEE Transactions on Evolutionary Computations*, Vol. 1, No. 1, 67-82.

West, D. B. (1996). *Introduction to Graph Theory*. Prentice Hall, Upper Saddle River, NJ.

Wu, A. S. & DeJong, K. A. (1999). An Examination of Building Block Dynamics in Different Representations. In *Proceedings of the Congress on Evolutionary computation*, vol. 1 (pp. 715-721).

Zitzler, E., Deb, K., & Thiele, L. (2000). Comparison of Multiobjective Evolutionary Algorithms: Empirical Results. *Evolutionary Computation*, 8(2):173–195.

Chapter XVIII
Dynamics of Neural Networks as Nonlinear Systems with Several Equilibria

Daniela Danciu
University of Craiova, Romania

ABSTRACT

Neural networks—both natural and artificial, are characterized by two kinds of dynamics. The first one is concerned with what we would call "learning dynamics". The second one is the intrinsic dynamics of the neural network viewed as a dynamical system after the weights have been established via learning. The chapter deals with the second kind of dynamics. More precisely, since the emergent computational capabilities of a recurrent neural network can be achieved provided it has suitable dynamical properties when viewed as a system with several equilibria, the chapter deals with those qualitative properties connected to the achievement of such dynamical properties as global asymptotics and gradient-like behavior. In the case of the neural networks with delays, these aspects are reformulated in accordance with the state of the art of the theory of time delay dynamical systems.

INTRODUCTION

Neural networks are computing devices for *Artificial Intelligence* (AI) belonging to the class of learning machines (with the special mention that learning is viewed at the sub-symbolic level). The basic feature of the neural networks is the interconnection of some simple computing elements in a very dense network and this gives the so-called *collective emergent computing capabilities*. The simple computing element is here the neuron or, more precisely, the artificial neuron – a simplified model of the biological neuron.

Artificial Neural Networks structures are broadly classified into two main classes: recurrent and non-recurrent networks. We shall focus on the class of *recurrent neural networks* (RNN). Due to the cyclic interconnections between the neurons, RNNs are dynamical nonlinear systems displaying some very rich temporal and spatial qualitative behaviors: stable and unstable fixed points, periodic, almost periodic or chaotic behaviors. This fact makes RNN applicable for modeling some cognitive functions such as associative memories, unsupervised learning, self-organizing maps and temporal reasoning.

The mathematical models of neural networks arise both from the modeling of some behaviors of biological structures or from the necessity of Artificial Intelligence to consider some structures which solve certain tasks. None of these two cases has as primary aim stability aspects and a "good" qualitative behavior. On the other hand, these properties are necessary and therefore important for the network to achieve its functional purpose that may be defined as "global pattern formation". It thus follows that any AI device, in particular a neural network, has to be checked *a posteriori* (i.e. after the functional design) for its properties as a dynamical system, and this analysis is performed on its mathematical model.

A common feature of various RNN (automatic classifiers, associative memories, cellular neural networks) is that they are all *nonlinear dynamical systems with multiple equilibria*. In fact it is exactly the equilibria multiplicity that gives to all AI devices their computational and learning capabilities. As pointed out in various reference books, satisfactory operation of a neural network (as well as of other AI devices) requires its evolution towards those equilibria that are significant in the application.

Let us remark here that if a system has several isolated equilibria this does not mean that all these equilibria are stable – they may be also unstable. This fact leads to the necessity of a qualitative analysis of the system's properties. Since there are important both the local stability of each equilibrium point and also (or more) the global behavior of the entire network, we shall discuss here RNN within the frameworks of the *Stability Theory* and the *Theory of Systems with Several Equilibria*.

The chapter is organized as follow. The *Background* section starts with a presentation of RNN from the point of view of those dynamic properties (specific to the systems with several equilibria), which make them desirable for modeling the associative memories. Next, there are provided the definitions and the basic results of the *Theory of Systems with Several Equilibria*, discussing these tools related to the *Artificial Intelligence* domain requirements. The main section consists of two parts. In the first part it is presented the basic tool for analyzing the desired qualitative properties for RNN as systems with multiple equilibria. The second part deals with the effect of time-delays on the dynamics of RNN. Moreover, one will consider here the time-delay RNN under forcing stimuli that have to be "reproduced"(synchronization). The chapter ends with *Conclusions* and comments on *Future trends* and *Future research directions*. *Additional reading* is finally suggested.

BACKGROUND

A. The state space of RNN may display stable and unstable fixed points, periodic, almost periodic or even chaotic behaviors. A concise survey of these behaviors and their link to the activity patterns of obvious importance to neuroscience may be found in (Vogels, Rajan & Abbott, 2005). From the above mentioned behaviors, the fixed-point dynamics means that the system evolves from an initial state toward a state (a *stable fixed-point equilibrium* of the system) in which the variables of the system do not change over the time. If that *stable fixed-point* is used to retain a specific pattern then, given a distorted or noisy pattern of it as an initial condition, the system evolution will be such that the stable equilibrium point will be eventually attained to. This process is called the *retrieving of the stored pattern*. Since an associative memory has to retain several different patterns, the system which models it has to have *several stable equilibrium points*. More general, according to the concise but meaningful description of Noldus *et al.* (1994, 1995):

- when a neural network is used as a classification network, system's equilibria represent the "*prototype*" *vectors* that characterize the different classes: the ith class consists of those vectors x which, as an initial state for network's dynamics, generate a trajectory converging to the ith "*prototype*" *equilibrium state*.
- when the network is used as an *optimizer*, the *equilibria represent optima*.

The mathematical model of a RNN with m neurons consists of m-coupled first-order ordinary differential equations, which describe the time-evolution of the state of the dynamical system. The state x of the entire system is given by the states x_i, $i = 1, ..., m$ of each neuron in the network. On other words, the state of the RNN is described by m state variables; $x = [x_1 \dots x_m]^T$ is the state vector of RNN, where T denote the transpose. Using the notation $\dot{x}$ for the derivative of x with respect to the time variable t, the time evolution of the dynamical system is described by the first-order vector differential equation

$$\dot{x} = h(x), \quad x \in \mathbb{R}^m \tag{1}$$

where $h(x) = [h_1(x) \dots h_m(x)]^T$ is a nonlinear vector function which satisfies the sufficient conditions for the existence and uniqueness of the solution $x(t, x_0, t_0)$, for a given initial state x_0 at a given initial time t_0. In order to describe the functions $h_i(x)$, $i = 1, ..., m$ we briefly discuss now the mathematical model of the artificial neuron, without insisting on the biological aspects; these may be found in several references (see, for instance Cohen & Grossberg, 1983; Bose & Liang, 1996; Vogels *et al.*, 2005).

Consider the standard equations for the Hopfield neuron

$$\dot{x}_i = -a_i x_i - \sum_{j=1}^{m} c_{ij} y_j + I_i,$$
$$y_i = f(x_i) \tag{2}$$

The first equation describes the state evolution—the state x of the neuron at time t is given by its potential or short-term memory activity at that moment. It depends on its inputs (the second and third terms of the first equation) and on the passive decay of the activity at rate $-a_i$ (first term). The input of a neuron has two sources: external input I_i from sources outside the network and the weighted sum of the outputs y_j, $j = \overline{1,m}$ of an arbitrary set of neurons within the network (the pre-synaptic neurons). The weights are some real numbers c_{ij}, which describe the strength and type of the synapse from the pre-synaptic neurons j to the post-synaptic neuron i; positive weights $c_{ij} > 0$ model excitatory synapses, whereas negative weights $c_{ij} < 0$ model inhibitory synapses.

The second equation gives the output of the neuron i as a function of its state. Generally, the input/output (transfer) characteristic of the neuron $f(\cdot)$ is described by sigmoidal functions of the bipolar type with the shape shown in Figure 1. These are bounded monotonically non-decreasing functions that provide a graded nonlinear response within the range $[-1,1]$. Sigmoidal functions are globally Lipschitz, what means they satisfy inequalities as

$$0 \le \frac{f(\sigma_1) - f(\sigma_2)}{\sigma_1 - \sigma_2} \le L, \quad \forall \sigma_1 \ne \sigma_2 \tag{3}$$

Let us remark that $f(0) = 0$, what means that sigmoidal functions also satisfy inequalities as

$$0 \le \frac{f(\sigma)}{\sigma} \le L. \tag{4}$$

We can write now the mathematical model of the Hopfield neural network, which is representative for RNN

$$\dot{x}_i = -a_i x_i - \sum_{j=1}^{m} c_{ij} f_j(x_j) + I_i, \quad i = \overline{1,m} \tag{5}$$

and identify from (1) and (5) the nonlinear functions

$$h_i(x) = -a_i x_i - \sum_{j=1}^{m} c_{ij} f_j(x_j) + I_i.$$

It is important to say here that the presence of many nonlinear characteristics $f_j(\cdot)$, $j = \overline{1,m}$ leads to the existence of several equilibria, whereas the Lipschitz sufficient conditions (3) and (4) ensure the existence and the uniqueness of the solution of system (5) i.e. for a given initial condition there is only one way that the RNN can evolve in time (the reader is sent to any book which treats ordinary differential equations or nonlinear systems; see for instance: Hartman, 1964; Khalil, 1992).

B. Dynamical systems with several equilibria occur in many fields of science and engineering where the stable equilibria are possible "operating points" of the man-made systems or represent a "good behavior solution" of some natural physical systems (Răsvan, 1998). For instance, such systems there exist in the fields of biology, economics, recurrent neural networks, chemical reactions, and electrical machines. The standard stability properties (Lyapunov, asymptotic and exponential stability) are defined for a single equilibrium. Their counterparts for several equilibria are: mutability, global asymptotics and gradient behavior.

The basic concepts of the theory of systems with several equilibria are introduced by Kalman (1957) and Moser (1967) and have been developed by Gelig, Leonov & Yakubovich (1978, 2004), Hirsch (1988) and Leonov, Reitmann & Smirnova (1992). From the last reference we introduce here the basic concepts of the field.

Consider the system of n ordinary differential equations

$$\dot{x} = h(t, x) \tag{6}$$

with $h : \mathbb{R}_+ \times \mathbb{R}^n \to \mathbb{R}^n$ continuous, at least in the first argument.

Definition 1:

a. *Any constant solution of (6) is called equilibrium. The set of equilibria **S** is called stationary set.*

Figure 1. Sigmoidal function

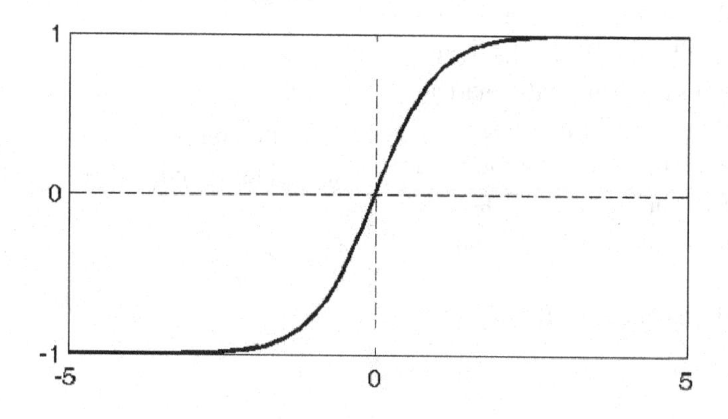

b. *A solution of (6) is called convergent if it approaches asymptotically some equilibrium:*

$$\lim_{t\to\infty} x(t) = c \in \boldsymbol{S} \tag{7}$$

c. *A solution is called quasi-convergent if it approaches asymptotically the stationary set* ***S****:*

$$\lim_{t\to\infty} d(x(t), \boldsymbol{S}) = 0 \tag{8}$$

where $d(\cdot, S)$ *denotes the distance from a point to the set* ***S****.*

Definition 2: *System (6) is called monostable if every bounded solution is convergent; it is called quasi-monostable if every bounded solution is quasi-convergent.*

Definition 3: *System (6) is called gradient-like if every solution is convergent; it is called quasi-gradient-like if every solution is quasi-convergent.*

From the above definitions we have:

a. An equilibrium is a constant solution $\bar{x} = [\bar{x}_1 \ldots \bar{x}_n]^T$ of the equation $h(t, x) = 0$. The set of equilibria may consist of both stable and unstable equilibria.
b. For RNN, the stable equilibria are used to retain different memories; unstable equilibria will never be recalled.
c. An initial condition for the RNN is any pattern that is presented at the input of the network. If for an input pattern x_0 the evolution $x(t, x_0, t_0)$ of the network is such that a stored pattern $\bar{x}$ is attained in an asymptotical manner, then the initial pattern is within the *asymptotic stability (attraction) region* of the equilibrium pattern and the asymptotical convergence of the solution $x(t,x_0,t_0)$ means a "good" remembering process. The attraction region of a stored pattern may be viewed as the set of all its distorted or noisy patterns, which still allow recalling it. The problem of computing regions of asymptotic stability for nonlinear systems and also for analogue neural classification networks is widely discussed in the literature, but is beyond the scope of this chapter; the reader is send to the papers of Noldus *et al.* (1994) and Loccufier *et al.* (1995).
d. Since there exists also other terms for the above introduced notions some comments are necessary. *Monostability* has been introduced by Kalman (1957) and sometimes it is called *strict mutability* (term introduced by Popov in 1979), while *quasi-monostability* is called by the same author *mutability* and by others *dichotomy* (Gelig, Leonov & Yakubovich, 1978). The *quasi-gradient-like* property is called sometimes *global asymptotics*.
e. In the sequel we shall use the following terms:
    - **Dichotomy:** All bounded solutions tend to the equilibrium set;
    - **Global asymptotics:** All solutions tend to the equilibrium set;
    - **Gradient-like behavior:** The set of equilibria is stable in the sense of Lyapunov and any solution tends asymptotically to some equilibrium point.
f. Dichotomy signifies some genuine nonoscillatory behavior in the sense that there may exist unbounded solutions but no oscillations are allowed. In general, the gradient-like behavior represents the desirable behavior for almost all systems. If the equilibria are isolated then global asymptotics and gradient-like behavior are equivalent (any solution may not approach the stationary set otherwise than approaching some equilibrium).

C. All the properties of the multiple equilibria dynamical systems are analyzed using Lyapunov-like results. Before introducing them let us present briefly the basic idea of the Lyapunov stability tool, together with its strengths and weakness. The so-called method of the Lyapunov functions has its background originating from Mechanics: the energy E of an isolated dissipative system decreases in time until the system reach the resting state, where $dE/dt = 0$ (the Lagrange-Dirichlet theorem); in fact, at that point the energy E reach its minimum and the corresponding state is an equilibrium state for the system.

In 1892, A.M. Lyapunov—a Russian mathematician, proposed some generalized state functions, which are energy-like: they are of the constant sign and decrease in time along the trajectories of the system of differential equations. The importance of the Lyapunov method resides in reducing the number of the equations needed in order to analyze the behavior of a system: from a number of equations equal to the number of the system's state variables to only one function—the Lyapunov function. A Lyapunov function describes completely the system behavior; at each moment a single real number gives information on the entire system. Moreover, since these functions may have not any connection with physics and the system's energy, they can be used for a wide class of systems to determine the stability of an equilibrium point.

Let us mention here that the method gives only sufficient conditions for the stability of an equilibrium what means that on one hand, if one can not find a Lyapunov function for a system then one can not say anything about the qualitative behavior of its equilibrium (if it is stable or unstable); on the other hand, it is a quasi-permanent task to find sharper, less restrictive sufficient conditions (i.e. closer to the necessary ones) on system's parameters (in the case of neural networks—on synaptic weights) by improving the Lyapunov function. For an easy-to-understand background on Lyapunov stability see for instance, Chapter III of the course by Khalil (1992); for applications of Lyapunov methods in stability the reader is sent to Chapter V (*op. cit.*) and the book of Halanay and Răsvan (1993).

For a linear system, if the equilibrium at the origin $\bar{x} = 0$ is globally asymptotically stable (hence exponentially stable), then that equilibrium represents the global minimum for the associated Lyapunov function. A vertical section on the Lyapunov surface for a globally asymptotically stable fixed point may have the allure in Figure 2. For a nonlinear system with multiple equilibria, the Lyapunov surface will have multiple local minima, as in Figure 3.

In general, there are no specific procedures for constructing Lyapunov functions for a given system. For a linear m-dimensional system, a quadratic Lyapunov function $V : \mathbb{R}^m \to \mathbb{R}$ described by $V(x) = x^T Px$, with P a symmetric positive definite matrix—may provide necessary and sufficient conditions for exponential stability. Unfortunately, for nonlinear systems or even more—for time-delay systems, the sharpest most general quadratic Lyapunov function(al) is rather difficult to manipulate. In these cases, an approach is to associate the Lyapunov function in a natural way i.e. as an energy of a certain kind, but "guessing" a Lyapunov function(al) which gives sharper sufficient criteria for a given system "remains an art and a challenge" (Răsvan, 1998).

For systems with several equilibria in the time invariant case (and this is exactly the case in the AI devices, in particular for the case of neural networks) the following Lyapunov-like results are basic (Leonov, Reitmann & Smirnova, 1992: statements in Chapter I, proofs in Chapter IV):

Lemma 1. *Consider the nonlinear system (1) and assume existence of a continuous function* $V : D \subset \mathbb{R}^m \to \mathbb{R}$ *such that:*

i) *for any solution x of (1) $V(x(t))$ is nonincreasing;*

Figure 2. A Lyapunov function allure with a global minimum

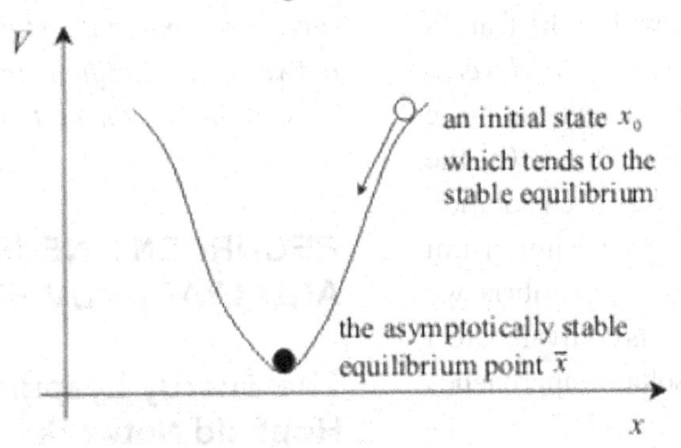

Figure 3. A Lyapunov function allure with multiple local minima

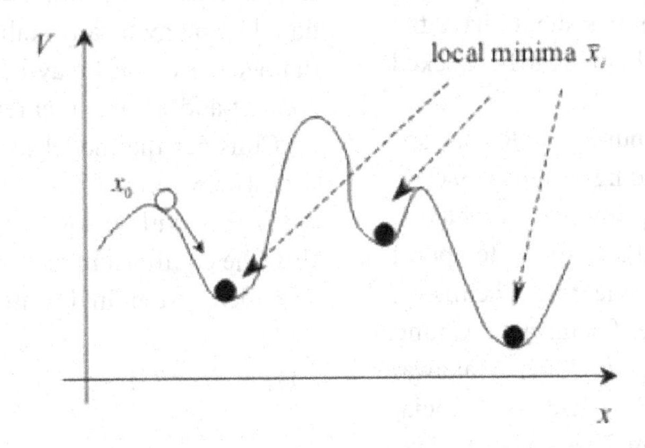

ii) *if $x(t)$ is a bounded solution of (1) on $\mathbb{R}^+$ for which there exists a $\tau > 0$ such that $V(x(\tau)) = V(x(0))$, then $x(t)$ is an equilibrium*

Then the system (1) is quasi-monostable (dichotomic).

Lemma 2. *If the assumptions of Lemma 1 hold and, additionally, $V(x) \to \infty$ as $|x| \to \infty$ (it is radially unbounded), then system (1) is quasi-gradient-like (has global asymptotics).*

Lemma 3. *If the assumptions of Lemma 2 hold and the set **S** is discrete (i.e. it consists of isolated equilibria only) then system (1) is gradient-like.*

Remark 1. (Moser, 1967): Consider the rather general nonlinear autonomous system

$$\dot{x} = -f(x), \quad x \in \mathbb{R}^m, \tag{9}$$

where $f(x)$ = *grad* $G(x)$ and $G : \mathbb{R}^m \to \mathbb{R}$ is such that: (a) it is radially unbounded, i.e.,

$$\lim_{|x| \to \infty} G(x) = \infty$$

and (b) the number of its critical points is finite. Under these assumptions any solution of (9) approaches asymtotically one of the equilibria (which is also a critical point of *G*—where its gradient, i.e., *f* vanishes) and thus the system's behavior is gradient-like.

At this point let us redefine the features of a well-designed neural network within the framework of the *Theory of Systems with Several Equilibria*: (a) it has several (and many, but finite number) fixed-point equilibrium states; (b) the network has to be convergent, i.e., each solution of the dynamics has to converge to an equilibrium state. From the finite number of equilibria we deduce that the second property is equivalent to a *gradient-like behavior* (every solution approaches an equilibrium).

D. The RNN have a dynamics induced *a posteriori* by the learning process that had established the synaptic weights. It is not at all compulsory that this *a posteriori* dynamics should have the required properties, hence they have to be checked separately.

In the last decades, the number of RNNs' applications increased, they being designed for classification, identification, optimization problems, and complex image, visual and spatio-temporal processing in fields as engineering, chemistry, biology and medicine (see, for instance Chung & Lee, 1991; Fortuna *et al.*, 2001; Arakawa, Hasegawa, & Funatsu, 2003; Fink, 2004; Atencia, Joya, & Sandoval, 2004; Jing-Sheng Xue, Ji-Zhou Sun, Xu Zhang, 2004; Maurer, Hersch, & Billard, 2005; Iwahori, Kawanaka, Fukui & Funahashi, 2005; Guirguis, & Ghoneimy, 2007). All these applications are mainly based on the existence of several equilibria, requiring the "good behavior" properties above discussed.

Neural networks with several equilibria (Bidirectional Associative Memories, Hopfield, cellular, Cohen-Grossberg) have a rather special systemic structure that allows associating Lyapunov functions in a natural way. However, since the properties of the Lyapunov function and its derivative give sufficient conditions on systems parameters, *one of the current stability problems in neural networks studies is to improve Lyapunov functions in order to obtain sharper, less restrictive stability conditions.* These conditions will be verified *a posteriori* on the mathematical model of RNN. *Another problem, but of theoretical nature, is to embed the stability analysis of neural networks in a unified stability theory for a wide class of nonlinear systems.*

RECURRENT NEURAL NETWORKS AND LYAPUNOV FUNCTIONS

The Energy Lyapunov Function for Hopfield Network

J.J. Hopfield (1982) showed that a physical system whose dynamics within the state space is dominated by many locally stable equilibrium points (named *attractors*) may be viewed as a general content-addressable memory.

Consider the model for the analog Hopfield neural network (5). If $\bar{x}_i, i = \overline{1,m}$ is some equilibrium of (5), then without loss of the generality one can shift the equilibrium to the origin, obtaining the so-called system in deviations

$$\dot{z}_i(t) = -a_i z_i(t) - \sum_{j=1}^{m} c_{ij} g_j(z_j), \quad i = \overline{1,m} \tag{10}$$

Here $z_i = x_i - \bar{x}_i$ and the nonlinearities defined by

$$g_i(z_i) = f_i(z_i + \bar{x}_i) - f_i(\bar{x}_i) \tag{11}$$

are also sigmoidal functions satisfying a condition of the type (4).

The Hopfield network model is based on the idea of defining the global behavior of the network related to the energy E of a physical system—which is in fact a natural Lyapunov function. Supposing that the potential energy of the origin equals zero, if one denotes by z the coordinates vector of some point as a measure of its displacement from the equilibrium in origin and takes into account that the energy, as a physical quantity, is smooth enough to allow a multivariable Taylor-series expansion about the origin (it is analytic), then the function $E:\mathbb{R}^m \to \mathbb{R}$ with the form

$$E(z) = \frac{1}{2} z^T C z \tag{12}$$

where the matrix C is symmetric, has the significance of the potential energy of the system. The negative of this function may be a Lyapunov function candidate for the system (Kosko, 1992). In this case the Lyapunov function was obtained in a natural way and is of the quadratic form.

Qualitative Behaviors for General Competitive Cohen-Grossberg Neural Networks

Consider the general case of the Cohen-Grossberg (1983) competitive neural network

$$\dot{x}_i = a_i(x_i)\left[b_i(x_i) - \sum_{j=1}^{m} c_{ij} d_j(x_j)\right], \qquad i = \overline{1,m}, \tag{13}$$

with $c_{ij} = c_{ji}$ (the symmetry condition). Worth mentioning that equations (13) account not only for the model of the neural networks just mentioned but they also include: Volterra-Lotka equations of population biology, Gilpin and Ayala system of growth and competition, Hartline-Ratliff equations related to the Limulus retina, Eigen and Schuster equation for the evolutionary selection of macromolecular quasi-species (*op. cit.*).

To system (13) it is associated the Lyapunov function (Cohen-Grossberg, 1983) $V:\mathbb{R}^m \to \mathbb{R}$

$$V(x) = \frac{1}{2}\sum_{1}^{m}\sum_{1}^{m} c_{ij} d_i(x_i) d_j(x_j) - \sum_{1}^{m}\int_{0}^{x_i} b_i(\lambda) d_i'(\lambda) d\lambda \tag{14}$$

Its derivative along the solutions of the system (13) is

$$W(x) = -\sum_{i=1}^{m} a_i(x_i)\, d_i'(x_i)\left[b_i(x_i) - \sum_{j=1}^{m} c_{ij}\, d_j(x_j)\right]^2 \tag{15}$$

The condition that the Lyapunov function (14) to be nonincreasing in time along the systems trajectories implies $W(x) \leq 0$. This inequality holds provided $a_i(\lambda) > 0$ and $d_i(\lambda)$ are monotone non-decreasing. If additionally $d_i(x_i)$ are strictly increasing, then the set where $W = 0$ consists of equilibria only; according to *Lemma 2* in this case the system has *global asymptotics* (is *quasi-gradient like*) i.e. every solution approaches asymptotically the stationary set. As already said, the above conditions on the system's parameters are only sufficiently for this behavior.

Let us remark that if functions $d_i(x_i)$ are strictly increasing, then (13) may be written as

$$\dot{x} = -A(x)\, grad V(x) \tag{16}$$

which makes system (16) a *pseudo-gradient system* – compare to (9); here $A(x)$ is a diagonal matrix with the entries (*op. cit.*)

$$A_{ij}(x) = \frac{a_i(x_i)}{d_i'(x_i)}\delta_{ij} \tag{17}$$

For a *gradient-like behavior* of system (13) one has to find the conditions ensuring that the stationary set consists of equilibrium points only and more, that these equilibria are isolated. Since the model of system (13) is more general and the stationary set consists of a large number of equilibria, the study of these qualitative properties is done on particular cases.

Qualitative Behaviors for a General Network with Several Sector-Restricted Nonlinearities

Consider the general model of a RNN (Noldus *et al.*, 1994, 1995)

$$\dot{x} = Ax - \sum_{1}^{m} b_k \varphi_k (c_k^* x) - h \,, \qquad (18)$$

where $x \in \mathbb{R}^n$ is the state vector; $A \in \mathbb{R}^{n \times n}$ is a constant matrix; B and C are matrices with b_k and c_k, $k = \overline{1,m}$ as columns; $h \in \mathbb{R}^n$ is a constant vector and the nonlinear functions $\varphi_k(\sigma)$ are differentiable, slope restricted, bounded functions; the matrix transfer function of the linear part of (18) is $T(s) = C^*(sI - A)^{-1}B$, where the asterix denote the transpose and complex conjugate. System (18) includes neural networks (13) provided the functions involved there have special forms that are quite easy to deduce (*op. cit.*).

The study of the qualitative properties of (18) has different approaches (Noldus *et al.*, 1994, 1995; Danciu & Răsvan, 2000). We shall present here the approach which places the problem within the field of *The Theory of Systems with Several Equilibria* (Danciu & Răsvan, 2000). Two cases have to be considered

a. The case of restricted slope nonlinearities verifying

$$\underline{\varphi}_k < \frac{d\varphi_k(\sigma_k)}{d\sigma_k} < \overline{\varphi}_k, \quad k = \overline{1,m} \qquad (19)$$

Denote

$$\underline{\Phi} = diag(\underline{\varphi}_1, ..., \underline{\varphi}_m),\ \overline{\Phi} = diag(\overline{\varphi}_1, ..., \overline{\varphi}_m),$$
$$Q = diag(q_1, \ldots, q_m),\ \Theta = diag(\theta_1, \ldots \theta_m)$$

Theorem 1: *Consider system (18) under the assumptions:*

i) $\det A \neq 0$;
ii) *(A, B) is a controllable pair and* (C^*, A) *is an observable pair;*
iii) $\det C^* A^{-1} B = \det T(0) \neq 0$

If for $k = \overline{1,m}$ *there exists the sets of parameters* $\theta_k \geq 0$, $\underline{\varphi}_k > 0$, $\overline{\varphi}_k > 0$, $q_k \neq 0$ *such that the frequency domain inequality of Popov*

$$\Theta\overline{\Phi}^{-1} + \mathrm{Re}\left\{\left[\Theta\left(I + \underline{\Phi}\,\overline{\Phi}^{-1}\right) + (i\omega)^{-1} Q\right] T(i\omega)\right\}$$
$$+ T^*(-i\omega)\Theta\underline{\Phi}\,T(i\omega) \geq 0 \qquad (20)$$

hold and matrix $Q(C^* A^{-1} B)^{-1}$ *is symmetric, then system (18) is dichotomic for all slope restricted nonlinear functions satisfying (19). If additionally, all equilibria are isolated, then each bounded solution approaches an equilibrium point.*

b. The case of the bounded nonlinearities verifying

$$\left|\varphi_k(\sigma) - \tilde{\varphi}_k \sigma\right| \leq p_k \leq p \qquad (21)$$

Theorem 2. *Consider system (18) under the assumptions of Theorem 1. Assume additionally that it is minimally stable, the nonlinear functions satisfy (21) and the equilibria are isolated. Then each solution of (18) approaches asymptotically an equilibrium state (the system (18) is gradient-like).*

For systems with sector restricted nonlinearities, the required *minimal stability property* (notion introduced by V.M. Popov (1966, 1973) means that in order to obtain stability for *all* nonlinear (and linear) functions from some sector, it is (minimally) necessary to have this property for *a single linear function* within this sector. In our case system (18) has to be internally stable for a linear function of the class: there exist the numbers $\tilde{\varphi}_k \in (\underline{\varphi}_k, \overline{\varphi}_k)$ such that

$$\left(A - \sum_{1}^{m} b_k \tilde{\varphi}_k c_k^*\right)$$

is a Hurwitz matrix (has all its eigenvalues in the open complex left half plane, Re $s < 0$).

Consider now the case of the Hopfield classification neural network (1984) described by

$$\dot{v}_k(t) = \frac{-1}{R_k C_k} v_k + \frac{1}{C_k}\left[\sum_{j=1}^{n}(\varphi_j(v_j) - v_k)/R_{kj} + I_k\right], \quad k = \overline{1,n} \tag{22}$$

where, C_k models the membrane capacitance, R_k - the transmembrane resistance, I_k - the external input current, v_k - the instantaneous transmembrane potential, $1/R_{kj}$, $k, j = \overline{1,n}$ represent the interconnection weight conductances between different neurons. The identically sigmoidal functions φ_k, $k = \overline{1,n}$ satisfying (19) are introduced in the electrical circuit by nonlinear amplifiers with bounded, monotone increased characteristics.

It can be verified that equations (22) are of the type (18) when

$$A = diag\left(-\frac{1}{C_k}\left(\frac{1}{R_k} + \sum_{1}^{n}\frac{1}{R_{kj}}\right)\right)_{k=1}^{n},$$

$$f(v) = col(\varphi_k(v_k))_{k=1}^{n}, \; h = -col(I_k/C_k)_{k=1}^{n},$$

$$C^* = I, \; B = -\Gamma\Lambda, \Gamma = diag(1/C_k)_{k=1}^{n},$$

$$\Lambda = (1/R_{kj})_{k,j=1}^{n}.$$

Chosing $\theta_k = C_k$ and

$$q_k = 1/R_k + \sum_{j=1}^{n}(1/R_{kj}),$$

the frequency domain inequality (20) holds provided the matrix of synaptic weights Λ is symmetric.

One obtained that the system (22) is in the case of *Theorem 2,* thus gradient-like if the symmetry condition $R_{ij} = R^{ji}$ is verified. This condition, mentioned also in (Noldus *et al.*, 1994, 1995) is quite known in the stability studies for neural networks and it is a standard design condition since the choice of the synaptic parameters is controlled by network adjustment in the process of "learning".

RECURRENT NEURAL NETWORKS AFFECTED BY TIME-DELAYS

A more realistic model for RNN has to take into account the dynamics affected by time delays due to the signal propagation at the synapses level or to the reacting delays in the case of the artificial neural network. Since these delays may introduce oscillations or even may lead to the instability of the network, the dynamical behaviors of the delayed neural networks have been investigating starting with the years of '90 (Marcus & Westervelt, 1989; Gopalsamy & He, 1994; Driessche & Zou, 1998; Danciu & Răsvan, 2001, 2005, 2007).

Investigating the dynamics of time delayed neural networks require some mathematical preliminaries. Let us recall that for systems modeled by ordinary differential equations (6) a fundamental presumption is that the future evolution of the system is completely determined by the current value of the state variables, i.e., one can obtain the value of the state $x(t)$ for $t \in [t_0, \infty)$ once the initial condition $x(t_0) = x_0$ is known. For systems with time delays the future evolution of the state variables $x(t)$ not only depends on their current value $x(t_0)$, but also on their past values $x(\xi)$, where $\xi \in [t_0 - \tau, t_0]$, $\xi > 0$. In this case the initial condition is not a simple point $x_0 \in \mathbb{R}^m$ in the state space, but a real valued vector function defined on the interval $[t_0 - \tau, t_0]$, i.e., $\phi : [t_0 - \tau, t_0] \to \mathbb{R}^m$. For this reason one choice for the state space of the time delay systems is the space $C(-\tau, 0; \mathbb{R}^m)$of continuous $\mathbb{R}^m$-valued mappings defined on the interval $[-\tau, 0]$. Time delay systems are modeled by *functional diferential equations* with the general form

$$\dot{x} = f(t, x_t) \tag{23}$$

where $x \in \mathbb{R}^m$, $f : \mathbb{R} \times C \to \mathbb{R}^m$ and generally, one denotes by ψ_t a segment of the function ψ defined as $\psi_t(\theta) = \psi_t(t + \theta)$ with $\theta \in [-\tau, 0]$. Equation (23) indicates that in order to determine the future evolution of the state, it is necessary to specify the initial state variables $x(t)$ in a time interval of length τ, i.e., $x_{t_0} = x(t_0 + \theta) = \phi(\theta), -\tau \le \theta \le 0$ (Gu *et al.*, 2003).

Global Asymptotic Stability for Time Delay Hopfield Networks via Lyapunov-Like Results

If we consider the standard equations for the Hopfield network (5) and the delay at the interconnection level as being associated with each neural cell, the following model is obtained (Gopalsamy & He, 1994)

$$\dot{x}_i(t) = -a_i\, x_i - \sum_{j=1}^{m} c_{ij} f_j\left(x_j(t-\tau_{ij})\right) + I_i, \quad i = \overline{1,m} \tag{24}$$

where the time delays τ_{ij} are positive real numbers.

Let $\bar{x}$ be an equilibrium for (24). In order to study the equilibrium at the origin one uses the system in deviations

$$\dot{z}_i(t) = -a_i\, z_i - \sum_{j=1}^{m} c_{ij} g_j\left(z_j(t-\tau_{ij})\right) \quad i = \overline{1,m} \tag{25}$$

As previously shown, if $f_j : \mathbb{R} \mapsto \mathbb{R}$ satisfy the usual sigmoid conditions then g_j defined by (11) are such. Denoting $\tau = \max_{i,j} \tau_{ij}$, one chooses as state space $\mathcal{C}(-\tau, 0; \mathbb{R}^m)$.

For time delay systems the Lyapunov-Krasovskii functional is the analogue of the Lyapunov function of the case without delays. One considers the Lyapunov-Krasovskii functional suggested by some early papers (Kitamura *et al.*, 1967; Nishimura *et al.*, 1969), $V : \mathcal{C} \mapsto \mathbb{R}_+$ as

$$V(z(\cdot)) = \sum_{i=1}^{m}\left[\frac{1}{2}\pi_i z_i^2(0) + \lambda_i \int_0^{z_i(0)} g_i(\theta)\, d\theta + \sum_{j=1}^{m} \int_{-\tau_{ij}}^{0} \left(\rho_{ij} z_j^2(\theta) + \delta_{ij} g_j^2(z_j(\theta))\right) d\theta\right]$$

with $\pi_i \ge 0$, $\lambda_i \ge 0$, $\rho_{ij} \ge 0$, $\delta_{ij} \ge 0$ some free parameters.

Differentiating it with respect to t along the solutions of (25) we may find the so-called derivative functional $W : \mathcal{C} \mapsto \mathbb{R}$ as below

$$\begin{aligned} W(z(\cdot)) = \sum_{i=1}^{m}\Big[& -a_i\pi_i z_i^2(0) - \lambda_i a_i g_i\left(z_i(0)\right) z_i(0) - \\ & -\left[\pi_i z_i(0) + \lambda_i g_i\left(z_i(0)\right)\right] \sum_{j=1}^{m} c_{ij} g_j\left(z_j(-\tau_{ij})\right)\Big] + \\ & + \sum_1^m \sum_1^m \left[\rho_{ij} z_j^2(0) + \delta_{ij} g_j^2\left(z_j(0)\right) - \rho_{ij} z_j^2(-\tau_{ij}) - \delta_{ij} g_j^2\left(z_j(-\tau_{ij})\right)\right]. \end{aligned} \tag{27}$$

The problem of the sign for W gives the following choice of the free parameters in (26) (see, Danciu & Răsvan, 2007)

$$\lambda_i > 0 \text{ and } \sigma_i = a_i^2 - \left(\sum_{j=1}^{m} \frac{c_{ij}^2}{\delta_{ji}}\right) \sum_{j=1}^{m} (\rho_{ji} + \delta_{ji}) > 0$$

$$2\left(\sum_{j=1}^{m} \frac{c_{ij}^2}{\delta_{ji}}\right)^{-1} (a_i - \sqrt{\sigma_i}) < \pi_i < 2\left(\sum_{j=1}^{m} \frac{c_{ij}^2}{\delta_{ji}}\right)^{-1} (a_i + \sqrt{\sigma_i}) \tag{28}$$

The application of the standard stability theorems for time delay systems (Hale & Verduyn Lunel, 1993) will give the asymptotic stability of the equilibrium $z = 0$ ($x = \bar{x}$). We thus obtain the following result (Danciu & Răsvan, 2007)

Theorem 3. *Consider system (24) with* $a_i > 0$ *and* c_{ij} *such that it is possible to choose* $\rho_{ij} > 0$ *and* $\delta_{ij} > 0$ *in order to satisfy* $\sigma_i > 0$ *with* σ_i *defined in (28). Then the equilibrium is globally asymptotically stable.*

Remark that the choice of the neural networks' parameters is not influenced here by the values of the delays, i.e. one obtains *delay-independent* sufficient *conditions* that characterize the local behavior of an equilibrium, expressed as previously in the language of the weights of the RNN.

Qualitative Behaviors for Time Delay Neural Networks via Popov-Like Results Using Comparison

Consider the time delay Hopfield network (24) and its system in deviations (25). From the point of view of the qualitative behavior of neural networks as systems with multiple equilibria, the results of V.M. Popov (1979) of interest here, are concerned with *dichotomy, global asymptotics* and *gradient behavior* for two systems that form the so-called comparison system (SM):

(M) – the "model" – which corresponds to a delayless neural network, e.g., described by (10) with $a_i = 1, \forall i$

$$\dot{u}_i = -u_i - \sum_{j=1}^{m} \pi_{ij} h_j(u_j), \quad i = \overline{1,m} \tag{29}$$

where $\pi_{ij} = \pi_{ji}$ define a non-singular symmetric matrix P and the functions $h_j(\cdot) : \mathbb{R} \mapsto \mathbb{R}$ are strictly increasing and uniformly Lipschitz verifying

$$0 < \varepsilon \le \frac{h_j(\sigma_1) - h_j(\sigma_2)}{\sigma_1 - \sigma_2} < L_j \ , \quad h_j(0) = 0 \, . \tag{30}$$

In this case every bounded solution of (29) approaches the set of equilibria (dichotomy). Let $L = \max_i L_i > 1$ and assume that $1 - m\varepsilon \mid P \mid > 0$, i.e., some small gain condition holds. Then *u(t)* is bounded and if the set of equilibria is finite (like in the case of neural networks) then system of (29) is gradient-like (for the proof see V.M. Popov, 1979).

(S) – the "system" – which is described by the following equation

$$x + \varphi + P\kappa * h(x) = 0 \tag{31}$$

where * denotes convolution with the diagonal matrix kernel κ.

In order to obtain the time delay system of the form considered by Popov (31) (*op. cit.*), it is necessary to introduce another assumption about (25), namely $\tau_{ij} = \tau_j$, $i = \overline{1,m}$ for each *j*. The neural network with delay will thus be described by

$$\dot{x}_i(t) = -x_i(t) - \sum_{1}^{m} c_{ij} h_j\big(x_j(t - \tau_j)\big), \quad i = \overline{1,m} \tag{32}$$

Note that the equations of the neural network with delay (32) may be reduced to (31) with κ defined by

$$\kappa_j(t) = \begin{cases} e^{-(t-\tau_j)} & , \quad t \ge \tau_j \\ 0 & , \quad elsewhere \, . \end{cases} \tag{33}$$

In this case it is easy to prove (see Răsvan & Danciu, 2004; Danciu, 2006) that the time-delayed neural network described by (32) inherits the global properties of (29): dichotomy, global asymptotics or gradient behavior (the best property for the neural networks).

Stability Results for Cellular Neural Networks with Time-Delays via "Exact" Lyapunov-Krasovskii Functional

Cellular Neural Networks (CNN) introduced in 1988 by Chua & Yang, are artificial RNN displaying multidimensional arrays of identical nonlinear dynamical systems (the so-called cells) and only local interconnections between the cells (allowing VLSI implementation). CNN have been successfully applied to complex image processing, shape extraction and edge detection, but also for nonlinear partial differential equations solution, spatio-temporal system modeling and space-distributed structures control as is shown in (Fortuna *et al.*, 2001).

The purpose here is to give sufficient conditions for the *absolute stability* (*robustness*) of CNN with time delay feedback and zero control templates defined by

$$\dot{x}_i = -a_i x_i(t) + \sum_{j \in N} c_{ij} f_j\left(x_j(t-\tau_j)\right) + I_i, \quad i = \overline{1,m} \tag{34}$$

where j is the index for the cells of the nearest neighborhood N of the ith cell and τ_j are positive delays. The nonlinearities for the cellular neural networks are of the bipolar ramp type

$$f(x_i) = \frac{1}{2}\left(|x_i + 1| - |x_i - 1|\right), \tag{35}$$

hence they are bounded, nondecreasing and globally Lipschitzian with $L_i = 1$ (Figure 4). Worth mentioning that other sigmoidal nonlinear functions, which are to be met in neural networks, may also be considered.

We shall use the system in deviations

$$\dot{z}_i(t) = -a_i z_i(t) + \sum_{j \in N} c_{ij} g_j\left(z_j(t-\tau_j)\right), \quad i = \overline{1,m} \tag{36}$$

where functions g_j are also subject to sector restrictions being nondecreasing and globally Lipschitzian verifying the inequalities (3) and (4) with $L = 1$.

Using the technique due to Kharitonov & Zhabko (2003) one obtains a more complex Lyapunov functional—the so-called "exact" Krasovskii-Lyapunov functional, which gives improved (less restrictive) conditions on system's parameters. When applied to neural networks with time delays this approach has to be completed with some robustness conditions suggested by Malkin (1952) for globally Lipschitz nonlinearities (as the sigmoid functions of the RNN are).

Denoting $A_0 = diag(-a_i)_1^m$, the Krasovskii-Lyapunov functional is (see Danciu & Răsvan, 2005)

$$V(x_t) = x^T(t)U(0)x(t) + \sum_{j=1}^{m} \int_{-\tau_j}^{0} x^T(t+\theta)\left[P_j + (\tau_j + \theta)R_j\right]x(t+\theta)\,d\theta \tag{37}$$

The sign condition on the derivative W of (37) along the solutions of system (36)

$$W(x_t) = x^T(t)P_0 x(t) + \sum_{j=1}^{m} x^T(t-\tau_j)P_j x(t-\tau_j) + \sum_{j=1}^{m} \int_{-\tau_j}^{0} x^T(t+\theta)R_j x(t+\theta)\,d\theta \tag{38}$$

Figure 4. The bipolar ramp function

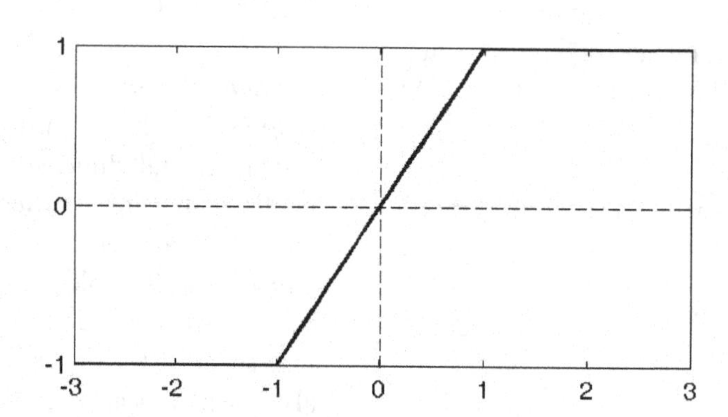

gives the following result (Danciu & Răsvan, 2005)

Theorem 4. *System (34) is exponentially stable for sigmoidal functions, in particular for (35), provided there exist positive definite matrices,* P_0, P_i, R_j, *and a positive value* μ, *such that*

$$P_0 > \mu U^T(0)U(0)\ ,\ R_j > 0\ ,$$
$$P_j - 2\left(\bar{k}_j^2\left(\sum_{i=1}^{m} c_{ij}^2\right)\Big/\mu\right)I > 0. \tag{39}$$

The *Linear Matrix Inequalities* (39) may be computed using the available commercial software on this topic. Cleary (39) gives a condition on the weights c_{ij} and the nonlinearity constants $\bar{k}_j$. The symmetry requirement for c_{ji} is no longer necessary, what gives a sensible relaxation of the stability restrictions since symmetry is not a result of the learning process.

Remark that the conditions (39) are valid regardless the values of the delays hence we obtained what is usually called *delay-independent stability.* Also stability is *exponential* since both the Krasovskii-Lyapunov functional and its derivative are quadratic functionals. Stability is also *global* since the functional and the inequality for the derivative are valid globally.

Synchronization Problems for Neural Networks with Time Delays

In the specific literature of the neuroscience domain are reported studies on the rhythmic activities in the nervous system (see for instance Kopell, 2000) and on the synchronization of the oscillatory responses with the external time-varying inputs (König & Schillen, 1991). Both rhythmicity and synchronization suggest some recurrence and this implies coefficients and stimuli being periodic or almost periodic in the mathematical model of the artificial neural networks. Let us mention that a time-varying periodic signal repeats in time with some periodicity, whereas an almost periodic signal is composed of two or more periodic signals with incommensurable frequencies; here the term incommensurable means that the ratio of the frequencies is an irrational number, which implies that the phase relationship between different periodic signals changes on every cycle forever.

The model for Hopfield-type neural networks with time delays and time-varying stimuli has the form

$$\dot{x}_i(t) = -a_i x_i(t) - \sum_{1}^{m} c_{ij} f_j\left(x_j(t-\tau_{ij})\right) + d_i(t)\ ,\quad i=\overline{1,m}. \tag{40}$$

The forcing stimuli $d_i(t)$ are periodic or almost periodic (see for instance the allures in Figure 5 and Figure 6) and the main mathematical problem is to find conditions on the systems to ensure existence and exponential stability of a unique global (i.e. defined on $\mathbb{R}$) solution that has the features of a limit regime, i.e., it is not defined by the initial conditions and is of the same type as the stimulus—periodic or almost periodic, respectively.

We shall present here for comparison some results for the two approaches of the absolute stability theory for the estimates of systems' solutions: Lyapunov function(al) and the frequency domain inequality of Popov.

The first result (Danciu & Răsvan, 2007) is based on the application of the Lyapunov functional (26) but restricted to be only quadratic in the state variables ($\lambda_i = 0$, $\delta_{ij} = 0$)

$$V(x(\cdot)) = \sum_{i=1}^{m}\left[\frac{1}{2}\pi_i x_i^2(0) + \sum_{j=1}^{m}\rho_{ij}\int_{-\tau_{ij}}^{0} x_j^2(\theta)\,d\theta\right] \tag{41}$$

with $\pi_i > 0$, $\rho_{ij} > 0$, $i,j=\overline{1,m}$. The derivative functional corresponding to $d_i(t)\equiv 0$ in (40) is

$$W(x(\cdot)) = \sum_{i=1}^{m}\left[-a_i\pi_i x_i^2(0) - \pi_i x_i(0)\sum_{j=1}^{m} c_{ij} f_j\left(x_j(-\tau_{ij})\right)\right] + \sum_{1}^{m}\sum_{1}^{m}\rho_{ij}\left[x_j^2(0) - x_j^2(-\tau_{ij})\right] \quad (42)$$

Theorem 5. *Consider system (40) with $a_i > 0$, the Lipschitz constants $L_i > 0$ and c_{ij} such that it is possible to choose $\pi_i > 0$ and $\rho_{ij} > 0$ in order to have the derivative functional (42) negative definite with a quadratic upper bound. Then the system (40) has a unique global solution $\bar{x}_i(t)$, $i = \overline{1,m}$ which is bounded on $\mathbb{R}$ and exponentially stable. Moreover, this solution is periodic or almost periodic according to the character of $d_i(t)$—periodic or almost periodic, respectively.*

The approach of the second result (Danciu, 2002) is based on the frequency domain inequality of Popov. Denoting

$$\mathbf{K}(s) = diag\left\{(s + a_i)^{-1}\right\}_{i=1}^{m} \cdot \left\{c_{ij} e^{-s\tau_{ij}}\right\}_{i,j=1}^{m}$$

$$\mathbf{L}^{-1} = diag\left\{L_i^{-1}\right\}_{i=1}^{m}, \quad \mathbf{\Theta} = diag\left\{\theta_i\right\}_{i=1}^{m} \quad (43)$$

the basic result here is (see the proof in Danciu, 2002):

Theorem 6. *Consider system (40) under the following assumptions:*

i) $a_i > 0, i = \overline{1,m}$;
ii) The nonlinear functions $f_i(\sigma)$ are globally Lipschitz satisfying conditions (3) and (4) with Lipschitz constants L_i;
iii) There exist $\theta_i \geq 0$, $i = \overline{1,m}$ such that the following frequency domain condition holds:

$$\mathbf{\Theta L}^{-1} + \frac{1}{2}\left[\mathbf{\Theta K}(i\omega) + \mathbf{K}^*(i\omega)\mathbf{\Theta}\right] > 0, \quad \omega > 0 \quad (44)$$

(the star denoting transpose and complex conjugation).

iv) $|d_i(t)| \leq M$, $i = \overline{1,m}$, $t \in \mathbb{R}$, i.e., boundedness of the forcing stimuli.

Then there exists a bounded on the whole real axis solution of (40), which is periodic, almost periodic respectively if $d_i(t)$ are periodic, almost periodic respectively. Moreover this solution is exponentially stable.

Remark that the two theorems give some relaxed sufficient conditions on the network's parameters, which do not impose the symmetry restriction for the matrix of weights.

In the sequel one considers that the "learning" process gave some structure for a Hopfield neural network with two neurons and that we have to check *a posteriori* the network behavior. If there are satisfied the assumptions of ***Theorem 6***, then the system has the desired properties. The purpose of the simulation is twofold: (a) we show that for periodic and almost periodic stimuli the solution is bounded on the whole real axis (for the time variable $t \in \mathbb{R}$) and it is of the same type as the stimuli, which means *the synchronization of the response of the network with the time-varying periodic or almost periodic external inputs*; (b) the solution is exponentially stable which means that it attains with exponential speed the same type of behavior as the stimuli.

The equations of the Hopfield neural network are

$$\dot{x}_1(t) = -2x_1(t) - 2f_1\left(x_1(t-0,5)\right) - 3f_2\left(x_2(t-0,4)\right) + \sin(t) + \sin(\pi t)$$

$$\dot{x}_2(t) = -5x_2(t) - f_1\left(x_1(t-0,7)\right) - 3f_2\left(x_2(t-0,6)\right) + 2\cos(t)$$

where the matrix of the weights

$$C = \begin{bmatrix} 2 & 3 \\ 1 & 3 \end{bmatrix}$$

is not symmetric and the values for the delays

$$T = \begin{bmatrix} 0,5 & 0,4 \\ 0,7 & 0,6 \end{bmatrix} \text{sec.}$$

are reasonable taking into account that the artificial neurons operate in times about 10^{-9} seconds. The simulation was made using the MATLAB-Simulink tool.

We verify the theorem's conditions:

1. $a_1 = 2 > 0, a_2 = 5 > 0$;
2. the nonlinear activation functions f_i, $i = 1, 2$ of the neurons are sigmoidal: the hyperbolic tangent function from the MATLAB-Simulink library
$$f(t) = \tanh(t) = \frac{2}{1+e^{-2t}} - 1 = \frac{e^t - e^{-t}}{e^t + e^{-t}}$$
its values are within the interval $[-1,1]$ and the Lipschitz constant is $L = 1$;
3. For second order systems, as in our example, the fulfilling of (44) gives: a small gain condition on system's parameters
$$c_{ii} < \frac{16a_i^2\tau_{ii}^2 + 9\pi^2}{2\sqrt{2}(4a_i\tau_{ii} + 3\pi)}$$
and for the free parameters θ_1, θ_2 the condition
$$\frac{\theta_1}{\theta_2} \in (0.0267, 2.03106)\cdot$$
It is easy to check that in our case, $c_{11} = 2 < 2.76$, $c_{22} = 3 < 11.81$ and taking $\theta_1 = \theta_2 = 1$, the frequency domain condition of Popov is fulfilled.
4. the external stimuli are bounded functions on the whole real axis as shown in Figure 5 and Figure 6: $d_1(t) = \sin(t) + \sin(\pi t)$ —an almost periodic signal with $|d_1(t)| < 2$ and $d_2(t) = 2\cos(t)$—a periodic signal with $|d_2(t)| < 3$.

Figure 7 and Figure 8 represent the time evolution of the state for the neural network with external stimuli; one can see that the two components of the state are almost periodic (as an effect of the first stimulus) and bounded on the whole real axis. Given an initial condition (in our example, $x_0 = [1,2]$), the internal behavior of the unforced system (the neural network without stimuli) can be discuss as time evolution of the state's components (Figure 9) and the evolution of the trajectory within the state space (Figure 10). Figure 9 shows an exponential convergence in time of the state toward the equilibrium in origin, whereas Figure 10 shows that this very good convergence is ensured by a stable focus in the origin of the state space.

Relating to the two theoretical results, let us remark that the system may be quite large from the dimensional point of view and the transmission delays may introduce additional difficulties in what concern the direct check of the conditions in theorems. Within this context, ***Theorem 5*** leads to computer tractable Linear Matrix Inequalities, while the frequency domain inequality in ***Theorem 6*** is difficult to verify.

The Estimation of the Admissible Delay for Preserving the Gradient-Like Behavior

The purpose here is to give an estimate of the admissible delays that preserves the asymptotic stability for each asymptotically stable equilibrium of a RNN with multiple equilibria. Consider the Hopfield-type neural networks (5) together with its model affected by time-delays (24) and theirs systems in deviations (10), respectively (25). Obviously, both systems (5) and (24) have the same equilibria and we assume the *gradient like-behavior* for the delayless Hopfield-type neural network described by (5).

The following result is based on the estimate for the state of system in deviation (25) using a consequence of rearrangement inequality No. 368 from the classical book of Hardy, Littlewood and Polya (1934) and a follows from a direct consequence of a technical Lemma belonging to Halanay (1963) (see the proof in Danciu & Răsvan, 2001); it reads as follows

Figure 5. The time-varying almost periodic stimulus $d_1(t)$

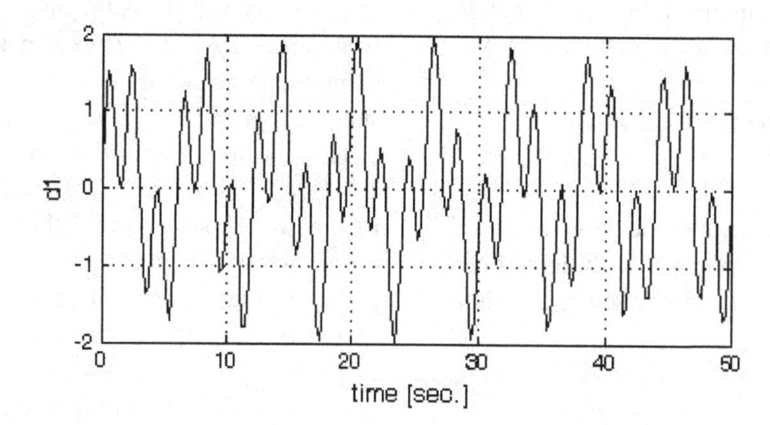

Figure 6. The time-varying periodic stimulus $d_2(t)$

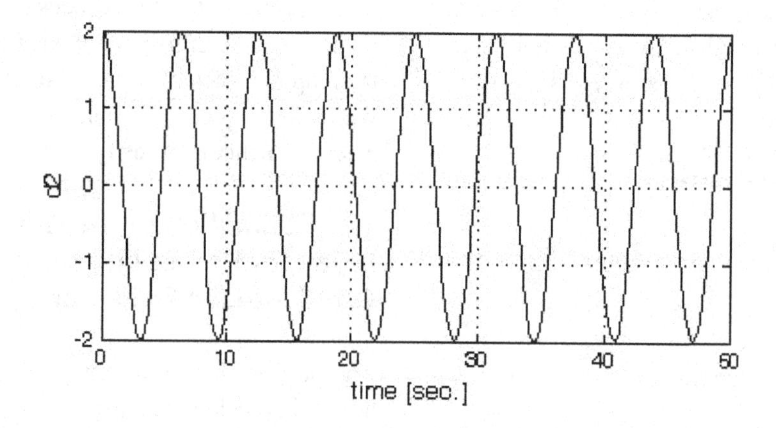

Figure 7. The first component of the state $x_1(t)$, for the forced system

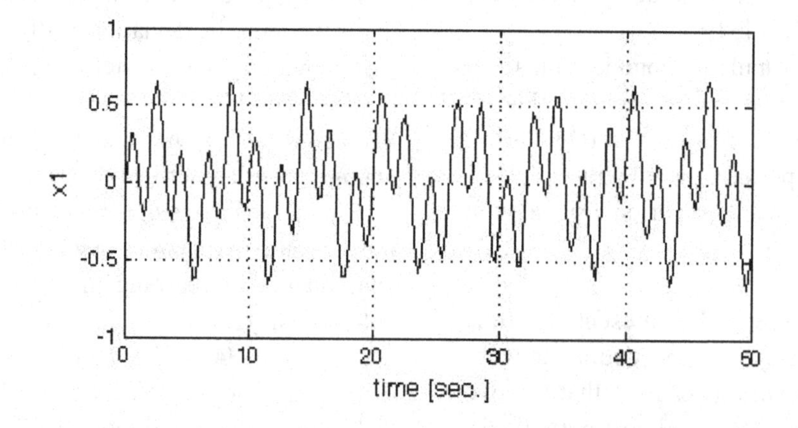

Figure 8. The second component of the state $x_2(t)$, for the forced system

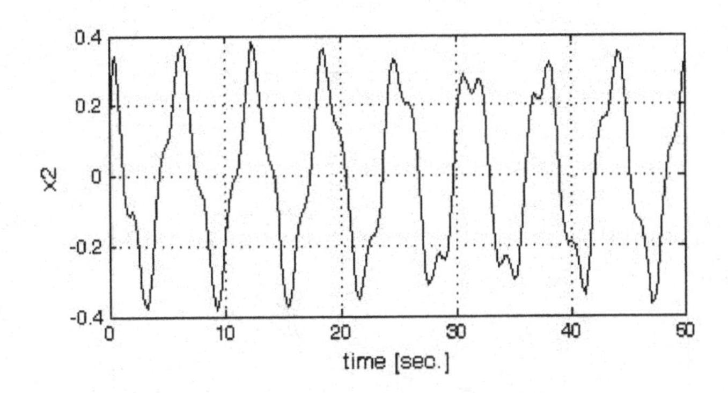

Figure 9. Time evolution of the state's components for the unforced system

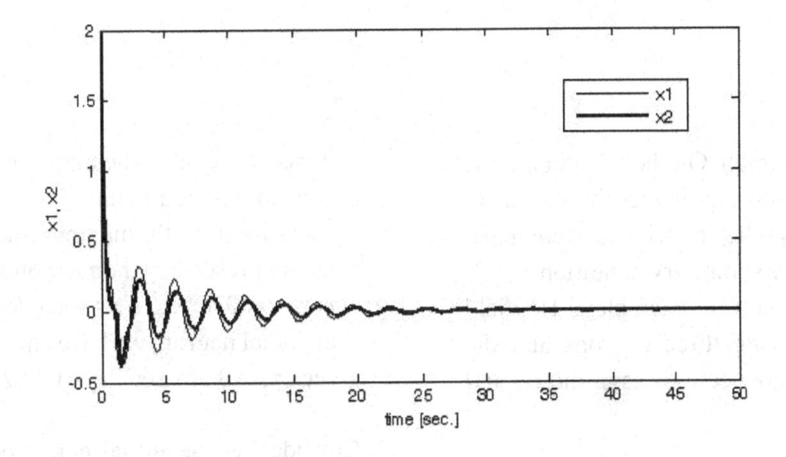

Theorem 7. *Consider the Hopfield-type neural network with time-delays (24) under the following assumptions:*

1. the nonlinearities f_i are of the sigmoidal type, what means they are non-decreasing, odd and restrict to the first and third quadrants;
2. the synaptic weight c_{ij} are such that they describe a doubly dominant matrix, i.e., they satisfy

$$c_{ii} \geq \sum_{k \neq i} |c_{ki}|, \; c_{kk} \geq \sum_{i \neq k} |c_{ki}|;$$

3. the delays are sufficiently small satisfying

$$\tau = \max_i \tau_i \leq \frac{\min(a_i)}{\left(1+\sum_1^m L_k\right)\left(\sum_1^m \max_j |c_{ij}|\right)\left(\sum_1^m L_j\left(a_j + \sum_1^m |c_{jk}|\right)\right)} \quad (45)$$

Then the network (24) has a gradient-like behavior as well as the network (5).

"Small delays don't matter" is an occasional remark of Jaroslav Kurzweil regarding the preservation of various properties of the dynamical systems for small delays—and it is also the con-

Figure 10. The evolution of the trajectory within the state space

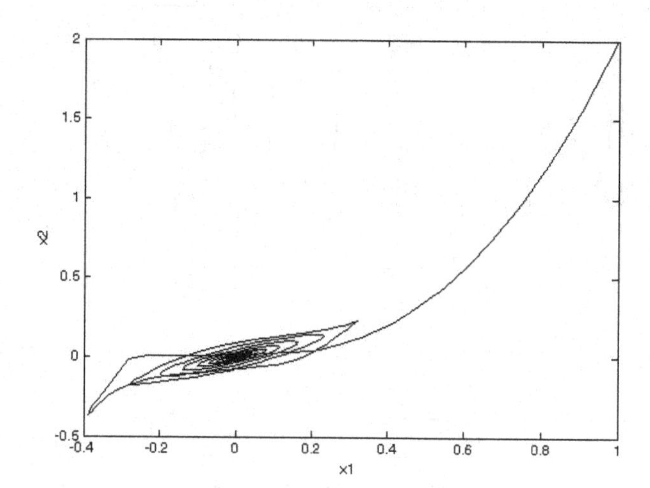

clusion of this paragraph. On the other hand, the condition of the doubly dominance for the matrix of interconnections is less restrictive in comparison with the usually symmetry condition.

Let us consider now an example: a Hopfield neural network having three neurons and the matrix of the interconnections asymmetric and doubly dominant

$$C = \begin{bmatrix} 3 & 1 & 0 \\ 1 & 4 & 2 \\ -2 & -1 & 3 \end{bmatrix}.$$

The system reads as

$$\dot{x}_1 = -6x_1(t) - 3f_1(x_1(t-\tau_1)) - f_2(x_2(t-\tau_2))$$
$$\dot{x}_2 = -5x_2(t) - f_1(x_1(t-\tau_1)) - 4f_2(x_2(t-\tau_2)) - 2f_3(x_3(t-\tau_3))$$
$$\dot{x}_3 = -9x_3(t) + 2f_1(x_1(t-\tau_1)) + f_2(x_2(t-\tau_2)) - 3f_3(x_3(t-\tau_3))$$

We verify the theorem's assumptions:

1. the nonlinearities f_i, $i = \overline{1,3}$ are of sigmoidal type: the hyperbolic tangent function from the MATLAB-Simulink library;
2. it easy to see that the matrix of the weights C is doubly dominant;
3. according (45), the maximum admissible delay for preserving the asymptotical stability is $\tau = 0.0038$ sec., a reasonable value for the artificial neuron; we have chosen $\tau_1 = 0.003$ sec., $\tau_2 = 0.001$ sec., $\tau_3 = 0.002$ sec.

Considering the initial condition $x_0 = [1 \;\; 6 \;\; 3]$ and comparing Figure 11 and Figure 12 one can observe an identical time evolution of the state for the delayless RNN and the RNN with admissible delays.

CONCLUSION

The above study resulted from the view of several sub-symbolic AI devices (mainly learning devices, in particular neural networks with feedback structure) as nonlinear dynamical systems with several (many) equilibria. The neural networks owe their "emergent computational capabilities" to the nonlinearities, to the high number of equilibria and to connectedness degree.

Figure 11. The time evolution of the state's components for the delayless neural network

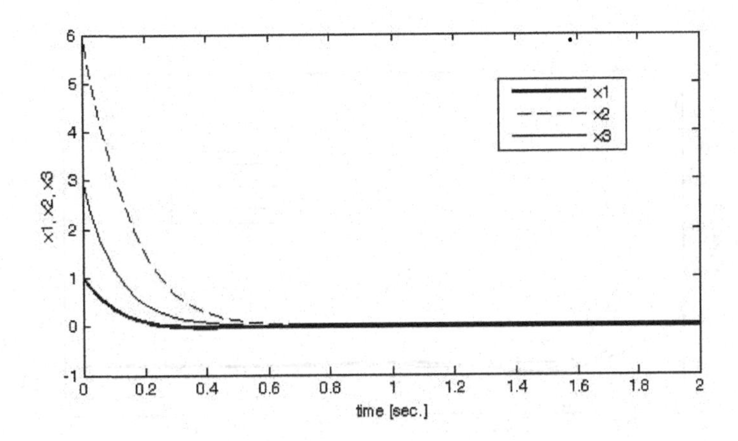

Figure 12. The time evolution of the state's components for the neural network with admissible delays for preserving the asymptotical stability $\tau_1 = 0.003$ *sec,* $\tau_2 = 0.001$ *sec,* $\tau_3 = 0.2$ *sec*

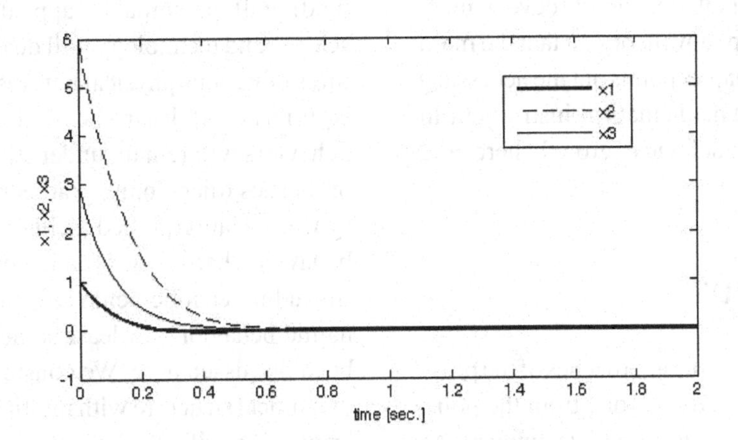

It is mentioned first that design and training of this devices is somehow independent of the necessary dynamical properties (in the sense that the networks are designed and trained for other purpose than, e.g., to ensure their own qualitative capabilities). From here the necessity of the dynamics studies as some kind of *a posteriori* analysis. This analysis uses all achievements and results of the theory of dynamical systems with several equilibria and globally Lipschitz nonlinearities.

These systems have in most cases an associated "natural" Lyapunov function. This fact allowed obtaining sufficient (but natural) conditions of a "good behavior" for the neural networks viewed as dynamical systems. This theory is less developed for systems with time delays; for this reason there was applied the technique of V. M. Popov

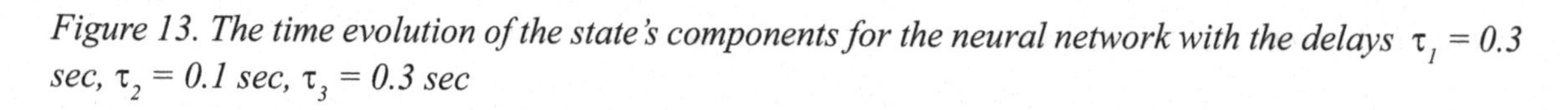

Figure 13. The time evolution of the state's components for the neural network with the delays $\tau_1 = 0.3$ *sec,* $\tau_2 = 0.1$ *sec,* $\tau_3 = 0.3$ *sec*

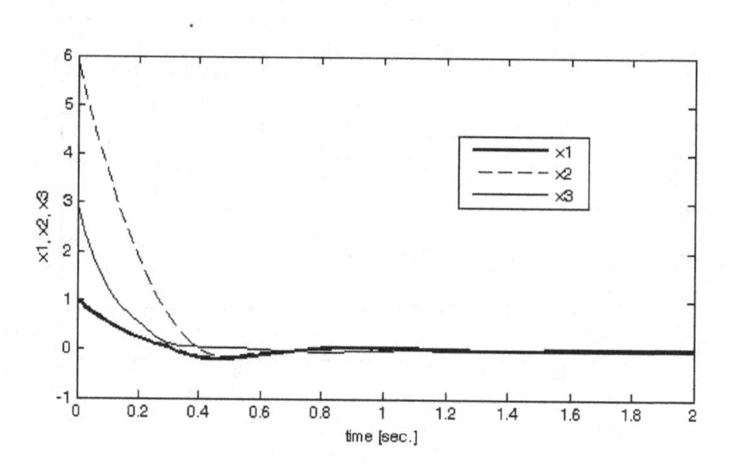

of frequency domain inequalities combined with some comparison principle. The results have been encouraging and stimulating for new development of the standard Lyapunov theory. In fact the main conclusion of the chapter points out the necessary development of the mathematical instrument in parallel to the technological growth here and elsewhere.

FUTURE TRENDS

The content, as well as the approaches of the present contribution, is a frontier one from the point of view of *Artificial Intelligence* development. At the same time the field tackled here has its own dynamics and trends, which are "ignited" by factors that may be independent of AI. We should thus analyze the trends of the theory of systems with several equilibria independently but also in connection with the field of AI.

Supposing the field of AI has its own dynamics, one has to admit that the neural networks and their structures will evolve in order to improve the imitative behavior i.e. more of the "natural" intelligence will be transferred to AI. Under the circumstances, new specific structures (with "emergent computing capabilities", to cite Hopfield) will presumably appear. Consequently, science and technology will deal with new structures of various physical natures having multiple equilibria. At least the following qualitative behaviors will remain under study: stability-like properties (dichotomy, gradient behavior a.s.o.), synchronization (forced oscillations, almost linear behavior, chaos control) and complex dynamics (including chaotic behavior). Such kinds of dynamic behavior – at least some of them – have been discussed here. We consider that each new dynamical structure with multiple equilibria will bring its specific features to be discussed within the qualitative theory.

Under the circumstances one has to foresee the development of the mathematical instrument. The main tool will remain, as presumable, the method of the Lyapunov function whose applications to the stability and oscillations is but well known. Actually we have to remark here the viability of several paradigms that underline this contribution as well as many other (some being cited in the references). One of them is that sub-symbolic AI systems are modeled by dynamical systems with several equilibria having adequate dynamical properties to ensure correct operation. The other one, which is borrowed from the theory of Hamiltonian and gradient systems, is that a "natural" Lyapunov function is associated to define structures of systems (Halanay & Răsvan, 1993; Răsvan, 1998). The other methods i.e. the frequency domain inequality Popov type and the comparison principles are useful substitutes for those cases where the Lyapunov function(al) is not available. Even in such cases we should expect development of the Lyapunov method since the Popov like inequalities are of matrix type for systems with several nonlinear functions – the case of AI devices, in particular of the neural networks.

From this point of view we hopefully consider that involvement of new mathematical development will bring new qualitative results for the dynamical systems of AI e.g. in the case of systems with time delays.

FUTURE RESEARCH DIRECTIONS

The interaction of the two paradigms – *emergent capabilities of neural networks* and *gradient-like behavior* of the dynamical systems as best achievable dynamic behavior for normal operation of this network – has far going implications. One may discover two corresponding research directions: field extension i.e. consideration of new network structures for AI but also for other applications and next, instrument development: here we include Lyapunov theory, generalized (in the La Salle sense) Lyapunov functions, new Lyapunov functionals for time delay systems.

Since conditions given by Lyapunov methods are only sufficient ones, obtaining sharper criteria is a quasi-permanent task. This goal is achievable by improving sharpness of the estimates, i.e. by making use of sharper inequalities. The researchers are thus in position to apply the entire set of procedures and results of the Lyapunov methods, also to complete and to extend it. Here it is necessary to discuss more about the method of V. L. Kharitonov of constructing Lyapunov functionals for time delay systems. One may find here the old idea of stating from a given derivative for the Lyapunov functional in order to reach via a suitable and not very abstract construction, a good Lyapunov functional satisfying suitable estimates ensuring the required qualitative behavior. The Kharitonov-like constructions for various types of delay equations and various behaviors define an interesting research direction. It appears that such research could be rewarding in both fields of dynamical systems and AI devices.

REFERENCES

Arakawa, M., Hasegawa, K. & Funatsu, K. (2003). Application of the Novel Molecular Alignment Method Using the Hopfield Neural Network to 3D-QSAR, *Journal of Chemical Information Computer Science,* 43 (5), 1396 -1402.

Atencia, M., Joya, G. & Sandoval, F. (2004). Parametric identification of robotic systems with stable time-varying Hopfield networks. *Neural Computing and Applications*, Springer London, *13*(4), 270-280.

Bose, N.K. & Liang, P. (1996). *Neural Network Fundamentals with Graphs, Algorithms and Applications*, U.S.A.: McGraw-Hill, Inc.

Chua, L. & Yang, L. (1988). Cellular neural networks: theory and applications, *IEEE Transactions on Circuits and Systems, CAS-35*, 1257-1290.

Chung, C.H. & Lee, K.S. (1991). Hopfield network application to optimal edge selection. *Neural Networks, IEEE International Joint Conference on*, vol.2, 1542 – 1547.

Cohen, M. A. & Grossberg, S. (1983). Absolute stability of pattern formation and parallel storage by competitive neural networks. *IEEE Transactions of Systems, Man & Cybernetics, 13*, 815-826.

Danciu, D. & Răsvan, V. (2000). On Popov-type stability criteria for neural networks. *Electronic Journal on Qualitative Theory of Differential Equations EJQTDE,* Proc. 6th Coll. QTDE 2000, *23.* http://www.math.uszeged.hu/ejqtde/6/623.pdf_

Danciu, D. & Răsvan, V. (2001). Gradient-like behaviour for Hopfield–type neural networks with delay. In I. Dumitrache & N. Constantin (Eds.), *Proceedings of the Third International Workshop on Intelligent Control Systems, ICS'2001,* (pp.20-24). Romania: Bucharest, Printech.

Danciu, D. (2002). Qualitative behavior of the time delay Hopfield type neural networks with time varying stimulus. *Annals of The University of Craiova,* Series: Electrical Engineering (Automatics, Computers, Electronics) *26,* 72–82.

Danciu, D. & Răsvan, V. (2005). Stability Results for Cellular Neural Networks with Time Delays. In J. Cabestany, A. Prieto, F. Sandoval (Eds.), *Computational Intelligence and Bioinspired Systems.* Lectures Notes in Computer Science, *3512* (pp. 366-373), Springer-Verlag. http://dx.doi.org/10.1007/11494669_45

Danciu, D. (2006). *Systems with several equilibria. Applications to the neural networks.* (in Romanian) Romania: Craiova, Universitaria Publishers House.

Danciu, D. & Răsvan, V. (2007). Dynamics of Neural Networks – Some Qualitative Properties. In F. Sandoval, A. Prieto, J. Cabestany, M. Graña (Eds.), *Computational and Ambient Intelligence.* Lectures Notes in Computer Science, *4507* (pp. 8-15), Springer-Verlag.

Driessche van den, P. & Zou, X. (1998). Global attractivity in delayed Hopfield neural network models. *SIAM Journal of Applied Mathematics, 58,* 1878-1890.

Fink, W. (2004). Neural attractor network for application in visual field data classification. *Physics in Medicine and Biology 49*(13), 2799-2809.

Fortuna, L., Arena, P., Balya, D. & Zarandy, A. (2001). Cellular Neural Networks. *IEEE Circuits and Systems Magazine, 4,* 6–21.

Gelig, A. Kh., Leonov, G.A. & Yakubovich, V.A. (1978). *Stability of nonlinear systems with non-unique equilibrium state.* (in Russian) U.R.S.S.: Moscow, Nauka Publishers House.

Gelig, A. Kh., Leonov, G.A. & Yakubovich, V.A. (2004). *Stability of Stationary sets in Control Systems with Discontinuous Nonlinearities.* Series on Stability, Vibration and Control of Systems, Series A, *14,* World Scientific Publishing House.

Gopalsamy, K. & He, X.Z. (1994). Stability in asymmetric Hopfield nets with transmission delays, *Physica D., 76,* 344-358.

Gu, K., Kharitonov, V.L. & Chen, J. (2003). *Stability of Time- Delay Systems.* U.S.A.: Boston, Birkhäuser.

Guirguis, L.A., Ghoneimy, M.M.R.E. (2007). Channel Assignment for Cellular Networks Based on a Local Modified Hopfield Neural Network. *Wireless Personal Communications,* Springer US, *41*(4), 539-550.

Halanay, A., (1963). *Differential Equations. Stability. Oscillations. Time Lags.* (in Romanian), Romania: Bucharest, Academia Publishers House.

Halanay, A. & Răsvan, V. (1993). *Applications of Lyapunov Methods to Stability,* The Netherlands: Dordrecht, Kluwer Academic Publishers.

Hale J. K. & Verduyn Lunel, S. M. (1993). *Introduction to Functional Differential Equations.* Springer-Verlag.

Hardy G.H., Littlewood J.E. & Polya, G. (1934). *Inequalities.* Cambridge University Press.

Hartman, P. (1964). *Ordinary Differential Equations,* Wiley, U.S.: NewYork.

Hirsch, M.W. (1988). Stability and convergence in strongly monotone dynamical systems. *Journal für Reine Angewandte Mathematik, 383,* 1-53.

Hopfield, J. J. (1982). Neural networks and physical systems with emergent collective computational abilities. *Proceedings of National Academic Science U.S.A., 79*, 2554-2558.

Hopfield, J. J. (1984). Neurons with graded response have collective properties like those of two state neurons. *Proceedings of National Academic Science U.S.A., 81*, 3088-3092.

Iwahori, Y., Kawanaka, H., Fukui, S. & Funahashi, K. (2005). Obtaining Shape from Scanning Electron Microscope using Hopfield Neural Network. *Journal of Intelligent Manufacturing*, Springer US, *16*(6), 715-725.

Jing-Sheng Xue; Ji-Zhou Sun; Xu Zhang (2004). Recurrent network in network intrusion detection system. *Proceedings of International Conference on Machine Learning and Cybernetics, 5*, 26-29 Aug. 2004, 2676 – 2679.

Kalman, R.E. (1957). Physical and mathematical mechanisms of instability in nonlinear automatic control systems. *Transactions of American Society Mechanical Engineering, 79*(3).

Khalil, H. (1992). *Nonlinear systems.* J. Griffin (Ed.) U.S.: Macmillan Publishing Company.

Kharitonov V.L. & Zhabko, A.P. (2003). Lyapunov-Krasovskii approach to the robust stability analysis of time-delay systems. *Automatica 39*, 15–20.

Kitamura, S., Hirai, K & Nishimura, M. (1967). Stability of a control system with several nonlinear elements and time lags. Technological Reports, Japan: Osaka University, *17*, 93-102.

König, P. & Schillen, J.B. (1991). Stimulus dependent assembly formation of oscillatory responses: I. Synchronization. *Neural Computation, 3*, 155-166.

Kopell, N. (2000). We got rhythm: dynamical systems of the nervous system. *Notices of American Mathematical Society, 47*, 6-16.

Kosko, B. (1992). *Neural networks and fuzzy systems. A dynamical systems approach to machine intelligence.* Prentice-Hall.

Leonov, G.A., Reitmann, V. & Smirnova, V.B. (1992). *Non-local methods for pendulum-like feedback systems*, Germany: Leipzig, Teubner Verlag.

Loccufier, M. & Noldus, E. (1995). On the estimation of asymptotic stability regions for autonomous nonlinear systems. *IMA Journal of Mathematical Control & Information, 12*, 91-109.

Malkin, I.G. (1952). *Stability of Motion.* (in Russian) U.R.S.S.: Moscow, Gostekhizdat.

Marcus, C.M. & Westervelt, R.M. (1989). Stability of analog neural networks with delay. *Physical Review A, 39*, 347-359.

Maurer, A., Hersch, M. & Billard, A. G. (2005). Extended Hopfield Network for Sequence Learning: Application to Gesture Recognition, *Proceedings of 15th International Conference on Artificial Neural Network*, 493-498.

Moser, J. (1967). On nonoscillating networks. *Quarterly Applied Mathematic, 25*, 1-9.

Nishimura, M., Kitamura, S. & Hirai, K. (1969). A Lyapunov Functional for Systems with Multiple Non-linearities and Time Lags, Technological Reports, Japan: Osaka University, *19*(860), pp. 83-88.

Noldus, E. Vingerhoeds, R & Loccufier, M. (1994). Stability of analogue neural classification networks. *International Journal of Systems Science 25*(1), 19-31.

Noldus, E. & Loccufier, M. (1995). An application of Lyapunov's method for the analysis of neural networks. *Journal of Computation and Applied Mathematics, 50*, 425-432.

Popov, V. M. (1966). *Hyperstability of Control Systems.* (in Romanian) Romania: Bucharest, Academia Publishers House, (English improved version by Springer-Verlag, 1973).

Popov, V.M. (1979). Monotonicity and Mutability. *Journal of Differential Equations, 31*(3), 337-358.

Răsvan, V. (1998). Dynamical Systems with Several Equilibria and Natural Lyapunov Functions, *Archivum mathematicum, 34*(1), [EQUADIFF 9], 207-215.

Răsvan, V. & Danciu, D. (2004). Neural networks - global behavior versus delay. In *Periodica Politechnica,* University of Timişoara, Transactions on Automatic Control and Computer Science, *49* (63), No. 2, (pp. 11-14). Romania: Timişoara, Politehnica Publishers House.

Vogels, T.P., Rajan, K. & Abbott L.F. (2005). Neural Network Dynamics. *Annual Reviews of Neuroscience, 28*, 357-376.

ADDITIONAL READING

Danciu, D. (1998). Stability of a Bidirectional Associative Memory System, *Proceedings of International Symposium on System Theory, Robotics, Computers & Process Informatics, SINTES 9*, vol. 1, System Theory, (pp. 54-59).

Danciu, D. & Răsvan, V. (2001). Steady state "almost linear" behavior of delayed Hopfield type neural networks. In I. Dumitrache & C. Buiu (Eds.), *Proceedings of the 13th International Conference on Control Systems and Computer Science, CSCS13,* (pp. 210-213). Romania: Bucharest, Politehnica Press.

Danciu, D. (2002). Time Delays and Oscillations in Neural Networks. In S. Holban (Ed.), *Periodica Politechnica,* University of Timişoara, Transactions on Automatic Control and Computer Science, *47*(61), No. 1, (pp. 131-134), Romania: Timişoara, Politehnica.

Danciu, D. & Răsvan, V. (2004). On the Stability of the Cellular Neural Networks with Time-Lags. In I. Dumitrache (Ed.), *Control Engineering and Applied Informatics, 6*(3), (pp. 11-15). Romania: Bucharest, Romanian Society of Control Engineering and Technical Informatics.

Halanay, A. (1967). *Invariant manifolds for systems with time lag.* In Hale & La Salle (Eds.) *Differential and dynamical systems.* (pp. 199–213), U.S.A.: New York, Academic Press.

Halanay, A. (1969). Almost periodic solutions for a class of linear systems with time lag, *Revue Roumaine de Mathematiques Pures et Appliquees, 14*(9), 1269-1276.

Halanay, A. (1971). For and against Lyapunov functions, *Symposia Mathematica, 6*, 167-175.

Halanay, A. & Răsvan, V. (1991). Absolute stability of feedback systems with several differentiable non-linearities, *International Journal of Systems Science, 22*(10), 1911-1927.

La Salle, J.P. (1967). An invariance principle in the theory of stability. In J.P. La Salle & J.K. Hale (Ed.) *Differential Equations and Dynamical Systems* (pp. 277-286), USA: New York, Academic Press.

La Salle, J.P. (1968). Stability Theory for Ordinary Differential Equations. *Journal of Differential Equations 4*(1), 57-65.

Noldus, E. & Loccufier, M. (1995). A new trajectory reversing method for the estimation of asymptotic stability regions. *International Journal on Control, 61*(4), 917-932.

Răsvan, V. (2002). Popov Theories and Qualitative Behavior of Dynamic and Control Systems, *European Journal on Control, 8*(3), 190-199.

Popov, V. M. (1981). Monotone-Gradient Systems. *Journal of Differential Equations, 41*(2), 245-261.

Timme M., Geisel, T. & Wolf, F. (2006). Speed of synchronization in complex networks of neural oscillators: Analytic results based on Random Matrix Theory. *Chaos, 16*(1), 015108.

Yi, Z. (1996). Global exponential stability and periodic solutions of delay Hopfield neural networks. *International Journal on Systems Science, 27*(2), 227-231.

Chapter XIX
A Genetic Algorithm–Artificial Neural Network Method for the Prediction of Longitudinal Dispersion Coefficient in Rivers

Jianhua Yang
University of Warwick, UK

Evor L. Hines
University of Warwick, UK

Ian Guymer
University of Warwick, UK

Daciana D. Iliescu
University of Warwick, UK

Mark S. Leeson
University of Warwick, UK

Gregory P. King
University of Warwick, UK

XuQin Li
University of Warwick, UK

ABSTRACT

In this chapter a novel method, the Genetic Neural Mathematical Method (GNMM), for the prediction of longitudinal dispersion coefficient is presented. This hybrid method utilizes Genetic Algorithms (GAs) to identify variables that are being input into a Multi-Layer Perceptron (MLP) Artificial Neural Network (ANN), which simplifies the neural network structure and makes the training process more efficient. Once input variables are determined, GNMM processes the data using an MLP with the back-propagation algorithm. The MLP is presented with a series of training examples and the internal weights are adjusted in an attempt to model the input/output relationship. GNMM is able to extract regression rules from the trained neural network. The effectiveness of GNMM is demonstrated by means of case study data, which has previously been explored by other authors using various methods. By comparing the results generated by GNMM to those presented in the literature, the effectiveness of this methodology is demonstrated.

INTRODUCTION

An important application of environmental hydraulics is the prediction of the fate and transport of pollutants that are released into watercourses, either as a result of accidents or as regulated discharges. Such predictions are primarily dependent on the water velocity, longitudinal mixing, and chemical/physical reactions etc, of which longitudinal dispersion coefficient is a key variable for the description of the longitudinal spreading in a river.

The concept of longitudinal dispersion coefficient was first introduced in Taylor (1954). Based on this work, the following integral expression was developed (Fischer, List, Koh, Imberger, & Brooks, 1979; Seo & Cheong, 1998) and generally accepted:

$$K = -\frac{1}{A}\int_0^B hu' \int_0^y \frac{1}{\varepsilon_t h}\int_0^y hu'\,dy\,dy\,dy \tag{1}$$

where K = longitudinal dispersion coefficient; A = cross-sectional area; B = channel width; h = local flow depth; u' = deviation of local depth mean flow velocity from cross-sectional mean; y = coordinate in the lateral direction; and ε_t = local (depth averaged) transverse mixing coefficient. An alternative approach utilises field tracer measurements and applies the method of moments. It is also well documented in the literature (Guymer, 1999; Rowinski, Piotrowski, & Napiorkowski, 2005; Rutherford, 1994) and defines K as

$$K = \frac{U_c^2}{2}\left[\frac{\sigma_t^2(x_2) - \sigma_t^2(x_1)}{\bar{t}_2 - \bar{t}_1}\right] \tag{2}$$

where U_c = mean velocity, x_1 and x_2 denotes upstream and downstream measurement sites, $\bar{t}$ = centroid travel time, $\sigma_t^2(x)$ = temporal variance,

However, owing to the requirement for detailed transverse profiles of both velocity and cross-sectional geometry, equation (1) is rather difficult to use. Furthermore, equation (2), called the method of moments (Wallis & Manson, 2004), requires measurements of concentration distributions and can be subject to serious errors due to the difficulty of evaluating the variances of the distributions caused by elongated and/or poorly defined tails. As a result, extensive studies have been made based on experimental and field data for predicting the dispersion coefficient (Deng, Singh, & Bengtsson, 2001; Jobson, 1997; Seo & Cheong, 1998; Wallis & Manson, 2004).

For example, employing 59 hydraulic and geometric data sets measured in 26 rivers in the United States, Seo and Cheong (1998) used dimensional analysis and applied the one-step Huber method, a nonlinear multi-regression method, to derive the following equation:

$$K = 5.915(Hu^*)\left(\frac{B}{H}\right)^{0.62}\left(\frac{U}{u^*}\right)^{1.428} \tag{3}$$

in which u^* = shear velocity. This technique uses the easily measureable hydraulic variables of B, H and U, together with a frequently used parameter, extremely difficult to accurately quantify in field applications, u^*, to estimate the dimensionless dispersion coefficient K from equation (3). Another empirical equation developed by Deng et al. (2001) is a more theoretically based approximation of equation (1), which not only includes the conventional parameters of (B/H) and (U/u^*) but also the effects of the transverse mixing ε_{t0}, as follows:

$$K = 0.15\left(\frac{Hu^*}{8\varepsilon_{t0}}\right)\left(\frac{B}{H}\right)^{5/3}\left(\frac{U}{u^*}\right)^{2} \tag{4}$$

where

$$\varepsilon_{t0} = 0.145 + \left(\frac{1}{3520.0}\right)\left(\frac{B}{H}\right)^{1.38}\left(\frac{U}{u^*}\right) \quad (5)$$

These equations are easy to use, assuming measurements or estimates of the bulk flow parameters are available. However, they may be unable to capture the complexity of the interactions of the fundamental transport and mixing mechanisms, particularly those created by non-uniformities, across the wide range of channels encountered in nature. In addition, the advantage of one expression over another is often just a matter of the selection of data and the manner of their presentation. Regardless of the expression applied, one may easily find an outlier in the data, which definitely does not support the applicability of a particular formula. An expectation that, in spite of the complexity of the river reach, the dispersion coefficient may be represented by one of the empirical formulae seems exaggerated (Rowinski et al., 2005).

Furthermore, most of the studies have been carried out based on specific assumptions and channel conditions and therefore the performance of the equations varies widely for the same stream and flow conditions. For instance, Seo and Cheong (1998) used 35 of the 59 measured data sets for establishing equation (3) and the remaining 24 for verifying their model. While the model of Deng et al. (equation (4) - (5)) (2001) is limited to straight and uniform rivers. They also assume that the river is straight and uniform with a width-to-depth ratio greater than 10. Therefore, a model that has greater general applicability is desirable.

Recently Artificial Neural Network (ANN) modelling approaches have been embraced enthusiastically by practitioners in water resources, as they are perceived to overcome some of the difficulties associated with traditional statistical approaches, e.g. making assumptions with regard to stream geometry or flow dynamics. ANNs are systems that are deliberately constructed to make use of some organizational principles resembling those of the human brain (Lin & Lee, 1996). They offer an effective approach for handling large amounts of dynamic, non-linear and noisy data, especially when the underlying physical relationships are not fully understood (Cannas, Fanni, See, & Sias, 2006; Hagan, Demuth, & Beale, 1996; Haykin, 1994).

In specific terms, several authors (Kashefipour, Falconer, & Lin, 2002; Piotrowski, Rowinski, & Napiorkowski, 2006; Rowinski et al., 2005; Tayfur, 2006; Tayfur & Singh, 2005) have reported successful applications of ANNs to the prediction of dispersion coefficient. For example, in the case of Tayfur and Singh (2005) the ANN was trained and tested using 71 data samples of hydraulic and geometric variables and dispersion coefficients measured on 29 streams and rivers in the United States, with the result that 90% of the dispersion coefficient was explained. However, there is always a lack of a suitable input determination methodology for ANN models in these applications. Moreover, without further interpretation of the trained network, their results are not easily transferable.

The aim of the present work is to present and demonstrate a data driven method (hereafter called GNMM, the Genetic Neural Mathematical Method) based on a Multi-Layer Perceptron (MLP) neural network, for the prediction of longitudinal dispersion coefficient. By utilizing a Genetic Algorithm (GA), GNMM is able to optimise the number of inputs to the ANN. As we will show this simplifies the network structure and also accelerates the training process. Employing a Mathematical Programming (MP) method (Tsaih & Chih-Chung, 2004), GNMM is also capable of identifying regression rules extracted from the trained MLP.

BACKGROUND TECHNIQUES

Before describing the specific details of GNMM, some relevant Intelligence Systems concepts are reviewed below.

Genetic Algorithms (GAs)

GAs are a powerful optimisation technique, based upon the underlying principles of natural evolution and selection (Holland, 1992). GAs are well suited to the task of selecting an appropriate combination of inputs to a model as they have the ability to search through large numbers of combinations, where interdependencies between variables may exist (Michalewicz, 1996; Reeves & Rowe, 2003; Rothlauf, 2006).

The main procedures in the GA optimisation process are shown in Figure 1. The basic idea is to maintain a population of chromosomes, representing candidate solutions to the concrete problem being solved. The possible solutions are generally coded as binary strings and these strings are equivalent to biological chromosomes. Other non-binary codings have proven to be useful in some applications (Gardner, Boilot, & Hines, 2005; Srinivasa, Venugopal, & Patnaik, 2007). Each bit of the binary string (chromosome) is referred to as a gene. A GA starts off with a population of randomly generated chromosomes and advances towards better chromosomes by applying genetic operators that are based on genetic processes occurring in nature (i.e., selection, crossover and mutation) (Haupt & Haupt, 2004; Mitchell, 1996).

The search is initialized with a random population of chromosomes, each representing a possible solution. Next, each chromosome in the population is decoded into a solution and its fitness is evaluated using an objective function. During successive iterations, or *generations*, the adaptation or associated fitness of chromosomes in the population is quantified by means of fitness functions. Chromosomes for the new population are selected with a probability proportional to their fitness, related to the purpose of the study. Once the chromosomes have been selected, a crossover procedure is used to partially exchange genetic information between two parent chromosomes. Chromosomes from the parent pool are randomly paired up and are tested to determine if an exchange will take place based on a crossover probability. If an exchange is to take place, a crossover site is selected at random for the two chromosomes and the genetic material (genes) after the crossover site is then exchanged between the two parent strings. In so doing, two child chromosomes are produced, which form the members of a new population. If an exchange is not to take place (i.e. the crossover probability is less than the crossover probability parameter), then the two parents enter the new population unchanged. Mutation has the purpose of keeping the population diverse and preventing the GA from prematurely converging onto a local minimum. Each chromosome is tested on a probability basis to determine if it will be mutated. In the most commonly used form of mutation, the probability that each bit in the chromosome will be mutated is determined by the mutation probability parameter. If a bit is to be mutated, then this occurs by flipping its value (i.e. a '0' will become a '1' and vice versa). The application of the mutation operator marks the end of one GA cycle. The GA is usually allowed to run for a specified number of generations, or until some stopping criterion is met, such as convergence of the population to a single solution.

A GA differs from many other optimisation methods by virtue of the fact that a population, or collection of possible solutions, is used rather than a single solution. It does not need knowledge of the problem domain, but it requires a fitness function to evaluate the fitness of a solution. The GA techniques have a solid theoretical foundation, which is based on the Schema Theorem. A comprehensive description of GAs can be found in Goldberg (1989) and Holland (1992).

Artificial Neural Networks (ANNs)

ANNs are a powerful set of tools for solving problems in pattern recognition, data processing, non-linear control and time series prediction. They offer an effective approach for handling large amounts of dynamic, non-linear and noisy data, especially when the underlying physical relationships are not fully understood (Cannas et al., 2006).

At heart, ANNs are distributed, adaptive, generally nonlinear learning machines comprised of different processing elements called neurons (Bishop, 1995). Each neuron is connected with other neurons and/or with itself. The interconnectivity defines the topology of the ANN. The connections are scaled by adjustable parameters called weights. Each neuron receives either external inputs or inputs from other neurons to which the neuron is connected and produces an output that is a nonlinear static function of the weighted sum of these inputs.

Although there are now a significant number of neural network types and training algorithms (Hagan et al., 1996; Haykin, 1994; Jang, Sun, & Mizutani, 1997), this chapter will focus on the MLP. Figure 2 provides an overview of the structure of this ANN. In this case, the ANN has three layers of neurons—an input layer, a hidden layer and an output layer. Each neuron has a number of inputs (from outside the neural network or the previous layer) and a number of outputs (leading to the subsequent layer or out of the neural network). A neuron computes its output response based on the weighted sum of all its inputs according to an activation function. Data flows in one direction through this kind of neural network starting from external inputs into the first layer, which are transmitted through the hidden layer, and then passing to the output layer from which the external outputs are obtained.

The ANN is trained by adjusting the weights that connect the neurons using a procedure called back-propagation (Bryson & Ho, 1975). In this procedure, for a given input-output pair, the back-propagation algorithm performs two phases of data flow (see Figure 2). First, the input pattern is propagated from the input layer to the output layer and, as a result of this forward flow of data, it produces an actual output. Then the error sig-

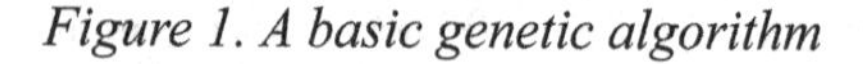
Figure 1. A basic genetic algorithm

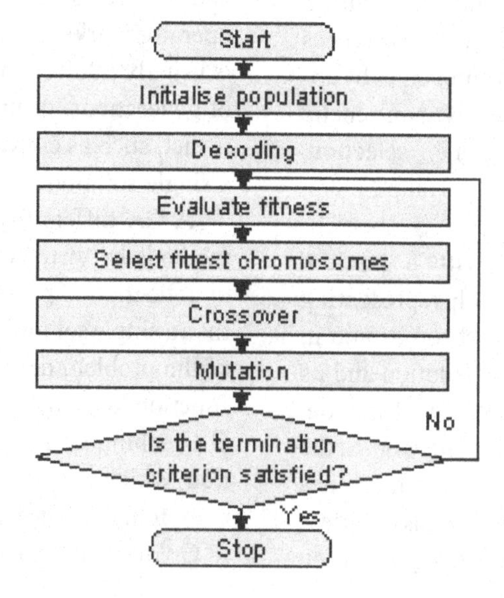

nals resulting from the difference between the expected and actual outputs are back-propagated from the output layer to the previous layers for them to update their weights.

It has been shown (Cybenko, 1989) that MLPs can approximate virtually any function with any desired accuracy, provided that there are enough hidden neurons in the network and that a sufficient amount of data is available.

METHODOLOGY

Briefly speaking, in GNMM GAs are used as a variable selection tool. Those variables identified by GAs are then redirected as inputs to an MLP. Finally, GNMM extracts regression rules from the trained network. Let us assume there are two data sets $\boldsymbol{X} = \{x_{(1,1)}, \ldots, x_{(a,b)}\}$ and $\boldsymbol{Y} = \{y_1, \ldots, y_a\}$, where $\boldsymbol{X}$ is the hydraulic/geometric measurements, $\boldsymbol{Y}$ is the corresponding longitudinal dispersion coefficient, a is the number of measurements that have been made and b denotes available variables. The implementation of GNMM can be described in 3 main stages as follows:

Step 1: Variable Selection

Randomly generate an initial population of chromosomes of size N_p ($N_p >> 0$ and ideally $N_p \approx 2^b$). A chromosome consists of b genes, each representing an input variable. The encoding of a gene is binary, meaning that a particular variable is considered as an input variable (represented by '1') or not (represented by '0'). The assessment of the fitness of a chromosome is the Mean Squared Error (MSE) when a three-layer MLP is being trained with the input variable subset $\boldsymbol{X}_i$ and output target $\boldsymbol{Y}$ for a certain number of epochs N_e. The number of neurons in the hidden layer is the same as the number of the input variables.

As mentioned previously, provided that there are enough hidden neurons in the network and that a sufficient amount of data is available, MLPs can approximate virtually any function with any desired accuracy. However, in the current stage the

Figure 2. Three-layer back-propagation neural network

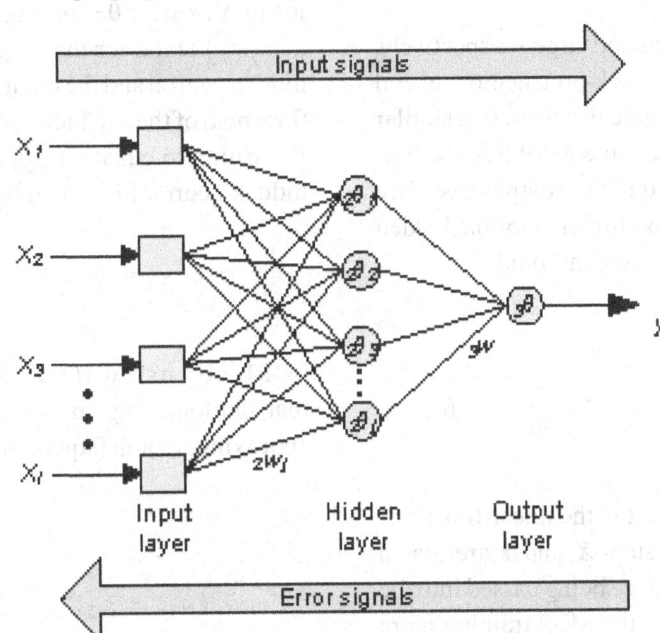

number of neurons in the hidden layer is selected to be the same as the number of the input variables. The reason for doing this is that the purpose of the current stage is only to explore the effectiveness of different input parameter combinations; such settings simplify the GA implementation and reduce the computational burden.

It should be noted that a small randomly selected portion of $\boldsymbol{X}$ and $\boldsymbol{Y}$ (e.g. 10%×a) is set aside for validation beforehand in order to avoid over fitting (Lin & Lee, 1996), thus the MLP does not necessarily reach its final epoch N_e. Also of note is that because the weights $\boldsymbol{w}$ and thresholds $\boldsymbol{\theta}$ in an MLP are randomly generated, it is not always true that offspring populations yield better results (smaller MSE) than parental ones.

The GA optimisation process is then realised by altering N_p, N_e, generation size N_g, crossover probability p_c and mutation probability p_m. As a result, the input variables which occur most frequently throughout all the populations can therefore be identified. The final subset formed by these variables $\boldsymbol{X}_f$ is the subset that produces the minimal error within a given number of epochs.

Step 2: MLP Training

Taking $\boldsymbol{X}_f$ and $\boldsymbol{Y}$ as inputs and targets respectively, an MLP is trained using back-propagation several times in order to minimize the error. In a similar way to the previous step, the MLP has one hidden layer and one neuron in the output layer (see Figure 2). The activation function for the hidden layer is the 'hyperbolic tangent' (tanh)

$$f(t) = \frac{1 - e^{-2t}}{1 + e^{-2t}} = \frac{2}{1 + e^{-2t}} - 1 \qquad (6)$$

and for the output layer it is the linear function.

As in the previous step, $\boldsymbol{X}_f$ and $\boldsymbol{Y}$ are scaled into the range [−1, 1] before being passed into the neural network to make the MLP training more efficient. For example, consider x_n as an element of the n-th column vector ($\mathbf{x}_n$) in $\boldsymbol{X}_f$, the mapping is carried out as follows:

$$x_n' = \frac{2 \times (x_n - \boldsymbol{x}_{min})}{(\boldsymbol{x}_{max} - \boldsymbol{x}_{min})} - 1 \qquad (7)$$

where x_n' denotes the mapped value, $\mathbf{x}_{\text{min}}$ and $\mathbf{x}_{\text{max}}$ are the minimum and maximum values in $\mathbf{x}_n$. After the neural network has been trained, the settings of equation (7) are used to transform any future inputs that are applied to the network. Thus, $\mathbf{x}_{\text{min}}$ and $\mathbf{x}_{\text{max}}$ effectively become a part of the network, just like the network weights and biases.

However, unlike in the case of the previous step, where the learning rate α has to be sufficiently small (e.g. 0.01) in order to avoid ANN oscillation, in the current step α can have a relatively greater value (e.g. 0.04) to accelerate the learning process.

Step 3: Rule Extraction

Let $\mathbf{x}_m = \{x_{(m,1)}, \ldots, x_{(m,i)}\}$ denote the m-th row vector in $\boldsymbol{X}_f$, where $0 < m < a$, $0 < i < b$, ${}_2\mathbf{w}_j = \{{}_2w_{1j}, \ldots, {}_2w_{ij}\}$ stand for the weights between the j-th hidden neuron and the input layer, ${}_2\theta_j$ stand for the threshold of the j-th hidden neuron (see Figure 2). Returning to equation (6), the output of the j-th hidden neuron for $\mathbf{x}_m$ can be written as

$${}_m h_j = f(\boldsymbol{x}_m \boldsymbol{w}_j' + {}_2\theta_j) \qquad (8)$$

It has been shown (Tsaih & Chih-Chung, 2004) that the following function $g(t)$ can be used to approximate tanh (equation (6)).

$$g(t) = \begin{cases} 1 & t \geq \kappa \\ \beta_1 t + \beta_2 t^2 & 0 \leq t \leq \kappa \\ \beta_1 t - \beta_2 t^2 & -\kappa \leq t \leq 0 \\ -1 & t \leq -\kappa \end{cases} \qquad (9)$$

in which $\beta_1 = 1.0020101308531$, $\beta_2 = -0.251006075157012$, $\kappa = 1.99607103795966$. Letting $t_j = \mathbf{x}_m\mathbf{w}_j$' and substituting $t = t_j + {}_2\theta_j$ into equation (9), we get the following

$$g(t_j + {}_2\theta_j) = \begin{cases} 1 & t_j \geq \kappa - {}_2\theta_j \\ (\beta_1\,{}_2\theta_j + \beta_2\,{}_2\theta_j^2) & \\ \quad +(\beta_1 + 2\beta_2\,{}_2\theta_j)t_j & -{}_2\theta_j \leq t_j \leq \kappa - {}_2\theta_j \\ \quad +\beta_2 t_j^2 & \\ (\beta_1\,{}_2\theta_j - \beta_2\,{}_2\theta_j^2) & \\ \quad +(\beta_1 - 2\beta_2\,{}_2\theta_j)t_j & -\kappa - {}_2\theta_j \leq t \leq -{}_2\theta_j \\ \quad -\beta_2 t_j^2 & \\ -1 & t_j \leq -\kappa - {}_2\theta_j \end{cases} \tag{10}$$

Thus for the j-th hidden neuron, the output value is approximated with a polynomial form of single variable t_j in each of four separate polyhedrons in the $\mathbf{x}_m$ space. For example, if $\mathbf{x}_m \in\in \{\mathbf{x}_m$: $-{}_2\theta_j \leq \mathbf{x}_m\mathbf{w}_j' \leq \kappa - {}_2\theta_j\}$, then tanh $(\mathbf{x}_m\mathbf{w}_j' + {}_2\theta_j)$ is approximated with $\beta_{12}\theta_j + \beta_{22}\theta_j^2 + (\beta_1 + 2\beta_{22}\theta_j)t_j + \beta_2 t_j^2$. Because the activation function for the output layer is a linear function, a comprehensible regression rule associated with a trained ANN with i hidden neurons is thus:

- *IF* $\mathbf{x}_m \in\in \{\mathbf{x}_m: -{}_2\theta_j \leq \mathbf{x}_m\mathbf{w}_j' \leq \kappa - {}_2\theta_j$ for all

$$y' = {}_3\theta + \textstyle\sum_{j=1}^{i} {}_3w_j(\beta_1\,{}_2\theta_j + \beta_2\,{}_2\theta_j^2 + (\beta_1 + 2\beta_2\,{}_2\theta_j)t_j + \beta_2 t_j^2)$$

$$y' = {}_3\theta + \textstyle\sum_{j=1}^{i} {}_3w_j(\beta_1\,{}_2\theta_j + \beta_2\,{}_2\theta_j^2 + (\beta_1 + 2\beta_2\,{}_2\theta_j)t_j + \beta_2 t_j^2)$$

Thus, once the training is done the neural network simulated output for $\mathbf{x}$ can be easily obtained. In other words, regression rules associated with the trained MLP can be derived.

CASE STUDY APPLICATION

In this case study, GNMM is applied to a very well studied set of data. By comparing the results generated by GNMM to those in the literature, we aim to demonstrate the effectiveness of GNMM.

Data Set

The whole data set, denoted by $\boldsymbol{S}_1$, contains 71 sets of measurements from 29 rivers in the United States (Deng et al., 2001; Rowinski et al., 2005; Seo & Cheong, 1998; Tayfur & Singh, 2005). Apart from the longitudinal dispersion coefficient K, a reach mean parameter, there are five independent variables. These are channel width (B), flow depth (H), velocity (U), shear velocity (u^*) and river sinuosity (α = channel length/valley length). Dependant variables are width-to-depth ratio (B/H), relative shear velocity (U/u^*) and channel shape variable ($\beta = ln(B/H)$); a total of eight variables are available.

The range for the dispersion coefficient (K) varies from 1.9 to 892 m^2/s, and K is greater than 100 m^2/s in 21 cases, which represents about 30% of the total measured coefficient values. The range for the width-to-depth ratio (B/H) of the data sets varies from 13.8 to 156.5 and (B/H) is greater than 50 in 26 cases. Other statistical information about $\boldsymbol{S}_1$ can also be found in Table 1.

$\boldsymbol{S}_1$ is divided into two sub-sets, $\boldsymbol{S}_{1t}$ and $\boldsymbol{S}_{1v}$ for training and validation/testing respectively. $\boldsymbol{S}_{1t}$ contains 49 data sets out of 71, while $\boldsymbol{S}_{1v}$ consists of the remaining 22. When forming these two sub-sets, the present work follows that of Tayfur and Singh (2005), in order to compare results. However, as mentioned in Tayfur and Singh (2005): "In choosing the data sets for training and testing, special attention was paid to ensure that we have representative sets so as to avoid bias in model prediction" (p. 993). Statistics for $\boldsymbol{S}_{1t}$ and $\boldsymbol{S}_{1v}$ are also listed in Table 1. As well as the statistics in Table 1, it is worth mentioning that the percentages of $K > 100$ m/s^2 and $B/H > 50$ are also comparable for both $\boldsymbol{S}_{1t}$ and $\boldsymbol{S}_{1v}$. For example, in $\boldsymbol{S}_{1v}$ 25% of K is greater than 100 m/s^2 (this ratio is 31% in $\boldsymbol{S}_{1t}$), also, in $\boldsymbol{S}_{1v}$ 40% of B/H is greater than 50 (this ratio is 31% in $\boldsymbol{S}_{1t}$).

Results and Discussion

GNMM is implemented in MATLAB (v7.2), mainly using the Genetic Algorithm and Direct Search Toolbox, as well as the Neural Network Toolbox. GAs are configured to run five times to explore different combinations of input variables. The configuration for each GA is shown in Table 2, along with CPU speed and CPU time. It should be noted that N_e in Table 2 stands for 'number of epochs per chromosome'. Also note that in all the cases $p_c = 0.8$, $p_m = 0.01$ and $\alpha = 0.01$ as these were empirically found suitable for the problem.

As mentioned previously, GAs serve as an input variable selection tool in GNMM. From this point of view, GNMM belongs to a broader concept of evolutionary ANNs (Yao, 1999), i.e. GAs evolve to an optimal set of input variables, and hence the structure of the ANN. A three-layer MLP is used to evaluate the fitness of each chromosome, with input/output neurons being the same as the number of input/target variables (the target is the longitudinal dispersion coefficient). For simplicity, the number of neurons in the hidden layer is made the same as of the input layer, i.e. the number of the input variables. The assessment of the fitness of a chromosome is the Mean Squared Error (MSE) when this MLP is being trained with the selected input variable combination and the corresponding longitudinal dispersion coefficient.

After running the GAs five times, a clear distinction was evident between variables. A variable's *appearance percentage* was defined as the accumulated appearance in the winning chromosome (minimizing MSE) of each population within a GA divided by the total number of generations. The average *appearance percentage* of each variable in the five GAs is illustrated in Figure 3. It can be seen that the most frequently appearing variables are U (99%), B (96%) and H (70%), followed by u^* (28%), U/u^* (26%), whereas β, α and B/H are all less than 2%. Thus U, B and H are used during the final MLP training.

The effect of such variable selection can clearly be seen in Figure 4. Within 200 epochs, the 'selected variable combination' achieved a much smaller MSE than using them all (0.067 vs. 0.115). This further demonstrates that not all of the potential input variables are equally informative since some may be correlated (U/u^*, B/H, β), noisy or have no significant relationship (α) with the longitudinal dispersion coefficient.

By setting $\alpha = 0.04$, number of neurons in the hidden layer = 3, running the MLP several times, the minimum Root Mean Square Error (RMSE) for the training data is 34.85, the coefficient of determination (R^2) is 0.96 at $N_e = 19887$. This implies that the MLP model is satisfactorily trained. A plot of the measured and predicted longitudinal dispersion coefficients can be seen in Figure 5. It is evident that the data is evenly distributed around the '$y=x$' line.

Comparing the above results to those in Tayfur and Singh (2005) (N_e = 20000, R^2 = 0.90), GNMM performs better. Although MLPs are being adopted in both applications, the difference lies in the fact that only a portion of available variables are used in GNMM instead of using them all as in Tayfur and Singh (2005). For the test data, a comparison has also been made with some other models, as in Table 3, which also shows that GNMM produces the best results, and ANN models (GNMM and MLP (Tayfur & Singh, 2005)) generally perform better.

The final weights and biases of the MLP that minimizes MSE are as follows: ${}_3\theta = -0.6031$, ${}_2\theta_1 = 1.4022$, ${}_2\theta_2 = -0.0143$, ${}_2\theta_3 = -4.1393$, ${}_3\mathbf{w} = (-1.7705, 0.8517, -1.2564)$, ${}_2\mathbf{w}_1 = (4.1222, 0.9600, -1.5078)$, ${}_2\mathbf{w}_2 = (5.7385, -4.3290, 1.1943)$, ${}_2\mathbf{w}_3 = (-0.7147, -6.7842, 0.3987)$. Returning to equation (10), we have:

$$t_1 = 4.1222B' + 0.9600H' - 1.5078U' \tag{11}$$

Table 1. Case study data statistics

	B (m)	H (m)	U (m/s)	u^* (m/ s)	B /H	U/u^*	β	α	K (m2/s)
Max	711.2	19.94	1.74	0.553	156.5	19.63	5.05	2.54	892.0
Min	11.9	0.22	0.03	0.002	13.8	1.29	2.62	1.08	1.9
$\bar{S}_1$	83.0	1.70	0.54	0.088	51.7	7.62	3.79	1.39	107.7
$\bar{S}_{1t}$	62.9	1.31	0.49	0.084	51.4	7.13	3.79	1.37	98.4
$\bar{S}_{1v}$	127.6	2.55	0.66	0.097	52.4	8.72	3.77	1.42	128.4

$$t_2 = 5.7385B' + 4.3290H' + 1.1943U' \tag{12}$$

$$t_3 = -0.7147B' - 6.7842H' + 0.3987U' \tag{13}$$

where B', H' and U' are scaled variables according to equation (7) and Table 1. Also, we have

$$g(t_1) = \begin{cases} 1 & t_1 \geq 0.5939 \\ 0.9115 + 0.2981t_1 - 0.2510t_1^2 & -1.4022 \leq t_1 \leq 0.5939 \\ 1.8985 + 1.7059t_1 + 0.2510t_1^2 & -3.3983 \leq t_1 \leq -1.4022 \\ -1 & t_1 \leq -3.3983 \end{cases} \tag{14}$$

$$g(t_1) = \begin{cases} 1 & t_2 \geq 2.0104 \\ -0.1444 + 1.0092t_2 - 0.2510t_2^2 & 0.0143 \leq t_2 \leq 2.0104 \\ -0.0143 + 0.9948t_2 + 0.2510t_2^2 & -1.9818 \leq t_2 \leq 0.0143 \\ -1 & t_2 \leq -1.9818 \end{cases} \tag{15}$$

$$g(t_1) = \begin{cases} 1 & t_3 \geq 6.1354 \\ -8.4483 + 3.0800t_3 - 0.2510t_3^2 & 4.1396 \leq t_3 \leq 6.1354 \\ 0.1531 - 1.0760t_3 + 0.2510t_3^2 & 2.1432 \leq t_3 \leq 4.1396 \\ -1 & t_3 \leq 2.1432 \end{cases} \tag{16}$$

Since the activation function in the output neuron is a linear function, we also get

$$y' = -1.7705g_1 + 0.8517g_2 - 1.2564g_3 - 0.6031 \tag{17}$$

Thus, regression rules are extracted from the trained MLP. Among these 64 (4^3) potential rules, some are null and will never execute. Null rules can be identified using the Simplex algorithm (Tsaih & Chih-Chung, 2004), which is not within the scope of the present work. However, rules fired for the training and test data are listed in Table 4, in which the number under the function name denotes the corresponding sub-function.

Table 2. GA parameters and CPU speeds/time

Case	N_p	N_g	N_e (/ chromosome)	CPU Speed (GHz)	CPU Time (s)
1	200	101	100	3.20	17830.33
2	200	101	50	2.66	31772.62
3	400	101	100	3.20	33600.17
4	200	201	20	2.66	45690.86
5	200	201	100	3.20	39280.16

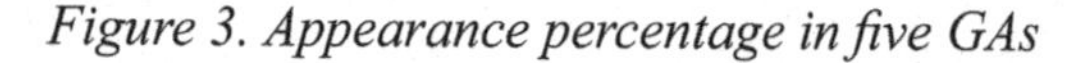

Figure 3. Appearance percentage in five GAs

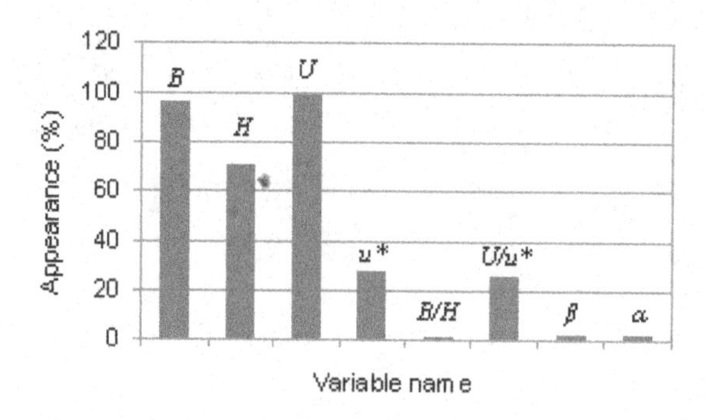

Figure 4. Comparison of performance using all the variables and selected variables

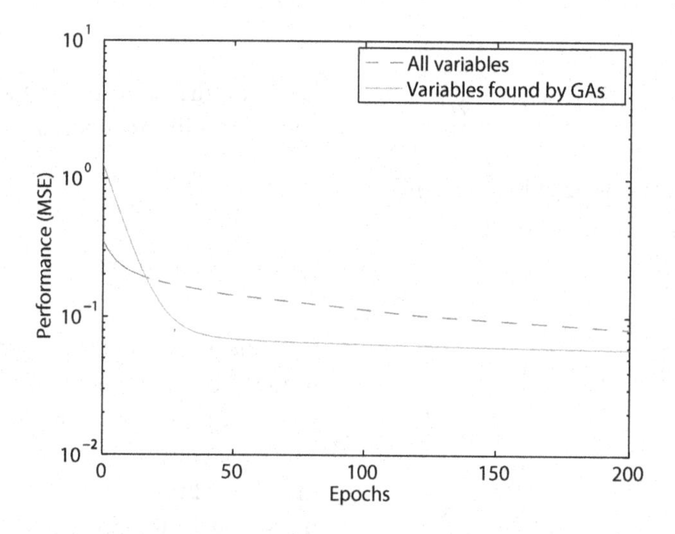

The number of training and test data samples associated with each rule is also listed in Table 4; this gives us an idea of the importance of each rule and the distribution of the data.

Regression rules summarised in Table 4 can also be written in the antecedent/consequent format. For example, Rule 2, which is executed most of the time for both the training and test data, can be rewritten as

- *IF* $t_1 \geq 0.5939$
- *AND* $t_2 \geq 2.0104$
- *AND* $2.1432 \leq t_3 \leq 4.1393$
- *THEN* $y' = -1.7143 + 1.3519t_3 - 0.3154\, t_3^2$

However, the above derived y' needs to be mapped back to the normal range using the reverse function of equation (7) to obtain the GNMM simulated dispersion coefficient y:

$$y = \frac{(y'+1)\times(K_{max}-K_{min})}{2} + K_{min} = 445.05 \times (y'+1) + 1.90 \quad (18)$$

These regression rules could provide environmental managers or field response teams with a rapid assessment tool for identifying values of the longitudinal dispersion coefficients required

Figure 5. Predicted and measured longitudinal dispersion coefficients

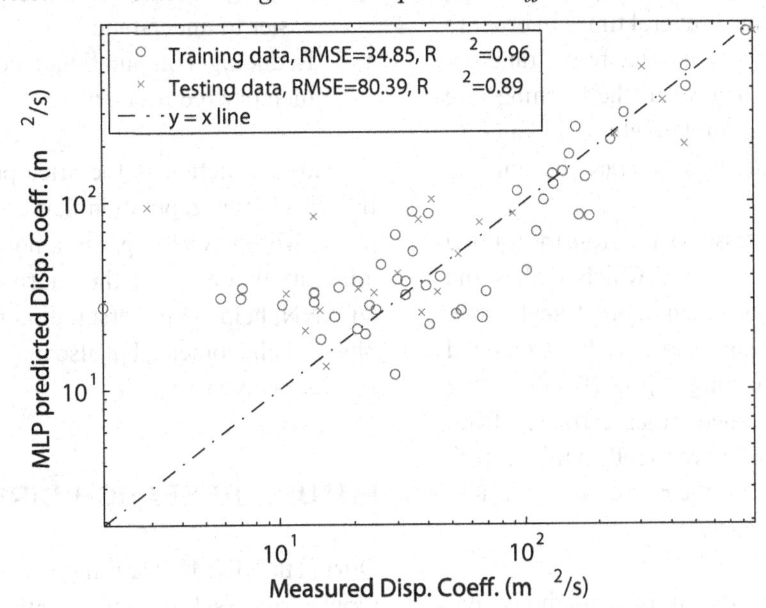

Table 3. Comparison of test data results when using 4 different methods

Model (Reference)	R^2	RMSE(m^2/s)
GNMM	0.89	80.4
MLP (Tayfur & Singh, 2005)	0.70	193.0
equation (4) (Deng et al., 2001)	0.55	610.0
equation (3) (Seo & Cheong, 1998)	0.50	812.0

for the prediction of contaminant spread and concentrations immediately following an accidental spill.

CONCLUSION

In this chapter, we have presented a new methodology, GNMM, for the prediction of longitudinal dispersion coefficient. Through a benchmarking case study, the effectiveness of GNMM has been demonstrated by comparing the results generated by GNMM to those presented in the literature.

To conclude, the scenario of GNMM can be summarized as follows:

- Utilizing GAs to optimize input variables, this simplifies the MLP structure in the GNMM, and makes the training process more efficient. In this stage, the population number (N_p) has to be sufficiently large ($N_p \approx 2^b$) to include most possible input variable combinations; while the learning rate α has to kept small ($\alpha = 0.01$) in order to avoid network oscillation. Furthermore, since weights and thresholds for each neuron are randomly generated, GAs have to be run several times until a clear distinction is evident between input variables.
- Using the input variables found by the GA with the associated targets to develop an

MLP. As in the previous step, the training has to be repeated several times in order to get satisfactory results due to its 'random' starting point. However, the learning rate α can be set to a relatively large value (α = 0.04) in order to accelerate the training process.

- Extracting regression rules from the trained MLP neural network, which makes the training results much more transferable. Since the original data have been mapped into a specific range ([−1, 1]) before the MLP being trained, rules extracted from the trained MLP have to reflect this feature (i.e. reversely map the rule executed results into normal ranges).

Compared with conventional methods that provides longitudinal dispersion prediction (e.g. equation (3) and equation (4)), GNMM as a data driven approach needs no *a priori* knowledge. Although *a priori* knowledge is widely used in many ANN applications, it is dependent on expert knowledge and hence very subjective and case dependent. This is particularly true for complex problems, where the underlying physical mechanism is not fully understood. Furthermore, GNMM is adaptive. This means that when new data samples are applied into the system, the system is capable of self-learning and thus adjusting its results and improving prediction accuracy. Another outstanding advantage of GNMM over conventional methods is that, due to its ANN feature, it can approximate virtually any function with any desired accuracy without making assumptions with regard to stream geometry or flow dynamics.

GNMM is distinct from other solely ANN-based methods by also incorporating variable selection and rule extraction. GA-based variable selection stage is capable of:

- Filtering out irrelevant and noisy variables, improving the accuracy of the model.
- Making ANN structure less complex and easier to understand.
- Reducing computational complexity and memory requirements.

Rule extraction is the attempt to overcome the 'black box' reputation that comes with neural networks. Such a process not only provides a facility that explains the internal behaviour of an ANN, helps in understanding the underlying physical phenomena, but also makes the training results easily applicable.

FUTURE RESEARCH DIRECTIONS

During the GNMM training process, either in the input variable selection stage or the MLP training stage, the back-propagation learning algorithm is being used throughout. However, this algorithm is not immune from problems. For example, the calculations are extensive and, as a result, training is slow. One of the most effective means to

Table 4. Rules fired for the training and test data

Rule	g_1	g_2	g_3	$\mathbf{S}_{1t}$	$\mathbf{S}_{1v}$
1	1	1	2	1	2
2	1	1	3	24	6
3	1	2	3	3	1
4	1	2	4	6	6
5	1	3	3		2
6	2	1	3	3	
7	2	1	4	2	1
8	2	2	3	5	
9	2	2	4	1	1
10	2	3	3	4	
11	4	3	1		1
12	4	4	1		1
13	4	4	2		1

accelerate the convergence of back-propagation is to adjust the learning rate parameter, as has been done during our case study. However, there exist some other means of improving the performance of MLP either with the back-propagation algorithm or using other algorithms instead. An example of the former is to set an adaptive learning rate, that is, a learning rate that keeps the learning step size as large as possible while keeping learning stable. In this case, the learning rate is made responsive to the complexity of the local error surface. An example of an alternative algorithm is the so-called 'Conjugate Gradient Algorithms' (Lin & Lee, 1996). The basic back-propagation algorithm adjusts the weights in the steepest descent direction (negative of the gradient); the direction in which the performance function is decreasing most rapidly. It turns out that, although the function decreases most rapidly along the negative of the gradient, this does not necessarily produce the fastest convergence. In Conjugate Gradient Algorithms, a search is performed along conjugate directions, which produces generally faster convergence than steepest descent directions.

In GNMM, rule extraction is based on the approximation of the hidden neurons' tanh activation function (equation (6) and equation (9)). Such an approximation is derived through the numerical analysis of Sequential Quadratic Programming. Despite the fact that the coefficients used in equation (9) are all kept to 14 significant digits (maximum in MATLAB v7.2), as in any approximation, there are always associated errors. Thus, methods that extract regression rules from ANN with high accuracy are desirable. Since neural networks are low-level computational structures that perform well when dealing with raw data, while fuzzy logic deals with reasoning on a higher level, using linguistic information acquired from domain experts, rule extraction from such a hybrid neuro-fuzzy system would be easier and more accurate. In particular, Kasabov (2001) proposed an Evolving Fuzzy Neural Network (EFuNN), which is a hybrid fuzzy ANN intelligent systems paradigm. In EFuNN, the system allows for the construction of fuzzy rules from the network weights, and hence knowledge extraction. Furthermore, it implements a strategy of dynamically growing and pruning the connectionist (i.e. ANN) architecture. Therefore, a system that integrates GNMM and EFuNN would offer a promising approach to the prediction of longitudinal dispersion and rule extraction.

Moreover, GNMM as a data driven method relies heavily on the quality of the data. Thus, future works may also include applications of GNMM to some more comprehensive data sets to improve the performance of GNMM.

REFERENCES

Bishop, C. M. (1995). *Neural networks for pattern recognition*. Oxford: Oxford University Press.

Bryson, A. E., & Ho, Y. C. (1975). *Applied optimal control*: John Wiley & Sons.

Cannas, B., Fanni, A., See, L., & Sias, G. (2006). Data preprocessing for river flow forecasting using neural networks: Wavelet transforms and data partitioning. *Time Series Analysis in Hydrology, 31*(18), 1164-1171.

Cybenko, G. (1989). Approximation by superpositions of a sigmoidal function. *Mathematics of Control, Signals, and Systems, 2*, 303-314.

Deng, Z. Q., Singh, V. P., & Bengtsson, L. (2001). Longitudinal dispersion coefficient in straight rivers. *Journal of Hydraulic Engineering, 127*(1), 919-927.

Fischer, H. B., List, E. J., Koh, R. C. Y., Imberger, J., & Brooks, N. H. (1979). *Mixing in inland and coastal waters*. New York: Academic Press.

Gardner, J. W., Boilot, P., & Hines, E. L. (2005). Enhancing electronic nose performance by sensor selection using a new integer-based genetic algorithm approach. *ISOEN 2003 - Selected Papers from the 10th International Symposium on Olfaction and Electronic Noses, 106*(1), 114-121.

Goldberg, D. E. (1989). *Genetic Algorithms in search, optimization and machine learning*. Boston, MA, USA: Addison-Wesley Longman Publishing Co., Inc.

Guymer, I. (1999). A national database of travel time, dispersion and methodologies for the protection of river abstractions. Environment Agency R&D Technical Report P346.

Hagan, M. T., Demuth, H. B., & Beale, M. H. (1996). *Neural network design* (1st ed.). Boston: PWS Pub.

Haupt, R. L., & Haupt, S. E. (2004). *Practical genetic algorithms* (2nd ed.). Hoboken, N.J.: John Wiley.

Haykin, S. S. (1994). *Neural networks: A comprehensive foundation*. New York: Maxwell Macmillan International.

Holland, J. H. (1992). *Adaptation in natural and artificial systems*. MA, USA: MIT Press Cambridge.

Jang, J.-S. R., Sun, C.-T., & Mizutani, E. (1997). Neuro-fuzzy and soft computing: A computational approach to learning and machine intelligence. Upper Saddle River, NJ: Prentice Hall.

Jobson, H. E. (1997). Predicting travel time and dispersion in rivers and streams. *Journal of Hydraulic Engineering, 123*(11), 971-978.

Kasabov, N. (2001). Evolving fuzzy neural networks for supervised/unsupervised online knowledge-based learning. *IEEE Transactions on Systems, Man, and Cybernetics, Part B: Cybernetics, 31*(6), 902-918.

Kashefipour, S. M., Falconer, R. A., & Lin, B. (2002). Modeling longitudinal dispersion in natural channel flows using ANNs. *River Flow 2002*, 111-116.

Lin, C. T., & Lee, C. S. G. (1996). *Neural fuzzy systems: A neuro-fuzzy synergism to intelligent systems*. NJ, USA: Prentice-Hall, Inc. Upper Saddle River.

Michalewicz, Z. (1996). *Genetic algorithms + data structures = evolution programs* (3rd rev. and extended ed.). Berlin: Springer-Verlag.

Mitchell, M. (1996). *An introduction to genetic algorithms*. Cambridge, Mass.: MIT Press.

Piotrowski, A., Rowinski, P. M., & Napiorkowski, J. J. (2006). *Assessment of longitudinal dispersion coefficient by means of different neural networks*. Paper presented at the 7th International Conference on Hydroinformatics, Nice, FRANCE.

Reeves, C. R., & Rowe, J. E. (2003). *Genetic algorithms: principles and perspectives: A guide to GA theory*. Boston: Kluwer Academic Publishers.

Rothlauf, F. (2006). Representations for genetic and evolutionary algorithms (2nd ed.). Heidelberg: Springer.

Rowinski, P. M., Piotrowski, A., & Napiorkowski, J. J. (2005). Are artificial neural network techniques relevant for the estimation of longitudinal dispersion coefficient in rivers? *Hydrological Sciences Journal, 50*(1), 175-187.

Rutherford, J. C. (1994). *River mixing*. Chichester: Wiley.

Seo, I. W., & Cheong, T. S. (1998). Predicting longitudinal dispersion coefficient in natural streams. *Journal of Hydraulic Engineering, 124*(1), 25-32.

Srinivasa, K. G., Venugopal, K. R., & Patnaik, L. M. (2007). A self-adaptive migration model genetic algorithm for data mining applications. *Information Sciences, 177*(20), 4295-4313.

Tayfur, G. (2006). Fuzzy, ANN, and regression models to predict longitudinal dispersion coefficient in natural streams. *Nordic Hydrology, 37*(2), 143-164.

Tayfur, G., & Singh, V. P. (2005). Predicting longitudinal dispersion coefficient in natural streams by artificial neural network. *Journal of Hydraulic Engineering, 131*(11), 991-1000.

Taylor, G. (1954). The dispersion of Matter in turbulent flow through a pipe. *Proceedings of the Royal Society of London. Series A, Mathematical and Physical Sciences, 223*(1155), 446-468.

Tsaih, R., & Chih-Chung, L. (2004). *The layered feed-forward neural networks and its rule extraction.* Paper presented at the Advances in Neural Networks - ISNN 2004. International Symposium on Neural Networks, Dalian, China.

Wallis, S. G., & Manson, J. R. (2004). Methods for predicting dispersion coefficients in rivers. *Water Management, 157*(3), 131-141.

Yao, X. (1999). Evolving artificial neural networks. *Proceedings of the IEEE, 87*(9), 1423-1447.

ADDITIONAL READING

Aitkenhead, M. J., McDonald, A. J. S., Dawson, J. J., Couper, G., Smart, R. P., Billett, M., et al. (2003). A novel method for training neural networks for time-series prediction in environmental systems. *Ecological Modelling, 162*(1-2), 87-95.

Baratti, R., Cannas, B., Fanni, A., Pintus, M., Sechi, G. M., & Toreno, N. (2003). River flow forecast for reservoir management through neural networks. *Evolving Solution with Neural Networks, 55*(3-4), 421-437.

Bazartseren, B., Hildebrandt, G., & Holz, K. P. (2003). Short-term water level prediction using neural networks and neuro-fuzzy approach. *Evolving Solution with Neural Networks, 55*(3-4), 439-450.

Bowden, G. J., Dandy, G. C., & Maier, H. R. (2005). Input determination for neural network models in water resources applications. Part 1--background and methodology. *Journal of Hydrology, 301*(1-4), 75-92.

Bowden, G. J., Maier, H. R., & Dandy, G. C. (2005). Input determination for neural network models in water resources applications. Part 2. Case study: forecasting salinity in a river. *Journal of Hydrology, 301*(1-4), 93-107.

Brath, A., Montanari, A., & Toth, E. (2002). Neural networks and non-parametric methods for improving realtime flood forecasting through conceptual hydrological models. *Hydrology and Earth System Sciences, 6*(4), 627-640.

Bruen, M., & Yang, J. (2005). Functional networks in real-time flood forecasting--a novel application. *Advances in Water Resources, 28*(9), 899-909.

Chang, F.-J., & Chen, Y.-C. (2001). A counter-propagation fuzzy-neural network modeling approach to real time streamflow prediction. *Journal of Hydrology, 245*(1-4), 153-164.

D'Heygere, T., Goethals, P. L. M., & De Pauw, N. (2003). Use of genetic algorithms to select input variables in decision tree models for the prediction of benthic macroinvertebrates. *Modelling the structure of acquatic communities: concepts, methods and problems., 160*(3), 291-300.

Dawson, C. W., Abrahart, R. J., Shamseldin, A. Y., & Wilby, R. L. (2006). Flood estimation at ungauged sites using artificial neural networks. *Journal of Hydrology, 319*(1-4), 391-409.

Dibike, Y. B., & Solomatine, D. P. (2001). River flow forecasting using artificial neural networks. *Physics and Chemistry of the Earth, Part B: Hydrology, Oceans and Atmosphere, 26*(1), 1-7.

Gardner, J. W., Craven, M., Dow, C., & Hines, E. L. (1998). The prediction of bacteria type and culture growth phase by an electronic nose with a multi-layer perceptron network. *Measurement Science and Technology, 9*(1), 120-127.

Gevrey, M., Dimopoulos, I., & Lek, S. (2003). Review and comparison of methods to study the contribution of variables in artificial neural network models. *Modelling the structure of acquatic communities: concepts, methods and problems, 160*(3), 249-264.

Guymer, I. (1998). Longitudinal dispersion in sinuous channel with changes in shape. *Journal of Hydraulic Engineering, 124*(1), 33-40.

Imrie, C. E., Durucan, S., & Korre, A. (2000). River flow prediction using artificial neural networks: generalisation beyond the calibration range. *Journal of Hydrology, 233*(1-4), 138-153.

Jobson, H. E. (2001). Predicting river travel time from hydraulic characteristics. *Journal of Hydraulic Engineering, 127*(11), 911-918.

Juha, V., & Esa, A. (2000). Clustering of the self-organizing map. *IEEE Transactions on Neural Networks, 11*(3), 586-600.

Kerh, T., & Lee, C. S. (2006). Neural networks forecasting of flood discharge at an unmeasured station using river upstream information. *Advances in Engineering Software, 37*(8), 533-543.

Kingston, G. B., Maier, H. R., & Lambert, M. F. (2006). A probabilistic method for assisting knowledge extraction from artificial neural networks used for hydrological prediction. *Application of Natural Computing Methods to Water Resources and Environmental Modelling, 44*(5-6), 499-512.

Maier, H. R., & Dandy, G. C. (1998). Understanding the behaviour and optimising the performance of back-propagation neural networks: an empirical study. *Environmental Modelling and Software, 13*(2), 179-191.

Maier, H. R., & Dandy, G. C. (2000). Neural networks for the prediction and forecasting of water resources variables: A review of modelling issues and applications. *Environmental Modelling and Software, 15*(1), 101-124.

Milanese, M., & Novara, C. (2004). Nonlinear Set Membership prediction of river flow. *Systems & Control Letters, 53*(1), 31-39.

Negnevitsky, M. (2001). *Artificial intelligence: A guide to intelligent systems*. Boston, MA, USA: Addison-Wesley Longman Publishing Co., Inc.

Príncipe, J. C., Euliano, N. R., & Lefebvre, W. C. (1999). *Neural and adaptive systems: fundamentals through simulations*. New York: Wiley.

Riad, S., Mania, J., Bouchaou, L., & Najjar, Y. (2004). Rainfall-runoff model usingan artificial neural network approach. *Mathematical and Computer Modelling, 40*(7-8), 839-846.

Saad, E. W., & Wunsch Ii, D. C. (2007). Neural network explanation using inversion. *Neural Networks, 20*(1), 78-93.

Saito, K., & Nakano, R. (2002). Extracting regression rules from neural networks. *Neural Networks, 15*(10), 1279-1288.

Seo, S., & Obermayer, K. (2004). Self-organizing maps and clustering methods for matrix data. *New Developments in Self-Organizing Systems, 17*(8-9), 1211-1229.

Sivakumar, B., Jayawardena, A. W., & Fernando, T. M. K. G. (2002). River flow forecasting: use of phase-space reconstruction and artificial neural networks approaches. *Journal of Hydrology, 265*(1-4), 225-245.

Sudheer, K. P. (2005). Knowledge extraction from trained neural network river flow models. *Journal of Hydrologic Engineering, 10*(4), 264-269.

Waldon, M. G. (2004). Estimation of average stream velocity. *Journal of Hydraulic Engineering, 130*(11), 1119-1122.

Zha, H., Ding, C., Gu, M., He, X., & Simon, H. (2002). Spectral relaxation for k-means clustering. *Advances in Neural Information Processing Systems, 14*, 1057-1064.

Chapter XX
The Exposition of Fuzzy Decision Trees and Their Application in Biology

Malcolm J. Beynon
Cardiff University, UK

Kirsty Park
University of Stirling, UK

ABSTRACT

This chapter employs the fuzzy decision tree classification technique in a series of biological based application problems. With its employment in a fuzzy environment, the results, in the form of fuzzy 'if.. then..' decision rules, bring with them readability and subsequent interpretability. The two contrasting applications considered concern, the age of abalones and the lengths of torpor bouts of hibernating Greater Horseshoe bats. Emphasis is on the visual results presented, including the series of membership functions used to construct the linguistic variables representing the considered attributes and the final fuzzy decision trees constructed. Technical details presented further offer the opportunity to readers to future employ the technique in other biological applications.

INTRODUCTION

Fuzzy set theory (FST) stands out as a general methodology that has contributed to the development of already established techniques used throughout many areas of science, including biology and medicine (recent examples include, Morato *et al.*, 2006; Mastorocostas and Theocharis, 2007). Since its introduction in Zadeh (1965), the utilisation of FST in such development has been with the inclusion of the acknowledgement of the presence of vagueness and ambiguity during its operation. Further, it has also invoked the ability to interpret the structuring and subsequent results from data analysis in a linguistic orientated language (Grzegorzewski and Mrówka, 2005). Indeed, artificial intelligence, with respect to FST, is closely associated with the mimicry of human cognition and linguistic language (Trillas and Guadarrama, 2005).

The issue of interpretability is particularly relevant in classification based problems, but overlooked, since so often the concomitant analysis is more oriented to the resultant classification accuracy, rather than interpretability. Indeed, Breiman (2001), in an informative discussion on the cultures of statistical modelling, comes down heavily on the need to accommodate the ability to interpret results in analysis undertaken. Their discussion offers a pertinent illustrative argument, describing the example of a medical doctor, with experimental data, in a choice between accuracy and interpretability they would choose interpretability. This issue of interpretability over accuracy is also pertinent in general biology.

This chapter considers fuzzy decision trees (FDTs), an example of an already established technique that has been developed using FST. The fundamentals of the decision tree technique, within a crisp or fuzzy environment, is concerned with the classification of objects described by a number of attributes, with concomitant decision rules derived in the constructed decision tree. The inherent structure a consequence of the partitioning algorithm used to discern the classification impact of the attributes. An early FDT reference is attributed to Chang and Pavlidis (1977). In the area of medicine, for example, Podgorelec *et al.* (2002) offer a good review of decision trees, with a most recent employment of FDTs presented in Armand *et al.* (2007), which looked into gait deviations. Relative to this, there is a comparative dearth of their employment in a biological setting, one exception being Beynon *et al.* (2004a).

An important feature of FDTs is the concomitant sets of fuzzy '*if* .. *then* ..' decision rules constructed, whose condition and decision parts, using concomitant attributes, can be described in linguistics terms (such as low, medium or high). The suggested FDT approach employed here was presented in Yuan and Shaw (1995) and Wang *et al.* (2000), and attempts to include the cognitive uncertainties evident in the data values. This FDT approach has been used in Beynon *et al.* (2004b) and Beynon *et al.* (2004a), the latter investigating the songflight of the Sedge Warbler, expositing the relationship between the birds' characteristics like, repertoire size and territory size, against their song flight duration.

Central to the utilisation of FDTs is the fuzzification of the considered data set, through the employment of FST related membership functions (MFs), which further enable the linguistic representation of the attributes considered (Kecman, 2001), present also in the subsequent decision rules constructed. Alongside the exposition of FDTs in this chapter, the results from two contrasting biology based applications are considered; the first using a well known data set from the UCI data repository and relates to the prediction of the age of abalones (Waugh, 1995), the second is a more contemporary application looking at the torpor bouts of hibernating Greater Horseshoe bats (Park *et al.*, 2000).

The contribution of this book chapter is the clear understanding of the potential advantages of the utilization of FDTs in biological processes. The small hypothetical example allows the reader to clearly follow the included analytical rudiments, and the larger applications demonstrate the interpretability allowed through the use of this approach. It expected a reader will gain effective insights into the type of technical details necessary for a FDT analysis, and its analysis within biological problems.

BACKGROUND

The contents of this section are partitioned into three subsections; the first introduces fuzzy set theory including the fuzzification of a small example data set, the second outlines the FDT technique used to exposit the concomitant general methodology, and the third presents a tutorial FDT analysis on the previously described small example data set.

Fuzzy Set Theory

In classical set theory, an element (value) either belongs to a certain set or it does not. It follows, the definition of a set can be defined by a two-valued membership function (MF), which takes values 0 or 1, defining membership or non-membership to the set, respectively. In fuzzy set theory (Zadeh, 1965), a grade of membership exists to characterise the association of a value x to a set S. The concomitant MF, defined $\mu_S(x)$, has range [0, 1]. The relevance and impact of the notion of a MF is clearly stated in Inuiguchi *et al.* (2000, p. 29);

In fuzzy systems theory, a membership function plays a basic and significant role since it characterizes a fuzzy object and all treatments of those fuzzy objects are made in terms of their membership functions. The membership function elicitation is the first and an important stage that allows us to deal with the fuzzy objects.

It is clear from this statement that the types of MFs used will have an impact on the analysis undertaken with them. In this chapter, the use of MFs is with respect to their association with defining attributes describing objects as linguistic variables. Briefly, a linguistic variable is made up of a number of linguistic terms, which are each defined by a MF. Garibaldi and John (2003) consider an important surrounding issue, namely choosing the type of MFs to define the linguistic terms. Recent research studies that consider the issue of the types of MFs used include; Pedrycz and Vukovich (2002) and Grzegorzewski and Mrówka (2005). In the latter study, Grzegorzewski and Mrówka (2005, p. 115) clearly state that;

... the crucial point in fuzzy modeling is to assign membership functions corresponding to fuzzy numbers that represent vague concepts and imprecise terms expressed often in a natural language.

This issue of the expression in a natural language is pertinent to the results from analysis when using FDTs (see later). Here a series of piecewise linear MFs are used to define linguistic terms to describe attribute values in a data set. Dombi and Gera (2005) identifies that fuzzy applications use piecewise linear MFs because of their easy handling (see also, Yu and Li, 2001). The general form of a piecewise linear MF, in the context of the jth linguistic term T_j^k of a linguistic variable A_k, is shown in Equation 1 with the respective *defining values* in list form are [$\alpha_{j,1}$, $\alpha_{j,2}$, $\alpha_{j,3}$, $\alpha_{j,4}$, $\alpha_{j,5}$]. This definition is restricted in the sense that the resultant MF is made up of four 'piecewise' parts only, a visual example of the kind of structure this type of MF produces, is given in the fuzzification of a small example data set next described, including also the use of the defining values.

The small example data set considered here consists of four objects, u_1, u_2, u_3 and u_4, described by three condition (T1, T2 and T3) and one decision (C) attribute, see Table 1.

The fuzzification of this data set starts with the fuzzification of the individual attribute values. Moreover, each attribute in Table 1 needs to be viewed in terms of them being a linguistic variable, which is itself described by a number of linguistic terms. Here, two linguistic terms are used to describe a linguistic variable, so two MFs are used to fuzzify a single attribute, see Figure 1 where the linguistic variable form associated with the decision attribute C is shown.

In Figure 1, two MFs, $\mu_L(C)$ (labelled C_L) and $\mu_H(C)$ (C_H), are shown to cover the domain of the decision attribute C, the concomitant defining values are, for C_L: [−∞,−∞, 7, 9, 11] and C_H: [7, 9, 11, ∞, ∞]. An interpretation could then simply be the associations of a decision attribute value to the two linguistic terms denoted by the words, low (L) and/or high (H). For example, a value of C = 8, would mean, C_L = 0.75 and C_H = 0.25,

Equation 1.

$$\mu_{T_j^k}(x) = \begin{cases} 0 & \text{if } x \le \alpha_{j,1} \\ 0.5\dfrac{x-\alpha_{j,1}}{\alpha_{j,2}-\alpha_{j,1}} & \text{if } \alpha_{j,1} < x \le \alpha_{j,2} \\ 0.5+0.5\dfrac{x-\alpha_{j,2}}{\alpha_{j,3}-\alpha_{j,2}} & \text{if } \alpha_{j,2} < x \le \alpha_{j,3} \\ 1 & \text{if } x = \alpha_{j,3} \\ 1-0.5\dfrac{x-\alpha_{j,3}}{\alpha_{j,4}-\alpha_{j,3}} & \text{if } \alpha_{j,3} < x \le \alpha_{j,4} \\ 0.5-0.5\dfrac{x-\alpha_{j,4}}{\alpha_{j,5}-\alpha_{j,4}} & \text{if } \alpha_{j,4} < x \le \alpha_{j,5} \\ 0 & \text{if } \alpha_{j,5} < x \end{cases}$$

indicating this value of C is more associated with being low rather than high.

Two MFs are also used to fuzzify each condition attribute, here only T1 and T2 are shown, see Figure 2.

The MFs described in Figure 2 are each found from a series of defining values, in this case for; T1: [[−∞,−∞, 6, 10, 11], [6, 10, 11, ∞, ∞]], T2: [[−∞,−∞, 12, 16, 19], [12, 16, 19, ∞, ∞]] and T3: [[−∞,−∞, 16, 20, 22], [16, 20, 22, ∞, ∞]]. Applying these MFs on the example data set achieves a fuzzy data set, see Table 2.

In Table 2, each condition attribute, T1, T2 and T3, is described by two values associated with two linguistic terms, with the value in bold the larger fuzzy value describing each condition attribute.

Description of a Fuzzy Decision Tree Technique

The idea of FDTs has been around since the late 1970s (Chang and Pavlidis, 1977), since then a number of different techniques have been developed to undertake a decision tree analysis in a fuzzy environment. For information on the types of FDTs introduced studies like, Yuan and Shaw (1995), Janikow (1998) Chiang and Hsu. (2002), Podgorelec *et al.* (2002) and Olaru and Wehenkel (2003), describe the pertinent issues.

This section of the chapter outlines the FDT approach first presented in Yuan and Shaw (1995) and later references (Beynon *et al.*, 2004a, 2004b). For simplicity the notation used in the example data set, and its fuzzification, are continued here. The underlying knowledge within a FDT, related to a linguistic term C_j for a decision attribute C, can be represented as a set of fuzzy '*if .. then ..*' decision rules, each of the form;

If (A_1 is $T^1_{i_1}$) and (A_2 is $T^2_{i_2}$) … and (A_k is $T^k_{i_k}$) then C is C_j,

where A_1, A_2, .., A_k and C are linguistic variables, and $T(A_k) = \{T^k_1, T^k_2, .. T^k_{S_i}\}$ and $\{C_1, C_2, …, C_L\}$ are their respective linguistic terms. Each linguistic term T^k_j is defined by the MF $\mu_{T^k_j}(x)$, which transforms a value in its associated domain to a grade of membership value to between 0 and 1. The MFs, $\mu_{T^k_j}(x)$ and $\mu_{C_j}(y)$, represent the grade of membership of an object's attribute value for A_j being T^k_j and C being C_j, respectively.

Table 1. Example small data set

Object	T1	T2	T3	C		Object	T1	T2	T3	C
u_1	13	15	26	7		u_3	8	18	19	5
u_2	10	20	28	8		u_4	15	12	11	10

Figure 1. Fuzzification of C with two linguistic terms defined by two MFs (labeled CL and CH)

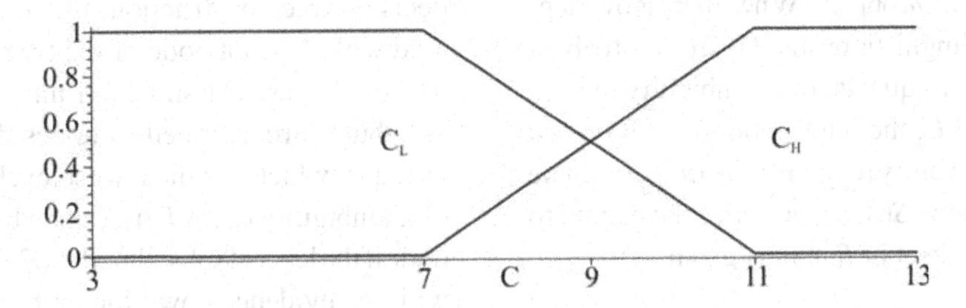

Figure 2. Fuzzification of condition attributes using two MFs

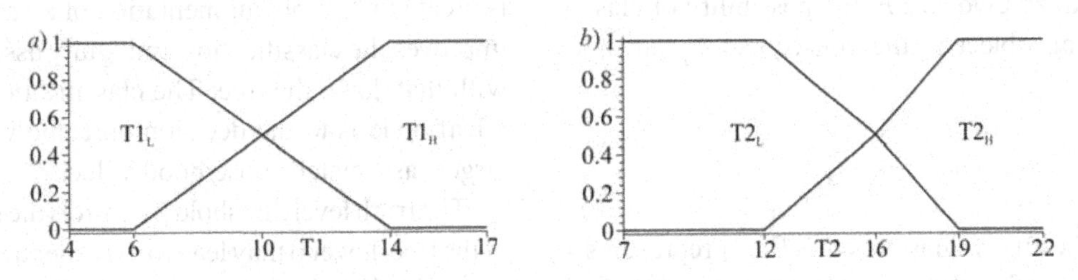

Table 2. Fuzzy data set using two MFs for each condition attribute

Object	T1 = [T1$_L$, T1$_H$]	T2 = [T2$_L$, T2$_H$]	T3 = [T3$_L$, T3$_H$]	C = [C$_L$, C$_H$]
u_1	[0.000, **1.000**]	[**0.625**, 0.375]	[0.000, **1.000**]	[**1.000**, 0.000]
u_2	[**0.500, 0.500**]	[0.000, **1.000**]	[0.000, **1.000**]	[**0.750**, 0.250]
u_3	[**0.750**, 0.250]	[0.167, **0.833**]	[**0.625**, 0.375]	[**1.000**, 0.000]
u_4	[0.000, **1.000**]	[**1.000**, 0.000]	[**1.000**, 0.000]	[0.250, **0.750**]

A MF $\mu(x)$ from the set describing a fuzzy linguistic variable Y defined on X, can be viewed as a possibility distribution of Y on X, that is $\pi(x) = \mu(x)$, for all $x \in X$ the values taken by the objects in U (also normalized so $\max_{x \in X} \pi(x) = 1$). The possibility measure $E_\alpha(Y)$ of ambiguity is defined by

$$E_\alpha(Y) = g(\pi) = \sum_{i=1}^{n} (\pi_i^* - \pi_{i+1}^*) \ln[i],$$

where $\pi^* = \{\pi_1^*, \pi_2^*, \ldots, \pi_n^*\}$ is the permutation of the normalized possibility distribution $\pi = \{\pi(x_1), \pi(x_2), \ldots, \pi(x_n)\}$, sorted so that $\pi_i^* \geq \pi_{i+1}^*$ for $i = 1, .., n$, and $\pi_{n+1}^* = 0$ (see Zadeh, 1978). In the limit, if $\pi_2^* = 0$, then $E_\alpha(Y) = 0$, indicates no ambiguity, whereas if $\pi_n^* = 1$, then $E_\alpha(Y) = \ln[n]$, which indicates all values are fully possible for Y, representing the greatest ambiguity.

The ambiguity of attribute A (over the objects $u_1, .., u_m$) is given as:

$$E_\alpha(A) = \frac{1}{m} \sum_{i=1}^{m} E_\alpha(A(u_i)),$$

where

$$E_\alpha(A(u_i)) = g\left(\mu_{T_s}(u_i) \Big/ \max_{1 \leq j \leq s}(\mu_{T_j}(u_i))\right),$$

with $T_1, \ldots, T_s$ the linguistic terms of an attribute (antecedent) with m objects. When there is overlapping between linguistic terms (MFs) of an attribute or between consequents, then ambiguity exists.

For all $u \in U$, the intersection $A \cap B$ of two fuzzy sets is given by $\mu_{A \cap B} = \min[\mu_A(u), \mu_B(u)]$. The fuzzy subsethood $S(A, B)$ measures the degree to which A is a subset of B, and is given by:

$$S(A, B) = \sum_{u \in U} \min(\mu_A(u), \mu_B(u)) \Big/ \sum_{u \in U} \mu_A(u).$$

Given fuzzy evidence E, the possibility of classifying an object to the consequent C_i can be defined as:

$$\pi(C_i|E) = S(E, C_i) / \max_j S(E, C_j),$$

where the fuzzy subsethood $S(E, C_i)$ represents the degree of truth for the classification rule ('if E then C_i'). With a single piece of evidence (a fuzzy number for an attribute), then the classification ambiguity based on this fuzzy evidence is defined as: $G(E) = g(\pi(C| E))$, which is measured using the possibility distribution $\pi(C| E) = (\pi(C_1| E), \ldots, \pi(C_L| E))$.

The classification ambiguity with fuzzy partitioning $P = \{E_1, \ldots, E_k\}$ on the fuzzy evidence F, denoted as $G(P| F)$, is the weighted average of classification ambiguity with each subset of partition:

$$G(P| F) = \sum_{i=1}^{k} w(E_i | F) G(E_i \cap F),$$

where $G(E_i \cap F)$ is the classification ambiguity with fuzzy evidence $E_i \cap F$, and where $w(E_i| F)$ is the weight which represents the relative size of subset $E_i \cap F$ in F:

$$w(E_i| F) = \sum_{u \in U} \min(\mu_{E_i}(u), \mu_F(u)) \Big/ \sum_{j=1}^{k} \left(\sum_{u \in U} \min(\mu_{E_j}(u), \mu_F(u)) \right)$$

To summarize the underlying process for fuzzy decision tree construction, the attribute associated with the root node of the tree is that which has the lowest classification ambiguity ($G(E)$). Attributes are assigned to nodes down the tree based on which has the lowest level of classification ambiguity ($G(P| F)$). A node becomes a leaf node if the level of subsethood ($S(E, C_i)$) associated with the evidence down the path, is higher than some truth value β assigned across the whole of the fuzzy decision tree. Alternatively a leaf node is created when no augmentation of an attribute improves the classification ambiguity associated with that down the tree. The classification from a leaf node is to the decision outcome with the largest associated subsethood value.

The truth level threshold β controls the growth of the tree; lower β may lead to a smaller tree (with lower classification accuracy), higher β may lead to a larger tree (with higher classification accuracy). The construction process can also constrain the effect of the level of overlapping between linguistic terms that may lead to high classification ambiguity. Moreover, evidence is strong if its membership exceeds a certain significant level, based on the notion of α-cuts (Yuan & Shaw, 1995).

Tutorial FDT Analysis of Example Data Set

Beyond the description of the FDT technique presented in Yuan and Shaw (1995) and Beynon *et al.* (2004b), here the small example problem previously described, with its data fuzzified, is used to offer a tutorial on the calculations needed to be made in a FDT analysis. Before the construction process commences, a threshold value of $\beta = 0.75$ for the minimum required truth level was used throughout.

The construction process starts with the condition attribute that is the root node. Hence it is necessary to calculate the classification ambiguity $G(E)$ of each condition attribute. The evaluation of a $G(E)$ value is shown for the first attribute T1 (i.e. $g(\pi(\text{C}|\ \text{T1}))$), where it is broken down to the fuzzy labels L and H, for L;

$$\pi(\text{C}|\ \text{T1}_L) = S(\text{T1}_L, \text{C}_i) / \max_j S(\text{T1}_L, \text{C}_j),$$

considering C_L and C_H with the information in Table 1

$S(\text{T1}_L, \text{C}_L)$

$$= \sum_{u \in U} \min(\mu_{\text{T1}_L}(u), \mu_{\text{C}_L}(u)) \Big/ \sum_{u \in U} \mu_{\text{T}_L}(u)$$

$$= \frac{(\min(0, 1) + \min(0.5, 0.75) + \min(0.75, 1) + \min(0, 0.25))}{(0 + 0.5 + 0.75 + 0)}$$

$$= \frac{(0 + 0.5 + 0.75 + 0)}{1.25} = \frac{1.25}{1.25} = 1.000,$$

whereas,

$S(\text{T1}_L, \text{C}_H)$

$$= \frac{(\min(0, 1) + \min(0.5, 0.25) + \min(0.75, 0) + \min(0, 0.75))}{(0 + 0.5 + 0.75 + 0)}$$

$$= \frac{(0 + 0.25 + 0 + 0)}{1.25} = \frac{0.25}{1.25} = 0.200.$$

Hence $\pi = \{1.000, 0.200\}$, giving the ordered normalized form of $\pi^* = \{1.000, 0.200\}$, with $\pi_3^* = 0$, then;

$$G(\text{T1}_L) = g(\pi(\text{C}|\ \text{T1}_L)) = \sum_{i=1}^{2} (\pi_i^* - \pi_{i+1}^*) \ln[i]$$

$$= (1.000 - 0.200)\ln[1] + (0.200 - 0.000)\ln[2] = 0.139,$$

along with $G(\text{T1}_H) = 0.347$, then $G(\text{T1}) = (0.139 + 0.347)/2 = 0.243$. Compared with $G(\text{T2}) = 0.294$ and $G(\text{T2}) = 0.338$, the condition attribute T1, with the least classification ambiguity, forms the root node for the desired fuzzy decision tree.

The subsethood values in this case are; for T1: $S(\text{T1}_L, \text{C}_L) = \mathbf{1.000}$ and $S(\text{T1}_L, \text{C}_H) = 0.200$, and $S(\text{T2}_H, \text{C}_L) = \mathbf{0.727}$ and $S(\text{T2}_H, \text{C}_H) = 0.363$. For T2_L and T2_H, the larger subsethood value (in bold), defines the possible classification for that path. In the case T1 = L, the subsethood value $S(\text{T1}_L, \text{C}_L) = \mathbf{1.000}$ is above the threshold truth value 0.75 employed, so a leaf node is created. For the path T1 = H the largest subsethood value is less then 0.75 so it is not a leaf node, instead further augmentation of this path is considered.

With three condition attributes included in the example data set, the possible augmentation to T1_H is with either T2 or T3. Concentrating on T2, where with $G(\text{T1}_H) = 0.347$, the ambiguity with partition evaluated for T2 ($G(\text{T1}_H$ and $\text{T2}|\ \text{C})$) has to be less than this value, where:

$$G(\text{T1}_H \text{ and T2}|\ \text{C}) = \sum_{i=1}^{k} w(\text{T2}_i\ |\ \text{T1}_H) G(\text{T1}_H \cap \text{T2}_i).$$

Starting with the weight values, in the case of T1_H and T2_L, it follows:

$w(\text{T2}_L|\ \text{T1}_H) =$

$$\sum_{u \in U} \min(\mu_{\text{T2}_L}(u), \mu_{\text{T1}_H}(u)) \Big/ \sum_{j=1}^{k} \left(\sum_{u \in U} \min(\mu_{\text{T2}_j}(u), \mu_{\text{T1}_H}(u)) \right)$$

where

$$\sum_{u \in U} \min(\mu_{\text{T2}_L}(u), \mu_{\text{T1}_H}(u)) = 1.792$$

and

$$\sum_{j=1}^{k} \left(\sum_{u \in U} \min(\mu_{\text{T2}_j}(u), \mu_{\text{T1}_H}(u)) \right) = 2.917,$$

so $w(\text{T2}_L|\ \text{T1}_H) = 1.792/2.917 = 0.614$. Similarly $w(\text{T2}_H|\ \text{T1}_H) = 0.386$, hence

$G(\text{T1}_H$ and T2| C)

$= 0.614 \times G(\text{T1}_H \cap \text{T2}_L) + 0.386 \times G(\text{T1}_H \cap \text{T2}_H)$

$= 0.614 \times 0.499 + 0.\ 86 \times 0.154$

$= 0.366,$

A concomitant value for $G(\text{T1}_H$ and T3| C) = 0.250, the lower of these ($G(\text{T1}_H$ and T3| C)) is lower than the concomitant $G(\text{T1}_H) = 0.347$, so less ambiguity would be found if the T3 attribute was augmented to the path T1 = H. The subsequent subsethood values in this case for each new path

Figure 3. FDT for example data set with two MFs describing each condition attribute

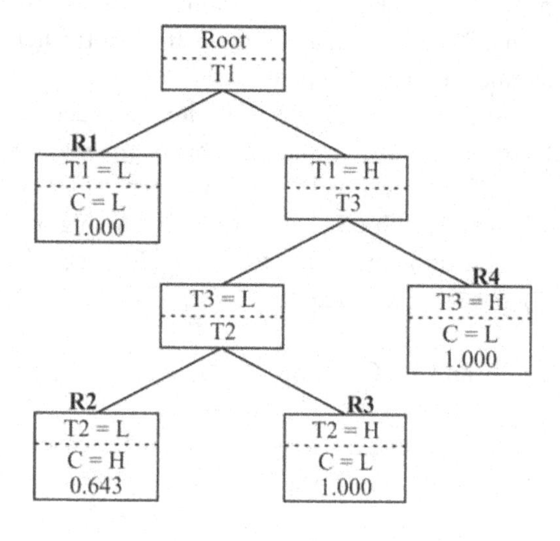

are; $T3_L$; $S(T1_H \cap T3_L, C_L) = 0.400$ and $S(T1_H \cap T3_L, C_H) = \mathbf{0.600}$; $T3_H$: $S(T1_H \cap T3_H, C_L) = \mathbf{1.000}$ and $S(T1_H \cap T3_H, C_H) = 0.143$. With each suggested classification path, the largest subsethood value is below (for $S(T1_H \cap T3_L, C_H) = 0.600$) and above ($S(T1_H \cap T3_H, C_L)$) the defined truth level threshold, hence only one of the paths becomes a leaf node (T1 = H then T3 = H), the other path would need checking for further augmentation with the last of the attributes T2. The construction process continues as before for this last check, the resultant FDT in this case is presented in Figure 3.

The tree structure in Figure 3 clearly demonstrates the visual form of the results using FDTs. Each path (branch) from the root node (using T1) describes a fuzzy decision rule using conditions from the condition attributes to a specified decision class. Only shown in each leaf node box is the truth level associated with the highest subsethood value to the represented decision category. There are four levels of the tree showing the use of all the considered condition attributes. There are four leaf nodes which each have a defined decision rule associated with them. For example, the rule **R4** can be written as,

"If T1 = H and T3 = H then C = L, with truth level 1.000".

Following this exposition of one of the fuzzy decision rules constructed from the analysis, its presence as a path in tree structure shown in Figure 3. Indeed, further inspection of Figure 3 can familiarize the reader with the tree based representation of all the fuzzy decision rules constructed

FUZZY DECISION TREE ANALYSES OF TWO BIOLOGY PROBLEMS

The main thrust of this chapter is the demonstration of the previously described FDT technique employed in two animal biology based studies. The first application of the FDT technique investigates the prediction of the age of abalones through an analysis of their physical measurements (Waugh, 1985). The second application considers the torpor bouts of hibernating Greater Horseshoe bats, based on associated characteristics (Park *et al.*, 2000). In each study the FDTs constructed should be viewed as attempting to exposit the relationship between the condition attributes describing the objects and the decision attribute which categorises them in some way. Due to the availability of the respective data, from these two applications are contrasting, which further exposits the possible use of FDT based analyses.

FDT Analysis of Abalones

The first FDT analysis concerns the subject of abalones, in particular, the prediction of their age using information regarding physical measurements (the data is included in the UCI data repository, with the original source is from Waugh, 1995). As a classification problem, there continues to be concerns on the inaccuracy or bias associated with current techniques used to obtain age and growth data for abalones, such as age from

Table 3. Information on the initial intervalisation of the AGE decision attribute

Description (AGE - years)	Low (L)	Medium (M)	High (H)
Interval	L ≤ 8.5 (333)	8.5 < M ≤ 11.5 (334)	11.5 < H (333)
Mean	6.937	9.823	14.634

the shell layers, prompting consideration of alternative techniques (see for example the decision presented in Gurney, 2005).

Here, a sample of 1000 abalones was analysed, which offered a balance in numbers over the three age categories considered on them. In this study, the three ordered categories of age are described as (in linguistic terms); low (L), medium (M) and high (H), see Table 3.

In Table 3, the intervals shown define the majority support sub-domains of the AGE linguistic terms, L, M and H, where the values in brackets are the numbers of abalones included in each interval (supporting the previously stated balanced numbers desired with each category). Also shown are the means of the AGE values of the abalones included in each interval. The interval details shown in Table 3 are used to construct the respective MFs that define each of the linguistic terms L, M and H, in a fuzzy sense, so used as the necessary defining values [$\alpha_{j,1}$, $\alpha_{j,2}$, $\alpha_{j,3}$, $\alpha_{j,4}$, $\alpha_{j,5}$] for each term, previously defined.

To demonstrate the assignment of values to the defining values, with the linguistic term AGE = M (2nd interval) its mean value is 9.823 years, hence $\alpha_{2,3}$ = 9.823. The neighbouring defining values, $\alpha_{j,2}$ and $\alpha_{j,4}$, around $\alpha_{j,3}$ are the left and right boundary values of that interval, respectively (so here they are $\alpha_{2,2}$ = 8.5 and $\alpha_{2,4}$ = 11.5). The final outer defining values, $\alpha_{j,1}$ and $\alpha_{j,5}$, are the middle (modal) defining values of its neighbouring intervals (so here $\alpha_{2,1}$ = 6.937 and $\alpha_{2,5}$ = 14.634). The left and right end intervals need special consideration, often necessitating the use of $-\infty$ and ∞, respectively. In the case of the constructed intervals for the AGE attribute given in Table 3, Figure 4 elucidates the sequence of linear piecewise (triangular) MFs which describe the linguistic terms associated with the AGE decision attribute.

The MFs presented in Figure 4 clearly show how the defining values have been employed, as described earlier, for example the MF $\mu_{AGE,M}(\cdot)$ has the associated defining values [6.937, 8.5, 9.823, 11.5, 14.634].

Attention then turns to the description of the condition attributes used to describe the abalones and which may be used in the prediction of the abalones' respective AGE values (when using FDTs), see Table 4.

Included in the eight condition attributes described in Table 4 are seven continuous attributes, which quantify different variations on their size and weight of the abalone, and a nominal attribute,

Figure 4. Membership functions of the linguistic terms, L, M and H, describing the linguistic

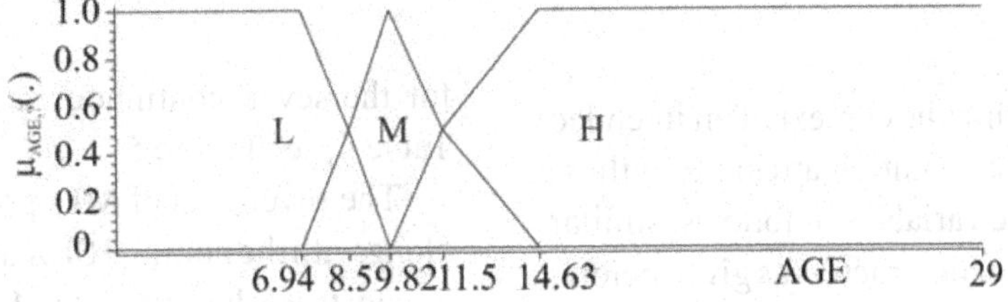

Table 4. Descriptions of the independent condition attributes considered

Attribute	Function
1: Length (LTH)	Longest shell length (mm)
2: Diameter (DMT)	Perpendicular to length (mm)
3: Height (HGT)	With meat in shell (mm)
4: Whole weight (WLW)	Whole abalone (grams)
5: Shucked weight (SKW)	Weight of meat (grams)
6: Viscera weight (VSW)	Gut weight after bleeding (grams)
7: Shell weight (SLW)	After being dried (grams)
8: Sex (SEX)	M, F and I (infant)

Figure 5. Membership functions of the linguistic terms, L, M and H, defining the linguistic variables of the seven continuous condition attributes describing the abalones

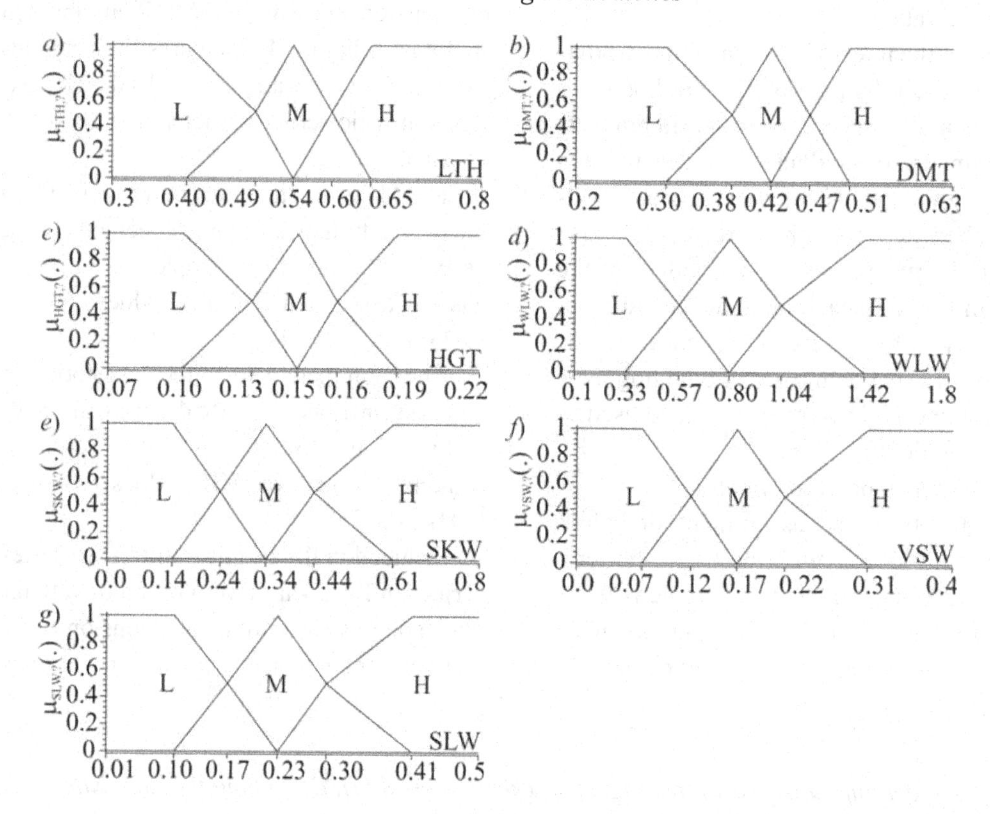

namely its sex. Within the context of an intended FDT analysis, these condition attributes will be viewed as linguistic variables. It follows, similar sets of three MFs are constructed, as given before, for the seven continuous attributes outlined in Table 4, see Figure 5.

The seven sets of MFs presented in Figure 5, along with the nominal SEX attribute, can be used to construct the associated FDT for this abalone

Table 5. Subsethood values of SLW linguistic terms to AGE classes

SLW	AGE = L	AGE = M	AGE = H
L	**0.730**	0.322	0.109
M	0.332	**0.597**	0.374
H	0.100	0.476	**0.594**

Table 6. Classification ambiguity values for the possible augmentation of condition attributes down the paths SLW = M and H

SLW		G(· | SLW = ·)						
G(SLW = ·)		LTH	DMT	HGT	WLW	SKW	VSW	SEX
M: 0.659	M	0.669	0.667	0.671	0.666	**0.658**	0.670	0.690
H: 0.624	H	0.713	0.723	0.651	0.691	0.706	0.709	**0.625**

Table 7. Subsethood values of SLW = M then SKW paths to AGE classes

SLW = M then SKW	AGE = L	AGE = M	AGE = H
L	0.431	**0.611**	0.452
M	0.370	**0.625**	0.391
H	0.347	**0.797**	0.366

age problem. In this construction process a necessary subsethood threshold value of $\beta = 0.72$ was adopted, termed a truth level, upon which a node is deemed a leaf node. The subsequent construction process is comparable to the tutorial analysis given in the previous section.

The first stage of the construction process is the evaluation of the classification ambiguity values $G(E)$ associated with the eight condition attributes describing the considered abalones, found to be, G(LTH) = 0.649, G(DMT) = 0.646, G(HGT) = 0.577, G(WLW) = 0.626, G(SKW) = 0.670, G(VSW) = 0.638, G(SLW) = 0.550 and G(SEX) = 0.626. Inspection of these values, which are all around the 0.6 value, shows the attribute SLW (Shell weight) possesses the least classification ambiguity, so forms the root node. With the root node identified (SLW), the subsethood S(SLW = ·, AGE = ·) values of the paths represented by the SLW linguistic terms, L, M and H, to the classes of the decision attribute AGE, also with the terms L, M, H, are investigated, see Table 5.

In each row presented in Table 5, the largest value identifies the category each of the paths from the root node SLW should take. In the case of the path SLW = L, the largest value is 0.730 (highlighted in bold associated with AGE = L). Since this value is above the minimum truth level of 0.72 earlier defined, the path SLW = L ends at a leaf node which categorisesabalones to AGE =

Figure 6. Fuzzy decision tree for abalone age problem

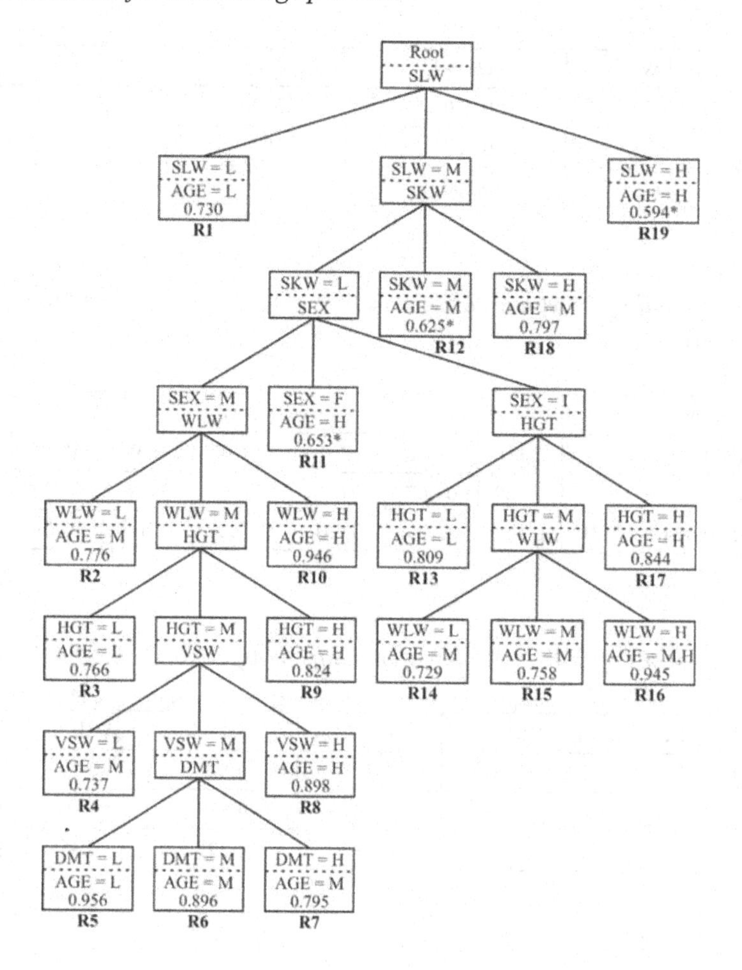

L with truth level 0.730. For the other two paths from the root node, SLW = M and SLW = H, the largest of the values in the sets of subsethood values are each below the minimum truth level of 0.72, previously defined, which means further checks for the lessening of ambiguity are made for these two paths, see Table 6.

In Table 6, the G(SLW = ·) column of values identifies the level of classification ambiguity with each subset partition from the SLW root node (only M and H considered). These values are important since they benchmark the required value below which a weighted average of classification ambiguity with each subset of partition $G(P| F)$ must be to augment a node to a particular path.

For the path defined by SLW = M, with G(SLW = M) = 0.659, the subsequently evaluated lowest G(· | SLW = M) value was found to be associated with the condition attribute SKW (value in bold), with G(SKW | SLW = M) = 0.658. It follows, the path SLW = M is augmented with the attribute SKW and the linguistic terms it is described by (L, M and H). In the case of the path defined by SLW = H, with G(SLW = H) = 0.624, the lowest G(· | SLW = ·) value is found to be, G(SEX | SLW = H) = 0.625. Hence, with G(SEX | SLW = H) = 0.625 > 0.624 G(SLW = H), there is no further possible augmentation possible that would reduce the considered ambiguity for that path, so it is a leaf node with less than the desired truth level.

For the path, SLW = M, the subsethood values associated with SKW (the paths L, M and H), to the possible classification to AGE = L, M or H, are reported in Table 7.

The results in Table 7 relate to the progression from the main SLW = M path to the SKW = · paths. The only largest concomitant subsethood value greater than the 0.720 minimum truth threshold is associated with SKW = H and consequent AGE = M (value 0.797). It follows, the defined path is SLW = M then SKW = H, which classifies to ADT = L with 0.797 level of truth. For the other two paths from SLW, there would be further checks needed on the possibility of minimising ambiguity by further augmenting other previously unconsidered condition attributes to the respective path.

This described process for FDT construction is repeated until all paths end with a leaf node, found from either achieving the required truth level of subsethood defined, or that no further augmentation is possible. The resultant FDT for the abalone problem is reported in Figure 6, which as in the tree found from the tutorial analysis, shows all the fuzzy decision rules constructed (each rule denoted by a path down from the root node to a leaf node).

The reported FDT in Figure 6, constructed from the analysis given previously, is shown to have eight levels, indicating a maximum of seven condition attributes are used in the concomitant condition parts of the fuzzy decision rules derived. Indeed, only the attribute LTH (longest shell length) is not used in the FDT shown. There are in total 19 fuzzy decision rules, **R1**, **R2**, .., **R19**, shown to be constructed. Of these 19 rules, three of them have an associated truth level labelled with a *, which indicates they have a truth level bellow that desired, since no further augmentation of other condition attributes was able to still lower the classification ambiguity. One of the decision rules is next explained more fully, namely **R11**, directly from the FDT the rule can be written down as;

'If SLW = M, SKW = L and SEX = F then AGE = H with truth level 0.653.'

Re-written as in a more linguistic form,

'If Shell weight is medium, Shucked weight is low and Sex is female then Age of abalone is high, with truth level 0.653.'

Also through using the defining values associated with each condition attribute,

'If Shell weight is between 0.17 grams and 0.30 grams, Shucked weight is below 0.24 grams and Sex is female then Age of abalone is above 11.5 years, with truth level 0.653.'

Further interpretation can be given to any of the fuzzy rules defined in the FDT presented, as well as the whole FDT itself.

FDT Analysis of Greater Horseshoe Bats

This section presents a second biology based FDT analysis. Motivation for this analysis comes from a study investigating patterns of torpor and arousal from hibernation in free-ranging bats (Park *et al.* 2000). Moreover, temperature sensitive radiotags were used to monitor the body temperature and movements of Greater Horseshoe bats, *Rhinolophus ferrumequinum* for an average of 87±29 days (mean±SD). These measurements, amongst others, were used as the characteristics associated with the torpor bouts undertaken by the bats.

Many species of small, temperate zone mammals spend much of the winter in hibernation, minimizing their energy expenditure when there is little or no food, and when ambient temperatures are generally low (Lyman *et al.*, 1982). All hibernating mammals studied, however, arouse periodically throughout winter and much attention has focused on why arousals occur since it is energetically expensive to do so (Thomas *et*

Table 8. Information on the initial intervalisation of the BOUT decision attribute

Description (BOUT - hours)	Low (L)	Medium (M)	High (H)
Interval	L ≤ 22.403 (43)	22.403 < M ≤ 94.630 (43)	94.630 < H (44)
Mean	13.927	60.840	172.001

Figure 7. Membership functions of the linguistic terms, L, M and H, describing the linguistic variable version of the BOUT decision attribute

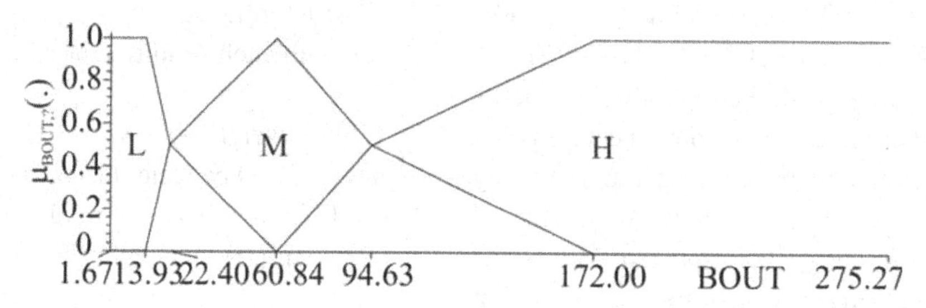

al., 1990). Hence arousals are assumed either have direct functional significance, or to permit important events, such as drinking, feeding, urinating or sleeping, to occur (Thomas and Cloutier, 1992; Willis, 1982; Daan *et al.*, 1991; Ransome, 1971).

Previous research has suggested that that euthermic activity after arousal functions to restore metabolic imbalances developed during the previous torpor bout. If this is correct, torpor bout duration should decrease with increasing temperature, as metabolic processes are faster at higher temperatures, and a critical metabolic imbalance will be reached faster (see for example, Twente and Twente, 1965). Models assuming that evaporative water loss determines torpor bout duration make similar predictions (Thomas and Geiser, 1997).

The data considered here was made up of a sample of 130 continually monitored topor bouts - BOUT (measured in hours), undertaken by the bats, with a range of BOUT that goes from 1.67 hours to 257.27 hours. As in the previous FDT analysis, three intervals are identified to offer an initial partitioning of the BOUT decision attribute, which are later associated with the linguistic terms, low (L), medium (M) and high (H). An equal frequency based intervalisation of the BOUT values was employed, see Table 8.

In Table 8, the intervals shown define the majority support sub-domains of the BOUT linguistic terms, L, M and H, where the values in brackets are the numbers of bouts included in each interval. Also shown are the means of the BOUT values included in each interval. The interval details shown in Table 8 are used to construct the respective MF for each of the linguistic terms L, M and H, so used as the necessary defining values [$\alpha_{j,1}$, $\alpha_{j,2}$, $\alpha_{j,3}$, $\alpha_{j,4}$, $\alpha_{j,5}$] for each term, previously defined.

To demonstrate the assignment of values to the defining values, with the linguistic term BOUT = M (2nd interval) its mean value is 60.840, hence $\alpha_{2,3}$ = 60.840. The neighbouring defining values, $\alpha_{j,2}$ and $\alpha_{j,4}$, around $\alpha_{j,3}$ are the left and right boundary values of that interval, respectively (so here they are $\alpha_{2,2}$ = 22.40 and $\alpha_{2,4}$ = 94.63). The final outer defining values, $\alpha_{j,1}$ and $\alpha_{j,5}$, are the middle

Figure 8. Membership functions of the linguistic terms, L, M and H, defining the linguistic variables of the two continuous condition attributes, TEMP and AGE

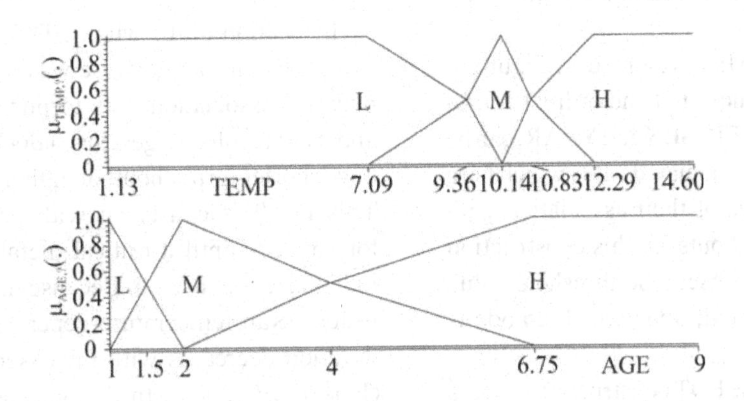

Figure 9. Fuzzy decision tree for torpor bout problem

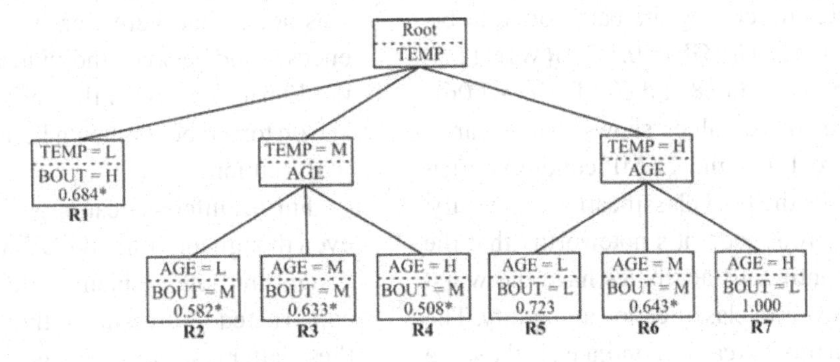

(modal) defining values of its neighbouring intervals (so here $\alpha_{2,1}$ = 13.93 and $\alpha_{2,5}$ = 172.00). In the case of the constructed intervals for the BOUT attribute given in Table 8, Figure 7 elucidates the sequence of linear piecewise (triangular) MFs which describe the linguistic terms associated with the BOUT decision attribute.

The MFs presented in Figure 7 clearly show how the defining values have been employed, as described earlier, for example the MF $\mu_{BOUT,M}(\cdot)$ has the associated defining values [13.93, 22.40, 60.84, 94.63, 172.00].

Five possible condition attributes, which may be associated with torpor bout duration, were investigated; 1 - TEMP, the average ambient temperature over the duration of the torpor bout; 2 - WINTER, the period of winter in which the torpor bout took place from early winter to late winter/spring (see Park et al. 2000 for details); 3 - YEAR, the year in which the bat was monitored (1995/6 and 1996/7); 4 - AGE (from 1 to 9 years); and 5 - SEX (M or F). According to different hypotheses as to why bats arouse during winter, these variables, with the exception of YEAR, might be expected to influence torpor bout durations in different ways (see Park *et al.*, 2000).

Similar sets of MFs can be constructed, as given before, for the two continuous attributes that are included in those that describe the bats and environment at the time of the torpor bouts, namely TEMP and AGE. Again, three MFs are used to define the linguistic terms which make up the linguistic variable versions of these attributes, see Figure 8.

The two sets of MFs presented in Figure 8, along with the other nominal and ordinal condition attributes, WINTER, SEX and YEAR, can be used to construct the associated FDT, which here assesses the relevance of their association with the duration of topor bouts. In this construction process a necessary subsethood threshold value of $\beta = 0.72$ was adopted, upon which a node is deemed a leaf node.

The first stage of the FDT construction process was the evaluation of the classification ambiguity $G(E)$ values associated with the five condition attributes describing the bats, found to be, G(TEMP) = 0.519, G(AGE) = 0.733, G(WINTER) = 1.078, G(SEX) = 1.068 and G(YEAR) = 1.068. Inspection of these values shows, which range from 0.519 to 1.078, the TEMP condition attribute possesses the least classification ambiguity, so forms the root node, it s noteworthy that the three non-continuous attributes have the worst associated levels of classification ambiguity. The construction process can be continued in the same way as previously described, the resultant FDT is reported in Figure 9.

The reported FDT in Figure 9 is shown to have three levels, indicating a maximum of two attributes are used in the concomitant condition parts of the fuzzy decision rules derived. Indeed, only two attributes are used in the FDT shown, namely TEMP and AGE, with seven fuzzy decision rules, **R1**, **R2**, .., **R7**, shown to be constructed. Of these seven rules five of them have an associated truth level, labelled with a *, that indicates they have a truth level bellow that desired, since no further augmentation of other condition attributes was worthwhile. One of these rules is next explained more fully, namely **R7**, expressing it with the most linguistic version of the rule;

'If average temperature is high and age is high then the torpor bout is low with truth level 1.000.'

Inspection of the whole FDT produced indicates ambient temperature was found to have a negative association with torpor bout duration, and these results suggest that older bats especially have shorter torpor bouts at higher temperatures. This result is consistent with predictions that torpor lasts until a critical metabolic or water imbalance is achieved. Because metabolism and water loss are temperature-dependent, torpor bout duration decreases with increased temperature (Park *et al.*, 2000). In the original analyses by Park *et al.* (2000), the influence of age on torpor bout duration was not investigated, although it was noted that there were no significant differences found between individuals. Since only 3 of the 12 bats were older than one year, the effect of age on torpor bouts shown here should be treated with caution.

Further inference can be gauged from this FDT, even though it is a small FDT with only a limited insight into the problem, unlike the larger FDT constructed to investigate the abalones. Perhaps these latter results may suggest further condition attributes should be found that may, through an FDT analysis, elucidate further interpretability into the problem.

FUTURE RESEARCH DIRECTIONS

Within a biological context, the rudiments of any analysis are to achieve a greater understanding of the issue being investigated. When analysing a data set of objects, such as the abalones and bats considered in this chapter, the precision of the attributes describing them may not be certain, as well as the a level of uncertainty with their association to the decision problem in question. The fuzzy decision tree (FDT) technique offers a novel approach to the wide range of classification problems that may have inherent uncertainties as those described.

There are a number of future issues that are associated with the general decision tree methodology which are pertinent with the work in biological applications. These include the generality of the results accrued from the FDT analyses, which technically moves the issue to the complexity of the FDT constructed. This complexity issue includes the, number, size and structure, of the membership functions used to fuzzify the data considered. This is a common issue with all techniques that operate within a fuzzy environment. Further, with respect to FDTs, there is also the role of pruning which refers to the cutting back of the established paths in a FDT.

Perhaps the future trends for the impact of FDTs in biology based applications is the ability of the biology orientated analysts to acquire the required level of understanding surrounding FDTs that make their employment practical and pertinent.

CONCLUSION

The interpretative power of fuzzy decision trees (FDTs) and their ability to handle uncertain information, may make it an appropriate technique to be employed in biological application analysis. That is, biologists are regularly investigating the physiology/relationship of objects (animals etc.) with their environment. The interpretation issue outlined, due to the readability of fuzzy decision rules constructed, could change the emphasis of what is wanted from an analysis.

Within classification problems, the standard notion of performance of a technique, like multivariate discriminant analysis, is classification accuracy. However, their interpretative power may be limited. The FDT analyses outlined does allow for both interpretation and accuracy.

The material presented in this chapter has shown FDTs with emphasis on the role of readability in the results found.

REFERENCES

Armand, S., Watelain, E., Roux, E., Mercier, M., & Lepoutre, F.-X. (2007). Linking clinical measurements and kinematic gait patterns of toe-walking using fuzzy decision trees. *Gait & Posture, 25*(3), 475-484.

Beynon, M. J., Buchanan, K., & Tang, Y.-C. (2004a). The Application of a Fuzzy Rule Based System in an Exposition of the Antecedents of Sedge Warbler Song Flight. *Expert Systems, 21*(1), 1-10.

Beynon, M.J., Peel, M.J., & Tang, Y.-C. (2004b). The Application of Fuzzy Decision Tree Analysis in an Exposition of the Antecedents of Audit Fees. *OMEGA - International Journal of Management Science, 32*(2), 231-244.

Breiman, L. (2001). Statistical Modeling: The Two Cultures. *Statistical Science, 16*(3), 199-231.

Chang, R.L.P., & Pavlidis, T. (1977). Fuzzy decision tree algorithms. *IEEE Transactions Systems Man and Cybernetics*, SMC-7(1), 28-35.

Chiang, I-J., & Hsu, J.Y-J. (2002). Fuzzy classification trees for data analysis. *Fuzzy Sets and Systems, 130*, 87-99.

Daan, S., Barnes, B.M., & Strijkstra, A.M. (1991). Warming up for sleep? Ground squirrels sleep during arousals from hibernation. *Neuroscience Letters, 128*, 265-268.

Dombi, J., & Gera, Z. (2005). The approximation of piecewise linear membership functions and Łukasiewicz operators. *Fuzzy Sets and Systems, 154*, 275-286.

Garibaldi, J.M., & John, R.I. (2003). Choosing Membership Functions of Linguistic Terms, *Proceedings of the 2003 IEEE International Conference on Fuzzy Systems*, St. Louis, USA, pp. 578-583.

Grzegorzewski, P., & Mrówka, E. (2005). Trapezoidal approximations of fuzzy numbers. *Fuzzy Sets and Systems, 153*, 115-135.

Gurney, L.J., Mundy, C., & Porteus, M.C. (2005). Determining age and growth of abalone using stable oxygen isotopes: a tool for fisheries management. *Fisheries Research, 72*, 353-360.

Inuiguchi, M., Tanino, T., & Sakawa, M. (2000). Membership function elicitation in possibilistic programming problems. *Fuzzy Sets and Systems, 111*, 29-45.

Janikow, C.Z. (1998). Fuzzy decision trees: Issues and methods. *IEEE Transactions of Systems, Man and Cybernetics Part B, 28*(1), 1-14.

Kecman, V. (2001). *Learning and Soft Computing: Support Vector Machines, Neural Networks, and Fuzzy Logic*. MIT Press, London, UK.

Lyman, C.P., Willis, J.S., Malan, A., & Wang L.C.H. (Eds.). (1982). *Hibernation and Torpor in Mammals and Birds*. San Diego: Academic Press.

Olaru, C., & Wehenkel, L. (2003). A complete fuzzy decision tree technique. *Fuzzy Sets and Systems, 138*, 221-254.

Mastorocostas, P., & Theocharis, J. (2007). A dynamic fuzzy neural filter for separation of discontinuous adventitious sounds from vesicular sounds. *Computers in Biology and Medicine, 37*(1), 60-69.

Morato, T., Cheung, W.W.L., & Pitcher, T.J. (2006). Vulnerability of seamount fish to fishing: fuzzy analysis of life-history attributes. *Journal of Fish Biology, 68*(1), 209-221.

Park, K. J., Jones, G., & Ransome, R.D. (2000). Torpor, arousal and activity of hibernating Greater Horseshoe Bats (*Rhinolophus ferrumequinum*). *Functional Ecology, 14*, 580-588.

Pedrycz, W., & Vukovich, G. (2002). On Elicitation of Membership Functions, *IEEE Transactions on Systems, Man, and Cybernetics - Part A: Systems and Humans, 32*(6), 761- 767

Podgorelec, V., Kokol, P., Stiglic, B., & Rozman, I. (2002). Decision trees: an overview and their use in medicine. *Journal of Medical Systems, 26*, 445-63.

Ransome, R.D. (1971). The effect of ambient temperature on the arousal frequency of the hibernating greater horseshoe bat, *Rhinolophus ferrumequinum*, in relation to site selection and the hibernation state. *Journal of Zoology, London, 164*, 353-371.

Thomas, D.W., & Cloutier, D. (1992). Evaporative water-loss by hibernating little brown bats, *Myotis lucifugus. Physiological Zoology, 65*, 443-456.

Thomas, D.W., & Geiser, F. (1997). Periodic arousals in hibernating mammals: Is evaporative water loss involved?, *Functional Ecology, 11*(5), 585-591.

Thomas, D.W., Dorais, J., & Bergeron, J.-M. (1990). Winter energy budgets and cost of arousals for hibernating little brown bats, *Myotis lucifugus. Journal of Mammalogy, 71*, 475-479.

Trillas, E., & Guadarrama, S. (2005). What about fuzzy logic's linguistic soundness?. *Fuzzy Sets and Systems, 156*, 334-340.

Twente, J.L., & Twente, J.A. (1965). Effects of core temperature upon duration of hibernation of Citellus lateralis. *Journal of Applied Physiology, 20*, 411-416.

Wang, X., Chen, B., Qian, G., & Ye, F. (2000). On the optimization of fuzzy decision trees. *Fuzzy Sets and Systems, 112*(1), 117-125.

Waugh, S. (1985). *Extending and benchmarking Cascade-Correlation*, PhD Thesis, University of Tasmania.

Willis, J.S. (1982). The mystery of the periodic arousal. In C.P. Lyman, J.S. Willis, A. Malan & L.C.H. Wang (Eds.), *Hibernation and Torpor in*

Mammals and Birds (pp. 92-103). San Diego: Academic Press.

Yu, C-S., & Li, H-L. (2001). Method for solving quasi-concave and non-cave fuzzy multi-objective programming problems. *Fuzzy Sets and Systems, 122*(2), 205-227.

Yuan, Y., & Shaw, M.J. (1995). Induction of fuzzy decision trees. *Fuzzy Sets and Systems, 69*(2), 125-139.

Zadeh, L.A. (1965). Fuzzy Sets. *Information and Control, 8*(3), 338-353.

Zadeh, L. A. (1978). Fuzzy sets as a basis for a theory of possibility. *Fuzzy Sets and Systems, 1*, 3-28.

ADDITIONAL READING

Abdel-Galil, T.K., Sharkawy, R.M., Salama, M.M.A, & Bartnikas, R. (2005). Partial Discharge Pattern Classification Using the Fuzzy Decision Tree Approach, *IEEE Transactions on Instrument and Measurement, 54*(6), 2258-2263.

Adamo, J. M. (1980). Fuzzy Decision Trees, *Fuzzy Sets and Systems, 4*(3), 207-219.

Akay, M., Cohen, M., & Hudson, D. (1997). Fuzzy Sets in Life Sciences, *Fuzzy Sets and Systems, 90*, 219-224.

Chen, R.Y., Sheu, D.D., & Liu, C. M. (2007). Vague Knowledge Search in the Design for Outsourcing using Fuzzy Decision Tree, *Computers and Operations Research, 34*(12), 3628-3627.

Cohen, H., Zohar, J., Matar, M., Kaplan, Z, & Geva, A. (2005). Unsupervised Clustering Analysis Supports behavioural Cutoff Criteria in an Animal Model of Postraumatic Stress Disorder, *Animal Psychiatry, 58*(8), 640-650.

Crockett, K., Bandar, Z., & O'Shea, J. (2006). On Producing Balanced Fuzzy decision Tree Classifiers, *IEEE Conference on Fuzzy Systems*, Vancouver, Canada, 1756-1762.

Dumitrescu, D., Lazzerini, B., & Jain, L. C. (2000). *Fuzzy Sets and their Application to Clustering and Training*, CRC Press.

Demicco, R.V., & Klir, G. J. (2004). *Fuzzy Logic in Geology*, Elsevier.

Dubitzky, W., & Azuaje, F. (2004). *Artificial Intelligence Methods and Tools for Systems Biology*, Springer.

Haskell, R.E., Lee, C., & Hanna, D. M. (2004). Geno-fuzzy classification trees, *Pattern Recognition*, 37(8), 1653-1659.

Jafelice, R.M., De Barros, L.C., Bassanezi, R.C., & Gomide, F. (2004). Fuzzy Modeling in Symptomatic HIV Virus Infected Population, *Bulletin of Mathematical Biology, 66*, 1597-1620.

Kurgan, L. A., Stach, W., & Ruan, J. (2007). Novel scales based on hydrophobicity indices for secondary protein structure, *Journal of Theoretical Biology, 248*(2), 354-366.

Nieto, J.J., Torres, A., Georgiou, D.N., & Karakasidis, T.E. (2006). Fuzzy Polynucleotide Spaces and Metrics, *Bulletin of Mathematical Biology, 68*, 703-725.

Pedrycz, W., & Sosnowski, Z. A. (2005). C-Fuzzy Decision Trees, *IEEE Transactions on Systems, Man and Cybernetics-Part C: Applications and Reviews, 35*(4), 498-511

Setzkorn, C. & Paton, R. C. (2005) On the use of multi-objective evolutionary algorithms for the induction of fuzzy classification rules systems, *BioSystems, 81*(2), 101-112.

Wang, X., Nauck, D.D., Spott, M., & Kruse, R. (2007). Intelligent Data Analysis using Fuzzy Decision Trees, *Soft Computing, 11*(5), 439-457.

Wolkenhauer, O. (2000). Fuzzy Relational Biology: A Factor-Space Approach to genome Analysis, *Working document*, Available at: http://www.umist.ac.uk/csc/people/wolkenhauer.htm.

KEY TERMS

Condition Attribute: An attribute that describes an object. Within a decision tree it is part of a non-leaf node, so performs as an antecedent in the decision rules used for the final classification of an object.

Decision Attribute: An attribute that characterises an object. Within a decision tree is part of a leaf node, so performs as a consequent, in the decision rules, from the paths down the tree to the leaf node.

Decision Tree: A tree-like structure for representing a collection of hierarchical decision rules that lead to a class or value, starting from a root node ending in a series of leaf nodes.

Induction: A technique that infers generalizations from the information in the data.

Leaf Node: A node not further split, the terminal grouping, in a classification or decision tree.

Linguistic Term: One of a set of linguistic terms, which are subjective categories for a linguistic variable, each described by a membership function.

Linguistic Variable: A variable made up of a number of words (linguistic terms) with associated degrees of membership.

Path: A path down the tree from root node to leaf node, also termed a branch.

Membership Function: A function that quantifies the grade of membership of a variable to a linguistic term.

Node: A junction point down a path in a decision tree that describes a condition in an if-then decision rule. From a node, the current path may separate into two or more paths.

Root Node: The node at the tope of a decision tree, from which all paths originate and lead to a leaf node.

Compilation of References

Abbott, L. F. & Regehr W.G. (2004). Synaptic Computation. *Nature*, Vol. 431, pp. 796-803.

Abeles, M., Bergman, H., Margalit, E., & Vaadia E. (1993). Spatiotemporal firing patterns in the frontal cortex of behaving monkeys. *Journal Neurophysiology*, 70, 1629-1638.

Acosta, G. G., Alonso, C., & Pulido, B. (2001). Basic Tasks for Knowledge Based Supervision in Process Control. *Engineering Applications of Artificial Intelligence*, Vol. 14, N° 4, Elsevier Science Ltd/IFAC, August 2001, pp. 441-455.

Acosta, G.G., Curti, H.J., & Calvo, O.A. (2005). *Autonomous Underwater Pipeline Inspection in AUTOTRACKER PROJECT: the Navigation Module*. Proc. of the IEEE/Oceans'05 Europe Conference, Brest, France, June 21-23, 2005, pp. 389-394.

Acosta, G.G., Curti, H.J., Calvo, Evans, J., & Lane, D. (2007). *Path Planning in an autonomous vehicle for deep-water pipeline inspection,* submitted to the Journal of Field Robotics.

Acosta, G.G., Curti, H.J., Calvo, O.A, & Mochnacs, J. (2003). An Expert Navigator for an Autonomous Underwater Vehicle. *Proceedings of SADIO/Argentine Symposium on Artificial Intelligence ASAI '03*, Buenos Aires, Argentina, Sept. 3-5 2003, in CD.

Adam, J. (2005). *Designing Emergence*. Doctoral dissertation, University of Essex.

Adleman, L. M. (1994). Molecular computation of solutions to combinatorial problems. *Science*, 226, 1021-1024.

Adrianantoandro, E., Basu, S., Karig, D. K., & Weiss, R. (2006). Synthetic biology: new engineering rules for an emerging discipline. *Molecular Systems Biology*, doi: 10.1038/msb4100073.

Aktas, M. S., Fox, G. C., & Pierce, M. (2007). Fault tolerant high performance Information Services for dynamic collections of Grid and Web services. *Future Generation Computer Systems*, 23(3), 317-337.

Alegre, M., Labarga, A., Gurtubay, I., Iriarte, J., Malanda, A., & Artieda, J. (2003). Movementrelated changes in cortical oscillatory activity in ballistic, sustained and negative movements. *Experimental Brain Research*, 148,17–25.

Alenda, A., & Nuñez, A. (2004). Sensory-interference in rat primary somatosensory cortical neurons. *European Journal Neuroscience*, 19, 766-70.

Alenda, A., & Nuñez, A. (2007). Cholinergic modulation of sensory interferente in rat primary somatosensory cortical neurons. *Brain Research* 1133,158-167.

Alexander, R.McN. (1984). The gait of bipedal and quadrupedal animals. *International Journal of Robotics Research,* 3, 49-59.

Almeida, F., Mateus, L., Costa, E., Harvey, I. & Coutinho, A. (2007). *Advances in Artificial Life. LNAI 4648*. Springer, Berlin.

Alon, U. (2007). *An introduction to systems biology: design principles of biological circuits.* Chapman & Hall/CRC - Taylor & Francis Group.

Amaral, L. A. N., Díaz-Guilera, A., Moreira, A. A., Goldberger, A. L., & Lipsitz, L. A. (2004). Emergence of complex dynamics in a simple model of signaling networks. *Proceedings of the National Academy of Sciences of the United States of America*, 101, 15551-15555.

Andreasen, M., & Nedergaard, S. (1996). Dendritic electrogenesis in rat hippocampal CA1 pyramidal neurons: functional aspects of Na+ and Ca2+ currents in apical dendrites. *Hippocampus*, 6, 79-95.

Angeline, P.J., Saunders, G.M. & Pollack, J.B. (1994). An evolutionary algorithm that constructs recurrent neural networks. *IEEE Transactions on Neural Networks*, 5(1), 54–65.

Antonelli, G., Chiaverini, S., Finotello, R., & Schiavon, R. (2001). Real-Time Path Planning and Obstacle Avoidance for RAIS: an Autonomous Underwater Vehicle. April 2001, *IEEE Journal of Oceanic Engineering*, Vol. 26, N°2, pp. 216-227.

Antonio, A., Ramírez, J. , Imbert, R., Méndez, G. & Aguilar, R. A. (2005). A Software Architecture for Intelligent Virtual Environments Applied to Education. *Revista Facultad de Ingeniería - Universidad de Tarapacá*, Arica – Chile, 13 (1), 47-55.

Antsaklis P.J., & Passino K.M. (1993). *An Introduction to Intelligent and Autonomous Control.* Boston, Dordrecht, London, 1993, Kluwer Academic Pub.

Arakawa, M., Hasegawa, K. & Funatsu, K. (2003). Application of the Novel Molecular Alignment Method Using the Hopfield Neural Network to 3D-QSAR, *Journal of Chemical Information Computer Science,* 43 (5), 1396 -1402.

Araque, A., Carmignoto, G., & Haydon, P. G. (2001). Dynamic Signaling Between Astrocytes and Neurons. *Annu. Rev. Physiol*, 63, 795-813.

Araque, A., Púrpura, V., Sanzgiri, R., & Haydon, P. G. (1999). Tripartite synapses: glia, the unacknowledged partner. *Trends in Neuroscience*, 22(5).

Aristotle. De Partibus Animalium, de Incessu Animalium, de Motu Animalium. In A.S. Peek & E.S. Forster (Translators), *In Parts of Animals, Movement of Animals, Progression of Animals.* Cambridge, MA: Harvard University Press, 1936.

Arkin, R.C. (1998). *Behaviour-Based Robotics.* The MIT Press, Cambridge, Massachusetts, 1998.

Armand, S., Watelain, E., Roux, E., Mercier, M., & Lepoutre, F.-X. (2007). Linking clinical measurements and kinematic gait patterns of toe-walking using fuzzy decision trees. *Gait & Posture*, *25*(3), 475-484.

Armstrong-James, M., & George, M. J. (1988). Bilateral receptive fields of cells in rat Sm1 cortex. *Experimental Brain Research*, 70, 155-65.

Aron, S., Deneubourg, J.L., Goss, S., & Pasteels, J.M. (1990), *Functional Self-Organization Illustrated by Inter-Nest Traffic in the Argentine Ant Iridomyrmex Humilis.* Lecture Notes in BioMeathematics, 89, 533-547.

Ashlock, D. & Lathrop, J. I. (1998). A Fully Characterized Test Suite for Genetic Programming. In *Evolutionary Programming VII*, (pp. 537-546). New York, NY: Springer-Verlag.

Ashlock, D. & Schonfeld, J. (2005). Nonlinear Projection for the Display of High Dimensional Distance Data. In *Proceedings of the 2005 Congress on Evolutionary Computation*, vol. 3 (pp. 2776-2783). Piscataway, NJ: IEEE Press.

Ashlock, D. A. (2006). *Evolutionary Computation for Modeling and Optimization.* New York, NY: Springer Science+Business Meida, Inc.

Ashlock, D., Guo, L., & Qiu, F. (2002). Greedy Closure Genetic Algorithms. In *Proceedings of the 2002 Congress on Evolutionary Computation* (pp. 1296-1301). Piscataway, NJ: IEEE Press.

Ashlock, D.A., Bryden, K.M., & Corns, S.M. (2005). Graph Based Evolutionary Algorithms Enhance the Location of Steiner Systems. In *Proceedings of the 2005 Congress on Evolutionary Computation* (pp. 1861-1866). Piscataway, NJ: IEEE Press.

Atencia, M., Joya, G. & Sandoval, F. (2004). Parametric identification of robotic systems with stable time-varying Hopfield networks. *Neural Computing and Applications*, Springer London, *13*(4), 270-280.

Athithan, G., & Dasgupta, C. (1997). *On the Problem of Spurious Patterns in Neural Associative Models*, IEEE Transactions on Neural Networks, vol. 8, no. 6, 1483-1491.

Aylett, R.& Cavazza, M. (2001). Intelligent Virtual Environments – A state of the art report. *Eurographics Conference*, Manchester, UK.

Aylett, R.& Luck, M. (2000). Applying Artificial Intelligence to Virtual Reality: Intelligent Virtual Environments. *Applied Artificial Intelligence*, 14(1), 3-32.

Bair, W., Koch, C. (1996). Temporal precision of spike trains in extrastriate cortex of the behaving macaque monkey. *Neural Computation*, 8, 1185-202.

Balbín, A. & Andrade, E. (2007). Designing a Methodology to Estimate Complexity of Protein Structures. *Advances in Artificial Life. LNAI 4648.* Springer, Berlin.

Balch, T. & Arkin, R. (1998). Behavior-based formation control for multi-robot teams. *IEEE Transaction on Robotics and Automation.* 14:6, 926-939.

Baldassarre, G.; Nolfi, S., & Parisi, D. (2002). *Evolving mobile robots able to display collective behaviour.* In Helmelrijk C.K. (ed.). International Workshop on Self-Organization and Evolution of Social Behaviour, pp. 11-22. Zurich: Swiss Federal Institute of Technology.

Ball, P. (2005). Synthetic biology for nanotechnology. *Nanotechnology*, 16, R1-R8.

Ballester, P.J., & Carter, J.N. (2003). Real-Parameter Genetic Algorithm for Finding Multiple Optimal Solutions in Multimodel Optimization. *Proceedings of Genetic and Evolutionary Computation* (pp. 706-717)

Banzhaf, W. G., Nordin, P., Keller, R. E., & Francone, F. D. (1998). *Genetic Programming: An Introduction*. San Francisco, CA: Morgan Kaufmann.

Barbosa, V.C. (1996). *An introduction to distributed algorithms*. Cambridge, MA: The MIT Press.

Barbosa, V.C., & Gafni, E. (1989). Concurrency in heavily loaded neighbourhood-constrained systems. *ACM Transactions on Programming Languages and Systems*, 11(4), 562-584.

Barbosa, V.C., Benevides, M.R.F., & França, F.M.G. (2001). Sharing resources at nonuniform access rates. *Theory of Computing Systems*, 34(1), 13-26.

Barlow, H.B. (1972) Single units and sensation: a neuron doctrine for perceptual psychology? *Perception, 1*, 371-394.

Barros, L., Silva, A. & Musse, S. (2004) "PetroSim: An Architecture to Manage Virtual Crowds in Panic Situations", in *CASA 2004 – Computer Animation and Social Agents*, Geneva, Switzerland.

Bässler, U. (1986). On the definition of central pattern generator and its sensory control. *Biological Cybernetics*, 54, 65-69.

Bath, J., & Turberfield, A. J. (2007). DNA nanomachines. *Nature Nanotechnology*, 2, 275-284.

Baum, E. B. (1995). Building an associative memory vastly larger than the brain. *Science*, 268, 583-585.

Bay, J.S., Hemami, H. (1987). Modeling of a neural pattern generator with coupled nonlinear oscillators. *IEEE Transactions on Biomedical Engineering*, 34, 297-306.

Beckers, R., Deneubourg, J. L. & Goss, S. (1992). Trails ant u-turns in the selection of a path by the and lasius niger. *Journal of Theoretical Biology*, 159, 397-415.

Beer, B.D. (1990). *Intelligence as adaptive behavior: An experiment in computational neuroethology*. San Diego, CA: Academic Press.

Benenson, Y., Gil, B., Ben-Dor, U., Adar, R., & Shapiro, E. (2004). An autonomous molecular computer for logical control of gene expression. *Nature*, 429, 423-429.

Benenson, Y., Paz-elizur, T., Adar, R., Keinan, E., Liben, Z., & Shapiro, E. (2001). Programmable and autonomous computing machine made of biomolecules. *Nature*, 414, 430-434.

Beni, G. (1988). The concept of cellular robotic system. In *Proceedings of the IEEE International Symposium on Intelligent Control* (pp. 57-62).

Benner, S. A., & Sismour, A. M. (2005). Synthetic biology. *Nature Reviews Genetics*, 6, 533-543.

Bentley, P.J. (2002) *Digital Biology*. Simon and Schuster, New York.

Bermudez, E. & Seth A.K. (2007). Simulations of Simulations in Evolutionary Robotics. *Advances in Artificial Life. LNAI 4648*. Springer, Berlin.

Bernstein, N.A. (1967) *The co-ordination and regulation of movements*. Pergamon Press, London.

Beyer, S., & Simmel, F. C. (2006). A modular DNA signal translator for the controlled release of a protein by an aptamer. *Nucleic Acids Research*, 34, 1581-1587.

Beynon, M.J., Peel, M.J., & Tang, Y.-C. (2004b). The Application of Fuzzy Decision Tree Analysis in an Exposition of the Antecedents of Audit Fees. *OMEGA - International Journal of Management Science*, *32*(2), 231-244.

Bishop, C. M. (1995). *Neural networks for pattern recognition*. Oxford: Oxford University Press.

Blaize, M., Knight, D., & Rasheed, K. (1998). Automated Optimal Design of Two-Dimensional Supersonic Missile Inlets. *Journal of Propulsion and Power* 14, 890-898.

Bonabeau, E., & Cogne, F. (1997). Self-organization in social insects. *Trends in Ecology and Evolution*, 12, 188-193.

Bonabeau, E., Dorigo, M., & Theraulaz, G. (Eds.). (1999). *Swarm Intelligence: From Natural to Artificial Systems*, New York, USA: Oxford University Press.

Borenstein, J., & Koren, Y. (1991). The vector field histogram-fast obstacle avoidance for mobile robots. June 1991, *IEEE Transactions on Robotics and Automation*, Vol. 7, Issue 3, pp. 278 – 288.

Bose, N.K. & Liang, P. (1996). *Neural Network Fundamentals with Graphs, Algorithms and Applications*, U.S.A.: McGraw-Hill, Inc.

Box, G. E. & Jenkins, G. M. (1976). *Time Series Analysis Forecasting and Control*. Ed. Holden Day.

Braga, L. A. F. (2006). *Escape Route Signage Simulation Using Multiagents and Virtual Reality.* D.Sc Thesis (in Portuguese), COPPE -Universidade Federal do Rio de Janeiro.

Braga, R.R., Yang, Z., & França, F.M.G. (in press). Implementing an artificial centipede CPG: Integrating appendicular and axial movements of the scolopendromorph centipede. *International Conference on Bio-inspired Systems and Signal Processing*, Madeira, Portugal, January, 2008.

Braun, A., Musse, S. R., de Oliveira, L. P. L. & Bodmann, B. E. J. (2003). Modeling individual behaviors in crowd simulation. *International Conference on Computer Animation and Social Agents.* (pp. 143-148).

Brecht, M., Preilowski, B., & Merzenich, M. M. (1997) Functional architecture of the mystacial vibrissae. Behav *Brain Research*, 84, 81-97.

Breiman, L. (2001). Statistical Modeling: The Two Cultures. *Statistical Science, 16*(3), 199-231.

Brillinger, D. R. (1978). Developments in statistics. In *Comparative aspects of the study of ordinary time series and of point processes* (pp. 33–129). Orlando, FL: Academic Press.

Brion, G.M. and Lingireddy, S. (1999). A neural network approach to identifying nonpoint sources of microbial contamination. Water Res.,33(14), 3099-3106.

Brooks, R.A. (1986). A Robust Layered Control System for a Mobile Robot. March 1986, *IEEE Journal of Robotics and Automation*, Vol. RA-2, N° 1, pp. 14-23.

Bryden, K. M., Ashlock, D. A., Corns, S. M., & Willson, S. J. (2006). Graph Based Evolutionary Algorithms. *IEEE Transactions on Evolutionary Computations*, Vol. 10:5, 550-567.

Bryden, K. M., Ashlock, D. A., McCorkle, D. S., & Urban, G. L. (2003). Optimization of Heat Transfer Utilizing Graph Based Evolutionary Algorithms. *Computer Methods in Applied Mechanics and Engineering*, 192(44-46), 5021-5036.

Bryson, A. E., & Ho, Y. C. (1975). *Applied optimal control*: John Wiley & Sons.

Buono, P.L., & Golubitsky, M. (2001) Models of central pattern generators for quadrupedal locomotion: I. Primary gaits. *Journal of Mathematical Biology*, 42, 291-326.

Burton, J. L. & Franks, N. R. (1985). The foraging ecology of the army ant eciton rapax : An ergonomic enigma? *Ecological Entomology*, 10, 131–141.

Butler, S. (1982). *Erewhon.* Barcelona: Bruguera.

Calabretta, R, Nolfi, S., Parisi, D., & Wagner, G. P. (2000). *Duplication of modules facilitates functional specialization.* Artificial Life, 6(1), 69-84.

Calvo, O., Rodríguez, G., & Picos, R. (1999). Real Time Navigation Of Autonomous Robot With Fuzzy Controller And ultrasonic Sensors. Palma de Mallorca, Spain, September 1999, *Proc. of the European Society of Fuzzy Logic and Technology Symposium EUSFLAT-ESTYLF99*, pp 95-98.

Camazine, S., Deneubourg, J.L., Franks, N., Sneyd, J., Theraulaz, G., & Bonabeau, E. (2001). *Self-Organization in Biological Systems.* Princeton University Press.

Canals, S., López-Aguado, L., & Herreras O. (2005). Synaptically-recruited apical currents are required to initiate axonal and apical spikes in hippocampal pyramidal cells: modulation by inhibition. *Journal of Neurophysiology*, 93, 909-918.

Cannas, B., Fanni, A., See, L., & Sias, G. (2006). Data preprocessing for river flow forecasting using neural networks: Wavelet transforms and data partitioning. *Time Series Analysis in Hydrology, 31*(18), 1164-1171.

Caro G. D. & Dorigo, M. (1998). Antnet: Distributed stigmergetic control for communications networks. *Journal of Artificial Intelligence Research*, 9, 317–365.

Carvell, G. E., & Simons, D. J. (1990). Biometric analyses of vibrissal tactile discrimination in the rat. *Journal Neuroscience*, 10, 2638-2648.

Castellanos, N. P. & Makarov, V. A. (2006). Recovering EEG brain signals: artifact suppression with wavelet enhanced independent component analysis. *Journal Neuroscience Methods*, 158, 300-312.

Castellanos, N. P., Malmierca, E., Nuñez, A., & Makarov, V. A. (2007). Corticofugal modulation of the tactile response coherence of projecting neurons in the gracilis nucleus. *Journal Neurophysiology*, 98, 2537-2549.

Cattaert, D., & Bévengut, M. (2002) Effects of antidromic discharges in crayfish primary afferents. *Journal of Neurophysiology, 88*, 1753-1765.

Cattaert, D., & El Manira, A. (1999) Shunting versus inactivation: analysis of presynaptic inhibitory mechanisms in primary afferents of the crayfish. *Journal of Neuroscience, 19*, 6079-6089.

Cattaert, D., & Le Ray, D. (1998) Direct glutamate-mediated presynaptic inhibition of sensory afferents by the postsynaptic motor neurons. *European Journal of Neuroscience, 10*, 3737-3746.

Cattaert, D., & Le Ray, D. (2001) Adaptive motor control in crayfish. *Progress in Neurobiology, 63*, 199-240.

Cattaert, D., Araque, A., Buño, W., & Clarac, F. (1994a) Motor neurones of the crayfish walking system possess TEA+-revealed regenerative electrical properties. *Journal of Experimental Biology, 188*, 339-345.

Cattaert, D., Araque, A., Buño, W., & Clarac, F. (1994b) Nicotinic and muscarinic activation of motoneurons in the crayfish locomotor network. *Journal of Neurophysiology, 72*, 1622-1633.

Cattaert, D., El Manira, A., & Clarac, F. (1992) Direct evidence for presynaptic inhibitory mechanisms in crayfish sensory afferents. *Journal of Neurophysiology, 67*, 610-624.

Cattaert, D., El Manira, A., & Clarac, F. (1994c) Chloride conductance produces both presynaptic inhibition and antidromic spikes in primary afferents. *Brain Research, 666*, 109-112.

Cattaert, D., El Manira, A., Marchand, A., & Clarac, F. (1990) Central control of the sensory afferent terminals from a leg chordotonal organ in crayfish in vitro preparation. *Neuroscience Letters, 108*, 81-87.

Cattaert, D., Le Bon, M., & Le Ray, D. (2002) Efferent controls in crustacean mechanoreceptors. *Microscopy Research and Techniques, 58*, 312-324.

Cattaert, D., Libersat, F., & El Manira, A. (2001) Presynaptic inhibition and antidromic spikes in primary afferents of the crayfish: a computational and experimental analysis. *Journal of Neuroscience, 21*, 1007-1021.

Cavaliere, M., Egecioglu, O., Ibarra, O.H., Ionescu, M., Păun, Gh., &. Woodworth, S. (2007). Unsynchronized spiking neural P systems: decidability and undecidability. *Proceedings of 13rd DNA Based Computers Conference, Memphis, USA.*

Cello, J., Paul, A. V., & Wimmer, E. (2002). Chemical synthesis of poliovirus cDNA: generation of infectious virus in the absence of natural template. *Science*, 297, 1016-1018.

Chaimowicz, L., Kumar, V., & Campos, M.F. (2001). Framework for coordinating multiple robots in cooperative manipulation tasks. In *Proceedings SPIE Sensor Fusion and Decentralized Control in Robotic Systems IV* (pp. 120-127).

Chang, R.L.P., & Pavlidis, T. (1977). Fuzzy decision tree algorithms. *IEEE Transactions Systems Man and Cybernetics*, SMC-7(1), 28-35.

Chen, H., Freund, R., Ionescu, M., Păun, Gh., & Perez-Jimenez, M.J. (2007). On string languages generated by spiking neural P systems. *Fundamenta Informaticae, 75*(1-4), 141–162.

Chen, H., Ionescu, M., & Ishdorj, T.-O. (2006c). On the efficiency of spiking neural P systems. *Proceedings of the 8th International Conference on Electronics, Information, and Communication*, Ulanbator, Mongolia, June 2006, 49–52.

Chen, H., Ishdorj, T.-O., Păun, Gh., & Perez-Jimenez, M.J. (2006a). Spiking neural P systems with extended rules. In (Gutierrez-Naranjo et al., eds., (2006)), vol. I, 241–265.

Chen, H., Ishdorj, T.-O., Păun, Gh., Perez-Jimenez, M.J. (2006b). Handling languages with spiking neural P systems with extended rules. *Romanian Journal of Information Sciience and Technology, 9*(3), 151–162.

Chiang, I-J., & Hsu, J.Y-J. (2002). Fuzzy classification trees for data analysis. *Fuzzy Sets and Systems, 130*, 87-99.

Chittaro, L., Ieronutti, L. & Rigutti, S. (2005. Supporting Presentation Techniques based on Virtual Humans in Educational Virtual Worlds. *Proceedings of CW 2005: 4th International Conference on Cyberworlds*, (pp.245-252), IEEE Press, Los Alamitos, CA, US.

Chiva, E., Devade, J. Donnart, J.-Y. Maruéjouls, S. (2004). Motivational Graphs: A New Architecture for Complex Behavior Simulation. In Rabin, S. *AI Game Programming Wisdom 2*, 361-372 Charles River Media, Inc.

Chrachri, A., & Clarac, F. (1989) Synaptic connections between motor neurons and interneurons in the fourth thoracic ganglion of the crayfish, *Procambarus clarkii. Journal of Neurophysiology, 62*, 1237-1250.

Chua, L. & Yang, L. (1988). Cellular neural networks: theory and applications, *IEEE Transactions on Circuits and Systems, CAS-35*, 1257-1290.

Chung, C.H. & Lee, K.S. (1991). Hopfield network application to optimal edge selection. *Neural Networks, IEEE International Joint Conference on*, vol.2, 1542 – 1547.

Church, G. M. (2005). From systems biology to synthetic biology. *Molecular Systems Biology*, doi: 10.1038/msb4100007.

Ciobanu, G., Păun, Gh., &. Perez-Jimenez, M.J. (Eds.) (2006). *Applications of Membrane Computing*. Berlin: Springer.

Ciobanu, G., Pérez-Jiménez, M. J., & Paun, G. (2006). Applications of Membrane Computing Series: Natural Computing Series. Berlin, Springer.

Clarac, F., Cattaert, D., & Le Ray, D. (2000) Central control components of a 'simple' stretch reflex. *Trends in Neuroscience, 23*, 199-208.

Cohen, D. (1994). *Los Genes de la Esperanza*. Barcelona: Seix Barral.

Cohen, M. A. & Grossberg, S. (1983). Absolute stability of pattern formation and parallel storage by competitive neural networks. *IEEE Transactions of Systems, Man & Cybernetics, 13*, 815-826.

Colbert, C. M., & Johnston, D. (1998). Protein kinase C activation decreases activity-dependent attenuation of dendritic Na+ current in hippocampal CA1 pyramidal neurons. *Journal of Neurophysiology* 79, 491-495.

Collins, J.J. (1995). Gait transitions. In M.A. Arbib (Ed.), *The handbook of brain theory and neural networks* (pp.420-423). New York: The MIT Press.

Collins, J.J., & Richmond, S.A. (1994). Hard-wired central pattern generators for quadrupedal locomotion. *Biological Cybernetics,* 71, 375-385.

Collins, J.J., & Stewart, I.N. (1993a). Coupled nonlinear oscillators and the symmetries of animal gaits. *Journal of Nonlinear Science,* 3, 349-392.

Collins, J.J., & Stewart, I.N. (1993b). Hexapodal gaits and coupled nonlinear oscillator models. *Biological Cybernetics,* 68, 287-298.

Colombo, G. & Cugini , U. (2006). Virtual humans and prototypes to evaluate ergonomics and safety. *Journal of Engineering Design*,.16(.2), 195-203.

Condon, A. (2004). Automata make antisense. *Nature*, 429, 351-352.

Conte, G., Zanoli, S., Perdon, A., & Radicioni, A. (1994). *A system for the automatic survey of underwater structures*. Brest, France, 13-16 Sept. 1994, Proc. of OCEANS '94, Vol. 2, pp. 92-95.

Conway J.H. (1971) *Regular Algebra and Finite Machines*. Chapman and Hall, Ltd., London.

Corns, S. M., Ashlock, D. A., McCorkle, D. S. & Bryden, K. M. (2006a). Improving Design Diversity Using Graph Based Evolutionary Algorithms. In *Proceedings of the 2006 IEEE World Congress on Computational Intelligence* (pp. 1037-1043). Piscataway, NJ: IEEE Press.

Corns, S. M., Bryden, K. M., & Ashlock, D. A. (2003). Evolutionary Optimization Using Graph Based Evolutionary Algorithms. In *Proceedings of the 2003 International Mechanical Engineering Congress and Exposition* (pp. 315-320). ASME Press.

Corns, S. M., Bryden, K. M., & Ashlock, D. A. (2005a). Solution Transfer Rates in Graph Based Evolutionary Algorithms. In *Proceedings of the 2005 Congress on Evolutionary Computation* (pp. 1699-1705). Piscataway, NJ: IEEE Press.

Corns, S. M., Bryden, K. M., & Ashlock, D. A. (2005b). The Impact of Novel Connection Topologies on Graph Based Evolutionary Algorithms. In C. H. Dagli (Ed.), *Smart Engineering System Design: Neural Networks, Evolutionary Programming, and Artificial Life*, vol. 15 (pp. 201-209). ASME Press.

Corns, S. M., Hurd, H. S., Hoffman, L. J. & Bryden, K. M. (2006b). Evolving Antibiotic Regimens to Minimize Bacterial Resistance in Swine. In *11th Annual AIAA/ISSMO Multidisciplinary Analysis and Optimization Conference*, Portsmouth, VA, On-line proceedings.

Crespo, A., & Garcia-Molina, H. (2002). *Routing indices for peer-to-peer systems*. Paper presented at the 22nd International Conference on Distributed Computing Systems (ICDCS 2002), Wien, Austria.

Crick, F., & Mitchinson, G. (1983). *The Function of Dream Sleep*, Nature, vol. 304, 111-114.

Crick, F.H.G.; Barnett, L.; Brenner, S. & Watts-Tobin, R.J. (1961) General Nature of the Genetic Code for Proteins. *Nature*, 192, 1277-1232.

Cruse, H. (1991). Coordination of leg movement in walking animals. *Proceedings of 1st International Conference on Simulation of Adaptive Behaviour* (pp. 105-119), Paris, France.

Csete, M., & Doyle J. (2004). Bow ties, metabolism and disease. *Trends in Biotechnology*, 22, 446-450.

Cybenko, G. (1989). Approximation by superpositions of a sigmoidal function. *Mathematics of Control, Signals, and Systems, 2*, 303-314.

Daan, S., Barnes, B.M., & Strijkstra, A.M. (1991). Warming up for sleep? Ground squirrels sleep during arousals from hibernation. *Neuroscience Letters, 128*, 265-268.

Danciu, D. & Răsvan, V. (2000). On Popov-type stability criteria for neural networks. *Electronic Journal on Qualitative Theory of Differential Equations EJQTDE*, Proc. 6th Coll. QTDE 2000, *23*. http://www.math.uszeged.hu/ejqtde/6/623.pdf

Danciu, D. & Răsvan, V. (2001). Gradient-like behaviour for Hopfield–type neural networks with delay. In I. Dumitrache & N. Constantin (Eds.), *Proceedings of the Third International Workshop on Intelligent Control Systems, ICS'2001*, (pp.20-24). Romania: Bucharest, Printech.

Danciu, D. & Răsvan, V. (2005). Stability Results for Cellular Neural Networks with Time Delays. In J. Cabestany, A. Prieto, F. Sandoval (Eds.), *Computational Intelligence and Bioinspired Systems*. Lectures Notes in Computer Science, *3512* (pp. 366-373), Springer-Verlag. http://dx.doi.org/10.1007/11494669_45

Danciu, D. & Răsvan, V. (2007). Dynamics of Neural Networks – Some Qualitative Properties. In F. Sandoval, A. Prieto, J. Cabestany, M. Graña (Eds.), *Computational and Ambient Intelligence*. Lectures Notes in Computer Science, *4507* (pp. 8-15), Springer-Verlag.

Danciu, D. (2002). Qualitative behavior of the time delay Hopfield type neural networks with time varying stimulus. *Annals of The University of Craiova*, Series: Electrical Engineering (Automatics, Computers, Electronics) *26*, 72–82.

Danciu, D. (2006). *Systems with several equilibria. Applications to the neural networks.* (in Romanian) Romania: Craiova, Universitaria Publishers House.

Darwin, C. (1859). *On the Origin of Species*. John Murray, London.

Darwin, C. (1864). *On the Origin of Species by means of Natural Selection or the Preservation of Favoured Races in the Struggle for Life*. Cambridge University Press, Cambridge, UK, sixth edition. Originally published in 1859.

Dasgupta, P. (2004). *Intelligent Agent Enabled P2P Search Using Ant Algorithms*, Paper presented at the 8th International Conference on Artificial Intelligence, Las Vegas, USA.

Davidor, Y., Yamada, T. and Nakano, R. (1993). "The ECOlogical framework II: Improving GA Performance at Virtually Zero Cost. In *Proceedings of the Fifth International Conference on Genetic Algorithms* (pp. 171-175). Morgan Kaufman.

Davis, G.W., & Bezprozvanny, I. (2001) Maintaining the stability of neural function: a homeostatic hypothesis. *Annual Review of Physiology, 63*, 847-869.

De Lorenzo, V., Serrano, L., & Valencia, A. (2006). Synthetic biology: challenges ahead. *Bioinformatics*, 22, 127-128.

Deb, K. (2003). *A Population-Based Algorithm-Generator for Real-Parameter Optimization. (Tech. Rep. No.* KanGAL 2003003)

Deb, K., & Agrawal, S. (1995). Simulated binary crossover for continuous search space. *Complex Systems 9(2)* (pp. 115-148)

DeJong, K.A., (1975). An Analysis of the Behaviour of a Class of Genetic Adaptive Systems. Phd. Thesis, University of Michigan, Ann Arbor.

Dellaert F., Beer R.D. (1996) *A Developmental Model for the Evolution of Complete Autonomous Agent* In From animals to animats: Proceedings of the Forth International Conference on Simulation of Adaptive Behaviour, Massachusetts, 9-13 September 1996, pp. 394-401, MIT Press.

Deneubourg, J. L., Goss, S., Franks, N., Sendova-Franks, A., Detrain, C., & Chretien, L. (1990) *The dynamics of collective sorting robot-like ants and ant-like robots*. Paper presented at the 1st International Conference on Simulation of Adaptive Behavior on From Animals to Animats, Cambridge, USA.

Deneubourg, J.L., Goss, S., Franks, N. R., & Pasteels, J. M. (1989). The blind leading the blind: Modeling chemically mediated army ant raid patterns. *Insect Behavior*, 2(), 719-725.

Deng, Z. Q., Singh, V. P., & Bengtsson, L. (2001). Longitudinal dispersion coefficient in straight rivers. *Journal of Hydraulic Engineering, 127*(1), 919-927.

Di Paolo, E. (2000). Homeostatic adaptation to inversion of the visual field sensorimotor disruptions. *Proceedings of SAB 2000*, MIT Press.

Di Paolo, E. (2003). Evolving spike-timing dependent plasticity for single-trial learning in robots. *Philosophical Transactions of the Royal Society*, A 361, pp. 2299 - 2319.

Di Ventura, B., Lemerle, C., Michalodimitrakis, K., & Serrano, L. (2006). From *in vivo* to *in silico* biology and back. *Nature*, 443, 527-533.

Dijkstra, E.W. (1971). Hierarchical ordering of sequential processes. *Acta Informatica*, 1(2), 115-138.

Dimitrijevic, M.R., Gerasimenko, Y., & Pinter, M.M. (1998). Evidence for a spinal central pattern generator in humans. *Annals of the New York Academy of Sciences*, 860, 360-376.

Dittmer, W. U., & Simmel, F. C. (2004). Transcriptional control of DNA-Based Nanomachines. *Nano Letters*, 4, 689-691.

Dittmer, W. U., Kempter, S., Rädler, J. O., & Simmel, F. C. (2005). Using gene regulation to program DNA-based molecular devices. *Small*, 1, 709-712.

Dombi, J., & Gera, Z. (2005). The approximation of piecewise linear membership functions and Łukasiewicz operators. *Fuzzy Sets and Systems*, *154*, 275-286.

Dopazo, H., Gordon, M.B., Perazzo, R.S., & Risau-Gusman, S.A. (2003). Model for the Emergence of Adaptive Subsystems. *Bulletin of Mathematical Biology*, 65, Elsevier, pp. 27–56.

Dorado, J. (1999). *Modelo de un Sistema para la Selección Automática en Dominios Complejos, con una Estrategia Cooperativa, de Conjuntos de Entrenamiento y Arquitecturas Ideales de Redes de Neuronas Artificiales Utilizando Algoritmos Genéticos*. Phd Thesis. Facultad de Informática. University of A Coruña, Spain.

Dorado, J., Santos, A., Pazos, A., Rabuñal, J.R., & Pedreira, N. (2000). Automatic selection of the training set with Genetic Algorithm for training Artificial Neural Networks. In: *Proc. Genetic and Evolutionary Computation Conference GECCO'2000*, (pp. 64-67) Las Vegas, USA.

Dorigo, M., Caro, G. D. & Gambardella, L. (1999). Ant algorithms for discrete optimization. *Artificial Life*, 5(2), 137–172.

Dorigo, M., Maniezzo, V. & Colorni, A. (1996). The Ant System: Optimization by a colony of cooperating agents. *IEEE Transactions on Systems, Man, and Cybernetics Part B: Cybernetics*, 26(1), 29–41.

Driessche van den, P. & Zou, X. (1998). Global attractivity in delayed Hopfield neural network models. *SIAM Journal of Applied Mathematics*, *58*, 1878-1890.

Dunn R.C. Jr., & Tolbert, D. L. (1982). The corticotrigeminal projection in the cat. A study of the organization of cortical projections to the spinal trigeminal nucleus. Brain Research, 240, 13-25.

Eames, M.H.A., Cosgrove, A., & Baker, R. (1999). Comparing methods of estimating the total body centre of mass in three-dimensions in normal and pathological gaits. *Human Movement Science*, 18, 637-646.

Eberhart, R. & Kennedy, J. (1995). A new optimizer using particle swarm theory. In *Proceedings of the Sixth International Symposium on Micro Machine and Human Science* (pp. 39–43).

Eberhart, R. & Shi, Y. (2001). Particle swarm optimization: Developments, applications and resources. In *Proceedings of the 2001 Congress on Evolutionary Computation*, (pp. 81-86).

Eggenberger P. (1996) *Cell Interactions as a Control Tool of Developmental Processes for Evolutionary Robotics* In From animals to animats: Proceedings of the Forth International Conference on Simulation of Adaptive Behaviour, Massachusetts, 9-13 September 1996, pp. 440-448, MIT Press.

Egmont-Petersen, M., de Ridder, D., & Handels, H. (2002). *Image processing with neural networks—a review*, Pattern Recognition, vol. 35, 2279-2301.

El Manira, A., & Clarac, F. (1994) Presynaptic inhibition is mediated by histamine and GABA in the crustacean escape reaction. *Journal of Neurophysiology, 71*, 1088-1095.

El Manira, A., Cattaert, D., & Clarac, F. (1991) Monosynaptic connections mediate resistance reflex in crayfish (*Procambarus clarkii*) walking legs. *Journal of Comparative Physiology A, 168*, 337-349.

El Manira, A., Cattaert, D., Wallen, P., DiCaprio, R.A., & Clarac, F. (1993) Electrical coupling of mechanoreceptor afferents in the crayfish: a possible mechanism for enhancement of sensory signal transmission. *Journal of Neurophysiology, 69*, 2248-2251.

Elowitz, M. B., & Leibler, S. (2000). A synthetic oscillatory network of transcriptional regulators. *Nature*, 403, 335-338.

Endo K., Maeno T., Kitano H. (2003) *Co-evolution of morphology and walking pattern of biped humanoid robot using evolutionary computation -designing the real robot.* ICRA 2003: pp. 1362-1367

Erdem, M. H., & Ozturk, Y. (1996). *A New family of Multivalued Networks*, Neural Networks 9,6, 979-89.

Erkmen, I., & Ozdogan, A. (1997) Short term load forecasting using genetically optimized neural network cascaded with a modified Kohonen clustering process, in: *Proc. IEEE Int. Symp. Intelligent Control* (pp. 107–112).

Eshelman, L.J., & Schaffer, J.D. (1993). Real coded genetic algorithms and interval schemata. *Foundations of Genetic Algorithms (2)* (pp. 187-202)

ETC group. (2007). Extreme genetic engineering: An introduction to synthetic biology. Published online January, 2007, http://www.etcgroup.org/upload/publication/pdf_file/602

Eugster, P., Guerraoui, R., Kermarres, A.M., & Massoulieacute, L. (2004). *Epidemic Information Dissemination in Distributed System*, IEEE Computer, 37(5), 60-67

Fabbri, G. (1997). A Genetic Algorithm for Fin Profile Optimization. *International Journal of Heat and Mass Transfer*, Vol. 40, No. 9, pp. 2165-2172.

Farmer, J., Packard, N. & Perelson, A. (1986). *The Immune System, Adaptation and Machine Learning*. Physica D. 22. 1986. pp. 187-204.

Feliciano, M., Potashner, S. J. (1995). Evidence for a glutamatergic pathway from the guinea pig auditory cortex to the inferior colliculus. *Journal Neurochemistry*, 65, 1348-1357.

Fernández León, J.A, (2005). *Estudio de Neuro-Controladores Evolutivos para Navegación de Robots Autónomos*. Master in Systems Engineering. UNCPBA, Argentina.

Fernández León, J.A., Tosini, M., & Acosta G.G, (2004). Evolutionary Reactive Behaviour for Mobile Robots Navigation. Singapore, December 1-3, 2004, *Proc. of the 2004 IEEE Conference on Cybernetics and Intelligent Systems (CIS04)*, pp. 532-537.

Fernández, D., Fuenzalida, M., Porto, A., & Buño, W. (2007). Selective shunting of NMDA EPSP component by the slow after hyperpolarization regulates synaptic integration at apical dendrites of CA1 pyramidal neurons. *Journal of Neurophysiology*, 97, 3242-3255.

Fernandez-Blanco E., Dorado J., Rabuñal J.R., Gestal M., Pedreira N.(2007) *A New Evolutionary Computation Technique for 2D Morphogenesis and Information Processing*. WSEAS Transactions on Information Science & Applications Volume 4 pp. 600-607. April 2007. WSEAS Press.

Fernández-León, J.A., & Di Paolo, E. A. (2007). Neural uncertainty and sensorimotor robustness. *Proceedings of the 9th European Conference on Artificial life ECAL 2007*. Springer-Verlag. 2007.

Ferreira C. (2006) *Gene Expression Programming: Mathematical Modelling by an Artificial Intelligence* Springer, Berlin.

Fiacco, T.A., & McCarthy, K.D. (2004). Intracellular astrocyte calcium waves in situ increase the frequency of spontaneous ampa receptor currents in Ca1 pyramidal neurons. *Journal of Neuroscience*, 24, 722-732.

Fink, W. (2004). Neural attractor network for application in visual field data classification. *Physics in Medicine and Biology 49*(13), 2799-2809.

Fischer, H. B., List, E. J., Koh, R. C. Y., Imberger, J., & Brooks, N. H. (1979). *Mixing in inland and coastal waters*. New York: Academic Press.

Fisher, R.A. (1936). The use of multiple measurements in taxonomic problems, *Annals of Eugenics*, 7, 179-188.

Fogel, L. & Atmar, J.W (Eds.) (1992) *Proceedings First Annual Conference on Evolutionary Programming*.

Fogel, L.J., Owens A. J. and Walsh, M.A. (1966) *Artificial Intelligence through Simulated Evolution*. Wiley, New York.

Fontana, W. (2006). Pulling strings. *Science*, 314, 1552-1553.

Forde, S.E., Thompson, J., & Bohannan, B.J. (2004). M. *Adaptation varies through space and time in a coevolving host-parasitoid interaction*. Nature, vol. 431.

Forestiero, A., Mastroianni, C., & Spezzano, G. (2005). *A Multi Agent Approach for the Construction of a Peer-to-Peer Information System in Grids*, Paper presented at the International Conference on Self-Organization and Adaptation of Multi-agent and Grid Systems (SOAS 2005), Glasgow, United Kingdom.

Forestiero, A., Mastroianni, C., & Spezzano, G. (2007). *A Decentralized Ant-Inspired Approach for Resource Management and Discovery in Grids*, Multiagent and Grid Systems - An International Journal, 3(1), 43-63.

Forster, A. C., & Church, G. M. (2007). Synthetic biology projects *in vitro*. *Genome Research*, 17, 1-6.

Fortuna, L., Arena, P., Balya, D. & Zarandy, A. (2001). Cellular Neural Networks. *IEEE Circuits and Systems Magazine*, *4*, 6–21.

Fossen, T. (2002). *Marine Control Systems*. Trondheim, Norway, Marine Cybernetics Ed. 2002.

França, F.M.G. (1994). *Neural networks as neighbourhood-constrained systems*. Unpublished doctoral dissertation, Imperial College, London, England.

França, F.M.G., & Yang, Z. (2000). Building artificial CPGs with asymmetric hopfield networks. *Proceedings of International Joint Conference on Neural Networks* (pp. IV:290-295), Como, Italy.

Franklin, S.(1995). *Artificial Minds.* MIT Press.

Freeman, J. A., & Skapura, D.M. (1991). *Neural Networks: Algorithms, Applications, and Programming Techniques*, Addison-Wesley, Reading, MA.

Friedberg, M. H., Lee, S. M., & Ebner, F. F. (2004). The contribution of the principal and spinal trigeminal nuclei to the receptive field properties of thalamic VPM neurons in the rat. *Journal Neurocytology*, 33, 75-85.

Fujibayashi, K., Murata, S., Sugawara, K., & Yamamura, M. (2002). Self-organizing formation algorithm for active elements. In *21st IEEE Symposium on Reliable Distributed Systems* (pp. 416-421).

García-Rojas, A. Vexo, F. Thalmann, D. Raouzaiou, A. Karpouzis, K. Kollias, S. Moccozet L., Magnenat-Thalmann, N (2006). Emotional face expression profiles supported by virtual human ontology. *Computer Animation and Virtual Worlds*, 17(3-4) 259 – 269.

Gardner, J. W., Boilot, P., & Hines, E. L. (2005). Enhancing electronic nose performance by sensor selection using a new integer-based genetic algorithm approach. *ISOEN 2003 - Selected Papers from the 10th International Symposium on Olfaction and Electronic Noses, 106*(1), 114-121.

Gardner, T. S., Cantor, C. R, & Collins, J. J. (2000). Construction of a genetic toggle switch in *Eschiria coli. Nature*, 403, 339-342.

Garibaldi, J.M., & John, R.I. (2003). Choosing Membership Functions of Linguistic Terms, *Proceedings of the 2003 IEEE International Conference on Fuzzy Systems*, St. Louis, USA, pp. 578-583.

Gasparini, S., Migliore, M., & Magee, J.C. (2004). On the initiation and propagation of dendritic spikes in CA1 pyramidal neurons. *Journal of Neuroscience,* 24, 11046-11056.

Gaudiano, P., Shargel, B. & Bonabeau, E. (2003). Control of UAV swarms: What the bugs can teach us. In *2nd American Institute of Aeronautics and Astronautics "Unmanned Unlimited" Conf. and Workshop and Exhibit.*

Gehani, A., LaBean, T. H., & Reif, J. H. (1999). *DNA-based cryptography.* Paper presented at 5th DIMACS Workshop on DNA Based Computers, MIT, Cambridge.

Gelig, A. Kh., Leonov, G.A. & Yakubovich, V.A. (1978). *Stability of nonlinear systems with non-unique equilibrium state.* (in Russian) U.R.S.S.: Moscow, Nauka Publishers House.

Georgopoulos, A., Schwartz, A.B., & Kettner, R.E. (1986) Neuronal population coding movement direction. *Science, 233*, 1416-1419.

Gerstner, W., & Kistler, W. (2002). *Spiking Neuron Models. Single Neurons, Populations, Plasticity.* Cambridge Univ. Press.

Gestal, M., Gómez-Carracedo, M.P., Andrade, J.M., Dorado, J., Fernández, E., Prada , D. & Pazos, A. (2005). Selection of variables by Genetic Algorithms to Classify Apple Beverages by Artificial Neural Networks. *Applied Artificial Intelligence* (pp. 181-198)

Gestal, M., Gómez-Carracedo, M.P., Andrade, J.M., Dorado, J., Fernández, E., Prada , D. & Pazos, A. (2004). Classification of Apple Beverages using Artificial Neural Networks with Previous Variable Selection. *Analytica Chimica Acta* (pp. 225-234)

Getting, P.A. (1989). Emerging principles governing the operation of neural networks", *Annual Review of Neuroscience,* 12, 185-204.

Glass, J. I., Alperovich, N., Assad-Garcia, N., Baden-Tillson, H., Khouri, H., Lewis, M., Nierman, W. C., Nelson, W. C., Pfannkoch, C., Remington, K., Yooseph, S., Smith, H. O., & Venter, J. C. (2004). Estimation of the minimal mycoplasma gene set using global transposon mutagenesis and comparative genomics. *Genomics: GTL II*, 51-52.

Glazewski, S., Benedetti, B. L., & Barth, A. L. (2007). Ipsilateral whiskers suppress experience-dependent plasticity in the barrel cortex. *Journal Neuroscience*, 27, 3910-3920.

Goelz, H., Jones, R., & Bones, P. (2000). Wavelet analysis of transient biomedical signals and its application to detection of epileptiform activity in the EEG. *Clinical Electroencephalography*, 31, 181–191.

Goldberg, D. E. (1989). *Genetic Algorithms in Search, Optimization and Machine Learning.* Addison-Wesley Reading, MA.

Goldberg, D. E., & Richardson J. (1987). Genetic Algorithms with Sharing for Multimodal Function Optimization. *Proceedings of 2nd International Conference on Genetic Algorithms (ICGA)*, (pp. 41-49), Springer-Verlag.

Goldberg, D.E (1989) *Genetic Algorithms in Search, Optimization & Machine Learning.* Addison-Wesley.

Goldenfeld, N., & Kadanoff, L. A. (1999). Simple lessons from complexity. *Science*, 284, 87-89.

Golding, N. L. & Spruston, N. (1998). Dendritic sodium spikes are variable triggers of axonal action potentials in hippocampal CA1 pyramidal neurons. *Neuron*, 21, 1189-1200.

Goldstein, S. S., & Rall W. (1974). Changes of action potential shape and velocity for changing core conductor geometry. *Biophysical Journal,* 14, 731-757.

Golubitsky, M., Stewart, I., Buono, P.L., & Collins, J.J. (1999) Symmetry in locomotor central pattern generators and animal gaits. *Nature,* 401, 693-695.

Golubitsky, M., Stewart, I., & Schaeffer, D.G. (1988). *Singularities and groups in bifurcation theory, Volume II.* Springer-Verlag.

Golubitsky, M., Stewart, I., Buono, P.L., & Collins, J.J. (1998). A modular network for legged locomotion. *Physica D,* 115, 56-72.

Gopalsamy, K. & He, X.Z. (1994). Stability in asymmetric Hopfield nets with transmission delays, *Physica D., 76,* 344-358.

Grassè, P. (1959). *La reconstruction du nid et les coordinations inter-individuelles chez belicositermes natalensis et cubitermes sp. la theorie de la stigmergie: Essai d'interprtation du comportement des termites constructeurs,* Insectes Sociaux 6, 41–84.

Grillner, S. (1975). Locomotion in vertebrates: Central mechanisms and reflex interaction. *Physiological Reviews,* 55, 247-304.

Grillner, S. (1985). Neurobiological bases of rhythmic motor acts in vertebrates. *Science,* 228, 143-149.

Grinsted, A., Moore, J. C., & Jevrejeva, S. (2004). Application of the cross wavelet transform and wavelet coherence to geophysical time series. *Nonlinear Processes in Geophysics*, 11, 561–566.

Grushin, A., & Reggia, J.A. (2006). Stigmergic self-assembly of prespecified artificial structures in a constrained and continuous environment. *Integrated Computer-Aided Engineering*, 13(4), 289-312.

Grzegorzewski, P., & Mrówka, E. (2005). Trapezoidal approximations of fuzzy numbers. *Fuzzy Sets and Systems, 153*, 115-135.

Gu, K., Kharitonov, V.L. & Chen, J. (2003). *Stability of Time- Delay Systems.* U.S.A.: Boston, Birkhäuser.

Guirguis, L.A., Ghoneimy, M.M.R.E. (2007). Channel Assignment for Cellular Networks Based on a Local Modified Hopfield Neural Network. *Wireless Personal Communications*, Springer US, *41*(4), 539-550.

Gurney, L.J., Mundy, C., & Porteus, M.C. (2005). Determining age and growth of abalone using stable oxygen isotopes: a tool for fisheries management. *Fisheries Research, 72,* 353-360.

Gusfield, D. (1997). *Algorithms on Strings, Trees and Sequences: Computer Science and Computational Biology.* Cambridge, MA: Cambridge University Press.

Gustafson, J. W., & Felbain-Keramidas, S. L. (1977). Behavioral and neural approaches to the function of the mystacial vibrissae. *Psychological Bulletin*, 84, 477-488.

Gutierrez-Naranjo, M.A. et al., eds. (2006). *Proceedings of Fourth Brainstorming Week on Membrane Computing.* Sevilla: Fenix Editora.

Gutierrez-Naranjo, M.A. et al., eds. (2007). *Proceedings of Fifth Brainstorming Week on Membrane Computing.* Sevilla: Fenix Editora.

Guymer, I. (1999). A national database of travel time, dispersion and methodologies for the protection of river abstractions. Environment Agency R&D Technical Report P346.

Hagan, M. T., Demuth, H. B., & Beale, M. H. (1996). *Neural network design* (1st ed.). Boston: PWS Pub.

Halanay, A. & Răsvan, V. (1993). *Applications of Lyapunov Methods to Stability,* The Netherlands: Dordrecht, Kluwer Academic Publishers.

Halanay, A., (1963). *Differential Equations. Stability. Oscillations. Time Lags.* (in Romanian), Romania: Bucharest, Academia Publishers House.

Hale J. K. & Verduyn Lunel, S. M. (1993). *Introduction to Functional Differential Equations.* Springer-Verlag.

Hardy G.H., Littlewood J.E. & Polya, G. (1934). *Inequalities.* Cambridge University Press.

Harik, G. (1995). Finding multimodal solutions using restricted tournament selection. *Proceedings of the Sixth International Conference on Genetic Algorithms*, (ICGA) (pp. 24-31).

Harris, C.J., Moore, C.G., & M. Brown. (1993). *Intelligent Control: Aspects of Fuzzy Logic and Neural Nets*, World Scientific, 1993.

Hartman, P. (1964). *Ordinary Differential Equations*, Wiley, U.S.: NewYork.

Hartwell, L. H., Hopfield, J. J., Leibler, S., & Murray, A. W. (1999). From molecular to modular cell biology. *Nature*, 402, c47-c52.

Harvey, I. (1992). *The SAGA cross. The mechanics of recombination for species with variable length genotypes.* Amsterdam: North-Holland. R. Manner and B. Manderick (Eds.), Parallel Problem Solving from Nature 2.

Hassoun M.H. (1995) *Fundamentals of Artificial Neural Networks.* University of Michigan Press, MA, USA

Hasty, J., McMillen, D., & Collins, J. J. (2002). Engineered gene circuits. *Nature*, 420, 224-230.

Haupt, R. L., & Haupt, S. E. (2004). *Practical genetic algorithms* (2nd ed.). Hoboken, N.J.: John Wiley.

Haykin, S. (1999). *Neural Networks: a Comprehensive Foundation.* 2nd Edition, PrenticeHall.

Hebb, D. O. (1949) . *The Organization of Behavior.* New York, Wiley.

Hee Yeal & Sung Yan B. (1997). An Improved Time Series Prediction by Applying the Layer-by-Layer Learning Method to FIR Neural Networks. *Neural Networks.* 10(9), 1717-1729.

Helbing, D., Farkas, I. & Vicsek, T. (2000). Simulating dynamical features of escape panic. *Nature.* 407, 487-490.

Hemmen, J. L. van, & Kuhn, R. (1991). *Collective Phenomena in Neural Networks*, E. Domany, J. L. van Hemmen and K. Shulten, eds. Berlin: Springer-Verlag, 1-105.

Herreras, O. (1990). Propagating dendritic action potential mediates synaptic transmission in CA1 pyramidal cells in situ. *Journal of Neurophysiology*, 64, 1429-1441.

Herreras, O., & Somjen, G, G., (1993). Effects of prolonged elevation of potassium on hippocampus of anesthetized rats. *Brain Research,* 617, 194-204.

Hertz, J. A., Grinstein, G., & Solla, S. A. (1987). *Heidelberg Colloquium on Glassy Dynamics*, J. L. van Hemmen and I. Morgenstern, eds. Berlin: Springer-Verlag, 538-546.

Hertz, J., Krogh, A., & Palmer, R. G. (1991). *Introduction to the Theory of Neural Computation.* Lecture Notes, Volume 1, Addison Wesley.

Hirsch, M.W. (1988). Stability and convergence in strongly monotone dynamical systems. *Journal für Reine Angewandte Mathematik, 383*, 1-53.

Ho, C. Y., Sasase, I. and Mori, S. (1992). On *the Capacity of the Hopfield Associative Memory.* In Proceedings of IJCNN 1992, II196-II201.

Hodgkin, A.L., & Huxley, A.F. (1952). A quantitative description of membrane current and its application to conduction and excitation in nerve. *Journal of Physiology,* 117(4), 500-544.

Hoffman, D. A., Magee, J. C., Colbert, C. M., & Johnston, D. (1997). K+ channel regulation of signal propagation in dendrites of hippocampal pyramidal neurons. *Nature* 387, 869-875.

Holland J.H. (1975) *Adaptation in natural and artificial systems.* University of Michigan Press, Ann Arbor, MA, USA.

Holland, J. H. (1975) *Adaptation in Natural and Artificial Systems.* Ann Arbor, Michigan: University Michigan Press.

Hopfield, J. J. (1982). Neural networks and physical systems with emergent collective computational abilities. *Proceedings of National Academic Science U.S.A., 79*, 2554-2558.

Hopfield, J. J. (1984). *Neurons with graded response have collective computational properties like those of two-state neurons*, Proceedings of the National Academy of Sciences USA, 81, 3088-3092.

Hopfield, J.J., & Tank, D.W. (1985). Neural computation of decisions in optimization problems. *Biological Cybernetics,* 52, 141-152.

Hopfield, J.J., & Tank, D.W. (1986). Computing with neural circuits: A model. *Science,* 233, 625-632.

Hosokawa, K., Tsujimori, T., Fujii, T., Kaetsu, H., Asama, H., Kuroda, Y., & Endo, I. (1998). Self-organizing collective robots with morphogenesis in a vertical plane. In *IEEE International Conference on Robotics and Automation* (pp. 2858-2863).

Hu, S-T. (1965). *Elements of Modern Algebre.* San Francisco, Ca: Holden-Day, Inc.

Hutchison, C. A., Peterson, S. N., Gill, S. R., Cline, R. T., White, O., Fraser, C. M., Smith, H. O., & Venter J. C. (1999). Global transposon mutagenesis and a minimal mycoplasma genome. *Science*, 286, 2165-2169.

Hutson, K. A., & Masterton, R. B. (1986). The sensory contribution of a single vibrissa's cortical barrel. *Journal Neurophysiology* 56, 1196-1223.

Hyland, J. & Taylor, F. (1993). Mine avoidance techniques for underwater vehicles. *IEEE J. Oceanic Engineering*, Vol. 18, 1993, pp.340-350.

Iamnitchi, A., & Foster, I. (2005). *Interest-aware information dissemination in small-world communities*, Paper presented at the 14th IEEE International Symposium on High Performance Distributed Computing (HPDC 2005), Research Triangle Park, USA.

Ibarra, O.H., Păun, A., Păun, Gh., Rodriguez-Paton, A., Sosik, P., & Woodworth, S. (2007). Normal forms for spiking neural P systems. *Theoretical Computer Science, 372*(2-3), 196–217.

Ibarz, J. M., Makarova, I., & Herreras, O. (2006). Relation of apical dendritic spikes to output decision in CA1 pyramidal cells during synchronous activation: a computational study. *European Journal of Neuroscience,* 23, 1219-1233.

Ijspeert, A.J., Crespi, A., Ryczko, D., & Cabelguen, J.M. (2007) From swimming to walking with a salamander robot driven by a spinal cord model. *Science, 315*, 1416-1420.

Inuiguchi, M., Tanino, T., & Sakawa, M. (2000). Membership function elicitation in possibilistic programming problems. *Fuzzy Sets and Systems, 111*, 29-45.

Ionescu, M., Păun, Gh., & Yokomori, T. (2006). Spiking neural P systems. *Fundamenta Informaticae, 71*(2-3), 279–308.

Isaacs, F. J., Dwyer, D. J., & Collins, J. J. (2006). RNA synthetic biology. *Nature Biotechnology*, 24, 545-554.

Ishiguro, A., Kondo, T., Shirai, Y., & Uchikawa, Y. (1996). Immunoid: An Architecture for Behaviour Arbitration Based on the Immune Networks. *Proceedings of the 1996 IEEE/RSJ International Conference on Intelligent Robots and Systems*, volume 3, pp. 1730 - 1738.

Issa, F.A., Adamson, D.J., & Edwards, D.H. (1999) Dominance hierarchy formation in juvenile crayfish *Procambarus clarkii. Journal of Experimental Biology, 202*, 3497-3506.

Iwahori, Y., Kawanaka, H., Fukui, S. & Funahashi, K. (2005). Obtaining Shape from Scanning Electron Microscope using Hopfield Neural Network. *Journal of Intelligent Manufacturing*, Springer US, *16*(6), 715-725.

Izhikevich, E.M. (2007). *Dynamical systems in neuroscience: The geometry of excitability and bursting*. Cambridge, MA: The MIT Press.

Jacobi, N. (1997). Evolutionary robotics and the radical envelope of noise hypothesis. *Journal of Adaptive Behaviour.* 1997.

Jang, J.-S. R., Sun, C.-T., & Mizutani, E. (1997). Neuro-fuzzy and soft computing: A computational approach to learning and machine intelligence. Upper Saddle River, NJ: Prentice Hall.

Janikow, C.Z. (1998). Fuzzy decision trees: Issues and methods. *IEEE Transactions of Systems, Man and Cybernetics Part B, 28*(1), 1-14.

Jarvis, M. R. & Mitra, P. P. (2001). Sampling Properties of the Spectrum and Coherency of Sequences of Action Potentials. *Neural Computation*, 13, 717–749.

Jen, P. H. S., Zhou, X., Zhang, J., & Sun, X. (2002). Brief and short-term corticofugal modulation of acoustic signal processing in the bat midbrain. *Hear Research*, 168, 196-207.

Jing-Sheng Xue; Ji-Zhou Sun; Xu Zhang (2004). Recurrent network in network intrusion detection system. *Proceedings of International Conference on Machine Learning and Cybernetics, 5*, 26-29 Aug. 2004, 2676 – 2679.

Jobson, H. E. (1997). Predicting travel time and dispersion in rivers and streams. *Journal of Hydraulic Engineering, 123*(11), 971-978.

Johansson, E. M., Dowla, F. U. & Goodman, D. M. (1992). Backpropagation learning for multi-layer feed-forward neural networks using the conjugate gradient method. *International Journal of Neural Systems*, 2 (4), 291-301.

Johnston, D., Magee, J. C., Colbert, C. M, & Christie, B. R. (1996). Active properties of neuronal dendrites. *Annual Review of Neuroscience* 19, 165-186.

Jones, C., & Mataric, M. (2003). From local to global behavior in intelligent self-assembly. In *IEEE International Conference on Robotics and Automation* (pp. 721-726).

Kalman, R.E. (1957). Physical and mathematical mechanisms of instability in nonlinear automatic control systems. *Transactions of American Society Mechanical Engineering, 79*(3).

Kamondi, A., Acsady, L., & Buzsaki, G. (1998) Dentritic spikes are enhanced by cooperative network activity in the intact hippocampus. *Journal of Neuroscience,* 18, 3919-3928.

Kaneko K. (2006) *Life: An Introduction to Complex Systems Biology*. Springer Complexity: Understanding Complex Systems, Springer Press.

Kasabov, N. (2001). Evolving fuzzy neural networks for supervised/unsupervised online knowledge-based learning.

IEEE Transactions on Systems, Man, and Cybernetics, Part B: Cybernetics, 31(6), 902-918.

Kashefipour, S. M., Falconer, R. A., & Lin, B. (2002). Modeling longitudinal dispersion in natural channel flows using ANNs. *River Flow 2002*, 111-116.

Kauffman S.A. (1969) *Metabolic stability and epigenesis in randomly constructed genetic nets.* Journal of Theoretical Biology 22 pp. 437-467.

Kauffman, S. A. & Levin, S. (1987). Towards a General Theory of Adaptive Walks on Rugged Landscapes. *Journal of Theoretical Biology*, vol. 128, 11-45.

Kazadi, S., Abdul-Khaliq, A., & Goodman, R. (2002). On the convergence of puck clustering systems. *Robotics and Autonomous Systems*, 38(2), 93-117.

Keane, A.J. (1994). Experiences with Optimizers in Structural Design. In *Proceedings of the 1st Conference on Adaptive Computing in Engineering Design and Control* (pp. 14-27). Plymouth, UK: Published by the University of Plymouth.

Kecman, V. (2001). *Learning and Soft Computing: Support Vector Machines, Neural Networks, and Fuzzy Logic.* MIT Press, London, UK.

Kemeny, J. G. (1955). Man Viewed as a Machine. *Scientific American,* 192 (4), *58-67.*

Kennedy, J. & Eberhart, R. (1995). Particle swarm optimization. In *IEEE International Conference on Neural Networks Vol.* 4 (pp. 1942–1948). Piscataway, NJ.

Kennedy, J., & Eberhart, R., & Shi, Y. (2001). *Swarm Intelligence.* Morgan Kaufmann.

Khalil, H. (1992). *Nonlinear systems.* J. Griffin (Ed.) U.S.: Macmillan Publishing Company.

Kharitonov V.L. & Zhabko, A.P. (2003). Lyapunov-Krasovskii approach to the robust stability analysis of time-delay systems. *Automatica 39*, 15–20.

Killackey, H. P. (1973). Anatomical evidence for cortical subdivisions based on vertically discrete thalamic projections from the ventral posterior nucleus to cortical barrels in the rat. *Brain Research*, 51, 326-331.

Kim, J., White, K. S., & Winfree, E. (2006). Construction of an *in vitro* bistable circuit from synthetic transcriptional switches. *Molecular Systems Biology*, doi: 10.1038/msb4100099.

Kimura, M. & Crow, J. (1963). On the Maximum Avoidance of Inbreeding. *Genetic Research*, vol. 4, 399-415.

King, J. L. & Jukes, T. H. (1969). Non Darwinian Evolution. *Science*, 164, 788-798.

Kinnebrock, W. (1994). Accelerating the standard backpropagation method using a genetic approach. *Neurocomputing* 6, 583–588.

Kirstukas, S. J., Bryden, K. M., & Ashlock, D. A. (2004). A Genetic Programming Technique for Solving Systems of Differential Equations. In C. H. Dagli (Ed.), *Intelligent Engineering Systems Through Artificial Neural Networks*, vol. 14 (pp. 241-246). ASME Press.

Kitamura, S., Hirai, K & Nishimura, M. (1967). Stability of a control system with several nonlinear elements and time lags. Technological Reports, Japan: Osaka University, *17*, 93-102.

Kitano H. et al. (2005) *Using process diagrams for the graphical representation of biological networks*, Nature Biotechnology 23(8) pp. 961 - 966.

Kitano, H. (2002). Systems biology: a brief overview. *Science*, 295, 1662-1664.

Kitano, H. (2004). Biological robustness. *Nature Reviews Genetics*, 5, 826-836.

Klavins, E. (2002). Automatic synthesis of controllers for distributed assembly and formation forming. In *IEEE International Conference on Robotics and Automation.*

Klein, A., Sauer, T., Jedynak, A., & Skrandies, W. (2006). Conventional and wavelet coherence applied to sensory–evoked electrical brain activity. IEEE Trans. Biomed. Engineer. 53, 266-272.

Kleinberg, J. (2000). Navigation in a Small World , Nature 406(2000), 845.

König, P. & Schillen, J.B. (1991). Stimulus dependent assembly formation of oscillatory responses: I. Synchronization. *Neural Computation, 3*, 155-166.

Kopell, N. (2000). We got rhythm: dynamical systems of the nervous system. *Notices of American Mathematical Society, 47*, 6-16.

Kopp, S. , Jung, B. Lessmann N., Wachsmuth, I. (2003). Max - A Multimodal Assistant in Virtual Reality Construction. *KI-Küstliche Intelligenz* 4(03), 11-17, Bremen: arenDTap Verlag.

Korec, I. (1996). Small universal register machines. *Theoretical Computer Science*, 168, 267–301.

Kosko, B. (1992). *Neural networks and fuzzy systems. A dynamical systems approach to machine intelligence.* Prentice-Hall.

Koza J. et. al. (1999) *Genetic Programming III: Darwin Invention and Problem Solving.* MIT Press, Cambridge, MA.

Krautmacher, M., & Dilger, W. (2004). AIS based robot navigation in a rescue scenario. *Lecture Notes Computer Science*, 3239, pp. 106-118.

Ku, K.W.C., Man, W.M. & Wan, C.S. (1999). Adding learning to cellular genetic algorithms for training recurrent neural networks. *IEEE Transactions on Neural Networks*, 10(2), 239–252.

Kube, R.C., & Bonabeau, E. (2000). Cooperative transport by ants and robots. *Robotics and Autonomous Systems*, Volume 30(1,2), 85-101.

Kuffler, S.W., & Eyzaguirre, C. (1955). Synaptic inhibition in an isolated nerve cell. *Journal of General Physiology,* 39, 155-184.

Kuh, A., & Dickinson, B. W. (1989). *Information Capacity of Associative Memory.* IEEE Transactions on Information Theory, vol. IT-35, 59-68.

Kumar S. (2004). *Investigating Computational Models of Development for the Construction of Shape and Form.* PhD Thesis. Department of Computer Science, University Collage London.

Kumar S. and Bentley P.J. (editors). (2003) *On Growth, Form and Computers.* Elsevier Academic Press. London UK.

Lachaux, J. P, Lutz, A., Rudrauf, D., Cosmelli, D., Le Van Quyen, M., Martinerie, J., & Varela, F. J. (2002). Estimating the time-course of coherence between single-trial brain signals: an introduction to wavelet coherence. *Neurophysiology Clinic*, 32, 157-174.

Laird, J. E., Newell, A., & Rosenbloom, P. S. (1987). Soar: An architecture for general intelligence. *.Artificial Intelligence, 33*(1): 1-64.

Landgon, W. (1996). *Evolution & Genetic Programming Populations.* (Tech. Rep. No. RN/96/125). London: University College.

Lapizco-Encinas, G., & Reggia, J.A. (2005). Diagnostic problem solving using swarm intelligence. In *IEEE Swarm Intelligence Symposium* (pp. 365-372).

Larkum, M. E., Zhu, J. J., & Sakmann, B. (2001). Dendritic mechanisms underlying the coupling of the dendritic with the axonal action potential initiation zone of adult rat layer 5 pyramidal neurons. *Journal of Physiology (London),* 533, 447-466.

Le Ray, D., & Cattaert, D. (1999) Active motor neurons potentiate their own sensory inputs via glutamate- induced long-term potentiation. *Journal of Neuroscience, 19*, 1473-1483.

Le Van Quyen, M., Foucher, J., Lachaux, J. P., Rodriguez, E., Lutz, A., Martinerie, J., & Varela, F. J. (2001). Comparison of Hilbert transform and wavelet methods for the analysis of neuronal synchrony. *Journal Neuroscience Method,* 111: 83–98.

Le Bon-Jego, M., & Cattaert, D. (2002) Inhibitory component of the resistance reflex in the locomotor network of the crayfish. *Journal of Neurophysiology, 88*, 2575-2588.

Le Bon-Jego, M., Cabirol-Pol, M.-J., & Cattaert, D. Activity-dependent plasticity induced in a sensory-motor network by the loss of a proprioceptive afferent in crayfish. (in prep)

Le Bon-Jego, M., Cattaert, D., & Pearlstein, E. (2004) Serotonin enhances the resistance reflex of the locomotor network of the crayfish through multiple modulatory effects that act cooperatively. *Journal of Neuroscience, 24*, 398-411.

Le Bon-Jego, M., Masante-Roca, I., & Cattaert, D. (2006) State-dependent regulation of sensory-motor transmission: role of muscarinic receptors in sensory-motor integration in the crayfish walking system. *European Journal of Neuroscience, 23*, 1283-1300.

Le Ray, D., & Cattaert, D. (1997) Neural mechanisms of reflex reversal in coxo-basipodite depressor motor neurons of the crayfish. *Journal of Neurophysiology, 77*, 1963-1978.

Le Ray, D., Clarac, F., & Cattaert, D. (1997a) Functional analysis of the sensory motor pathway of resistance reflex in crayfish. I. Multisensory coding and motor neuron monosynaptic responses. *Journal of Neurophysiology, 78*, 3133-3143.

Le Ray, D., Combes, D., Dejean, C., & Cattaert, D. (2005) In vivo analysis of proprioceptive coding and its antidromic modulation in the freely behaving crayfish. *Journal of Neurophysiology, 94*, 1013-1027.

Lee, D. (2002). Analysis of phase-locked oscillations in multi-channel single-unit spike activity with wavelet cross-spectrum. *Journal Neuroscience Method*, 115, 67-75.

Lee, D., (2003). Coherent oscillations in neuronal activity of the supplementary motor area during a visuomotor task. *Journal Neuroscience,* 23, 6798-809.

Lee, S.W. (1996). Off-line recognition of totally unconstrained handwritten numerals using multilayer cluster neural network, *IEEE Trans. Pattern Anal. Machine Intell.* 18, 648–652.

Leonov, G.A., Reitmann, V. & Smirnova, V.B. (1992). *Non-local methods for pendulum-like feedback systems,* Germany: Leipzig, Teubner Verlag.

LeRay, D., Fernández, D., Porto, A., Fuenzalida, M., & Buño, W. (2004). Heterosynaptic Metaplastic Regulation of Synaptic Efficacy in CA1 Pyramidal Neurons of Rat Hippocampus. *Hippocampus,* 14, 1011-1025.

Li X, Yao X, Fox J, Jefferys JG. Interaction dynamics of neuronal oscillations analysed using wavelet transforms. J Neurosci Meth 160: 178–185, 2007.

Liedl, T., Sobey, T. L., & Simmel, F. C. (2007). DNA-based nanodevices. *Nanotoday,* 2, 36-41.

Lin, C. T., & Lee, C. S. G. (1996). *Neural fuzzy systems: A neuro-fuzzy synergism to intelligent systems.* NJ, USA: Prentice-Hall, Inc. Upper Saddle River.

Lindenmayer, A. (1968) *Mathematical models for cellular interaction in development: Part I and II.* Journal of Theorical Biology. Vol. 18 pp. 280-299, pp. 300-315.

Linkens, D.A., Taylor, Y., & Duthie, H.L. (1976). Mathematical modeling of the colorectal myoelectrical activity in humans. *IEEE Transactions on Biomedical Engineering,* 23, 101-110.

Lippmann, R. P. (1987). An Introduction to Computing with Neural Nets, *IEEE ASSP Magazine.*

Lipton, R. J. (1995). DNA solutions of hard computational problems. *Science,* 268, 542-545.

Liu, Q-S, Xu, Q., Arcuino, G., Kang, J., & Nedergaard, M. (2004). Astrocytes mdiated activation of neural kinate receptors. *Proceedings of the National Academy of Sciences,* Vol. 101 (pp. 3172-3177). USA

Lo, F. S,, Guido, W., Erzurumlu, R. S. (1999). Electrophysiological properties and synaptic responses of cells in the trigeminal principal sensory nucleus of postnatal rats. *Journal Neurophysiology,* 82, 2765-2775.

Loccufier, M. & Noldus, E. (1995). On the estimation of asymptotic stability regions for autonomous nonlinear systems. *IMA Journal of Mathematical Control & Information, 12,* 91-109.

López-Aguado, L., Ibarz, J. M., Varona, P. & Herreras, O. (2002). Structural inhomogeneities differentially modulate action currents and population spikes initiated in the axon or dendrites. *Journal of Neurophysiology,* 88, 2809-2820.

López-Rodríguez, D., Mérida-Casermeiro, E., Ortiz-de-Lazcano-Lobato, J. M., & López-Rubio, E. (2006). *Image Compression by Vector Quantization with Recurrent Discrete Networks.* Lecture Notes in Computer Science, 4132, 595-605.

Lorente de Nó, R. (1947). Action potential of the motoneurons of the hypoglossus nucleus. *Journal Cell Comparative Physiology,* 29, 207-288.

Lv, C., Cao, P. Cohen, E. Li, L., & Shenker, S. (2002). Search and *replication in unstructured peer-to-peer networks,* ACM, Sigmetrics.

Lyman, C.P., Willis, J.S., Malan, A., & Wang L.C.H. (Eds.). (1982). *Hibernation and Torpor in Mammals and Birds.* San Diego: Academic Press.

Ma, P. M. (1991). The barrelettes--architectonic vibrissal representations in the brainstem trigeminal complex of the mouse. I. Normal structural organization. *Journal Comparative Neurology,* 309, 161-199.

Maaref, H. & Barref, C. (2002). Sensor-based navigation of a mobile robot in an indoor environment. *Robotics and Autonomous Systems,* Elsevier, Vol. 38, pp. 1-18.

Maass, W., &. Bishop, C., eds. (1999). *Pulsed Neural Networks.* MIT Press.

Mackey M. & Glass L. (1977). Oscillation and chaos in physiological control systems. *Science,* 197-287.

Magee, J. C., & Cook, E. P. (2000). Somatic EPSP amplitude is independent of synapse location in hippocampal pyramidal neurons. *Nature Neuroscience,* 3, 895-903.

Magnenat-Thalmann, N. & Volino, P (2005). From early draping to haute couture models: 20 years of research. *Visual Computing,* 21:506–519,

Mainen, Z. F., & Sejnowski, T. J. (1995). Reliability of spike timing in neocortical neurons. *Science.* 268, 1503-1506.

Mainen, Z. F., & Sejnowski, T.J. (1996). Influence of dendritic structure on firing pattern in model neocortical neurons. *Nature* 382, 363-66.

Mainen, Z.F., Joerges, J., Huguenard, J.R., & Sejnowski, T.J. (1995). A model of spike initiation in neocortical pyramidal neurons. *Neuron* 15:1427-1439.

Malkin, I.G. (1952). *Stability of Motion.* (in Russian) U.R.S.S.: Moscow, Gostekhizdat.

Malmierca, E. & Nuñez, A. (1998). Corticofugal action on somatosensory response properties of rat nucleus gracilis cells. *Brain Research,* 810, 172-180.

Malmierca, E. & Nuñez, A. (2004). Primary somatosensory cortex modulation of tactile responses in nucleus gracilis cells of rats. *European Journal Neuroscience,* 19, 1572-1580.

Malmierca, E. & Nuñez, A. (2007). Corticofugal modulation of sensory information. *Advances Anatomy Embryology Cell Biology,* 187, 1-74.

Mandiau, R., Strugeon, G. (2001). Multi-agent Systems (in French). In Mandiau, R. and Strugeon, G. (Eds.), *Techniques De L'ingénieur* (Ed.), pp. 1-17, Paris.

Marcus, C.M. & Westervelt, R.M. (1989). Stability of analog neural networks with delay. *Physical Review A, 39,* 347-359.

Mariño, J., Canedo A. & Aguilar J. (2000) Sensorimotor cortical influences on cuneate nucleus rhythmic activity in the anesthetized cat. *Neuroscienc,e* 95, 657-673.

Martín, E.D., & Araque, A. (2006). Astrocytes and The Biological Neural Networks. In J. Rabuñal (Ed.), *Artificial Neural Networks in Real-Life Applications* (pp. 22-45) Hershey PA, USA: Idea Group Inc.

Martin, S. J., Grimwood, P. D., & Morris, R. G. (2000). Synaptic plasticity and memory: an evaluation of the hypothesis. *Annual Review Neuroscience,* 23, 649–711.

Martínez, A. & Goddard, J. (2001). Definición de una red neuronal para clasificación por medio de un programa evolutivo. *Revista Mexicana de Ingeniería Biomédica* 22 nº 1, 4-11.

Mastorocostas, P., & Theocharis, J. (2007). A dynamic fuzzy neural filter for separation of discontinuous adventitious sounds from vesicular sounds. *Computers in Biology and Medicine, 37*(1), 60-69.

Masukawa, L. M., & Prince, D. A. (1984). Synaptic control of excitability in isolated dendrites of hippocampal neurons. *Journal of Neuroscience,* 4, 217-227.

Mataric, M. (1995). Designing and understanding adaptive group behavior. *Adaptive Behavior. 4(1),* 51-80.

Matsuoka, K. (1987). Mechanisms of frequency and pattern control in the neural rhythm generators. *Biological Cybernetics,* 56, 345-353.

Maurer, A., Hersch, M. & Billard, A. G. (2005). Extended Hopfield Network for Sequence Learning: Application to Gesture Recognition, *Proceedings of 15th International Conference on Artificial Neural Network,* 493-498.

Mayr, E. & Ashlock, P. D. (1991). *Principles of Systematic Zoology.* New York, NY: McGraw-Hill.

McAdams, H. H., & Arkin, A. (1999). It's a noisy business! Genetic regulation at the nanomolar scale. *Trends in Genetic,* 15, 65-69.

McCulloch, W.S., & Pitts, W. (1943). A logical calculus and the ideas immanent in the nervous activity. *Bulletin of Mathematical Biophysics,* 5, 115-133.

McEliece, R. J., Posner, E. C., Rodemich, E. R., & Venkatesh, S. S. (1990). *The Capacity of the Hopfield Associative Memory.* IEEE Transactions on Information Theory, vol. IT-33, no. 4, 461-482.

Mel, B. (1999) Why have dendrites? A computational perspective. In G. Stuart, N. Spruston and M. Häusser (Eds.), *Dendrites* (pp 271-289). OUP, New York.

Menezes, T. & Costa, E. (2007). Designing for Surprise. *Advances in Artificial Life. LNAI 4648.* Springer, Berlin.

Mengshoel, O.J., & Goldberg, D.E. (1999). Probabilistic Crowding: Deterministic Crowding with Probabilistic Replacement. *Proceedings of Genetic and Evolutionary Computation* (pp. 409-416)

Mérida Casermeiro, E., Muñoz-Pérez, J. & García-Bernal, M.A. (2002). *An Associative Multivalued Recurrent Network,* IBERAMIA 2002, 509-518.

Mérida-Casermeiro, E., & López-Rodríguez, D. (2004). *Multivalued Neural Network for Graph MaxCut Problem,* ICCMSE, 1, 375-378.

Mérida-Casermeiro, E., & López-Rodríguez, D. (2005). *Graph Partitioning via Recurrent Multivalued Neural Networks.* Lecture Notes in Computer Science, 3512:1149-1156.

Mérida-Casermeiro, E., & Muñoz-Pérez, J. (2002). *MREM: An Associative Autonomous Recurrent Network.* Journal of Intelligent and Fuzzy Systems, 12 (3-4), 163-173.

Mérida-Casermeiro, E., Galán-Marín, G., & Muñoz-Pérez, J. (2001). *An Efficient Multivalued Hopfield Network for the Travelling Salesman Problem*. Neural Processing Letters, 14:203-216.

Mérida-Casermeiro, E., Muñoz-Pérez, J., & Benítez-Rochel, R. (2001). *A recurrent multivalued neural network for the N-queens problem*, Lecture Notes in Computer Science 2084, 522-529.

Mérida-Casermeiro, E., Muñoz-Pérez, J., & Domínguez-Merino, E. (2003). *An N-parallel Multivalued Network: Applications to the Travelling Salesman Problem*. Computational Methods in Neural Modelling, Lecture Notes in Computer Science, 2686, 406-413.

Meystel, A. (1991). *Autonomous Mobile Robots: Vehicles With Cognitive Control*, World Scientific, 1991.

Miao, Y., Hoppe, H. U. & Pinkwart, N. (2006). Naughty Agents Can Be Helpful: Training Drivers to Handle Dangerous Situations in Virtual Reality. In Kinshuk et al (Eds.), *Proceedings of the 6th IEEE International Conference on Advanced Learning Technologies* (p. 735-739).

Michalewicz, Z. (1996). *Genetic algorithms + data structures = evolution programs* (3rd rev. and extended ed.). Berlin: Springer-Verlag.

Michalewitz, Z. (1992). *Genetic Algorithms + Data Structures = Evolutionary Programs*. New York: Springer Verlag.

Miller, B., & Shaw, M., (1995). *Genetic Algorithms with Dynamic Niche Sharing for Multimodal Function Optimization*. (Tech. Rep. No. IlliGAL 95010). University of Illinois.

Miller, J. F. & Thomson, P. (2000). Cartesian Genetic Programming. *Proceedings of the European Conference on Genetic Programming* (pp. 121-132). London, UK: Springer-Verlag.

Miller, J. P., Rall, W., & Rinzel, J. (1985). Synaptic amplification by active membrane in dendritic spines. *Brain Research*, 325, 325-330.

Minar, N., Burkhart, R., Langton, C., & Askenazi, M. (1996). *The swarm simulation system, a toolkit for building multi-agent simulations*, from http://www.santafe.edu.

Minsky, M. (1986). *The Society of Mind*. Simon and Schuster, New York.

Mitchell, M. (1996). *An introduction to genetic algorithms*. Cambridge, Mass.: MIT Press.

Mjolsness E., Sharp D.H. and Reinitz J. (1995) *A Connectionist Model f Development*. Journal of Theoretical Biology 176 pp. 291-300.

Morato, T., Cheung, W.W.L., & Pitcher, T.J. (2006). Vulnerability of seamount fish to fishing: fuzzy analysis of life-history attributes. *Journal of Fish Biology*, *68*(1), 209-221.

Moser, J. (1967). On nonoscillating networks. *Quarterly Applied Mathematic, 25*, 1-9.

Murata, A. (2005). An attempt to evaluate mental workload using wavelet transform of EEG. *Human Factors*, 47, 498–508.

Musse, S., Silva, A., Roth, B., Hardt, K., Barros, L., Tonietto, L. And Borba, M. (2003) "PetroSim: A Framework to Simulate Crowd Behaviors in Panic Situations". In *MAS 2003 - Modeling & Applied Simulation*, Bergeggi, Italy.

Nadkarni, S., & Jung, P. (2007). Modeling synaptic transmission of the tripartite synapse. *Physical Biology*, 4, 1-9.

Nagl S.B., Parish J.H., Paton R.C. and Warner G.J. (1998) *Macromolecules, Genomes and Ourselves*. Chapter in *Computation in cells and tissues.Perspective and tools of thought*. Paton R., Bolouri H., Holcombe M., Parish J.H. and Tateson R. Springer Press

Nelson, A.L., Grant, E., Galeotti, J.M., & Rhody, S. (2004). Maze exploration behaviours using an integrated evolutionary robotic environment. *Robotic and Autonomous Systems*, 46, pp. 159-173.

Newell, A. (1990). *Unified Theories of Cognition*. Harvard Press: Cambridge, MA.

Nishimura, M., Kitamura, S. & Hirai, K. (1969). A Lyapunov Functional for Systems with Multiple Non-linearities and Time Lags, Technological Reports, Japan: Osaka University, *19*(860), pp. 83-88.

Noldus, E. & Loccufier, M. (1995). An application of Lyapunov's method for the analysis of neural networks. *Journal of Computation and Applied Mathematics, 50*, 425-432.

Noldus, E. Vingerhoeds, R & Loccufier, M. (1994). Stability of analogue neural classification networks. *International Journal of Systems Science 25*(1), 19-31.

Nolfi S. (2006). *Behaviour as a complex adaptive system: On the role of self-organization in the development of individual and collective behaviour*. Unpublished yet, internal communication with author.

Nolfi, S. & Floreano, D. (2000). *Evolutionary Robotics: The Biology, Intelligence, and Technology of Self-Organizing Machines*. MA: MIT Press/Bradford Books. Ref.: Phototaxis: pp. 121-152; Obstacle avoidance/Navigation: pp. 69-92; Learning: pp.153-188.

Nolfi, S. (1997). *Using emergent modularity to develop control systems for mobile robots.* Adaptive Behaviour (5) 3-4: 343-364.

Olaru, C., & Wehenkel, L. (2003). A complete fuzzy decision tree technique. *Fuzzy Sets and Systems, 138*, 221-254.

Olavarria, J., Van Sluyters, R. C., & Killackey, H. P. (1984). Evidence for the complementary organization of callosal and thalamic connections within rat somatosensory cortex. *Brain Research,* 291, 364-368.

Ono, I., & Kobayashi, S. (1999). A real-coded genetic algorithm for function optimization using unimodal normal distribution. *Proceedings of International Conference on Genetic Algorithms* (pp. 246-253)

Ortiz, D., Hervás, C., & García, N. (2005). CIXL2: A crossover operator for evolutionary algorithms based on population features. *Journal of Artificial Intelligence Research.*

Osório, F. S.; Musse, S. R.; Santos, C. T.; Heinen, F.; Braun, A. & Silva, A. T. (2005). Intelligent Virtual Reality Environments (IVRE): Principles, Implementation, Interaction, Examples and Practical Applications. In: Fischer, Xavier. (Org.). *Virtual Concept (Proceedings - Tutorials).* Biarritz, França, 1: 1-64.

Ozturk, Y., & Abut, H. (1997). *System of associative relationships (SOAR),* In Proceedings of ASILOMAR.

Panait, L. & Luke, S. (2005). *Cooperative Multi-Agent Learning: The State of the Art. Autonomous Agents and Multi-Agent Systems.* 11(3) : 387–434. Springer 2005.

Papagiannakis, G., Schertenleib, S., O'Kennedy, B., Arevalo-Poizat, M., Magnenat-Thalmann, N., Stoddart, A. & Thalmann, D. (2005). Mixing Virtual and Real Scenes in the site of ancient Pompeii. *Journal Of Computer Animation and Virtual Worlds,* 16(1)11 – 24. John Wiley and Sons Ltd. Chichester, UK.

Parisi, G. (1986). *Asymmetric Neural Networks and the Process of Learning,* J. Phys. A: Math. and Gen., vol 19, L675-L680.

Park, K. J., Jones, G., & Ransome, R.D. (2000). Torpor, arousal and activity of hibernating Greater Horseshoe Bats (*Rhinolophus ferrumequinum*). *Functional Ecology, 14*, 580-588.

Parnas, I., Shahrabany-Baranes, O., Feinstein, N., Grant, P., Adelsberger, H., & Dudel, J. (1998) Changes in the ultrastructure of surviving distal segments of severed axons of the rock lobster. *Journal of Experimental Biology, 201*, 779-791.

Parunak, H., Purcell, M. & O'Connell, R. (2002). Digital pheromones for autonomous coordination of swarming UAV's. In *Proceedings 1st UAV Conference.*

Pasti, L., Volterra, A., Pozzan, R., & Carmignoto, G. (1997). Intracellular calcium oscillations in astrocytes: a highly plastic, bidirectional form of communication between neurons and astrocytes in situ. *Journal of Neuroscience,* 17, 7817-7830.

Păun, A., & Păun, Gh. (2007). Small universal spiking neural P systems. *BioSystems, 90*(1), 48–60.

Păun, Gh. (2000). Computing with membranes. *Journal of Computer and System Sciences, 61*(1), 108–143, and Turku Centre for Computer Science (TUCS) Report 208, November 1998.

Păun, Gh. (2002). *Membrane Computing. An Introduction.* Berlin: Springer.

Păun, Gh. (2007). Twenty six research topics about spiking neural P systems. In (Gutierrez-Naranjo et al., eds., (2007)), 263–280.

Păun, Gh., Perez-Jimenez, M.J., & Rozenberg, G. (2006a). Spike trains in spiking neural P systems. *International Journal of Foundations of. Computer Science, 17*(4), 975–1002.

Păun, Gh., Perez-Jimenez, M.J., & Rozenberg, G. (2006b). Infinite spike trains in spiking neural P systems. Submitted.

Pavlov, A. N., Makarov, V. A., Mosekilde, E., & Sosnovtseva, O. V. (2006) Application of wavelet-based tools to study the dynamics of biological processes", *Briefings Bioinformatics,* 7, 375-389.

Pavlov, A. N., Tupitsyn, A. N., Makarov, V. A., Panetsos, F., Moreno, A., Garcia-Gonzalez, V., & Sanchez-Jimenez, A. (2007). Tactile information processing in the trigeminal complex of the rat. Proceedings of SPIE: Complex Dynamics and Fluctuations in Biomedical Photonics IV. Vol 6436, doi:10.1117/12.709155

Paxinos, G., & Watson, C. (1986). The Rat Brain in Stereotaxic Coordinates, Academic Publishing, New York.

Pazos, J. (2000). El Criptoanálisis en el Desciframiento del Código Genético Humano. In J. Dorado et al. (Eds.), *Protección y Seguridad de la Información* (pp. 267-351). Santiago de Compostela: Fundación Alfredo Brañas.

Pearlmutter, B. A. (1990). Dynamic Recurrent Neural Networks. *Technical Report CMU-CS*, School of Computer Science, Carnegie Mellon University, 88-191.

Pearson, K.G. (1976) The control of walking. *Scientific American,* 235, 72-86.

Pearson, K.G. (1993). Common principles of motor control in vertebrates and invertebrates. *Annual Review of Neuroscience,* 16, 265-297.

Pedrycz, W., & Vukovich, G. (2002). On Elicitation of Membership Functions, *IEEE Transactions on Systems, Man, and Cybernetics - Part A: Systems and Humans, 32*(6), 761- 767

Penrose, L. J. (1974). Máquinas que se Autorreproducen. In R. Canap, et al., *Matematicas en las Ciencias del Comportamiento* (pp: 270-289). Madrid: Alianza Editorial, S.A.

Percival, D. P. (1995). On estimation of the wavelet variance. *Biometrika,* 82, 619–631.

Perea, G., & Araque, A. (2002). Communication between astrocytes and neurons: a complex language. *Journal of Physiology.* Paris: Elsevier Science.

Perea, G., & Araque, A. (2005). Properties of synaptically evoked astrocyte calcium signal reveal synaptic information processing by astrocytes. *The Journal of Neuroscience,* 25(9), 2192-2203.

Perea, G., & Araque, A. (2007). Astrocytes Potentiate Transmitter Release at Single Hippocampal Synapses. *Science,* 317(5841), 1083 – 1086.

Perez-Otaño, I., & Ehlers, M.D. (2005) Homeostatic plasticity and NMDA receptor trafficking. *Trends in Neuroscience, 28,* 229-238.

Perkel, D. H., Gerstein, G. L., & Moore, G. P. (1967). Neuronal spike trains and stochastic point processes. II. Simultaneous spike trains. *Biophysical Journal,* 7, 419–440.

Perkel, D.H. (1976). A computer program for simulating a network of interacting neurons: I. Organization and physiological assumptions. *Computers and Biomedical Research,* 9, 31-43.

Peschanski, M. (1984). Trigeminal afferents to the diencephalon in the rat. *Neuroscience,* 12, 465-487.

Petersen, K., Spreitzer, M., Terry, D., Theimer, M., & Demers, A. (1997). *Flexible Update Propagation for Weakly Consistent Replication,* Paper presented at the 16th ACM Symposium on Operating System Principles, Saint-Malo, France.

Petrowski, A. (1996). A Clearing Procedure as a Niching Method for Genetic Algorithms. *International Conference on Evolutionary Computation.* IEEE Press. Nagoya, Japan.

Piotrowski, A., Rowinski, P. M., & Napiorkowski, J. J. (2006). *Assessment of longitudinal dispersion coefficient by means of different neural networks.* Paper presented at the 7th International Conference on Hydroinformatics, Nice, FRANCE.

Podgorelec, V., Kokol, P., Stiglic, B., & Rozman, I. (2002). Decision trees: an overview and their use in medicine. *Journal of Medical Systems, 26,* 445-63.

Popov, V. M. (1966). *Hyperstability of Control Systems.* (in Romanian) Romania: Bucharest, Academia Publishers House, (English improved version by Springer-Verlag, 1973).

Popov, V.M. (1979). Monotonicity and Mutability. *Journal of Differential Equations, 31*(3), 337-358.

Porto, A. (2004). *Computational Models for optimizing the Learning and the Information Processing in Adaptive Systems,* Ph.D. Thesis, Faculty of Computer Science, University of A Coruña, Spain.

Porto, A., & Pazos, A. (2006). Neuroglial behaviour in computer science. In J. Rabuñal (Ed.): *Artificial Neural Networks in Real-Life Applications* (pp. 1-21). Hershey, PA: Idea Group Inc.

Porto, A., Araque, A., & Pazos, A. (2005). Artificial Neural Networks based on Brain Circuits Behaviour and Genetic Algorithms. *LNCS.* 3512, 99-106.

Porto, A., Araque, A., Rabuñal, J., Dorado, J., & Pazos, A. (2007). A New Hybrid Evolutionary Mechanism Based on Unsupervised Learning for Connectionist Systems. *Neurocomputing,* 70(16-18), 2799-2808.

Qiu, F., Guo, L., Wen, T.J., Ashlock, D.A. & Schnable, P.S. (2003). DNA Sequence-Based Barcodes for Tracking the Origins of Ests from a Maize CDNA Library Constructed Using Multiple MRNA Sources. *Plant Physiology,* vol. 133, 475-481.

Querrec, R., Buche, C., Maffre, E. & Chevaillier, P. (2004). .Multiagents systems for virtual environment for training: application to fire-fighting. *Special issue Advanced Technology for Learning of International Journal of Computers and Applications* 1: p. 25-34. ACTA Press.

Quevedo, F. de. (1995). *Poesía Completa I* (pp: 43-44). Madrid: Turner Libros, S.A..

Quiroga, R., Garcia, H. (2003). Single-trial event-related potentials with wavelet denoising. *Clinical Neurophysiology*, 114, 376–390.

Rabuñal, J. (1998). *Entrenamiento de Redes de Neuronas Artificiales con Algoritmos Genéticos*. Tesis de Licenciatura. University of A Coruña. Spain.

Rabuñal, J., Dorado, J., Pazos, A., Pereira, J. & Rivero, D. (2004). A New Approach to the Extraction of Rules from ANN and their Generalization Capacity through GP, Neural Computation, 16(7), 1483-1523.

Ramón y Cajal, S. (Ed.). (1911). *Histologie du sisteme nerveux de l'homme et des vertebres*. Maloine, Paris.

Ransome, R.D. (1971). The effect of ambient temperature on the arousal frequency of the hibernating greater horseshoe bat, *Rhinolophus ferrumequinum*, in relation to site selection and the hibernation state. *Journal of Zoology, London*, *164*, 353-371.

Rao, S. A., & Georgeff. M. P. (1995) BDI Agents: From Theory to Practice. *Proc. of 1st international Conference on Multiple Agent System.*

Rapp, M., Yarom, Y., & Segev, I. (1996). Modeling back propagation action potential in weakly excitable dendrites of neocortical pyramidal cells. *P.N.A.S.* 93, 11985-11990.

Răsvan, V. & Danciu, D. (2004). Neural networks - global behavior versus delay. In *Periodica Politechnica,* University of Timişoara, Transactions on Automatic Control and Computer Science, *49* (63), No. 2, (pp. 11-14). Romania: Timişoara, Politehnica Publishers House.

Răsvan, V. (1998). Dynamical Systems with Several Equilibria and Natural Lyapunov Functions, *Archivum mathematicum*, *34*(1), [EQUADIFF 9], 207-215.

Rechenberg, I. (1973). *Evolutionsstrategie: Optimierung Technischer Systeme nach Prinzipien der Biologischen Evolution.* Stuttgart, Germany: Fromman-Holzboog Verland.

Reeves, C. R., & Rowe, J. E. (2003). *Genetic algorithms: principles and perspectives: A guide to GA theory.* Boston: Kluwer Academic Publishers.

Ressot, C., Collado, V., Molat, J. L., & Dallel, R. (2001). Strychnine alters response properties of trigeminal nociceptive neurons in the rat. Journal Neurophysiology 86, 3069-3072.

Reyes, A. (2001). Influence of dendritic conductances on the input-output properties of neurons. *Annual Review of Neuroscience,* 24, 653-675.

Reynolds, C. (1987). Flocks, herds, and schools: A distributed behavioral model. *Computer Graphics*, 21(4), 25-34.

Reynolds, C. (1992). An Evolved, Vision-Based Behavioral Model of Coordinated Group Motion. In J. Meyer, H. L. Roiblat, & S. Wilson (Eds.), *From Animals to Animats 2* (pp. 384-392). Cambridge, MA: The MIT Press.

Reynolds, C. (1999). Steering behaviors for autonomous characters. In *Proc. Game Developers Conference* (pp. 763-782).

Reynolds, J.H., Chelazzi, L. & Desimone, R. (1999). Competitive mechanisms subserve attention in macaque areas V2 and V4. *Journal Neuroscience*, 19, 1736-1753.

Rinaudo, K., Bleris, L., Maddamsetti, R., Subramanian, S., Weiss, R., & Benenson, Y. (2007). A universal RNAi-based logic evaluator that operates in mammalian cells. *Nature Biotechnology*, doi: 10.1038/nbt1307.

Rist, T., André, E. & Baldes, S. (2003). A flexible platform for building applications with life-like characters. *Proceedings of the 8th international conference on Intelligent user interfaces*, p. 158-165. ACM Press New York, NY, USA

Ro, D., Paradise, E. M., Ouellet, M., Fisher, K. J., Newman, K. L, Ndungu, J. M., Ho, K. A., Eachus, R. A., Ham, T. S., Kirby, J., Chang, M. C. Y., Withers, S. T., Shiba, Y., Sarpong, R., & Keasling, J. D. (2006). Production of the antimalarial drug precursor artemisinic acid in engineered yeast. *Nature*, 440, 940-943.

Roberts, A., & Tunstall, M.J. (1990). Mutual re-excitation with post-inhibitory rebound: A simulation study on the mechanisms for locomotor rhythm generation in the spinal cord of Xenopus embryos. *European Journal of Neuroscience,* 2, 11-23.

Rodriguez, A., & Reggia, J.A. (2004). Extending self-organizing particle systems to problem solving. *Artificial Life*, 10(4), 379-395.

Rodriguez, A., & Reggia, J.A. (2005). Collective-movement teams for cooperative problem solving. *Integrated computer-aided engineering*, 12(3), 217--235.

Rodríguez-Patón, A. (1999). *Variantes de la concatenación en computación con ADN.* PhD thesis, Universidad Politécnica de Madrid.

Roitt, I. (1997). Essential Immunology. Ninth Edition. *Pub. Blackwell Science.*

Roll, J.P., Bergenheim, M., & Ribot-Ciscar, E. (2000) Proprioceptive population coding of two-dimensional limb movements in humans: II. Muscle-spindle feedback during "drawing-like" movements. *Experimental Brain Research, 134*, 311-321.

Rossi-Durand, C. (1993) Peripheral proprioceptive modulation in crayfish walking leg by serotonin. *Brain Research, 632*, 1-15.

Rossignol, S., Dubuc, R., & Gossard, J.P. (2006) Dynamic sensorimotor interactions in locomotion. *Physiological Reviews, 86*, 89-154.

Rothlauf, F. (2006). Representations for genetic and evolutionary algorithms (2nd ed.). Heidelberg: Springer.

Rowinski, P. M., Piotrowski, A., & Napiorkowski, J. J. (2005). Are artificial neural network techniques relevant for the estimation of longitudinal dispersion coefficient in rivers? *Hydrological Sciences Journal, 50*(1), 175-187.

Rowland, B.A., Quessy, S., Stanford, T.R., & Stein, B.E. (2007) Multisensory integration shortens physiological response latencies. *Journal of Neuroscience, 27*, 5879-5884.

Rutherford, J. C. (1994). *River mixing.* Chichester: Wiley.

Saavedra, L., García, A. & Barbas, C. (2000). Development and validation of a capillary electrophoresis method for direct measurement of isocitric, citric, tartaric and malic acids as adulteration markers in orange juice. *Journal of Chromatography 881(1-2)* (pp. 395-401)

Sakamoto, K., Gouzu, H., Komiya, K., Kiga, D., Yokoyama, S., Yokomori, T., & Hagiya, M. (2000). Molecular computation by DNA hairpin formation. *Science* , 288, 1223-1226.

Schiff, S., Aldroubi, A., Unser, M., & Sato, S. (1994). Fast wavelet transformation of EEG. *Electroencephalography Clinical Neurophysiology*, 91, 442–455.

Schmitz, J., Dean, J., Kindermann, T., Schumm, M., & Cruse, H. (2001) A biologically inspired controller for hexapod walking: simple solutions by exploiting physical properties. *Biological Bulletin, 200*, 195-200.

Schöner, G., Jiang, W.Y., & Kelso, J.A.S. (1990). A synergetic theory of quadrupedal gaits and gait transitions. *Journal of Theoretical Biology,* 142, 359-391.

Schoonderwoerd, R., Holland, O.& Bruten, J.. Ant-like agents for load balancing in telecommunications networks. In *AGENTS '97: Proceedings of the first international conference on Autonomous agents*, (pp. 209–216). New York, NY.

Schreiber, J. V, & Schmitz, A. (2000). Surrogate time series. *Physica D* 142, 646-652.

Schrödinger, E. (1944). *What is Life?* Cambridge: Cambridge Universities Press. Spanish translation (1983). *¿Qué es la Vida?* (pp: 40-42). Barcelona: Tusquets Editores, S.A.

Schwab, B. (2004). *AI Game Engine Programming (Game Development Series).* Charles River Media.

Seelig, G., Soloveichik, D., Zhang, D. Y., & Winfree, E. (2006). Enzyme-free nucleic acid logic circuits. *Science*, 314, 1585-1588.

Segev, I., & London, M. A. (1999) Theoretical view of passive and active dendrites. In G. Stuart, N. Spruston and M. Häusser (Eds.), *Dendrites* (pp 205-230). OUP, New York.

Seo, I. W., & Cheong, T. S. (1998). Predicting longitudinal dispersion coefficient in natural streams. *Journal of Hydraulic Engineering, 124*(1), 25-32.

Seth, A.K. & Edelman, G. (2004). M. Theoretical neuroanatomy: Analyzing the structure and dynamics of networks in the brain. *Complex Networks.* E. Ben-Naim, H. Fraunfelder, & Z. Toroczkai (eds). p. 487-518, Springer-Verlag, Berlin. 2004.

Sevin, E., Thalmann, D. (2005a). A Motivational Model of Action Selection for Virtual Humans, In *Computer Graphics International (CGI)*, IEEE Computer, SocietyPress, New York.

Sevin, E., Thalmann, D. (2005b). An Affective Model of Action Selection for Virtual Humans, In *Proceedings of Agents that Want and Like: Motivational and Emotional Roots of Cognition and Action symposium at the Artificial Intelligence and Social Behaviors Conference (AISB'05)*, University of Hertfordshire, Hatfield, England.

Shadlen, M. N., & Newsome, W. T. (1998). The variable discharge of cortical neurons: implications for connectivity, computation, and information coding. *Journal Neuroscience*, 18, 3870-3879.

Shapiro, E., & Benenson, Y. (2006). Bringing DNA computers to life. *Scientific American*, 294, 44-51.

Shen, G. Y., Chen, W. R., Midtgaard, J., Shepherd, G. M., & Hines M. L. (1999). Computational analysis of action potential initiation in mitral cell soma and dendrites based on dual patch recordings. *Journal of Neurophysiology*, 82, 3006-3020.

Shen, X., & De Wilde, F. (2006) Long-term neuronal behaviour caused by two synaptic modification mechanisms.

Department of Computer Science, Heriot-Watt University, United Kingdom.

Shin, J. S., & Pierce N. A. (2004). A synthetic DNA walker for molecular transport. *Journal of the American Chemical Society*, 126, 10834-10835.

Shuler, M. G., Krupa, D. J., Nicolelis, M. A. (2001). Bilateral integration of whisker information in the primary somatosensory cortex of rats. *Journal Neuroscience*, 21, 5251-5261.

Simmel, F. C., & Dittmer, W. U. (2005). DNA nanodevices. *Small*, 1, 284-299.

Singh, M. P., Chengxiang, Z., & Dasgupta, C. (1995). *Analytic Study of the Effects of Synaptic Asymmetry*, Phys. Rev. E, vol. 52, 5261-5272.

Smith, A.T., Singh, K.D. & Greenlee, M.W. (2000). Attentional suppression of activity in the human visual cortex. *Neuroreport*, 7, 271-277.

Smith, R. L. (1973). The ascending fiber projections from the principal sensory trigeminal nucleus in the rat. *Journal Comparative Neurology*, 148, 423-436.

Sneath, P. H. A. & Sokal, R. R. (1973). *Numerical Taxonomy; the Principles and Practice of Numerical Classification.* San Francisco, CA: W.H. Freeman.

Softky, W. R., & Koch, C. (1993). The highly irregular firing of cortical cells is inconsistent with temporal integration of random EPSPs. *Journal of Neuroscience*, 13, 334-350.

Song, S., Miller, K. D., & Abbott, L. F. (2000). Competitive Hebbian learning through spike-timing-dependent synaptic plasticity. Nature Neuroscience 3, 919–926.

Spencer, W.A., & Kandel, E.R. (1961). Electrophysiology of hippocampal neurons. IV. Fast prepotentials. *Journal of Neurophysiology* 24, 272-285.

Spigelman, I., Zhang, L., & Carlen, P. L. (1992). Patch-clamp study of postnatal development of CA1 neurons in rat hippocampal slices: membrane excitability and K^+ currents. *Journal of Neurophysiology* 68, 55-69.

Srinivasa, K. G., Venugopal, K. R., & Patnaik, L. M. (2007). A self-adaptive migration model genetic algorithm for data mining applications. *Information Sciences, 177*(20), 4295-4313.

Stanley, K. & Miikkulainen, R. (2003) *A Taxonomy for Artificial Embryogeny.* In Proceedings Artificial Life 9, pp. 93-130. MIT Press.

Stauffer, A., Mange, D. & Rossier J. (2007). Self-organizing Systems Based on Bio-inspired Properties. *Advances in Artificial Life. LNAI 4648.* Springer, Berlin.

Stein, P.S.G. (1978). Motor systems with specific reference to the control of locomotion. *Annual Review of Neuroscience,* 1, 61-81.

Stelling, J., Sauer, U., Szallasi, Z., Doyle III, F., & Doyle, J. (2004). Robustness of cellular functions. *Cell*, 118, 675-685.

Stevens, C. F., & Zador, A. M. (1998). Input synchrony and the irregular firing of cortical neurons. *Nature Neuroscience*, 1, 210-217.

Stojanovic, M. N., & Kolspashchikov, D. M. (2004). Modular aptameric sensors. *Journal of the American Chemical Society*, 126, 9266-9270.

Stojanovic, M. N., & Stefanovic, D. (2003). A deoxyribozyme-based molecular automaton. *Nature Biotechnology*, 21, 1069-1074.

Stuart, G., & Sakmann, B. (1995). Amplification of EPSPs by axosomatic sodium channels in neocortical pyramidal neurons. *Neuron*, 15,1065-76.

Stuart, G., Schiller, J. & Sakmann, B. (1997). Action potential initiation and propagation in rat neocortical pyramidal neurons. *Journal of Physiology (London)*, 505, 617-632.

Sudd, J. H. (1965). The transport of prey by ant. *Behavior*, 25, 234-271.

Sudd, J.H. (1960). The transport of prey by an ant pheidole crassinoda. *Behavior*, 15, 295-308.

Sugitani, M., Yano, J., Sugai, T., & Ooyama, H. (1990). Somatotopic organization and columnar structure of vibrissae representation in the rat ventrobasal complex. *Experimental Brain Research*, 81, 346-352.

Sutton, R.S., & Barto, A.G. (1998). *Reinforcement Learning: An Introduction.* The MIT Press.

Sycara, K. (1998). *Multiagent systems*, Artificial Intelligence Magazine, Association for the Advancement of Artificial Intelligence, 10(2), 79–93.

Syswerda, G. (1991). A Study of Reproduction in Generational and Steady State Genetic Algorithms. *Foundations of Genetic Algorithms* (pp. 94-101). San Francisco, CA: Morgan Kaufmann.

Taft, R. (1955). An introduction: Eadweard Muybridge and his work. In E. Muybridge (Ed.), *The human figure in motion* (pp.7-14). New York: Dover Publications.

Taga, G. (1995). A model of the neuro-musculo-skeletal system for human locomotion – I. Emergence of basic gait. *Biological Cybernetics,* 73, 97-111.

Taga, G., Yamaguchi, Y., & Shimizu, H. (1991). Self-organized control of bipedal locomotion by neural oscillators in unpredictable environment. *Biological Cybernetics,* 65, 147-159.

Taha M.A. & Hanna A. S. (1995). Evolutionary neural network model for the selection of pavement maintenance strategy. *Transportation Res. Rec.* 1497, 70–76.

Takahashi, K., Yaegashi, S., Kameda, A., & Hagiya., M. (2005). *Chain reaction systems based on loop dissociation of DNA.* Paper presented at the DNA 11, Eleventh International Meeting on DNA Based Computers, Preliminary Proceedings, London, Ontario, Canada.

Tay, J. & Jhavar, A. (2005). CAFISS: A complex adaptive framework for immune system simulation. *2005 ACM Symposium on Applied Computing (SAC'05),* pp. 158-164.

Tayfur, G. (2006). Fuzzy, ANN, and regression models to predict longitudinal dispersion coefficient in natural streams. *Nordic Hydrology, 37*(2), 143-164.

Tayfur, G., & Singh, V. P. (2005). Predicting longitudinal dispersion coefficient in natural streams by artificial neural network. *Journal of Hydraulic Engineering, 131*(11), 991-1000.

Taylor, D.J., Green, N. P. & Stout, G.W. (2003). *Biological Sciences*, 3rd ed. United Kingdom, Cambridge University Press.

Taylor, G. (1954). The dispersion of Matter in turbulent flow through a pipe. *Proceedings of the Royal Society of London.Series A, Mathematical and Physical Sciences, 223*(1155), 446-468.

Téllez, R. & Angulo, C. (2004). Evolving cooperation of simple agents for the control of an autonomous robot. *Proceedings of the 5th IFAC Symposium on Intelligent Autonomous Vehicles (IAV04)*, Lisbon, Portugal.

Tena Ruiz, I., Petillot, Y., & Lane, D. (2003). Improved AUV navigation using side-scan sonar, *Proceedings of IEEE OCEANS 2003*, 3, 22-26 Sept. 2003, pp. 1261 – 1268.

Thalmann, D. (2003). Concepts and Models for Inhabited Virtual Worlds. *Proceedings of the First International Workshop on Language Understanding and Agents for Real World Interaction.*

Theiler, J., Eubank, S., longtin, A., Galdrikian, B., & Farmer, D. (1992). Testing for nonlinearity in time series: the method of surrogate data. *Physica D,* 58, 77–94.

Theraulaz, G., & Bonabeau, E. (1995). Coordination in distributed building. *Science*, 269(), 686-688.

Theraulaz, G., & Bonabeau, E. (1999). A brief history of stigmergy. *Artificial Life*, 5(2), 97-116.

Thomas, D.W., & Cloutier, D. (1992). Evaporative water-loss by hibernating little brown bats, *Myotis lucifugus. Physiological Zoology, 65*, 443-456.

Thomas, D.W., & Geiser, F. (1997). Periodic arousals in hibernating mammals: Is evaporative water loss involved?, *Functional Ecology, 11*(5), 585-591.

Thomas, D.W., Dorais, J., & Bergeron, J.-M. (1990). Winter energy budgets and cost of arousals for hibernating little brown bats, *Myotis lucifugus. Journal of Mammalogy, 71*, 475-479.

Timmis, J., Knight, T., de Castro, L., & Hart, E. (2004). *An overview of artificial immune systems.* In Paton et al., editors, Computation in Cells and Tissues: Perspective and Tools for Thought. Natural Computation Series, pp. 51-86. Ed. Springer.

Tipler, F. J. (1980). Extraterrestrial Intelligent Beings Do Not Exist. *Quarterly Journal of the Royal Astronomical Society,* 21, 267-281.

Tipler, F. J. (1982). We Are Alone In Our Galaxy, *New Scientist,* 7 October, 33-35.

Togelius, J. (2003). *Evolution of the Layers in a Subsumption Architecture Robot Controller.* Master of Science in Evolutionary and Adaptive Systems. University of Sussex, UK.

Togelius, J. (2004). Evolution of a Subsumption Architecture Neurocontroller. *Journal of Intelligent & Fuzzy Systems,* 15:1, pp. 15-21.

Torrence, C., & Compo, G. P. (1998). A Practical Guide to Wavelet Analysis. *Bulletin of the American Meteorological Society,* 79, 61-78.

Tortell, R., Morie, J..F. (2006). Videogame play and the effectiveness of virtual environments for training. *Interservice/Industry Training, Simulation, and Education Conference (I/ITSEC).* (Paper No. 3001, 1-9), Orlando , Florida, USA.

Treves, A. & Amit, D. J. (1988). *Metastable States in Asymmetrically Diluted Hopfield Networks,* J. Phys A: Math. and Gen., vol. 21, 3155-3169.

Trillas, E., & Guadarrama, S. (2005). What about fuzzy logic's linguistic soundness?. *Fuzzy Sets and Systems, 156*, 334-340.

Tsaih, R., & Chih-Chung, L. (2004). *The layered feed-forward neural networks and its rule extraction.* Paper presented at the Advances in Neural Networks - ISNN 2004. International Symposium on Neural Networks, Dalian, China.

Tsoumakos, D., & Roussopoulos, N. (2003). *A Comparison of Peer-to-Peer Search Methods*, Paper presented at the 6th International Workshop on the Web and Databases (WebDB), San Diego, USA.

Tsubokawa, H., & Ross, W. N. (1996). IPSPs modulate spike backpropagation and associated [Ca2+]i changes in the dendrites of hippocampal CA1 pyramidal neurons. *Journal of Neurophysiology*, 76, 2896-2906.

Tsutsumi, K., & Matsumoto, H. (1984). A synaptic modification algorithm in consideration of the generation of rhythmic oscillation in a ring neural network. *Biological Cybernetics,* 50, 419-430.

Tu, X., & Terzopoulos, D. (1994). Artificial fishes: Physics, locomotion, perception, behavior. In *Computer Graphics 28 Annual Conference Series* (pp. 43-50).

Tufte G. and Haddow P. C. (2005) *Towards Development on a Silicon-based Cellular Computing Machine.* Natural Computing 4(4): 387-416.

Turing, A.(1952) *The chemical basis of morphogenesis.* Philosofical Transactions of the Royal Society B, vol.237, pp. 37-72

Turing, A.M. (1952). The chemical basis of morphogenesis. *Philosophical Transactions of the Royal Society B,* 237, 37-72.

Turner, R. W., Meyers, D. E. R., Richardson, T. L., & Barker, J. L. (1991). The site for initiation of action potential discharge over the somatodendritic axis of rat hippocampal CA1 pyramidal neurons. *Journal of Neuroscience*, 11, 2270-2280.

Turrigiano, G. (2007). Homeostatic signaling: the positive side of negative feedback. *Current Opinion in Neurobiology*, 17, 1-7.

Twente, J.L., & Twente, J.A. (1965). Effects of core temperature upon duration of hibernation of Citellus lateralis. *Journal of Applied Physiology, 20*, 411-416.

Tzounopoulos, T., Kim, Y., Oertel, D., & Trussell, L. O. (2004). Cell specific, spike timing-dependent plasticities in the dorsal cochlear nucleus. *Nature Neuroscience,* 7, 719–725.

Ursem, R. K. (2002). Diversity-Guided Evolutionary Algorithms. *Proceedings of VII Parallel Problem Solving from Nature* (pp. 462-471). Springer-Verlag.

Urzelai, J. & Floreano, D. (1999). *Incremental Evolution with Minimal Resources*. Proceedings of IKW99. 796-803. Computation. Nature, Vol. 431. 1991.

Van Dyke Parunak, H., Brueckner, S. A., Matthews, R., & Sauter, J. (2005). *Pheromone Learning for Self-Organizing Agents*, IEEE Transactions on Systems, Man, and Cybernetics, Part A: Systems and Humans, 35(3).

Van Harreveld, A. (1980) L-proline as a glutamate antagonist at a crustacean neuromuscular junction. *Journal of Neurobiology, 11*, 519-529.

Vanputte, M., B. Osborn, J. Hiles, (2002). A Composite Agent Architecture for Multi-Agent Simulations. In: *Proceedings of the Eleventh Conference in Computer Generated Forces and Behavior Representation.* Orlando, Florida.

Vargas, P., de Castro, L., Michelan, R., & Von Zuben, F. (2003). *Implementation of an immuno-genetic network on a real Khepera II robot.* 2003 IEEE Congress on Evolutionary Computation, Volume 1, 8-12 Dec. 2003, pp. 420 – 426.

Varona, P., Ibarz, J. M., López-Aguado, L., & Herreras, O. (2000). Macroscopic and subcellular factors shaping CA1 population spikes. *Journal of Neurophysiology*, 83, 2192-2208.

Vavak, F., Jukes, K., & Fogarty, T. C. (1997). Adaptive Combustion Balancing in Multiple Burner Boiler Using a Genetic Algorithm with Variable Range of local Search. In *The proceedings of the 7th International Conference on Genetic Algorithms* (pp. 719-726). San Francisco, CA: Morgan Kaufmann.

Veinante, P. & Deschenes, M. (1999). Single- and multi-whisker channels in the ascending projections from the principal trigeminal nucleus in the rat. *Journal Neuroscience,* 19, 5085-095.

Vetter, P. H., Roth, A., & Haüsser, M. (2001). Propagation of action potentials in dendrites depends on dendritic morphology. *Journal of Neurophysiology* 85, 926-937.

Victor, J.D. (1999) Temporal aspects of neural coding in the retina and lateral geniculate. *Network, 10*, 1-66.

Viggiano, A., Monda, M., Viggiano, A., Chiefari, M., Aurilio, C., & De Luca, B. (2004). Evidence that GABAergic neurons in the spinal trigeminal nucleus are involved in the transmission of inflammatory pain in the rat: a microdialysis and pharmacological study. *European Journal Pharmacology*, 496, 87-92.

Vilela, I.M.O(1998). Integrated Approach of Visual Perception Computational Modeling (in Portuguese). MSc Dissertation, COPPE -Universidade Federal do Rio de Janeiro.

Vilela, I.M.O. (2000) An Integrated Approach of Visual Computational Modeling. *6th Brazilian Symposium on Neural Networks (SBRN)*, Rio de Janeiro, Brazil. IEEE Computer Society.

Vilela, I.M.O., P. M.V. Lima, 2001. Conjecturing the Cognitive Plausibility of an ANN Theorem-Prover. In: *Lecture Notes on Computer Science*, 2084 (1)- 822-829.

Vogels, T.P., Rajan, K. & Abbott L.F. (2005). Neural Network Dynamics. *Annual Reviews of Neuroscience, 28*, 357-376.
Volino, P, Magnenat-Thalmann, N. (2006) Real-Time Animation of Complex Hairstyles. - *IEEE Transactions on Visualization and Computer Graphics* 12(2) pp. 131-142.

Von Neumann, J. (1951). La Teoría Lógica y General de los Autómatas. In L.A. Jeffres (Ed), *Cerebral Mechanisms in Behaviour* (pp. 1-41). New York: John Willey and Sons.. And in, J. von Neumann (1963). *Collected Works* (pp 288-328). Oxford: Pergamon.

Von Neumann, J. (1966). *Theory of Self-Reproducing Automata*. Urbana. Illinois: Illinois University Press.

Wallis, S. G., & Manson, J. R. (2004). Methods for predicting dispersion coefficients in rivers. *Water Management, 157*(3), 131-141.

Wang, X., Chen, B., Qian, G., & Ye, F. (2000). On the optimization of fuzzy decision trees. *Fuzzy Sets and Systems, 112*(1), 117-125.

Wang, X.J., & Rinzel, J. (1992). Alternating and synchronous rhythms in reciprocally inhibitory model neurons. *Neural Computation,* 4, 84-97.

Warren, C. (1999). A technique for autonomous underwater vehicle route planning. *IEEE J. Oceanic Engineering*, Vol. 15, 1990, pp. 199-204.

Watson J.D. & Crick. F. H. (1953) *Molecular structure of Nucleic Acids*. Nature vol. 171, pp. 737–738.

Watson, J. D. & Crick, F. H. C. (1953 a). Molecular Structure of Nucleic Acid. A Structure for Deoxyribose Nucleic Acid. *Nature* 171, 737-738.

Watson, J. D. & Crick, F. H. C. (1953 b). Genetic Implications of the Structure of Deoxyribonucleic Acid. *Nature* 171, 964-967.
Waugh, S. (1985). *Extending and benchmarking Cascade-Correlation*, PhD Thesis, University of Tasmania.

Webb, B. (2001). Can robots make good models of biological behaviour? *Behavioural and Brain Sciences,* 24, 1033-1050.

Webb, B. (2002). Robots in Invertebrate Neuroscience. *Nature,* 417, 359-363.

Webb, B. (2004). Neural mechanisms for prediction: Do insects have forward models? *Trends in Neurosciences,* 27, 278-282.

Weiss, R., & Basu, S. (2002) *The device physics of cellular logic gates. First Workshop on Non-Silicon Computing.* Boston, MA.

Weiss, R., Basu, S., Hooshangi, S., Kalmbach, A., Karig, D., Mehreja, R., & Netravali, I. (2003). Genetic circuit building blocks for cellular computation, communications, and signal processing. *Natural Computing*, 2, 47-84.

Welker, C. (1971). Microelectrode delineation of fine grain somatotopic organization of SM1 cerebral neocortex in albino rat. *Brain Research*, 26, 259–275.

Werfel, J., & Nagpal, R. (2006). Extended stigmergy in collective construction. *IEEE Intelligent Systems*, 21(), 42-48.

Whitbrook, A. (2005). *An idiotypic immune network for mobile robot control.* Msc Thesis, University of Nottingham.

White, P., Zykov, V., Bongard, J., & Lipson, H. (2005). Three dimensional stochastic reconfiguration of modular robots. In *Proc. Robotics: Science and Systems*, (pp. 161-168).

Whitley, D. & Starkweather, T. (1990). GENITOR II: A Distributed Genetic Algorithm. *Journal of Experimental and Theoretical Artificial Intelligence*, vol. 2, 189-214.

Whitley, D. (1989). The Genitor Algorithm and Selection Pressure: Why Rank Based Allocation of Reproductive Trials is Best. In *Proceedings of the 3rd ICGA* (pp. 116-121). San Francisco, CA: Morgan Kaufmann.

Whitley, D. (1995). Genetic algorithms and neural networks. In *Genetic Algorithms in Engineering and Computer Science* (vol. 11, pp. 1-15). John Wiley & Sons Ltd.

Whitley, D., Mathias, K., Rana S., & Dzubera, J. (1996). Evaluating Evolutionary Algorithms. *Artificial Intelligence*, vol. 85(1-2): 245-276.

William W. S. (1990). *Time Series Analysis. Univariate and Multi*

Williams, R.J. & Zipser, D. (1989). A Learning Algorithm for Continually Running Fully Recurrent Neural Networks. *Neural Computation* 1, 270-280.

Williams, S. R. (2004). Spatial compartmentalization and functional impact of conductance in pyramidal neurons. *Nature Neuroscience*, 7, 961-967.

Willis, J.S. (1982). The mystery of the periodic arousal. In C.P. Lyman, J.S. Willis, A. Malan & L.C.H. Wang (Eds.), *Hibernation and Torpor in Mammals and Birds* (pp. 92-103). San Diego: Academic Press.

Wilson, H.R. (1999). *Spikes, decisions, and actions: dynamical foundations of neuroscience.* New York: Oxford University Press.

Winder, R., & Reggia, J.A. (2004). Using distributed partial memories to improve self-organizing collective movements. *SMC-B*, 34(4), 1697-1707.

Winter, D.A., Prince, F., & Patla, A.E. (1998). Stiffness control of balance during quiet standing. *Journal of Neurophysiology*, 80, 1211-1221.

Wong, R. K. S., & Stewart, M. (1992). Different firing patterns generated in dendrites and somata of CA1 pyramidal neurones in guinea-pig hippocampus. *Journal of Physiology (London)*, 457, 675–687.

Wooldridge, M., Jennings, N. (1995). Intelligent Agents: Theory and Practice. In: *Knowledge Engineering Review*, 10 (2), Cambridge University Press, U.K.

Woolston, D. C., La Londe, J. R., & Gibson, J. M. (1982). Comparison of response properties of cerebellar- and thalamic-projecting interpolaris neurons. Journal Neurophysiology 48, 160-173.

Wright, S. (1986). *Evolution.* Chicago, IL: University of Chicago Press. Edited and with Introductory Materials by William B. Provine.

Xiao, A., Bryden, K. M., Engelbrecht, J., Huang, G. & McCorkle, D. S. (2004). Acceleration Methods in the Interactive Design of a Hydraulic Mixing Nozzle Using Virtual Engineering Tools. In *Proceedings of the 2004 International Mechanical Engineering Congress and Exposition.* ASME Press.

Xu, M., S.Sun,, Y. Pan, 2006. Virtual human model for ergonomics workplace design. *Zhongguo Jixie Gongcheng (China Mechanical Engineering)*,17(80),836-840.

Yamashita, A., Arai, T., Ota, J., & Asama, H. (2003). Motion planning of multiple mobile robots for cooperative manipulation and transportation. *IEEE Transactions on Robotics and Automation*, 19(2), 223-237.

Yan, W., Zhu, Z. & Hu R. (1997). Hybrid genetic/BP algorithm and its application for radar target classification. In: Proc. *IEEE National Aerospace and Electronics Conf. NAECON.* Part 2 of 2. 981–984.

Yang, J.M., Kao, C.Y. & Horng, J.T. (1996). Evolving neural induction regular language using combined evolutionary algorithms. In: *Proc. 1st Joint Conf. Intelligent Systems/* ISAI/IFIS 162–169.

Yang, Z., & França, F.M.G. (2003). A generalized locomotion CPG architecture based on oscillator building blocks. *Biological Cybernetics*, 89, 34-42.

Yao, X. (1999). Evolving artificial neural networks. *Proceedings of the IEEE, 87*(9), 1423-1447.

Yoerger, D., Bradley, A., Walden, B., Singh, H., & Bachmayer, R. (1996). *Surveying a subsea lava flow using the autonomous bentic explorer (ABE).* Proc. 6th Int. Advanced Robotics Program, 1996, pp. 1-21.

Yu, C-S., & Li, H-L. (2001). Method for solving quasi-concave and non-cave fuzzy multi-objective programming problems. *Fuzzy Sets and Systems, 122*(2), 205-227.

Yuan, Y., & Shaw, M.J. (1995). Induction of fuzzy decision trees. *Fuzzy Sets and Systems, 69*(2), 125-139.

Yurke, B., Turberfield, A. J., Mills, Jr A. P., Simmel, F. C., & Neumann J. L. (2000). A DNA-fuelled molecular machine made of DNA. *Nature*, 406, 605-608.

Zadeh, L. A. (1978). Fuzzy sets as a basis for a theory of possibility. *Fuzzy Sets and Systems, 1*, 3-28.

Zadeh, L.A. (1965). Fuzzy Sets. *Information and Control, 8*(3), 338-353.

Zhang, P., Sankai, Y. & Ohta, M. (1995). Hybrid adaptive learning control of nonlinear system. In: *Proc. American Control Conf.* Part 4 of 6, 2744–2748.

About the Contributors

Ana B. Porto Pazos is an associate professor at University of A Coruña, Spain (Department of Information Technologies and Communications). She is currently coordinator of the Master in Computer Science at University of A Coruña. She received the MS and PhD degrees in computer science from the University of A Coruña, Spain in 1999 and 2004 respectively. She has worked with several research groups such as: "Artificial Neural Networks" in Computer Science Faculty of the University Politécnica de Madrid (Spain) and "Cellular Electrophysiology" in Cajal Institute of the "Consejo Superior de Investigaciones Científicas (CSIC)" in Madrid. She has published more than 20 papers and book chapters about artificial neural networks, brain computer modelling, evolutionary computation, telemedicine, etc.

Alejandro Pazos Sierra is a professor at University of A Coruña, Spain. He is currently the director of the Department of Information Technologies and Communications in University of A Coruña. He received the MS degree in medicine from University of Santiago de Compostela in 1987. He received the PhD degree in computer science from University Politécnica de Madrid (Spain) in 1989 and the PhD degree in medicine from University Complutense de Madrid (Spain) in 1996. He has worked with several research groups such as: "Visualization and Usability Laboratory College of Computing" in Georgia Institute of Technology, "Decision Systems Group" in Harvard Medical School, "Section on Medical Informatics" in Stanford University, "Artificial Neural Networks Laboratory" in Computer Science Faculty of the University Politécnica de Madrid (Spain), etc. He found and is currently the director of "Artificial Neural Networks and Adaptive Systems Laboratory" in the Faculty of Computer Science in A Coruña University. His current research interests include artificial neural network, medical image, evolutionary computation, adaptive systems, control medical systems and telemedicine. Dr. Pazos is author of more than 40 published papers and is a member of the IEEE and several honour societies, including ACM, International Neural Network Society, American Association for the Advancement of Science, Internet Society, etc.

Washington Buño Buceta is a research professor at Cajal Institute (CSIC), Madrid, Spain. He found and is currently the leader of the Cellular Electrophysiology Group (Department of Functional and Systems Neurobiology, Cajal Institute). He received the MS degree in medicine from Universidad Autónoma of Madrid in 1975. He received the PhD degree in medicine from Universidad de la República, Montevideo, Uruguay in 1964 and the PhD degree in medicine from Universidad Autónoma of Madrid in 1982. His research interests cover neural plasticity, cellular electrophysiology, etc. He has more than 50 published papers, more than 20 research projects and has worked with several prestigious research groups such as: "Laboratoire de Neurosciences Fonctionnelles" Unité de Sensori-Motricité Comparée et Comportements - Centre National de la Recherche Scientific, Marsella (France), Dept. of Zoology and Genetics, Iowa State University, Ames (Iowa), "Unidad de Neurofisiología del Instituto de Investigaciones Biológicas Clemente Estable" Faculty of Medicine, Montevideo (Uruguay), etc. He is a reviewer in journals such as *Brain Research*, *Epilepsia*, *Hippocampus*, *Journal of Neuroscience*, etc. and is a member of honour societies, including International Brain Research Organization (IBRO), Society for Neuroscience (USA), Academia de Ciencias de Latinoamérica, founder member of Sociedad Uruguaya de Electroencefalografía y Neurofisiología Clínica and Sociedad Española de Neurociencias, etc.

* * *

Gerardo G. Acosta (IEEE Member '95 – Senior Member '01) was born in General Roca, Río Negro, Argentina in 1964. He has graduated as an engineer in electronics at the National University of La Plata, Argentina (1988), and as a PhD in computer science, at the University of Valladolid, Spain (1995). He is currently an associated professor in control systems (electronic area) in the Faculty of Engineering, National University at the Centre of Buenos Aires Province (UNCPBA). He is a researcher of the Argentinean Scientific and Technical Research Council (CONICET) and director of the R+D Group INTELYMEC, at UNCPBA. Dr. Acosta is a member of the Argentinean Automatic Control Society (AADECA-IFAC), the Hispanic-American Fuzzy Systems Association (HAFSA-IFSA), and president of the Argentinean Chapter of the IEEE Computational Intelligence Society. His research topic comprises the use of AI techniques in control and supervision. He has more than 90 papers in this and related fields.

Alberto Alvarellos González is a researcher at University of A Coruña, Spain (Department of Information Technologies and Communications). He received an MS in computer science from the University of A Coruña, Spain in 2007. He collaborates with the research group of "Cellular physiology of astrocytes and neurons" in Cajal Institute of the "Consejo Superior de Investigaciones Científicas (CSIC)" in Madrid. His research lines are about artificial neural networks, evolutionary computation, etc.

Alfonso Araque Almendros was born in Madrid, Spain in 1965. He received the MS and PhD degrees in biology from "Complutense" University of Madrid in 1988 and 1993 respectively. Since 1990, he has worked with several research groups: "Laboratoire de Neurosciences Fonctionnelles" Unité de Sensori-Motricité Comparée et Comportements - Centre National de la Recherche Scientific, Marsella (France), Dept. of Zoology and Genetics, Iowa State University, Ames (Iowa), "Unidad de Neurofisiología del Instituto de Investigaciones Biológicas Clemente Estable" and Depto. of Phisiology of the Faculty of Medicine in Montevideo (Uruguay), etc. He is currently a "Titular" scientific of the "Consejo Superior de Investigaciones Científicas (CSIC)" in the Neural Plasticity Department of Cajal Institute in Madrid. His research interests covers neurophysiology, neural plasticity, glial system, etc. Dr. Araque is author of more than 40 published papers from 1991 and he is a member of the honour societies, such as: Society for Neuroscience (EE.UU.), European Neuroscience Association, "Sociedad Española de Neurociencias", etc.

Dan Ashlock is a professor of mathematics in the Department of Mathematics and Statistics, University of Guelph. He has a doctorate from Caltech in mathematics and holds the Bioinformatics Chain in his department. Dr. Ashlock's research is in evolutionary computation, particularly issues of representation, bioinformatics, and machine learning. He has a book, *Evolutionary Computation for Optimization and Modeling*, published by Springer.

José María Barreiro Sorrivas, associate professor of computer science and artificial intelligence, Universidad Politécnica de Madrid's School of Computing, graduated in medicine from the Universidad Santiago de Compostela, and holds a PhD in computing and an MSc in knowledge engineering from the Universidad Politécnica de Madrid. Prof. Sorrivas has lengthy teaching and research experience, in the course of which he has published several books, congress papers and articles in national and international journals on subjects within his field of expertise. Prof. Sorrivas now teaches undergraduate computing students at the School of Computing and is a research member of CETTICO, attached to the Fundación General de la UPM. He has participated in several nationally and internationally funded R&D projects. He now specializes and researches in the field of ICT (artificial intelligence, artificial neural networks, machine learning), medical computing and sports medicine (clinical practice guidelines, medical databases, genetics) and rehabilitation technologies (isokinetics, diagnosis and patient assessment).

Malcolm J. Beynon is a reader in Cardiff Business Cardiff, Cardiff University (UK). He gained his BSc and PhD in pure mathematics and computational mathematics, respectively, at Cardiff University. His research areas include the theoretical and application of uncertain reasoning methodologies, including Dempster-Shafer theory, fuzzy set theory and rough set theory. Also the introduction and development of multi-criteria based decision making and classification techniques, including the Classification and Ranking Belief Simplex. He has published over 120 research articles. He is a member of the International Rough Set Society, International Operations Research Society and the International Multi-Criteria Decision Making Society.

Kenneth "Mark" Bryden is an associate professor of mechanical engineering at Iowa State University. He has written more than 100 peer reviewed articles in the areas of modeling, complex systems, and decision science. He is associate professor of the Mechanical Engineering Department, Iowa State University, and program director for the Simulation, Modeling and Decision Science Program in the U.S. Department of Energy's Ames Laboratory. Dr. Bryden's primary research interests are in integration of high performance computing, new computational algorithms, and new user interfaces to support engineering decision analysis. Dr. Bryden is the recipient of numerous awards including a 2006 R&D 100 award for the development of the software package VE-Suite.

Oscar A. Calvo Ibáñez was born in Necochea, Argentina, in 1954. He received his BSEE in electrical engineering from the National University of La Plata (Argentina) in 1979, his MSEE in computer and electrical engineering from the Illinois Institute of Technology (Chicago) in 1988 and his PhD in electrical engineering from the Polytechnic University of Cataluña, (UPC Barcelona, Spain) in 2005. He has been a guest scientist at the Fermi National Accelerator Laboratory (Chicago), sponsored research staff at the Massachusetts Institute of Technology (Boston), and adjunct professor and researcher at the National University of La Plata (Argentina). He currently holds a position as professor at the University of the Balearic Islands (Palma de Mallorca, Spain). His research is in the areas of control, robotics and power electronics.

Nazareth P. Castellanos graduated in physics at the Universidad Autonoma de Madrid. She is currently a PhD student at the Universidad Complutense of Madrid (Spain). Her research concentrates on analysis and comparison of neural dynamics and circuitry at different scales from unitary recordings to EEG.

Daniel Cattaert is a neurophysiologist specialist of crayfish locomotion and sensory-motor interactions occurring in central networks. He did his first 20 years of research under the supervision of François Clarac on crustacean locomotion during which he developed the *in vitro* preparation of the crayfish locomotor network used to study sensory-motor interactions. More especially, he analyzed in collaboration with Abdeljabbar El Manira, the mechanisms of presynaptic inhibition occurring in primary afferents during fictive locomotion. He also demonstrated several mechanisms of modulation of the sensory-motor pathways with Didier Le Ray, and studied 5-HT neuromodulation in the same sensory-motor pathways with Morgane Le Bon-Jego.

Steven Corns received the BS in 2001 and MS in 2003, both in mechanical engineering from Iowa State University. He is currently working towards the PhD degree in mechanical engineering with the Virtual Engineering Research Group at Iowa State University. His main research interests are in the area of evolutionary computation applied to biological systems and the mechanics of information transfer in evolutionary algorithms. He is also interested in energy systems and engineering education.

Doina Daciana Iliescu graduated in 1991 from the Polytechnic Institute of Bucharest, Romania, Faculty of Electronics and Telecommunications, specialisation in Telecommunications and Data Networks. She received her PhD in engineering in 1998 from the University of Warwick, UK, in the field of optical engineering. Since then she has been a lecturer in the School of Engineering, University of Warwick and a research member of the Optical Engineering Laboratory and the Intelligent Systems Laboratory.

Daniela Danciu received the MS degree in 1990 and the PhD degree in automatic control in 2003, both from the University of Craiova, Faculty (School) of Automation and Computers, Romania. In 2000, after 10 years of industrial experience both in automatic control and computer applications, she joined the Staff (Faculty) of the University of Craiova where she is currently an associate professor with the Department of Automatic Control and also a member of the Research Centre of *Nonlinear Control. Stability. Oscillations.* Her scientific interests include the qualitative behavior of nonlinear systems with sector restrictions, time-delays, synchronization issues and oscillations and their applications to neural networks.

Julian Dorado was born in A Coruña, Spain in 1970. He is associate professor in the Department of Information and Communications Technologies of the University of A Coruña. He finished his studies on computer engineering in 1993, and on 1999 he became doctor in computer science with his thesis "Methodology for the Development of Knowledge Extraction Systems in ANNs." He has worked on several Spanish and European projects, and he has published many books and papers on several international journals. He is currently working on evolutionary computation, artificial neural networks and knowledge extraction systems.

Norberto Ezquerra is an associate professor in the College of Computing and a faculty member of the Graphics, Visualization and Usability (GVU) Center. His research experience and interests lie in the areas of interactive computer graphics and visualization, computer vision, knowledge-based systems, and 3D imaging and modeling techniques. In particular, Dr. Ezquerra is exploring approaches for extracting, representing, and interpreting dynamic processes and complex structures within a biomedical context. His interdisciplinary research is supported by the National Library of Medicine (NLM), the National Institutes of Health, the Georgia Research Alliance, the Emory-Georgia Tech Center, and the Georgia Tech- Medical Col. of Ga Center.

José A. Fernández León was born in La Plata, Argentina, in 1974. He received his System Engineer degree in the National University at the Centre of Buenos Aires Province (UNCPBA), Argentina, in 2001. Since 2004 he has a fellowship from the CONICET to develop his Master Sciences studies at UNCPBA (2006) and PhD studies at University of Sussex (US), UK, since then, co-sponsored with an AlBan scholarship. He is member of the INTIA Research Institute, UNCPBA, and the CCNR, US. He is Assistant Lecturer at UNCPBA, and Associate Tutor at US. His research topic comprises the exploration of the boundaries between evolutionary biology, neuroscience, and robotics in bio-inspired systems affected by noise and uncertainty in sensorimotor control. He has more than fifteen papers in this and related fields.

Enrique Fernández-Blanco was born in Frankfurt, Germany in 1981. He received his MS in computer science from the university of A Coruña in 2005. Since that date he has been working with the research group SABIA of the University of A Coruña. Actually, he is developing his PhD thesis in computer science under the direction of professor Julian Dorado in the Department of Information Technologies and Communications. His current research interest include artificial neural networks, artificial embryogeny, swarm Intelligence, image processing and evolutionary computation

Felipe M. G. França was born in Rio de Janeiro, Brazil. He received his electronics engineer degree from the Federal University of Rio de Janeiro (UFRJ), in 1981, the MSc in computer science from COPPE/UFRJ, in 1987, and his PhD from the Department of Electrical and Electronics Engineering of the Imperial College London, in 1994. Since 1996, he has been with the Systems Engineering and Computer Science (graduate) Program, COPPE/UFRJ, as associate professor, and he has research and teaching interests in asynchronous circuits, computational intelligence, computer architecture, cryptography, distributed algorithms and other aspects of parallel and distributed computing.

Agostino Forestiero received his Laurea degree in computer engineering from the University of Calabria, Cosenza, Italy, in 2002 and his PhD in computer engineering from the University of Calabria in 2007. Since 2002 he collaborates with the Institute of High Performance Computing and Networks of the Italian National Research Council (ICAR-CNR) in Cosenza. His research interests include Grid computing, peer-to-peer networks and swarm intelligence.

Marcos Gestal is an assistant professor in the Computer Science Faculty (University of A Coruña) and member of the research laboratory Artificial Neural Networks and Adaptative Systems. His actual research interests are focused on evolutionary computation (mainly genetic algorithms), artificial neural networks and their interaction to perform variable selection. He has participated in several research projects and published papers in many international journals and books.

Alexander Grushin holds a BS in computer and information sciences from the University of Delaware, an MS in computer science from the University of Maryland at College Park, and a PhD in computer science, which he recently received from the University of Maryland. His doctoral dissertation dealt with the distributed control of self-assembly processes. More generally, his research interests lie in areas that include artificial intelligence, artificial life, biologically-inspired computing, machine learning, and multi-agent systems. Presently, he is a research scientist at Intelligent Automation, Inc., located in Rockville, MD.

Ian Guymer BSc, DIS, PhD (Bham), CEng, MCIWEM holds the chair in civil engineering at the University of Warwick. His research interests focus on identifying and quantifying mixing processes, through laboratory and field studies, to determine the fate of soluble and fine suspended sediments within both the natural environment and engineering applications. Recent publications and funding have investigated urban drainage systems and rivers through to estuarine and coastal processes, with an emphasis on understanding the impact on the natural environment.

Oscar Herreras obtained his PhD in neurophysiology at the University Complutense of Madrid in 1987. Early career was devoted to electrophysiological and neurochemical studies in the hippocampus in vivo, exploring the input/output relations of pyramidal cells in relation to EEG patterns and epilepsy. Back in Madrid after a postdoc time at Duke University, he begun to combine experimental and computational studies covering a broad range of interests, from single cell (dendritic electrogenesis) to network and macroscopic (spreading depression) levels. At present, he continues these lines at the Cajal Institute where he also explores the cellular generators of the EEG.

Evor L. Hines BSc, GDMA, PhD (Brad), FHEA, CEng, FIET joined Warwick in 1984 and is a Reader in Electrical and Electronic Engineering. He has worked in intelligent systems (expert systems (ESs), artificial neural networks (ANNs), genetic algorithms (GAs), fuzzy logic/systems (FL/Ss), neuro-fuzzy systems (NFSs), genetic programming (GP) and similar techniques) and their application for some 20 years. Typical application areas include intelligent sensors (e.g., electronic nose); non-destructive testing of for example composite materials; computer vision; telecommunications; and medicine amongst others. He has co-authored some 190 articles and won numerous industrial and government grants to fund his research work.

José M. Ibarz is an electronic engineer by the University Politécnica of Madrid (1978). His early work was devoted to the implementation of recording devices and systems for brain electrophysiology, and later became engaged in computational biophysics. For most of his career at the Ramón y Cajal Hospital in Madrid, he has been interested in dendritic electrogenesis, and developed a number of biophysical models to study the computational properties of active dendrites and the role of local spikes on the elaboration of the output code.

Greg P. King, BS, PhD (University of Texas, Austin) joined Warwick in 1987. He has published papers on nonlinear methods of time series analysis and particle dispersal by chaotic advection. Current research involves the use of satellite data to investigate atmospheric turbulence and coastal upwelling.

Morgane Le Bon-Jego is a specialist of the functioning and the functional plasticity of biological neuronal networks. She obtained her PhD in 2003 working on the organization and the modulation of the crayfish locomotor circuitry. Then, she went to Columbia University where she acquired an expertise in the domain of calcium imaging applied to the analysis of a neocortical neural network. In 2006, she obtained a permanent position in the Université Bordeaux 1 where she explores, combining electrophysiological and imaging techniques, the mechanistic bases of the behavioral plasticity derived from an operant conditioning in *Aplysia.*

Didier Le Ray obtained his PhD in 1997 working on the sensory-motor networks that control walking leg movements in crayfish. After two post-doctoral positions (in Madrid and Montreal) during which he broadened his knowledge of synaptic plasticity in the rat hippocampus and of motor circuit operation in the lamprey, he entered the CNRS in 2000 as a junior research scientist to continue his work on sensory-motor integration on the crayfish model. In 2003, he turned towards the developmental aspect of neural networks and started a new project on the amphibian *Xenopus laevis* in which he and his colleagues study the developmental transformation of the central nervous system associated with the animal's change in locomotory behaviour during metamorphosis.

Mark Leeson received a PhD from the University of Cambridge, UK, in 1990 and then worked as a network analyst for a UK bank until 1992. Subsequently, he held several academic appointments before joining the University of Warwick in March 2000 where he is an associate professor. His major research interests are optical communication systems, communication protocols, ad hoc networking and evolutionary optimisation. To date he has published about 100 journal or conference papers in these fields. He is a chartered member of the UK Institute of Physics and a member of the EPSRC Peer Review College.

XuQin Li received his BS in educational technology and an MSc in pattern recognition and intelligent systems from HuaZhong Normal University and Chinese Academy of Sciences in 2003 and 2006 respectively. He is now pursuing his PhD degree and serving as a postgraduate research fellow in University of Warwick, United Kingdom. His research interests include pattern recognition, neural networks and intelligent system. Li is also an Overseas Research Students Awards Scheme (ORSAS) awards holder.

Domingo López-Rodríguez (born 1978) received his doctoral degree by University of Málaga, Spain in 2007, in the area of applied mathematics. His research interests include fields such as optimization, neural networks and combinatorial optimization problems. He is particularly interested in the application of multivalued recurrent networks to solve some optimization problems on graphs, and in the development of a general framework to study competitive learning. Web-based education and distance learning are also fields of his study. He has now a contractual relationship with the Department of Applied Mathematics in University of Málaga.

Valeri A. Makarov graduated in physics and mathematics at the Nizhny Novgorod State University (Russia) in 1992, and received the PhD degree in physics and mathematics from the same University in 1997. From 1998 he is with the Universidad Complutense ofMadrid (Spain). His research interests are in nonlinear dynamics, time-series analysis and their applications in neuroscience and bio-inspired robotics.

Julia Makarova studied physics at Nizhni Novgorod University in Russia and become interested in biophysics during her doctoral training in the Cajal Institute in Madrid, where she developed an integral model of spreading depression. Her biophysical models cover from membrane channels and subcellular electrogenesis in single neurons to extracellular potentials in a volume conductor. After earning her PhD in neurophysiology at the University Complutense of Madrid in 2007, she now combines experimental and computational approaches to the study of dendritic electrogenesis in spreading depression, and the cellular generators of hippocampal rhythms.

Eduardo Malmierca graduated in medicine at the Universidad Autónoma de Madrid (Spain) in 1997, and received a PhD in neuroscience from the same university in 2006. He is currently assistant professor at the Faculty of Medicine, Universidad Autónoma de Madrid (Spain). His research interests are electrophysiological studies on somatosensory system, specially corticofugal influences on the subcortical relay stations. Recently he has been collaborating in the development of new mathematical tools for spike processing.

Carlo Mastroianni is a researcher at the Institute of High Performance Computing and Networks of the Italian National Research Council (ICAR-CNR) in Cosenza, Italy, since 2002. He received his PhD in computer engineering from the University of Calabria in 1999. His research interests focus on distributed systems and networks, in particular on Grid computing, peer-to-peer networks, content distribution networks, multi agent systems. He published more than 60 scientific papers on international journals and conferences. He currently lectures on computer networks at the University of Calabria. He is serving as a program committee member for many international journals and conferences.

Miguel A. Mayosky (IEEE Member '97-Senior Member `98) was born in La Plata, Argentina, in 1960. He received the Engineer on Electronics degree (First Class Award) from the University of La Plata (UNLP), La Plata, Argentina, and the PhD in computer science from the Autonomous University of Barcelona, Spain (Cum Laude) in 1983 and 1990, respectively. He is currently a full professor of automatic control systems at the School of Engineering, UNLP, and also a member of the Research Staff of the Buenos Aires' Scientific Research Commission (CICpBA). Dr. Mayosky was president of the Argentinean Chapter of the IEEE Computational Intelligence Society. His research activities involve real-time data acquisition and control systems, neural networks, and embedded computer architectures.

Enrique Mérida Casermeiro (born 1952) obtained a degree in mathematics, University of Granada in 1975 and his doctoral degree in computer engineering, University of Málaga in 2000. His research interests include fields such as optimization, neural networks and combinatorial optimization problems. He is particularly interested in the application of multivalued recurrent networks to solve some optimization problems and its aplicattions to engineering problems. He is a full proffessor since 1989 with the Department of Applied Mathematics, University of Málaga.

Mónica Miguélez Rico was born in 1978 in Pontevedra, Spain. She has a research fellowship in the Department of Information and Communications Technologies, University of A Coruña (Spain). She finished her graduate degree in computer science in 2001, and in 2006 she finished her studies in computer engineering. She is currently working in the areas of evolutionary computation and artificial neural networks.

Jesús M. Miró is a physicist (specialized in applied physics) and holds a Master in Biophysics from the Universidad Autónoma de Madrid (UAM). He is currently a member of the Artificial Intelligence Group at the Universidad Politécnica de Madrid (UPM) where he is researching his PhD thesis on biomolecular computation. His research focuses on the design and analysis of genetic circuits and biomolecular automata.

Angel Nuñez graduated in biology at the Universidad Autónoma de Madrid (Spain) in 1979, and received a PhD in neuroscience from the same university in 1988. He is currently professor of cellular biology at the Medicine School of the Universidad Autónoma de Madrid (Spain). His research interest is about synaptic plasticity of the somatosensory system, using in vivo and in vitro electrophysiological recordings.

Iara M. Oberg Vilela received her Bachelor of Science (BS) degree in both biology and psychology from Rio de Janeiro Federal University (UFRJ), Brazil, and has worked in educational orientation and psychopedagogical counseling in schools and private practice. She holds a Master and Doctor of Science degrees in AI from System Engineering and Computer Science Program of Rio de Janeiro Federal University (COPPE-Sistemas, UFRJ). Her areas of expertise are cognitive neuropsychology and psychology, AI, psychopedagogy, and educational interactive virtual environments. Her current research focuses on modeling of virtual characters' autonomous behavior for educational games, interactive narrative and virtual storytelling.

Juan M. Ortiz de Lazcano Lobato was born in 1978. He obtained his MTech degree in computer science from the University of Málaga (Spain) in 2003. Then he received his PhD in software engineering and AI from the University of Málaga in 2007. Since 2003, he has been working as a researcher with the Soft Computing and Image Analysis Unit, University of Málaga. His research interests include fields such as multidimensional data analysis, pattern recognition and neural networks, especially competitive and self-organizing neural networks.

Kirsty Park is a lecturer in Ecology at Stirling University (UK). After a BSc at Leeds University she went on to study bat ecology for a PhD at Bristol University. Her primary research focus is animal ecology and conservation in managed environments (e.g., urban, agricultural, forestry), and human-wildlife conflicts. One particular interest is in the distribution and use of fundamental resources for roosting and foraging bats within agricultural and urban environments. Other interests include impact assessment of non-native species, public attitudes towards management of non-native species, and models to investigate control strategies for invasive species such as *Rhododendron ponticum*.

Gheorghe Paun received his degree in mathematics and his PhD in mathematics from the University of Bucharest. From 1990 he is at the Institute of Mathematics of the Romanian Academy, where he is currently a senior researcher. He has been member of the Romanian Mathematical Society, the Romanian Computer Science Society, the American Mathematical Society for various periods; member of the council of EATCS (the European Association for Theoretical Computer Science) from 1991 to 2003; member of the Romanian Writers Association. He has been corresponding member of the Romanian Academy from October 1997 as well; honorary member and Doctor Honoris Causa of the International Academy of Informatization from March1998; honorary professor of HUST University, Wuhan, China, from2005; finally, he has been member of Academia Europea from 2006. He has lectured at over 100 universities, and gave numerous invited talks at recognized international conferences. He has published 11 books in mathematics and computer science, has edited over 25 collective volumes, and also published many popular science books and books on recreational mathematics (games). He is author or co-author (in collaboration with about 100 researchers from many countries) of more than 450 research papers published in refereed international journals too. He is on the editorial boards for 19 international journals in mathematics, computer science, and linguistics, and was/is involved in the program/steering/organizing committees for many recognized conferences and workshops. He has given invited lectures at more than 70 international conferences, communications at more than 100 international conferences; he has been member of the program committee and referee for many conferences.

Juan Pazos Sierra (BSc, computer science, Technical University of Madrid, 1976; PhD, computer science, Technical University of Madrid, 1980) is an analyst at the Ministry of Education and Science's Data Processing Centre (1973-1978), director of the Computer Centre at the Ministry of Education and Science's Directorate General of Programming and Investments and Building Commission (1978-1988), knowledge engineering consultant for IBM (1988-1992), EU advisor for the 7th Scientific Research Framework Programme, associate professor of computer science at the Technical University of Madrid (1980-1988), full professor of computer science of the Technical University of Madrid (1988-); author or co-author of over 60 publications in high-ranking international journals, author or co-author of six books on artificial intelligence and knowledge engineering. His current research interests include the theoretical foundations of SE, establishing an information theory, the anthropic principle.

Mario J. Pérez-Jiménez received his degree in mathematics from the Barcelona University and doctor degree in mathematics from the Sevilla University. In the past, he was associated professor at the University of Barcelona. Currently, he is titular professor of Computer Science and Artificial Intelligence at the University of Sevilla, where is the head of the Research Group on Natural Computing. He has published twelve books in computer science and mathematics, and over 100 scientific papers in international journals (collaborating with many researchers worldwide). He has served as reviewer for 10 prestigious ISI journals. He is a member of editorial boards of the *Soft Computing* and *International Journal of Computers, Communications and Control*. He is the main researcher in various European, Spanish and Andalusia research grants. He has been and independent expert to the evaluation of NEST (New and Emergent Science and Technology) proposals under the Sixth Framework Programme of the European Community, and he is an European Science Foundation peer reviewer.

Nieves Pedreira was born in was born in A Coruña, Spain in 1969. She is an assistant professor in the Department of Information and Communications Technologies of the University of A Coruña. She received the University Degree in computer science from the University of A Coruña in 1993. This was followed by a master's degree on communications and real time systems. After having worked in private enterprises, she returned to the University in 1997 as a PhD student and on 2003 she became doctor in computer science with her thesis "A model for virtual learning to learn." She is also a tutor in the UNED (Spanish Distance Education National University) since 1998. Her research interests are focussed on distance learning and new technologies.

Javier Pereira Loureiro is an associate professor, Faculty of Health Sciences, University of A Coruña (Spain). He finished his PhD in 1995 and is a member of the research group RNASA-IMEDIR (Artificial Neural Networks and Adaptive Systems / Medical Informatics and Radiological Diagnosis Center) in the Department of Information and Communications Technologies. He has work in many different fields, including neural network development, evolutionary computation, image acquisition, and processing on medical environments. He is currently on several projects about computer science applied in medical environments.

Fausto Pupo received his Laurea Degree in computer engineering from the University of Calabria in 2007. He currently collaborates with the ICAR-CNR staff and is involved in research activities concerning Grid and multi-agent systems.

James A. Reggia, MD, PhD, is a professor of computer science, with joint appointment in the Institute for Advanced Computer Studies, at the University of Maryland. Dr. Reggia's research interests are in the general area of biologically-inspired computation, including evolutionary computation, artificial life, and neural computation; he has authored numerous research papers in these and related areas. His current research focuses on developing new methods for creating very large-scale integrated neurocomputational systems as a basis for machine intelligence, the use of genetic programming to evolve modular neural networks and self-replicating configurations, swarm intelligence systems for collective problem solving and self-assembly of artificial structures, and neural network learning methods.

Juan Ramón Rabuñal Dopico is an associate professor in the Department of Information and Communications Technologies, University of A Coruña (Spain). He finished his studies in computer engineering in 1996, and in 2002, he became a doctor of computer science with his thesis "Methodology for the Development of Knowledge Extraction Systems in ANNs." He has worked on several Spanish and European projects and has published many books and papers in several international journals. He is currently working in the areas of evolutionary computation, artificial neural networks, and knowledge extraction systems.

Alejandro Rodriguez obtained the degree of BS in computer science summa cum laude from the Universidad Simon Bolivar in Caracas, Venezuela in 2001, MS in computer science from the University of Maryland in 2006 and PhD in computer science from the University of Maryland in 2007. His doctoral research focused on the emergence of cooperative behavior in swarms of mobile agents. Dr. Rodriguez's research interests include swarm intelligence, artificial life, emergent behavior, evolutionary computation and machine learning. He currently works at the National Library of Medicine at NIH.

Alfonso Rodríguez-Patón is a physicist (specialized in electronic and computation) and has a PhD in computer science. He is associate professor at the Universidad Politécnica de Madrid (UPM). He is director of the Artificial Intelligence Group in the same university. His main research interest lies in the interplay of computer science, biology, and nanotechnology. His topics of research are formal languages an automata theory, DNA/biomolecular computing, membrane computing, and any bio-inspired or unconventional model of computation.

Giandomenico Spezzano is a research director at the Institute of High Performance Computing and Networking (ICAR) of the Italian National Research Council (CNR), where he manages the Group of Intelligent Grid and Peer-to-Peer Systems. He is also a contract professor at the University of Calabria, Italy, since 1994. Previously, he worked at CRAI (Consortium for Research and Applications of Information Technology), Italy, where he has led various research projects in the distributed and parallel computing area. He received the Laurea degree in Industrial Technologies Engineering from the University of Calabria, Italy, in 1980. He has published two books and about 130 papers in international journals and conference proceedings such as IEEE TEC, IEEE CSE, FGCS, PDCP, Parallel Computing, Concurrency: Practice and Experience. His current research interests cover models and tools for massively parallel architectures, grid computing, peer-to-peer computing, parallel and distributed data mining, parallel genetic programming, cellular automata and swarm intelligence. He is serving as a program committee member for many international conferences and he is member of ACM and IEEE Computer Society.

José M. Vázquez was born in A Coruña, Spain, in 1978. He graduated with a degree in Computer Science from University of A Coruña (UDC), in 2004. Since June 2000 he is a member of the Medical Computing and Radiological Diagnosis Center of the UDC. At present, he is an FPU researcher of the Spanish Ministry of Education and Science at the Information and Communications Technologies Department of the UDC. His research interests include information integration, ontologies, security and evolutionary computation. He has participated in several research projects and published papers in many international journals, books and conferences.

Jianhua Yang, BEng (CAU, Beijing), MSc (Warwick), is a PhD student and a member of the Warwick Intelligent Systems Engineering Laboratory. He currently holds the UK Overseas Research Students Awards and the Warwick Postgraduate Research Fellowship. He is also a student member of IEEE and the Society for the Study of Artificial Intelligence and Simulation of Behaviour (SSAISB). After rejoining Warwick in 2006, he has several research papers published in the field of intelligent systems' applications.

Zhijun Yang received a Bachelor of Engineering from Huazhong University of Science and Technology, Master of Science from Nanjing University in China, both in electrical engineering. He then worked as an analogue electrical engineer for four years. In 1999 he received the Doctor of Science from Federal University of Rio de Janeiro, Brazil, studying the theory and application of machine learning. Yang was then a senior software engineer of Motorola global software centre. In 2001 he worked as a research associate at physics department of Heriot-Watt University. Since 2002 he has worked at department of computer science of Stirling University and school of engineering and electronics of Edinburgh University as a research fellow. His research interests include VLSI circuit design, artificial intelligence and software development. Yang is a member of IEEE, and IEEE computer society.

Index

S

T

V

W